Coxiella burnetii: Recent Advances and New Perspectives in Research of the Q Fever Bacterium

For further volumes:
http://www.springer.com/series/5584

Rudolf Toman • Robert A. Heinzen
James E. Samuel • Jean-Louis Mege
Editors

Coxiella burnetii: Recent Advances and New Perspectives in Research of the Q Fever Bacterium

Springer

Editors
Rudolf Toman
Department of Rickettsiology
Institute of Virology
Slovak Academy of Sciences
Bratislava, Slovakia

Robert A. Heinzen
Coxiella Pathogenesis Section
Rocky Mountain Laboratories
NIH/NIAID
Hamilton, MT, USA

James E. Samuel
Department of Microbial and Molecular Pathogenesis
College of Medicine
Texas A&M Health Science Center
3112 Medical Research and Education Building
Bryan, TX, USA

Jean-Louis Mege
Unité de Recherche sur les Maladies Infectieuses Transmissibles et Emergentes
CNRS-IRD UMR 6236
Institut Fédératif de Recherche 48
Faculté de Médecine
Université de la Méditerranée
Marseille, France

ISSN 0065-2598
ISBN 978-94-007-4314-4 ISBN 978-94-007-4315-1 (eBook)
DOI 10.1007/978-94-007-4315-1
Springer Dordrecht Heidelberg New York London

Library of Congress Control Number: 2012941642

Printed on acid-free paper

Springer is part of Springer Science+Business Media (www.springer.com)

Preface

Over 20 years have elapsed since the publication of the seminal two-volume series entitled *Q Fever: The Biology of Coxiella burnetii* (edited by J.C. Williams and H.A. Thompson) and *Q Fever: The Disease* (edited by T.J. Marrie) that described the current state of *Coxiella burnetii* research. The ensuing years have brought the post-genomic era and accompanying technologies that have catalyzed major advances in the field, including milestone discoveries of genetic transformation and host cell-free growth of this former obligate intracellular bacterium. *Coxiella* has matured from a niche organism, investigated by a handful of laboratories worldwide, to a model system to study macrophage parasitism, developmental biology, host-pathogen interactions, and immune evasion/modulation. Further bolstering interest are recent bioterrorism concerns that have delegated *Coxiella* to the list of regulated microbial agents with potential for illegitimate use, a designation confounded by the organism's ubiquitous nature and abundant natural reservoirs. Indeed, illustrating the zoonotic character of Q fever is a large outbreak (>4,000 cases) associated with intensive dairy goat farming that is currently subsiding in the southern Netherlands. The outbreak has highlighted the public health impact and re-emergent nature of Q fever, in addition to the need for improved surveillance and control methods.

Coxiella is a fascinating example of intracellular parasitism. The organism has uniquely evolved to thrive in the most inhospitable of cellular compartments—the phagolysosome. Understanding how *Coxiella* resist the degradative functions of this vacuole, and the host cell functions co-opted for successful parasitism, is central to understanding Q fever pathogenesis. These topics are among the relevant and timely research areas comprehensively reviewed in this volume. Structure-function relationships are discussed in chapters on *Coxiella* proteomics, lipopolysaccharide, and lipids (bacterial and host). Improved knowledge in these subject areas is guiding development of enhanced detection schemes for *Coxiella* in addition to shedding light on the host immune response to the Q fever pathogen and how it survives within the host. The cellular microbiology of the pathogen-host relationship is explored in chapters detailing how *Coxiella* responds to oxidative stress, manipulates host cell pro-survival signaling, and directs biogenesis of a replication vacuole. These and other host cell processes are likely modulated by a repertoire of secreted proteins with specific effector functions examined in a thorough review of *Coxiella* secretory systems and verified secreted proteins. The biological relevance and molecular biology of *Coxiella's* bi-phasic developmental cycle is explored. Several chapters survey immune functions that control or potentially exacerbate *Coxiella* infection and delve into correlates of protective immunity elicited by vaccination. These topics are particularly relevant to development of a safe and effective Q fever vaccine that does not require pre-vaccination skin testing. *Coxiella* genetic diversity is reviewed with the aim of better understanding the importance of strain variation and pathogenetic potential. Comparative genomics is also the foundation of chapters discussing diagnostic antigen discovery and molecular typing of *Coxiella,* with significance for development of new clinical, epidemiologic, and forensic tools. A rather underexplored area with relevance to the evolution of *Coxiella* as human pathogen is discussed in a chapter on *Coxiella*-like endosymbionts of ticks. The Netherlands Q fever outbreak is described with subject areas including surveillance, challenges in laboratory diagnostics, post-acute Q fever sequela, risks for pregnant women, and mitigating veterinary measures. Finally, chapters on host cell-free (axenic) growth and genetic manipulation of *Coxiella* illustrate how one technological advance feeds another. Axenic growth has relieved the considerable experimental constraints imposed by *Coxiella's* previous obligate intracellular lifestyle and enabled new lines of investigation. The increasing genetic tractability of *Coxiella* will inevitably lead to novel insight into intracellular parasitism and disease pathogenesis.

In Chap. 1, Hechemy introduces *Coxiella* as a pathogen and the natural history of the disease Q fever. Events leading to the co-discovery of *Coxiella* by Australian and American researchers are covered. Focused overviews of disease pathogenesis, epidemiology, diagnosis, treatment, and vaccination provide a prospective of Q fever as a zoonosis of continuing concern.

In Chap. 2, van Schaik and Samuel summarize studies that have compared the variety of isolates of *C. burnetii* from clinical, animal, and environmental sources, which phylogenetically separate these isolates into groups of close relationship using molecular techniques with increasing resolution. At the most precise level of investigation, whole genome sequence analysis has predicted a variety of polymorphic loci that might contribute to pathotype differences established in animal models of acute disease and consistent with difference in acute and chronic human infections. The chapter then predicts genes that might mediate differences in virulence with various mechanisms reviewed in subsequent chapters.

In Chap. 3, Mertens and Samuel detail predictions based on the sensitivity to oxidative stress identified in *C. burnetii*. Reactive oxygen intermediates (ROI) and reactive nitrogen intermediates (RNI) are the principal host innate defense mechanisms employed to control infection. A principal strategy of pathogenesis is to avoid activating ROI and RNI responses, and extracellular growth restricted to low oxygen environment supports the hypothesis that *Coxiella* has evolved to occupy such an environment. Important genome reductions have resulted in reduced detoxification mechanisms and DNA repair enzymes for survival in oxidative environment.

In Chap. 4, Narasaki and Toman provide a summary of studies on the only confirmed virulence macromolecular complex, lipopolysaccharide, providing a summary of studies on its composition and structure, biological activities and biosynthetic pathways for the unique sugars (virenose and dihydrohydroxystreptose) found in the O-chain. The differential expression of O-chain and core LPS is the bases of serodiagnostic antigens (phase I and II), targets for vaccine-induced protective antibody, and alternate interactions with host cells. While the structural prediction for the lipid A moiety of LPS appears to support studies that demonstrate antagonistic activity for TLR4 host receptors, other studies suggest LPS involvement in a variety of cellular responses.

In Chap. 5, Zhang, Zhang and Samuel explore the literature for an understanding of which components of the host immune response are essential for protective immunity engendered after infection or as a result of vaccine-induced immunity. One novel discovery by the authors identified an important role for antibody in protection, which is distinct from traditional models of protective immunity against an intracellular pathogen but has become recently appreciated for a variety of other bacterium models. The target of antibody-specific immunity appears to include the O antigen of LPS and may provide a component of next generation subunit vaccines. But equally well documented are the principal mechanisms of clearance in infection and vaccine-induced immunity as T-cell-mediated effectors elicited by activating cytokines, especially interferon-γ.

In Chap. 6, Ihnatko, Shaw and Toman critically evaluate the rapidly expanding field of *Coxiella* proteomics. Sub-cellular and whole cell proteomic studies have provided new knowledge of *Coxiella* developmental biology, physiology, secreted proteins, and strain variation associated with different virulence potential. Immunoproteomics, wherein whole proteome microarrays are high-throughput screened for seroreactive proteins, are revealing candidate antigens with potential use in a new generation of Q fever diagnostic and vaccine antigens based on recombinant protein. The chapter also details how proteome studies dovetail with genomics to reveal conserved and unique *Coxiella* biomarkers useful as detection tools.

In Chap. 7, Hussain and Voth describe the adept ability of *Coxiella* to manipulate host cell pro-survival signaling and describe pathogen reliance on various cellular kinases for proper development of the parasitophorous vacuole (PV). Akt and Erk1/2 are two pro-survival kinases actively modulated by *Coxiella* to promote host viability. Interplay between autophagic and apoptotic pathways is also observed during *Coxiella* infection, as exemplified by the interaction of autophagy-related Beclin-1 and anti-apoptotic Bcl-2 proteins. The general anti-apoptotic activity of *Coxiella* is viewed as a virulence strategy to sustain the host for the duration the pathogen's lengthy infectious cycle.

In Chap. 8, Ghigo, Colombo and Heinzen examine the biogenesis and biological character of the *Coxiella* PV. The PV is the intracellular microenvironment of *Coxiella* from which it interfaces with host cells. With the exception of human macrophages/monocytes, studies agree that the PV completely matures through the endolysosomal pathway to acquire characteristic of a phagolysosome. *Coxiella* has the uncanny ability to resist the degradative functions of the vacuole and to exploit its acidic pH for metabolic activation. This chapter presents opposing trafficking models for virulent phase I and avirulent phase II *Coxiella* in human macrophages/monocytes and discusses the different behavior of phase variants in primary murine macrophages. Cumulative evidence indicates that the PV is a specialized compartment that is actively modified by the pathogen, with interactions extending to engagement with autophagic and early secretory pathways.

In Chap. 9, McDonough, Newton and Roy critically review *Coxiella* secretion systems. The ability of *Coxiella* to transport proteins with effector functions across its membranes is considered critical for conditioning of the PV lumen and manipulating host cell processes that benefit the pathogen. *Coxiella* encodes components of the Sec-dependent secretion pathway, and an export system used for type IV pilus assembly, that are predicted to deliver proteins to the PV lumen. On the other hand, a *Coxiella* type IVB secretion system that is functionally analogous to the *Legionella pneumophila* Dot/Icm secretion system has recently been verified to translocate proteins to the host cell cytoplasm where their specific activities manipulate a variety of host cell processes, including pro-survival signaling and PV biogenesis.

In Chap. 10, Gilk examines the minimally studied but critically important topic of lipid metabolism (pathogen and host) and *Coxiella* infection. *Coxiella* requires lipids for both normal bacterial functions and formation of its PV membrane. The organism acquires lipids through *de novo* synthesis, and potentially through subversion of host cell pools. Phospholipids comprising the *Coxiella* cell envelope contain a high percentage of branched fatty acids, a modification that may serve a protective role. The PV membrane is sterol-rich and *Coxiella* encodes two "orphan" eukaryotic-like sterol reductases predicted to catalyze late reduction steps in cholesterol biosynthesis. Indeed, at least one reductase is enzymatically active, leading to the hypothesis that *Coxiella* modifies host cell sterols. Other lipid modifying enzymes of *Coxiella* include phospholipases of unknown function. Elucidation of how *Coxiella* exploits host cell lipids will improve our understanding of the interplay between pathogen and host.

In Chap. 11, Omsland discusses the milestone advance of *Coxiella* host cell-free (axenic) growth. *Coxiella's* prior obligate intracellular lifestyle presented severe experimental hurdles to gaining an improved understanding of the pathogen lifestyle. This chapter summarizes how transcription microarrays, metabolic pathway reconstruction, and metabolite typing aided development of acidified citrate cysteine medium (ACCM), a medium that supports robust growth of *Coxiella* in liquid culture as well as colony formation in semi-solid agarose plates. Technical issues related to ACCM culture are discussed as are utilization of the media in genetic transformation and viability assays. The technique is currently facilitating new lines of investigation and dramatically aiding development of new *Coxiella* genetic tools. ACCM culture methods should also prove useful in a clinical setting.

In Chap. 12, Minnick and Raghavan provide an overview of the developmental biology of *Coxiella*. Pronounced environmental stability is a hallmark of *Coxiella* that promotes efficient aerosol transmission. *Coxiella* undergoes a biphasic developmental cycle that generates morphologically distinct small cell variants (SCV) and large cell variants (LCV). The transcriptional and translational properties of

SCV and LCV, and ultrastructural observations of intracellular *Coxiella*, support a model whereby non-replicating SCV differentiate into replicating LCV. Based on SCV structure (i.e., condensed chromatin and multi-layered cell envelope) and mechanical disruption experiments, the SCV is considered the extracellular survival form of *Coxiella* that initiates most natural infections. Interestingly, SCV and LCV appear equally infectious for cultured cells. Many interesting questions related to the molecular biology of *Coxiella* differentiation remain unanswered. For example, do two-component regulatory systems drive stringent response physiology that triggers LCV to SCV transition during stationary phase? A better understanding of the pathogenic roles of *Coxiella* developmental forms in the natural history of Q fever is an important research goal.

In Chap. 13, Beare provides a brief overview of successes in genetic transformation of obligate intracellular bacteria and discusses the current state of *Coxiella* genetic manipulation including practical aspects of genetic transformation protocols. The rescue of *Coxiella* from its host cell eliminated several technical obstacles associated with host cell-based transformation systems, such as difficulties in cloning and transformant expansion/characterization. Axenic culture has allowed the rapid expansion of the *Coxiella* genetics toolbox to currently include *mariner*-based *Himar1* transposon systems for random mutagenesis, a Tn7 system for site-specific single-copy, chromosomal gene integration and *in cis* complementation, and multiple permutations of RSF1010 *ori*-based shuttle vectors that allow multi-copy *in trans* complementation and heterologous gene expression. A reliable method of targeted gene inactivation remains a current, but not insurmountable, challenge. The ability to fulfill molecule Koch's postulates for putative virulence genes will shepherd a new era of *Coxiella* virulence factor discovery.

In Chap. 14, Capo and Mege provide a wide-ranging view of elements of both the innate and adaptive immune responses that are engaged in controlling infection in acute and chronic infections. This chapter also provides an excellent example of the strength of this book in presenting several of the important controversies associated with modeling Q fever infection and pathogenesis. The authors provide a model for innate responses essential for clearance that does not involve IFN-γ killing through generation of ROI or RNI but rather engagement of nutritional restriction and activation of TNF-mediated apoptosis. The authors also posit an important role for TLR4, although other studies have shown the lipid A moiety of *C. burnetii* LPS does not serve as an agonist for TLR4-mediated activation. A detailed and important review of the dysregulation of immune suppression noted in chronic Q fever patients provides a basis for understanding the host component to this disease. Additionally, the authors provide a summary of recent (some unpublished) results that may provide a basis for sex-related susceptibility to acute Q fever symptoms in males.

In Chap. 15, Amara, Bechah and Mege outline several novel models describing mechanisms *C. burnetii* may use to evade the host response, thereby allowing persistence. Adipose and placenta tissues provide unique sites within the host to persist and *C. burnetii* may utilize these tissues to avoid host immune surveillance and to provide a permissive replication environment. The characterization that *C. burnetii* exists in a non-replicating, viable state in M1 polarized monocytes and replicate well in M2 polarized macrophage forms the basis for the principle infection model and confirms the general appreciation that infected macrophages do not become active in vitro. The role of the immunomodulatory cytokine, IL-10, in establishing the M1/M2 polarization and how dysregulated overexpression is critical for chronic infection is carefully reviewed. Finally, the role that apoptotic cell engulfment may play in infection and new data suggesting regulatory T cell involvement in chronic patient infection provide exciting expansions to the model of persistence.

In Chap. 16, Hendrix and Chen provide a detailed review of the vaccines and diagnostic methods that have been applied to Q fever. With this summary of previous approaches to these related problems, the authors explore current technologies that are designed to identify novel reagents for both subunit vaccines and diagnostic methods. The limitations to these approaches require a clear identification of strain heterogeneity and studies that summarize this diversity of *C. burnetii* isolates are provided. The thrust of this review is how characterization of dominant protein antigens, through traditional gel separation, mass spectrometric description, and genome wide expression of the entire proteome, has each contributed to the current understanding of dominant B and T cell antigens. Accompanying this review is an extensive table cataloging all proteins identified in the literature as immunodominant proteins.

In Chap. 17, Van Der Hoek et al. present a summary of the recent Q fever outbreak in the Netherlands where more than 4,000 human cases were reported from 2007 to 2010. This important review provides a careful and extensive analysis of the measures taken to define and successfully mitigate this seasonal epidemic. Besides providing a chronological summary of the events between 2007 and 2010, this chapter also provides an interesting analysis of public health policy, surveillance strategies, zoonotic transmission and selective application of large-scale animal vaccination. In particular, the ability to survey such a large population of at-risk and validated infections provides the opportunity to explore the risk factors associated with pregnancy for infected individuals and various chronic manifestations including chronic fatigue syndromes. The careful definition and evaluation of various diagnostic strategies will be important for public health surveillance and policy worldwide. Finally, the opportunity to administer the human Q fever vaccine, manufactured and licensed in Australia, to at-risk individuals will provide additional safety and efficacy data for future licensure consideration in the Netherlands and possibly other countries.

In Chap. 18, Zhong probes the fascinating field of *Coxiella*-like endosymbionts. A diverse collection of *Coxiella*-like bacteria have recently been found in hard and soft ticks, as well as in a few vertebrate hosts. In some tick species, colonization rates approach 100%. *Coxiella*-like bacteria and virulent *C. burnetii* are highly homologous and form a monophyletic clade. This review discusses the techniques used to study *Coxiella*-like bacteria and delves into their putative functions. Improved understanding of the biological significance of *Coxiella*-like bacteria will shed light on the evolution of *C. burnetii* as a human pathogen.

In Chap. 19, Massung, Cutler and Frangoulidis provide an in-depth review and analysis of historic and current methods for providing phylogenetic linkage between isolates of *C. burnetii*. The uses include molecular epidemiologic and pathogenesis studies but may also be required in forensic analysis in the event of a bioterrorist release. The authors define the most powerfully discriminatory and reliable methods and targets including multispacer sequence typing (MST), multiple loci variable number tandem repeat analysis (MLVA), IS1111 and single nucleotide polymorphism (SNP) typing. Finally, the authors provide an extensive review and outlook for appropriate sample analysis for both epidemiology and forensic applications.

This book is not intended to offer a complete review of the latest findings in *C. burnetii* and Q fever research but rather to provide the reader interesting insights into this unusual, highly infectious intracellular pathogen, summarized by several leading groups in the field. These presentations include new data, techniques, and methodologies that have transformed our understanding of the agent. It is our hope that established scientists and students working with *Coxiella* and other intracellular bacteria will read this book with interest and that young scientists looking for exciting and demanding biological models will consider working with this pathogen in their career. Additionally, the book will be useful for microbiologists dealing with highly infectious pathogens as well as veterinarians, physicians, and health authorities involved in clinical medicine, emerging disease surveillance, and biodefense.

The editors here wish to thank all the contributors to this book and also the reviewers of the manuscripts who have generously given their time to assist the editors in the reviewing process.

January 2012

Rudolf Toman
Robert A. Heinzen
James E. Samuel
Jean-Louis Mege

Contents

Chapter 1
History and Prospects of *Coxiella burnetii* Research

Karim E. Hechemy

Contents

Abstract *Coxiella burnetii*, the causative agent of Q fever, is the only known intracellular pathogenic bacterium that replicates within the acidic, degradative confines of a eukaryotic phagolysosome. It appears that the bacterium nullifies or thwarts the toxic elements of this vacuole and evades both innate and adaptive immune responses. *C. burnetii* is ubiquitous in its geographic distribution with the disease prevalence more widespread than previously considered. Recent molecular and cell biological advances, along with improved instrumentation, have provided unique insight into the host-parasite interrelationship and revealed previously unknown virulence strategies of *C. burnetii*. An example is completion of the genome sequences of several strains of *C. burnetii* that has improved our understanding of pathogenic mechanisms used by the organism to cause disease in mammalian hosts.

K.E. Hechemy (✉)
Wadsworth center for Laboratory and Research,
New York State Department of Health, Albany, NY 12208, USA

5 Windham Court, Clifton Park, NY 12065, USA
e-mail: khechemy@nycap.rr.com

R. Toman et al. (eds.), *Coxiella burnetii: Recent Advances and New Perspectives in Research of the Q Fever Bacterium*, Advances in Experimental Medicine and Biology 984,
DOI 10.1007/978-94-007-4315-1_1,

Keywords Q fever • Vaccine • Virulence • Epidemiology • Clinical diagnosis • Pathogenesis • Outbreak

1.1 History

Coxiella burnetii is an intracellular, Gram-negative pathogenic bacterium. It is the causative agent of Q fever. "Q" stands for "query" and not for Queensland, Australia where the first outbreak of this hitherto unknown disease was observed (Derrick 1937). This febrile illness outbreak occurred in 1935 in nine patients who worked in an abattoir in Brisbane. Derrick transferred the disease to guinea pigs by injecting animals with patient blood. However, his attempts to visualize the infectious agent were unsuccessful (Derrick 1937). Derrick submitted infected guinea pig tissue to Frank Macfarlane Burnet who, along with Mavis Freeman, reproduced the characteristic febrile reaction in guinea pigs (Burnet and Freeman 1937). They observed that the agent was "filterable with difficulty" and that smears of infected spleen tissue contained organisms that appeared to be of "rickettsial nature." Furthermore, they observed that suspensions of clarified tissue from infected guinea pigs agglutinated when mixed with the serum of patients who had convalesced from Q fever. The suspensions, however, failed to agglutinate when mixed with sera from patients with typhus rickettsia.

Nearly simultaneously in Hamilton, Montana, USA, an unknown agent was discovered as part of field study on the ecology of Rocky Mountain spotted fever. *Dermacentor andersoni* ticks collected in Nine Mile, Montana, were fed on guinea pigs, one of which developed a febrile illness that did not mimic Rocky Mountain spotted fever in presentation (Davis and Cox 1938). The designated Nine Mile agent could be serially passaged in guinea pigs, was filterable, and did not grow in axenic culture, indicating that the agent was not a "true" bacterium, e.g., *Francisella tularensis*. The agent had, according to Davis and Cox (1938), both bacterial and viral characteristics.

The link between the Nine Mile and Q fever agents was discovered serendipitously due to a laboratory-acquired infection by the Nine Mile agent. The patient displayed signs and symptoms strikingly similar to Q fever (Dyer 1938), and subsequent cross protection studies confirmed that the Q fever and Nine Mile agents were very likely the same pathogen (Dyer 1939). Because the organism did not behave exactly like a typical rickettsia, the agent was placed in a new genus in the family *Rickettsiaceae*, the genus "*Coxiella*" and species "*burnetii*" in honor of Cox and Burnet, respectively. The early history of the discovery and past milestones of *C. burnetii* are described in detail by McDade (1990). *C. burnetii* was moved to the order *Legionellales* after molecular taxonomy studies indicated that it was phylogenetically more closely related to members this order (see below).

1.2 The Bacterium

Based on the 16S rRNA-encoding gene sequence, *C. burnetii* was reclassified from the order *Rickettsiales* to the phylum *Proteobacteria*, class γ-*Proteobacteria*, order *Legionellales*, family *Coxiellaceae*. (Weisburg et al. 1989; Waag and Thompson 2005). *C. burnetii* has a genomic 42.2% guanosine-plus-cytosine content, which is closer to that of members of the order *Legionallales* than to the order *Rickettsiales* (~29% G+C). The genome size of *C. burnetii* strains was originally estimated to range from ~1.5 to 2.4 Mb based on low resolution pulsed-gel electrophoresis (Willems et al. 1998), but genome sequencing now indicates a range of ~2.0 to 2.2 Mb (Beare et al. 2009). The *Coxiella* genome encodes more basic proteins (average pI of 8.25) than almost any other medically important genus group (Waag and Thompson 2005).

The bacterium has a small gram-negative pleomorphic coccobacilli shape and produces two morphologically distinct cell types that comprise a bi-phasic developmental cycle. A small cell variant, with its characteristic condensed chromatin, is thought to be an extracellular survival form with enhanced resistance to environmental stressors such as desiccation and heat (Waag 2007). When the small cell variant invades the host, it develops into a large cell variant that is metabolically and divisionally active. Differentiation of large cell to small cell variant occurs during the stationary phase of the organism's growth cycle. It involves changes in the surface proteins (Waag 2007), but changes in lipopolysaccharide (LPS) have not been studied thus far. The large cell variant was originally proposed to produce an endospore that served as a progenitor of the small cell variant (McCaul and Williams 1981), but this developmental form has since been discounted (Heinzen et al. 1999).

C. burnetii displays considerable genetic homogeneity when examined solely by 16S rRNA gene sequencing (Stein and Raoult 1993). However, restriction fragment-length polymorphism (RFLP) (Hendrix et al. 1991), and microarray whole genome comparisons (Beare et al. 2006), reveal considerable heterogeneity in banding and hybridization patterns, respectively, indicating genetic diversity between isolates. Thirty-two isolates were categorized into six distinct genomic groups (I to VI) (Hendrix et al. 1991). Human isolates within genomic groups I, II, and III were derived from acute disease patients, whereas human isolates within genomic groups IV and V were from chronic disease patients. Organisms in genomic group VI were isolated from rodents near Dugway, Utah. Most isolates harbor one of four autonomously-replicating plasmids termed QpH1, QpRS, QpDV, and QpDG (Beare et al. 2006; Ghigo et al. 2009). These plasmids range from 32 to 42 kb in size and share a common 25-kb "core" region along with unique regions. Strains that lack plasmids contain plasmid sequences integrated into their chromosomes.

The genetic diversity of *C. burnetii* is further evidenced by the presence of antigenically and structurally unique LPS molecules. Three distinct LPS chemotypes have been described that are associated with specific genomic groups (Toman et al. 2009).

A potential link between LPS chemotype and virulence potential has been proposed (Hackstadt 1990; Skultety et al. 1998). An LPS phase variation occurs in the laboratory. Virulent *C. burnetii* isolated from natural sources and infections all produce a full-length LPS that is serologically defined as "phase I" or the smooth variant, which is virulent to its host. Serial *in vitro* passage of phase I *C. burnetii* in embryonated eggs or tissue culture results in LPS molecules with decreasing molecular weights and distinct constituent sugar compositions, culminating in the truncated LPS of avirulent "phase II" organisms or the rough type. Phase II LPS contains a lipid A identical to that of phase I LPS, and has some core sugars. However, it is missing two unusual O-antigen sugars, namely virenose (6-deoxy-3-C-methyl-D-gulose) and dihydrohydroxystreptose [3-C-(hydroxymethyl)-L-lyxose] that are unique biomarkers of *C. burnetii* (Skultety et al. 1998; Toman et al. 2009). Two cloned LPS variants of the virulent Nine Mile phase I (NMI) isolate have been described: Nine Mile Crazy (NMC; intermediate virulence), producing an intermediate-length LPS (Toman et al. 2009), and Nine Mile phase II (NMII; avirulent), producing a severely truncated LPS (Toman and Skultety 1996). NMII and NMC have large chromosomal deletions that eliminate open reading frames involved in the biosynthesis of O-antigen sugars, including the rare sugar virenose (Hoover et al. 2002). However, at present, the genetic lesion(s) leading to the severely truncated LPS of phase II organisms is unknown.

1.3 The Disease

Q fever is a zoonotic infection that manifests in humans primarily as an acute flu-like syndrome with potential complications including pneumonia and hepatitis (Raoult et al. 2005). The symptoms may resolve or they may become chronic. Chronic disease, although rare, is serious as endocarditis is the most frequent clinical presentation. The majority (60%) of human infections result in asymptomatic seroconversion. Resolution of symptoms does not necessarily indicate that the patient is free of infection. DNA was detected in patients several months post onset of the disease (Harris et al. 2000). The outcome of the infection is dependent on the cell-mediated immune response and is inversely proportional to granuloma formation (Raoult and Marrie 1995). Chronic infection is mainly seen in patients with valvulopathy and to a lesser extent with immunocompromised patients with defective cell-mediated immunity.

Infection typically occurs by inhalation of the bacterium contained in contaminated dust particles. Sources include barnyards (Thomas et al. 1995), contaminated straw (van Woerden et al. 2004), and historically, facilities housing *C. burnetii* research programs (Huebner 1947). In our laboratory, we received a blood specimen from a painter who did not have contact with farm animals. The painter came down with Q fever-like symptoms after painting the walls of a science laboratory where a newborn lamb had been dissected. In addition, indirect accidental exposures have occurred with workers in offices near elevators used to transport pregnant sheep that were unknowingly infected with *C. burnetii* (Meiklejohn et al. 1981). It should be noted that Q fever as an occupational hazard associated with research facilities has largely been eliminated due to implementation of modern biosafety equipment and protocols.

1.4 Clinical/Laboratory Diagnosis and Treatment

The signs and symptoms of human Q fever are protean and non-specific, which can complicate and delay clinical diagnosis. The incubation period may vary from a few days to several weeks probably depending in part on the infectious dose and the immune status of the host. Typical presentations of acute disease include a flu-like syndrome (i.e., fever, headache, chills, and fatigue), pneumonia and granulomatous hepatitis. Acute disease is usually self-limited, even without antibiotic therapy. Atypical presentations of acute or chronic Q fever indicative of a systemic inflammatory disease can misleadingly suggest other diseases such as Crohn's disease, Goodpasture's syndrome, and polymyalgia rheumatica (Lefebvre et al. 2010). The most frequent and serious chronic presentation of Q fever is a culture-negative endocarditis. Without prompt recognition and appropriate antimicrobial therapy, this disease can be fatal. Pregnant women whose sera scored positive have been known to have adverse outcomes including neonatal death (Raoult et al. 2002).

It is estimated that 1–5% of asymptomatic or acute Q fever infections give rise to more serious chronic Q fever (Tissot-Dupont and Raoult 2008). Patients with an immunocompromised state and/or preexisting valvular heart disease are at greater risk of developing chronic disease. Chronic Q fever could result from the bacterium that migrated to the bone marrow and subsequently, at a later time, migrated to other tissues, e.g., the endocardium (Waag 2007). Although pre-existing host conditions are clearly involved in the evolution of chronic Q fever, a running debate in the field concerns the extent to which *C. burnetii* strains have the propensity to cause acute or chronic disease. Genetically distinct strains clearly display different virulence in animal models of Q fever (Russell-Lodrigue et al. 2009). Furthermore, correlations have been made between disease outcome and *C. burnetii* genomic composition (Hendrix et al. 1991; Glazunova et al. 2005).

Given the non-specific nature of Q fever, clinical suspicion when evaluating a fever of unknown origin is critical in correctly diagnosing the disease and initiating appropriate antibiotic therapy. Subsequent laboratory confirmation normally involves testing for the presence of *C. burnetii*-specific antibodies, which develop in patients 1–2 weeks after infection. The gold standard serological test for Q fever is an indirect immunofluorescence assay (IFA) that relies on serum reactivity with fixed whole cell phase I and phase II *C. burnetii*. A four-fold rise in IgG titer between paired samples taken 2–3 weeks apart, with antibodies primarily directed against phase II antigen, is diagnostic of acute Q fever. Conversely, in chronic Q fever, phase I titers are higher than phase II titers with an IgG phase I titer of ≥1:800 diagnostic (Maurin and Raoult 1999).

Significant advances have been made in nucleic acid based-detection of infectious agents using techniques like real-time PCR (Brennan and Samuel 2003; Waag and

Thompson 2005). The insertion sequence IS*1111* is the target of choice in PCR detection of *C. burnetii* as the repetitive element has multiple genomic copies, thereby increasing assay sensitivity. This technology is now largely restricted to reference and research laboratories. However, when a platform is developed for routine use, PCR should be the test of choice for STAT laboratory confirmation of a clinical diagnosis. It should be noted that PCR-based technology is sensitive mainly in the early disease state. As the disease progresses, the test sensitivity decreases, presumably because neutralizing antibodies clarify the patient's blood of infectious agent. Indeed, in Q fever, it was reported that *C. burnetii* DNA becomes undetectable in serum 17 days after infection (Schneeberger et al. 2010).

Several high-throughput approaches using whole proteome protein microarrays have recently been employed to systematically evaluate the human humoral immune response to *C. burnetii* infection (Beare et al. 2008; Vigil et al. 2010, 2011). In one study, an array containing 1,901 *C. burnetii* ORFs (84% of the entire proteome) was probed with Q-fever patient sera and naïve controls. Thirteen antigens specifically reacted with patient sera, and nine of these were validated using an immunostrip platform applicable to a clinical laboratory (Vigil et al. 2010). These studies are providing needed information for development of a new generation of diagnostic and vaccine antigens based on recombinant protein.

Although most acute infections spontaneously resolve, antibiotic therapy significantly reduces the duration of symptomatic disease and the likelihood of chronic infection. The current recommended therapy for acute Q fever is a 14-day course of doxycycline (200 mg/day) (Angelakis and Raoult 2010). Rifampin and fluoroquinolone antibiotics are alternate treatments. Combination trimethoprim and sulfamethoxazole therapy is recommended for pregnant women. For chronic Q fever, adults are treated with 200 and 600 mg per day of doxycycline and hydroxychloroquine, respectively, for at least 18 months (Angelakis and Raoult 2010). Hydroxychloroquine alkalinizes the acidic phagolysosome, thereby potentiating the activity of doxycycline (Raoult et al. 1990).

1.5 Epidemiology

Infection with this zoonotic pathogen is worldwide with the possible exception of New Zealand (Maurin and Raoult 1999). *C. burnetii* has a strong association with domestic ruminants, and contact with these animals and their products is the primary route of human exposure to the Q fever pathogen (Raoult et al. 2005). However, the range of natural reservoirs is extensive and includes both wild and domestic mammals, birds, and arthropods such as ticks (Maurin and Raoult 1999). In most animals, *C. burnetii* does not cause overt disease. Exceptions are sheep and goats where massive proliferation of the microorganism in the female reproductive system can result in late-term abortion. This is in contrast to the pathology in human, which mainly occurs in the respiratory system. Mammals transmit the bacteria through their urine, feces, milk and birth products, such as placenta and amniotic fluid. Consequently, desiccation of these substances can deposit tremendous numbers of infectious bacteria into barnyard dust. This material can be widely dispersed by air currents where it remains viable in the environment for years (Tissot-Dupont et al. 2004).

The infection is transmitted to humans primarily through inhalation of contaminated dust or aerosols. Transmission of *C. burnetii* through consumption of unpasteurized dairy products from infected sheep or cows occurs but is less efficient than aerosol transmission (Rodolakis 2009). Additionally, ticks, although noted reservoirs, do not vector the disease to humans in any significant way. Infection with *C. burnetii* is recognized as an occupational hazard for people who work with or around waste and birth products of livestock and may include farmers, veterinarians, zoo and slaughterhouse workers (Garner et al. 1997). The kinetics of excretion in a commercial dairy sheep flock after treatment with oxytetracycline showed that the bacteria were excreted in feces for 5 months after parturition, for 3 months in vaginal discharges and for 4 months in milk. These data suggests that treatment does not prevent shedding or limit the duration of bacterial excretion (Astobiza et al. 2010).

In the United States, *C. burnetii* has been shown to be widely distributed in the environment (Kersh et al. 2010). For example, 28% of samples from diverse locations (e. g., dairy farms, grocery stores) contained *C. burnetii* DNA as determined by quantitative PCR for the IS*1111* sequence. Complementing this study is a recent sero-survey indicating an overall seroprevalence of Q fever in the US of 3.1%. Prevalence is higher for men than for women (3.8%, vs. 2.5%, respectively) with antibody positivity increasing with age. Collectively, these data suggest that human exposure to *C. burnetii* in the US is much more common than the number of reported cases of Q fever. Subclinical or asymptomatic infections that go unreported are presumably important contributors to high seroprevalence rates. In Denmark, a recent serosurvey of occupational contacts in dairy farms showed that 39 of 359 individuals were seropositive, with the veterinarians surveyed having the highest positivity rate of 36% (Bosnjak et al. 2010). Q fever as an occupational risk associated with veterinary medicine is also illustrated by the high seropositivity rate of US veterinarians (~22%) (Whitney et al. 2009).

Several recent Q fever outbreaks exemplify how wide spread the pathogen is worldwide. In Israel, a large outbreak of Q fever in an urban school was possibly transmitted through an air conditioning system (Amitai et al. 2010). The conclusion from this outbreak is that *C. burnetii* infection should be considered when investigating point source outbreaks of influenza-like disease, especially outside the influenza season (Amitai et al. 2010). The largest Q fever outbreak ever recorded has occurred in an area of the southern Netherlands where the disease was formerly not prevalent (Enserink 2010; Schneeberger et al. 2010). Three thousand confirmed human cases have occurred in a span of 3 years from 2007 to 2009, with 2,300 cases in 2009 alone including six fatalities. The outbreak is clearly associated with high-density dairy goat farming. However, many sickened people had no contact with animal husbandry operations, providing new perspective on the epidemiology of *C. burnetii.*

C. burnetii can stay under the "radar" and reappear when conditions are favorable. This behavior is aided by the organism's resistance to environmental conditions, aerosol transmission, and low infectious dose approaching one bacterium, characteristics that also contribute to *C. burnetii's* potential as a biowarfare agent (US Centers for Disease Control and Prevention, category B) (Atlas 2003).

1.6 Vaccines

A vaccine for Q fever called Q-Vax is produced and licensed for use in Australia. It consists of formalin-inactivated whole-cell *C. burnetii* of the Henzerling phase I strain. The vaccine is considered 98% efficacious with protection lasting at least 5 years (Chiu and Durrheim 2007). Adverse local and systemic reactions, such as induration and shivering, respectively, can occur in immunized individuals previously infected with *C. burnetii* (Waag 2007). Therefore, serology and skin testing are conducted to assess immune status prior to vaccination. An important goal for Q fever preventatives is development of a safe and effective vaccine that does not require pre-vaccination skin testing. This countermeasure is particularly relevant in the event of an intentional release of *C. burnetii* where a susceptible population would need to be quickly vaccinated. Recent advances in axenic (host cell-free) culture and genetic manipulation of *C. burnetii* should facilitate development of such a vaccine (Omsland et al. 2009; Beare et al. 2011).

1.7 Prospective

Although the pace of research on *C. burnetii* has significantly increased during the last decade, many questions remain unanswered. The pathogen is highly infectious and able to colonize the inhospitable phagolysosome of a macrophage host cell. However, the virulence factors enabling successful host cell parasitism and disease pathogenesis are largely unknown. In fact, to date LPS is still the only verified virulence factor of *C. burnetii,* a finding that resulted from the testing of isogenic mutants in a guinea pig model of Q fever (Hackstadt 1990). New genetic tools should soon allow fulfillment of molecular Koch's postulates for putative *C. burnetii* virulence genes. Q fever disease pathogenesis is likely multifactorial, with one factor building upon another, creating a cascade of pathologic strikes that leads to clinical disease. It is essential to further investigate the immunomodulation occurring during Q fever to better understand immune correlates of protective immunity, information that is important for development of an efficacious vaccine with minimum side effects.

Conflicts of Interest The author declares no conflict of interest.

References

Amitai Z, Bromberg M, Bernstein M, Raveh D, Keysary A, David D, Pitlik S, Swerdlow D, Massung R, Rzotkiewicz S, Halutz O, Shohat T (2010) A large Q fever outbreak in an urban school in central Israel. Clin Infect Dis 50:1433–1438

Angelakis E, Raoult D (2010) Q fever. Vet Microbiol 140:297–309

Astobiza I, Barandika JF, Hurtado A, Juste RA, Garcia-Perez AL (2010) Kinetics of *Coxiella burnetii* excretion in a commercial dairy sheep flock after treatment with oxytetracycline. Vet J 184:172–175

Atlas RM (2003) Bioterrorism and biodefence research: changing the focus of microbiology. Nat Rev Microbiol 1:70–74

Beare PA, Samuel JE, Howe D, Virtaneva K, Porcella SF, Heinzen RA (2006) Genetic diversity of the Q fever agent, *Coxiella burnetii*, assessed by microarray-based whole-genome comparisons. J Bacteriol 188:2309–2324

Beare PA, Chen C, Bouman T, Pablo J, Unal B, Cockrell DC, Brown WC, Barbian KD, Porcella SF, Samuel JE, Felgner PL, Heinzen RA (2008) Candidate antigens for Q fever serodiagnosis revealed by immunoscreening of a *Coxiella burnetii* protein microarray. Clin Vaccine Immunol 15:1771–1779

Beare PA, Unsworth N, Andoh M, Voth DE, Omsland A, Gilk SD, Williams KP, Sobral BW, Kupko JJ 3rd, Porcella SF, Samuel JE, Heinzen RA (2009) Comparative genomics reveal extensive transposon-mediated genomic plasticity and diversity among potential effector proteins within the genus *Coxiella*. Infect Immun 77:642–656

Beare PA, Sandoz KM, Omsland A, Rockey DD, Heinzen RA (2011) Advances in genetic manipulation of obligate intracellular bacterial pathogens. Front Microbiol 2:97

Bosnjak E, Hvass AM, Villumsen S, Nielsen H (2010) Emerging evidence for Q fever in humans in Denmark: role of contact with dairy cattle. Clin Microbiol Infect 16:1285–1288

Brennan RE, Samuel JE (2003) Evaluation of *Coxiella burnetii* antibiotic susceptibilities by real-time PCR assay. J Clin Microbiol 41:1869–1874

Burnet FM, Freeman M (1937) Experimental studies on the virus of "Q" fever. Med J Aust 2:299–305

Chiu CK, Durrheim DN (2007) A review of the efficacy of human Q fever vaccine registered in Australia. NSW Public Health Bull 18:133–136

Davis GE, Cox HR (1938) A filter-passing infectious agent isolated from ticks. I. Isolation from *Dermacentor andersonii*, reactions with animals, and filtration experiments. Public Health Rep 53:2259–2276

Derrick EH (1937) "Q" fever, a new fever entity: clinical features, diagnosis, and laboratory investigation. Med J Aust 2:281–299

Dyer RE (1938) A filter-passing infectious agent isolated from ticks. IV. Human infection. Public Health Rep 53:2277–2282

Dyer RE (1939) Similarity of Austrailian "Q" fever and a disease cause by an infectious agent isolated from ticks in Montana. Public Health Rep 54:1229–1237

Enserink M (2010) Questions abound in Q-fever explosion in the Netherlands. Science 327:266–267

Garner MG, Longbottom HM, Cannon RM, Plant AJ (1997) A review of Q fever in Australia 1991–1994. Aust NZ J Public Health 21:722–730

Ghigo E, Pretat L, Desnues B, Capo C, Raoult D, Mege JL (2009) Intracellular life of *Coxiella burnetii* in macrophages. Ann N Y Acad Sci 1166:55–66

Glazunova O, Roux V, Freylikman O, Sekeyova Z, Fournous G, Tyczka J, Tokarevich N, Kovacava E, Marrie TJ, Raoult D (2005) *Coxiella burnetii* genotyping. Emerg Infect Dis 11:1211–1217

Hackstadt T (1990) The role of lipopolysaccharides in the virulence of *Coxiella burnetii*. Ann N Y Acad Sci 590:27–32

Harris RJ, Storm PA, Lloyd A, Arens M, Marmion BP (2000) Long-term persistence of *Coxiella burnetii* in the host after primary Q fever. Epidemiol Infect 124:543–549

Heinzen RA, Hackstadt T, Samuel JE (1999) Developmental biology of *Coxiella burnetii*. Trends Microbiol 7:149–154

Hendrix LR, Samuel JE, Mallavia LP (1991) Differentiation of *Coxiella burnetii* isolates by analysis of restriction-endonuclease-digested DNA separated by SDS-PAGE. J Gen Microbiol 137:269–276

Hoover TA, Culp DW, Vodkin MH, Williams JC, Thompson HA (2002) Chromosomal DNA deletions explain phenotypic characteristics of two antigenic variants, phase II and RSA 514 (crazy), of the *Coxiella burnetii* Nine Mile strain. Infect Immun 70:6726–6733

Huebner RJ (1947) Report of an outbreak of Q fever at the National Institute of Health. II. Epidemiological features. Am J Public Health 37:431–440

Kersh GJ, Wolfe TM, Fitzpatrick KA, Candee AJ, Oliver LD, Patterson NE, Self JS, Priestley RA, Loftis AD, Massung RF (2010) Presence of *Coxiella burnetii* DNA in the environment of the United States, 2006 to 2008. Appl Environ Microbiol 76:4469–4475

Lefebvre M, Grossi O, Agard C, Perret C, Le Pape P, Raoult D, Hamidou MA (2010) Systemic immune presentations of *Coxiella burnetii* infection (Q Fever). Semin Arthritis Rheum 39:405–409

Maurin M, Raoult D (1999) Q fever. Clin Microbiol Rev 12:518–553

Mccaul TF, Williams JC (1981) Developmental cycle of *Coxiella burnetii*: structure and morphogenesis of vegetative and sporogenic differentiations. J Bacteriol 147:1063–1076

Mcdade JE (1990) Historical aspects of Q fever. In: Marrie TJ (ed) Q fever, vol 1, The disease. CRC Press, Boca Raton

Meiklejohn G, Reimer LG, Graves PS, Helmick C (1981) Cryptic epidemic of Q fever in a medical school. J Infect Dis 144:107–113

Omsland A, Cockrell DC, Howe D, Fischer ER, Virtaneva K, Sturdevant DE, Porcella SF, Heinzen RA (2009) Host cell-free growth of the Q fever bacterium *Coxiella burnetii*. Proc Natl Acad Sci USA 106:4430–4434

Raoult D, Marrie T (1995) Q fever. Clin Infect Dis 20:489–495

Raoult D, Drancourt M, Vestris G (1990) Bactericidal effect of doxycycline associated with lysosomotropic agents on *Coxiella burnetii* in P388D1 cells. Antimicrob Agents Chemother 34:1512–1514

Raoult D, Fenollar F, Stein A (2002) Q fever during pregnancy: diagnosis, treatment, and follow-up. Arch Intern Med 162:701–704

Raoult D, Marrie T, Mege J (2005) Natural history and pathophysiology of Q fever. Lancet Infect Dis 5:219–226

Rodolakis A (2009) Q fever in dairy animals. Ann N Y Acad Sci 1166:90–93

Russell-Lodrigue KE, Andoh M, Poels MW, Shive HR, Weeks BR, Zhang GQ, Tersteeg C, Masegi T, Hotta A, Yamaguchi T, Fukushi H, Hirai K, Mcmurray DN, Samuel JE (2009) *Coxiella burnetii* isolates cause genogroup-specific virulence in mouse and guinea pig models of acute Q fever. Infect Immun 77:5640–5650

Schneeberger PM, Hermans MH, Van Hannen EJ, Schellekens JJ, Leenders AC, Wever PC (2010) Real-time PCR with serum samples is indispensable for early diagnosis of acute Q fever. Clin Vaccine Immunol 17:286–290

Skultety L, Toman R, Patoprsty V (1998) A comparative study of lipopolysaccharides from two *Coxiella burnetii* strains considered to be associated with acute and chronic Q fever. Carbohyd Polymers 35:189–194

Stein A, Raoult D (1993) Lack of pathotype specific gene in human *Coxiella burnetii* isolates. Microb Pathog 15:177–185

Thomas DR, Treweek L, Salmon RL, Kench SM, Coleman TJ, Meadows D, Morgan-Capner P, Caul EO (1995) The risk of acquiring Q fever on farms: a seroepidemiological study. Occup Environ Med 52:644–647

Tissot-Dupont H, Raoult D (2008) Q fever. Infect Dis Clin North Am 22:505–514

Tissot-Dupont H, Amadei MA, Nezri M, Raoult D (2004) Wind in November, Q fever in December. Emerg Infect Dis 10:1264–1269

Toman R, Skultety L (1996) Structural study on a lipopolysaccharide from *Coxiella burnetii* strain Nine Mile in avirulent phase II. Carbohydr Res 283:175–185

Toman R, Skultety L, Ihnatko R (2009) *Coxiella burnetii* glycomics and proteomics–tools for linking structure to function. Ann N Y Acad Sci 1166:67–78

Van Woerden HC, Mason BW, Nehaul LK, Smith R, Salmon RL, Healy B, Valappil M, Westmoreland D, De Martin S, Evans MR, Lloyd G, Hamilton-Kirkwood M, Williams NS (2004) Q fever outbreak in industrial setting. Emerg Infect Dis 10:1282–1289

Vigil A, Ortega R, Nakajima-Sasaki R, Pablo J, Molina DM, Chao CC, Chen HW, Ching WM, Felgner PL (2010) Genome-wide profiling of humoral immune response to *Coxiella burnetii* infection by protein microarray. Proteomics 10:2259–2269

Vigil A, Chen C, Jain A, Nakajima-Sasaki R, Jasinskas A, Pablo J, Hendrix LR, Samuel JE, Felgner PL (2011) Profiling the humoral immune response of acute and chronic Q fever by protein microarray. Mol Cell Proteomics 10:M110 006304

Waag DM (2007) *Coxiella burnetii*: host and bacterial responses to infection. Vaccine 25:7288–7295

Waag D, Thompson HA (2005) Pathogenesis of and immunity to *Coxiella burnetii*. In: Linder LE, Lebeda FK, Korch GW (eds) Biological weapons defense: infectious diseases and counter bioterrorism. Humana Press, Totowa

Weisburg WG, Dobson ME, Samuel JE, Dasch GA, Mallavia LP, Baca O, Mandelco L, Sechrest JE, Weiss E, Woese CR (1989) Phylogenetic diversity of the *Rickettsiae*. J Bacteriol 171:4202–4206

Whitney EA, Massung RF, Candee AJ, Ailes EC, Myers LM, Patterson NE, Berkelman RL (2009) Seroepidemiologic and occupational risk survey for *Coxiella burnetii* antibodies among US veterinarians. Clin Infect Dis 48:550–557

Willems H, Jager C, Baljer G (1998) Physical and genetic map of the obligate intracellular bacterium *Coxiella burnetii*. J Bacteriol 180:3816–3822

Chapter 2
Phylogenetic Diversity, Virulence and Comparative Genomics

Erin J. van Schaik and James E. Samuel

Contents

E.J. van Schaik • J.E. Samuel (✉)
Department of Microbial and Molecular Pathogenesis, College of Medicine,
Texas A&M Health Science Center, 3112 Medical Research and Education Building,
Bryan, TX 77807-3260, USA
e-mail: jsamuel@medicine.tamhsc.edu

R. Toman et al. (eds.), *Coxiella burnetii: Recent Advances and New Perspectives in Research of the Q Fever Bacterium*, Advances in Experimental Medicine and Biology 984,
DOI 10.1007/978-94-007-4315-1_2,

Abstract *Coxiella burnetii*, the causative agent of Q fever, has remained a public health concern since the identification of this organism in 1935 by E. H. Derrick in Australia and at the Rocky Mountain Laboratory in the USA by H.R. Cox and G. Davis. Human Q fever has been described in most countries where *C. burnetii* is ubiquitous in the environment except in New Zealand where no cases have been described. Most human infections are acquired through inhalation of contaminated aerosols that can lead to acute self-limiting febrile illness or more severe chronic cases of hepatitis or endocarditis. It is estimated that the actual incidence of human infection is under-reported as a result of imprecise tools for differential diagnosis. An intracellular lifestyle, low infectious dose, and ease of transmission have resulted in the classification of *C. burnetii* as a category B bio-warfare agent. The recent outbreaks in Europe are a reminder that there is much to learn about this unique intracellular pathogen, especially with the speculation of a hyper-virulent strain contributing to an outbreak in the Netherlands where over 4,000 human cases were reported. A new era in *C. burnetii* research has begun with the recent description of an axenic media making this an exciting time to study this bacterial pathogen.

Keywords *Coxiella burnetii* • Isolates • Phylogeny • Genomics • Pathogenecity • Virulence • Virulence factors

2.1 Phylogenetic Diversity

2.1.1 Taxonomy

Coxiella burnetii is the etiological agent of zoonotic Q fever, which has near world-wide distribution (Marrie 2010). This gram-negative, pleomorphic coccobacillus (0.2–0.4 μm wide, 0.4–1 μm long) is an obligate intracellular pathogen in nature with a doubling time estimated at 12–20 h (Zamboni et al. 2002). Though classified as a Gram-negative bacterium with a membrane similar to other Gram-negative bacteria, *C. burnetii* is not readily stainable by Gram techniques and instead is commonly visualized using Gimenez stain (Gimenez 1964). The early recovery of *C. burnetii* from ticks and its obligate intracellular lifestyle resulted in the original classification of this bacterium within the α-1 subdivision of the *Proteobacteria* to the *Rickettsiaceae* family and the *Rickettsiae* tribe along with the genera *Rickettsia* and *Rochalimaea* (now reclassified as genus *Bartonella* family *Baronellaceae* belonging to the α-2 subdivision of *Proteobacteria*) (Maurin and Raoult 1999). However, 16S rRNA sequence analysis resulted in the reclassification of *Coxiella* into a separate genus of the γ subdivision of *Proteobacteria* with the genera *Legionella*, *Francisella*, and *Rickettsiella* as its closest relatives (Weisburg et al. 1989). Sequencing of 16S rRNA from six *C. burnetii* isolates from different geographic areas revealed only three base substitutions, indicating that these strains are highly related (99%) supporting the phylogenetic homogeneity of the genus *Coxiella* into a single species (Stein et al. 1993). Furthermore, phylogenetic analysis demonstrated that *C. burnetii* shares a common line of decent with the genus *Legionella* and *Rickettsiella grylli* (Roux et al. 1997).

2.1.2 Genomic Organization

The genomes from several strains of *C. burnetii* have now been sequenced and range in size from approximately 1.9 to 2.2 Mb and are organized as circular chromosomes normally associated with a single autonomous plasmid or integrated plasmid-related sequences (Table 2.1; Seshadri et al. 2003). Currently there are four plasmid types designated QpH1, QpRS, QpDG, QpDV, as well as one plasmid without designation from a Chinese *C. burnetii* isolate (Jager et al. 2002). Sequence alignments and cross hybridization experiments defined a core region of ~16 kb that is found on all plasmid types (Willems et al. 1997; Savinelli and Mallavia 1990). The conserved presence of this sequence in all analyzed *C. burnetii* strains suggests that this plasmid sequence is an essential part of the genome (Table 2.1). The initial characterization of these plasmids, found that QpH1 associated with acute infection isolates, while plasmid QpRS associated with chronic infection isolates (Samuel et al. 1985). However, a subsequent study demonstrating isolates from chronic Q fever carry either QpH1 or QpRS, suggesting plasmid type was not a consistent predictor of disease presentation (Thiele and Willems 1994).

A high number of predicted pseudogenes suggests that genome reduction is ongoing in *C. burnetii*; reductive evolution is common among obligate intracellular bacteria (Seshadri et al. 2003). In addition, the genome of *C. burnetii* contains a high number of insertion sequences, which is unusual for obligate intracellular bacteria assuming there would be little opportunity for gene transfer, and thereby implies that the obligate intracellular lifestyle of *Coxiella* is a relatively recent adaptation (Seshadri et al. 2003; Raghavan et al. 2008). For example, a self-splicing intein in the genome, Cbu.DnaB, is very similar to inteins found in two extremophilic bacteria *Alkalilimnicola ehrlichei* and *Halorhodospira halophila*, which are distantly related to *C. burnetii*, making it difficult to determine whether Cbu.DnaB was acquired by horizontal gene transfer or inherited vertically from a common ancestor (Raghavan et al. 2008).

2.1.3 Differentiation of *C. burnetii* Isolates

Differentiation of *C. burnetii* strains was initially difficult as conventional serological methods failed to discriminate between isolates (Peacock et al. 1983). Additional attempts to discriminate isolates based on differences in LPS banding patterns or plasmid-type determination were limited (Hackstadt 1986; Samuel et al. 1985). Hence, *C. burnetii* strains were described by their source of isolation, geographic origin, and clinical manifestation. However, the extremely broad host range of *C. burnetii* as well as the wide spectrum of disease caused in humans makes differentiation of isolates extremely important for both epidemiology and diagnosis. The inability to grow *C. burnetii* in axenic media until recently has hampered the diagnosis of infections caused by this bacterium and made it difficult to conform to a single differentiation scheme (Omsland et al. 2009).

Table 2.1 Genotypic and phylogenetic information on select sequenced *C. burnetii* isolates

Isolate	Genome size	Plasmid	Source	Restriction group[a]	MST group[b]	MVLA group[c]	SNP group[d]
Nine Mile RSA 493	1,995,281	pQpH1	Tick, USA, 1935	I	16	E/3	3
Henzerling RSA 331	2,016,427	pQpH1	Human blood, Italy, 1945	II	18	A	8
Priscilla MSU Q177	2,090,565	pQpRS	Goat, USA, 1980	IV	8	F/21	6
CbuG_Q212	2,008,870	NP	Human heart valve, Canada 1981	V	21	G/3[e]	9
Dugway 5 J108–111	2,158,758	pQpDG	Rodent, USA, 1958	VI	20	I/21	N/A

[a] Hendrix et al. (1991)
[b] Multispacer sequence typing (MST) (Glazunova et al. 2005)
[c] Multiple locus variable number of tandem repeats analysis (MLVA) (Svraka et al. 2006; Arricau-Bouvery et al. 2006)
[d] Single nucleotide polymorphism (SNP) (Huijsmans et al. 2011)
[e] CbuG_Q212 was not used for analysis in these studies. But, CbuG_Q212 does group with Scurry Q217 using MST, and predicted to group with Scurry Q217 using MVLA

DNA-DNA hybridization showed a low degree of genetic heterogeneity among *C. burnetii* isolates (Vodkin et al. 1986). However, restriction fragment length polymorphism (RFLP) of DNA isolated from 32 *C. burnetii* strains resulted in the differentiation of isolates into six distinct genomic groups based on DNA fingerprinting (groups I–VI) (Table 2.1; Hendrix et al. 1991). Interestingly, isolates from groups I, II, and III were representative of acute infections whereas groups IV and V consisted of isolates representing those from chronic infections (Hendrix et al. 1991). These genotype groupings were later confirmed using pulse-field gel electrophoresis (PFGE) (Heinzen et al. 1990). Further studies using PFGE and DNA from 80 *C. burnetii* strains isolated from Europe, USA, Africa, and Asia identified 16 restriction groups and correlated the RFLP patterns to geographic origin, indicating that although *C. burnetii* is considered a homogenous species, there is substantial genetic diversity between isolates (Jäger et al. 1998). More recently, amplification and sequencing of the IS*1111* transposase elements from 14 *C. burnetii* isolates separated the strains into the original groupings from I to V, confirming the original groupings (Hendrix et al. 1991; Denison et al. 2007).

Other phylogenetic or differential analyses have been performed via sequencing of specific genes including *rpoB*, *icd*, and *com1/mucZ* (Nguyen and Hirai 1999; Mollet et al. 1998; Sekeyová et al. 1999). Sequencing of *rpoB* from 8 *C. burnetii* isolates revealed fewer than four base differences, confirming their phylogenetic homogeneity, however, distribution of these isolates did not correlate with the previously determined genotyping groups (Mollet et al. 1998). Furthermore, differentiation based on *com1/mucZ* found that strains were not delineated based on geographical origin (Sekeyová et al. 1999). On the other hand, sequencing of *icd* from 19 *C. burnetii* isolates separated them into three groups, where group I contained only isolates from acute disease, and groups II and III contained isolates from chronic disease, which did correlate with the original genotyping groups (Nguyen and Hirai 1999). In addition, the results of another study suggested that isolates could be separated into acute and chronic groups using the sequence of the *com1* gene (Zhang et al. 1997). The inability of these methods to develop a common typing method indicates that a fast and reliable method for typing *C. burnetii* isolates remains an important goal. These studies also indicate that the most common typing methods may not have the ability to distinguish between virulence potential, which is likely mediated by a complex accumulation of pathogenic factors.

2.1.4 Genotyping

More recent genotyping has taken advantage of the fully sequenced *C. burnetii* genome using such techniques as multispacer sequence typing (MST) or multiple locus variable number of tandem repeats analysis (MLVA). MST is based on intergenic region sequencing and screening of *C. burnetii* identified ten variable spacers, which were then used to separate 173 isolates into 30 different sequence types (MST) (Table 2.1) (Glazunova et al. 2005). Phylogenetic analysis of these genotypes revealed three monophyletic groups, which could be further subdivided into different clusters (Glazunova et al. 2005). Nine Mile RSA 493, Henzerling RSA 331, and Dugway 5 J108-111 all sectored into monophyletic group I, whereas, Priscilla MSU Q117 was found in group II, and CbuG_Q212 in group III (Table 2.1) (Glazunova et al. 2005). When *com1* typing was used in this study, six groups were generated and only one isolate was not in accordance with the MST typing results (Glazunova et al. 2005). Interestingly, acute disease was associated with several MST groups including ST18 ($p=10^{-2}$) which contains the Henzerling isolate and chronic disease associated with ST8 ($p=10^{-3}$) which contains the Priscilla isolate (Table 2.1) (Glazunova et al. 2005). The plasmid QpDV was associated with acute disease isolates and QpRS was correlated with chronic disease isolates, indicating some correlation between genotype and disease type (Glazunova et al. 2005). In addition, while most of the French isolates where found in monophyletic group I, other sequence types were not correlated to geographic distribution (Glazunova et al. 2005). This lack of correlation of geographic distribution is likely due to the movement of infected individuals, animals, or ticks in the environment. The advantage of MST analysis over PFGE, which had been the most discriminative approach, was the portability of sequence data, which allows different laboratories to compare their results without having to exchange strains (Glazunova et al. 2005).

MLVA takes advantage of repetitive DNA sequences that are found in all bacterial species and has been used to differentiate many bacterial isolates. MLVA is comparable to PFGE in its discriminatory power; however it is a much simpler method of genotyping bacterial strains. Variable number tandem repeats (VNTR) markers were used to identify nine MLVA types from 21 *C. burnetii* isolates (Svraka et al. 2006). This study, like the MST study, demonstrated that there was a difference between isolates from acute and chronic disease, where S and Pricilla isolates were found in two different MLVA groups (G and F, respectively) but clustered in the same monophyletic group (Table 2.1; Svraka et al. 2006; Glazunova et al. 2005). A more recent study that evaluated the systematic genotyping of *C. burnetii* isolates using MLVA confirmed the results of the Glazunova study (2005) and separated Nine Mile, Dugway, Scurry, and Priscilla isolates into the same monophyletic groups (Arricau-Bouvery et al. 2006). It is important to note that the results of the MLVA and MST analysis have been similar, suggesting there are potential genetic differences in strains based on geographic isolation and disease manifestation (Glazunova et al. 2005; Svraka et al. 2006). Even though both studies separated a pool of *C. burnetii* strains into three monophyletic groups, the relatedness of some isolates differed depending on the method used. For example, in the Glazunova study (2005) the Nine Mile and Henzerling isolates were assigned to the same monophyletic group, while in the Syraka study (2006) they were found in different monophyletic groups. The results of these two studies indicate that even though both support differences based on geographic isolation and disease manifestation, the predicted ancestral relationships of the strains depends on the method used (MST versus MLVA). Therefore, it will be clarifying to standardize the methodology used in the field for epidemiological reporting (Svraka et al. 2006; Glazunova et al. 2005).

Recently, single nucleotide polymorphism (SNP) genotyping has been applied directly to clinical samples from the outbreak in The Netherlands (Huijsmans et al. 2011). This technique has many advantages including sensitivity and rapid analysis. In addition, SNP genotyping has been used to determine that single haplotypes were dominant in *Salmonella* Typhi outbreaks in distinct geographic locations, supporting the notion of clonal expansion rather than successive replacement of the pathogen (Baker et al. 2010). Consequently, SNP analysis of *C. burnetii* isolates would be extremely valuable, especially during outbreak situations like in the Netherlands to determine the source and spread of strains. Using sequencing data, SNP analysis identified nine distinct SNP-genotypes (Table 2.1; Huijsmans et al. 2011). The SNP and MLVA genotype analysis separated *C. burnetii* isolates into similar phylogenetic groups with a concordance of 93% (Huijsmans et al. 2011; Svraka et al. 2006; Arricau-Bouvery et al. 2006). The advantages of SNP over MLVA include that it does not require expensive sequencing equipment and may be directly applied to serum samples (Huijsmans et al. 2011). However, the ability of SNP to distinguish differences between isolates is less powerful then MLVA as 9 and 21 different phylogenetic groups were discovered, respectively (Huijsmans et al. 2011; Arricau-Bouvery et al. 2006). As noted earlier, standardization in the field of *C. burnetii* epidemiology will be essential to determine sources of outbreaks and ancestral relationships accurately and reproducibly because of its worldwide distribution.

2.2 Comparative Genomics

The availability of the genome sequence from *C. burnetii* has allowed for rapid assessment of isolate diversity through comparative genomics (Beare et al. 2006). Comparative genomics can identify duplicated genes, deletions or insertions, gene-fusion events, and predict lateral gene transfer events (Bansal 2005). A major advantage of this technology for *C. burnetii* research is the ability to resolve differences between isolates that were not differentiated by serology or PCR (Beare et al. 2006). Comparative genomics between Nine Mile phase I RSA493 (NMI) and Nine Mile phase II *C. burnetii* RSA439 (NMII) confirm that the only major difference between the passage variant clone is the deletion of 20 ORFs (CBU0679 to CBU0698) associated with the production of LPS (Beare et al. 2006; Hoover et al. 2002). This indicates that the loss of virulence is associated with the truncation of the LPS and not the lack of production of other virulence factors. Interestingly, the Nine Mile passage variant clone "crazy" RSA514 (NMC) which produces an intermediate length LPS, has a slightly larger deletion then the NMII clone, suggesting that other point mutations in other regions might account for the severe truncation seen in NMII (Beare et al. 2006). In addition, the Australian QD isolate (Au) which has phase II LPS, does not contain major genome deletions indicating that our understanding of the biogenesis of LPS in *C. burnetii* is incomplete and requires renewed investigation. These results confirmed that large chromosomal deletions are not required for the avirulent phenotype of the phase variants, however, the inability to produce full length LPS is entirely responsible for this phenotype (Thompson et al. 2003; Beare et al. 2006; Hoover et al. 2002).

The genomes of 23 *C. burnetii* isolates obtained from different types of infections and diverse geographical regions were used to compare polymorphisms between open reading frames (ORFs) (Beare et al. 2006). This study confirmed that the original genomic groupings I through VI were valid based for phylogenetic tree prediction (Beare et al. 2009; Hendrix et al. 1991). Most of the polymorphic ORFs were generated via deletions, which is consistent with the previously proposed hypothesis that *C. burnetii* is undergoing reductive evolution (Seshadri et al. 2003; Beare et al. 2006; Seshadri and Samuel 2005). There were no polymorphic ORFs between genomic group I strains even though these strains were isolated from diverse temporal and geographic regions, suggesting a global spread of a common ancestral *C. burnetii* isolate that has undergone little evolution (Beare et al. 2006). In addition, all *C. burnetii* strains from all genotypes contain 65 conserved pseudogenes likely representing genes that were inactivated in a common ancestor (Beare et al. 2009). The large number of insertional sequences (IS) found among *C. burnetii* strains encompassing eight families of IS elements may drive the formation of pseudogene development, a scenario that was recently proposed for *F. tularensis*, a closely related intracellular pathogen (Beare et al. 2009; Rohmer et al. 2007). Formation of pseudogenes that are no longer transcribed removes the selective pressure to prevent mutations in these genes, which ultimately could result in their removal via genome reduction (Beare et al. 2009; Seshadri and Samuel 2005; Casadevall 2008). Genome reduction is a common occurrence for intracellular bacteria and it was proposed that the *C. burnetii* Dugway strain, with the largest chromosome and plasmid, is the most primitive of the sequenced strains (Beare et al. 2009). However, recent phylogenetic analysis revealed that although the Dugway strain may not be the common ancestor for the tree of all isolates, it does have the shortest distance to the root supporting the proposal that this isolate is more primitive (Beare et al. 2009).

There were no polymorphic ORFs between the isolates in genomic group I and the reference isolate NMI RSA493 (group I), however, 87 polymorphic ORFs were identified between group V isolates and NMI (RSA493) and at least half of these were annotated as hypothetical proteins (Beare et al. 2006). These results are intriguing as isolates from group V have been associated with chronic infections and suggest that novel virulence factors may contribute to acute versus chronic pathotypes (Hendrix et al. 1991). Cross-genome comparisons of the sequenced *C. burnetii* isolates revealed several differences that may account for the different virulence phenotypes observed between isolates. Pathotype specific virulence has been a long-standing hypothesis that could account for the difference in occurrence between acute and chronic *C. burnetii* isolates. Although several studies have suggested that different genotypes or plasmid types may account for different pathotypes, it only became clear recently that pathotype specific does virulence exists (Samuel et al. 1985; Thiele and Willems 1994; Glazunova et al. 2005; Svraka et al. 2006; Hendrix et al. 1991; Russell-Lodrigue et al. 2009).

Many of the polymorphic hypothetical proteins could function to dampen the immune response and shift the pathology of infection from acute to chronic.

Table 2.2 Potential virulence proteins that are polymorphic between reference strains that may provide pathotype specific virulence

ORF designation	Nine Mile RSA493	CbuG_Q212	Dugway 5 J108–111	Eukaryotic domain
AnkA	CBU_0072		CBUD_2034	Ankyrin repeats
AnkI	CBU_1213	CBUG_0798	CBUD_1298	Ankyrin repeats
AnkL	CBU_1608	CBUG_0406	CBUD_0382	Ankyrin repeats
AnkP	CBU_0069		CBUD_2035	Ankyrin repeats
EnhC	CBU_1136	CBUG_0874	CBUD_1234	Tetratrico Peptide Repeats (TPR)
PilC	CBU_0154	CBUG_2088[a]	CBUD_1952	Lipoprotein Type II secretion apparatus

The information used to create this table in Beare et al. (2006, 2009)
Shaded boxes identify the polymorphic genes, which are either frame-shifted pseudogenes or longer ORFs compared to the other isolates (Beare et al. 2009)
Blank box indicate that no homologue has been identified in the genome
[a] Indicates a mutated protein

For example, three ankyrin repeat proteins are polymorphic in isolates from genomic group V (Table 2.2; Hendrix et al. 1991). Ankyrin repeats are composed of a 33 amino acid residue motif that occurs in tandem arrays and are the most common protein-protein interaction domains found in eukaryotic proteins (Al-Khodor et al. 2010). Many of the ankyrin repeat containing proteins identified in *C. burnetii* are potential type IV secretion system effector proteins as they can be translocated in the host cytosol using the *L. pneumophila* Dot/Icm system (Voth et al. 2009). Table 2.2 lists the polymorphic ankyrin repeat proteins and EnhC, a tetratrice peptide repeat domain protein, that are polymorphic between potentially acute pathotypes such as NMI (RSA493 a group I isolate), potentially chronic pathotypes such as Graves (CbuG_Q212 a group V isolate) and the avirulent Dugway (5 J108–111 a group VI isolate). These polymorphisms may contribute to the virulence potential of these isolates by modulating host cell functions via their eukaryotic protein-protein interaction domains (Table 2.2). Several of the ankyrin repeat proteins are pseudogenes or encoded by longer OFRs in NMI, supporting the hypothesis that these proteins may be involved in dampening the immune response (Table 2.2; Beare et al. 2006). Another polymorphic difference between pathotypes is associated with genes required for the biogenesis of type IV pili (Beare et al. 2009). It is likely that the core pilus genes of *C. burnetii* form a functional type II secretion system similar to that found in *Francisella* species (Hager et al. 2006; Zogaj et al. 2008). Production of type IV pili on the surface of *C. burnetii* has never been observed. This may be because *C. burnetii* lack a homolog to PilT, an ATPase involved in pilus retraction (Beare et al. 2006). In support of this notion, PilT is required for pilus assembly and expression of surface fibers on *F. tularensis* LVS (Chakraborty et al. 2008). This is currently unique to LVS, as mutations affecting *pilT* in other species that express type IV pili such as *Pseudomonas aeruginosa* cause hyper-piliation (Chiang and Burrows 2003). One possible scenario is that the lack of *pilT* in the *C. burnetii* genome results in the inability to express surface piliation.

The type II secretion and type IV pilus systems are mediated by the same structure in *F. tularensis*, which is a unique adaptation since other bacteria express separate machinery for these functions (Forsberg and Guina 2007; Ayers et al. 2010). Since *C. burnetii* does not code for the pilus retraction protein *pilT*, it is likely the type IV pilus homologues function as a type II secretion system similar to *F. tularensis* (Forsberg and Guina 2007; Beare et al. 2006). Candidates for secretion include proteins with characteristic signal sequences including acid phosphatase, which is a recently characterized potential virulence factor (Hill and Samuel 2010; Beare et al. 2009). The genetic heterogeneity in the *pil* genes in *C. burnetii* isolates may account for differential virulence or disease outcome of these isolates (acute versus chronic) (Beare et al. 2009). For example, both chronic isolates K and G contain frameshift mutations in *pilC*, a gene that is necessary for secretion in *Francisella* subsp. *novicida* (Table 2.2; Beare et al. 2009; Zogaj et al. 2008).

Potential differences in pathogenic potential identified by comparative genomics must be validated using confirmatory methods to determine if differences in protein expression correlate with disease outcome or virulence potential. The recent description of an axenic media to propagate *C. burnetii* should facilitate such studies (Omsland et al. 2009, 2011a). The ability to grow *C. burnetii* in host-cell free culture conditions has also accelerated the ability to identify transposon mutations in this organism, a valuable approach to test the molecular Koch's postulates for virulence factors.

2.3 Virulence Factors and Host Response Remodeling

2.3.1 Phase Variation

C. burnetii does not produce any obvious cytotoxins and virulence factor identification has been hampered by the inability to genetically modify this organism until recently (Omsland et al. 2011b; Beare et al. 2011). Lipopolysaccharide (LPS) is the only genetically confirmed virulence factor of *C. burnetii* where phase I bacteria are virulent, but phase II bacteria are avirulent (Skultety et al. 1998). Phase variation in *C. burnetii* has been used to describe a phenomenon of passage in a non-immunologically competent host whereby phase I bacteria become phase II upon passage in tissue culture or embryonated eggs (Moos and Hackstadt 1987). The differences in phase I and phase II *C. burnetii* are similar but not identical to the smooth and rough LPS variation of *Enterobacteriaceae* (Amano et al. 1987). During passage rough mutants are favored in growth and the *C. burnetii* population shifts from smooth to predominately rough chemotypes. The shift from phase I to phase II occurs as a result of the loss of the methylated sugars virenose and dihydrohydroxystreptose, in addition to other sugars comprising the O-antigen core (Toman and Kazar 1991; Amano et al. 1987). Furthermore, the transition from phase I to II is often, but not always, associated with a large chromosomal deletion containing a number of genes predicted to be associated with LPS or LOS biosynthesis (in the Nine Mile I RSA493 isolate)

(Thompson et al. 2003). Since LPS variation is associated with a large chromosomal deletion, it is not phase variation in the classic sense as rough LPS variants are not able to switch back to the smooth chemotype. Furthermore, if a mixed population of phase I and phase II *C. burnetii* is used to infect an immuno-competent host (animal infection model) the rough mutants are removed from the population shifting the infection to "phase I".

There are several potential reasons that LPS is a virulence factor for *C. burnetii*. Phase II organisms are complement sensitive, whereas, phase I are resistant to the complement membrane attack complex (Vishwanath and Hackstadt 1988). Secondly, the LPS of phase I *C. burnetii* sterically inhibits the binding of antibodies to proteins on the bacterial surface (Hackstadt 1988). In addition, the LPS is likely involved in other forms of immune evasion that will be discussed in subsequent sections of this chapter.

Virulence factors in pathogenic microorganisms are normally defined by producing mutations in wild type infectious bacteria and looking for avirulent phenotypes *in vitro* or *in vivo*. The sequencing of the first *C. burnetii* isolate identified novel genes predicted to be involved in adhesion, invasion, and host modulation vastly expanded our model knowledge of the pathogenesis of this organism (Seshadri et al. 2003). Prediction of potential virulence factors as well as the characterization of a unique intracellular niche occupied by *C. burnetii* has supported the speculations that its major virulence strategy is immune evasion.

2.3.2 *C. burnetii* Metabolism and Microaerophilic Requirements

The first sequenced *C. burnetii* genome identified a number of enzymes potentially involved in glycolysis, the Embden-Meyerhof-Parnas pathway, the election transport chain, the Entner-Doudoroff pathways, gluconeogenesis, the pentose phosphate pathway, and the tricarboxylic acid (TCA) cycle (Seshadri et al. 2003). This large and seemingly complete biosynthetic capacity was not anticipated as most other obligate intracellular bacteria have an increased dependence on host-derived substrates and predicted reduced metabolic capabilities (Seshadri et al. 2003). In support of this robust biosynthetic capacity, studies have employed a cell-free growth medium for *C. burnetii* (Omsland et al. 2009, 2011a). Early *C. burnetii* metabolic studies in neutral pH buffers suggested that this bacterium had little to no metabolic activity outside of host cells (Ormsbee and Peacock 1964). A breakthrough in metabolic studies was the development of acid activation buffers (pH 4.5) that significantly enhanced *C. burnetii* metabolic potential *in vitro* (Hackstadt and Williams 1981). It is not surprising that acidic conditions support *C. burnetii* metabolism, as once internalized by host cells, *C. burnetii* promotes the development of the lysosome-like parasitophorous vacuole (PV) with a pH of ~4.5–5.3 where replication and growth occur (Maurin and Raoult 1999). It was proposed that the neutral pH of the extracellular environment restricted *C. burnetii* metabolism prolonging viability, whereas, the acidic pH of the intracellular PV triggered metabolic activation and replication (Hackstadt and Williams 1981). The cytosolic pH of *C. burnetii* was maintained near neutrality by both passive and active mechanisms even in highly acidic solutions, suggesting that it behaves like acidophilic bacteria (Hackstadt 1983). The ability to maintain neutral pH during intracellular infections was proposed to energize nutrient transport using proton motive force (Hackstadt 1983). In the development of extracellular growth sustained metabolism for >24 h *C. burnetii* was achieved using CCM (complex *Coxiella* medium) a citrate based acidic medium with complex nutrient sources (Omsland et al. 2008). It was this medium (CCM) that was used as a starting point to develop the current axenic media ACCM that supports *C. burnetii* growth of ~3 $\log_{10}$ over 3–5 days (Omsland et al. 2009). The major breakthrough was the identification of terminal oxidases associated with microaerophilic respiration, which lead to testing metabolic potential at low oxygen levels and the identification of 2.5% oxygen as optimal for *C. burnetii* growth and metabolism (Omsland et al. 2009). Glycolysis and TCA intermediates were most efficiently oxidized at 2.5% oxygen and subsequently incubation of *C. burnetii* for 6 days in ACCM at 2.5% oxygen resulted in an increase in genome equivalents from 8.71×10^4 to 5.49×10^7 genome equivalents (GE)/mL (Omsland et al. 2009). The development of axenic media for host-free growth of *C. burnetii* was a fundamental advancement in the field and will accelerate development of genetic tools and clinical isolate identification (Omsland et al. 2009, 2011a).

2.3.3 *C. burnetii* Uptake

C. burnetii is normally acquired by respiring infectious aerosols. Adherence to host cells is considered the first step in the establishment of an infection, in general, and for *C. burnetii* it is a pre-requisite for internalization. The interactions between host cells and phase I or phase II *C. burnetii* are significantly different, which may represent active virulence mechanisms. Alternatively, immune components may interact with phase I and phase II *C. burnetii* differently. The differences between these interactions likely impact virulence. Interestingly, avirulent phase II *C. burnetii* attach more readily to host cells than virulent phase I *C. burnetii* (Baca et al. 1993). Limited studies have evaluated the host cellular receptors that are involved in binding and uptake of *C. burnetii* leaving an incomplete model of how this bacterium mediates attachment and uptake into its primary target cells, monocytes and macrophages.

In the current model, virulent phase I *C. burnetii* bind to monocytes and macrophages using leukocyte response integrin ($\alpha_v\beta_3$ integrin) since RGD peptides inhibit binding and uptake (Fig. 2.1; Capo et al. 1999; Dellacasagrande et al. 2000). Integrins belong to a family of host cell proteins termed CAMs (cell adhesion molecules) that are involved in mediating cell-cell and cell-extracellular-matrix interactions. Mimicking host cell ligands and binding to CAMs is a common strategy for pathogenic bacteria as carefully demonstrated for the establishment of infections by *Yersinia enterolitica*, *Listeria monocytogenes*, and *Neisseria gonorrhoeae* (Boyle and Finlay 2003). The interaction between integrins and pathogenic bacteria normally

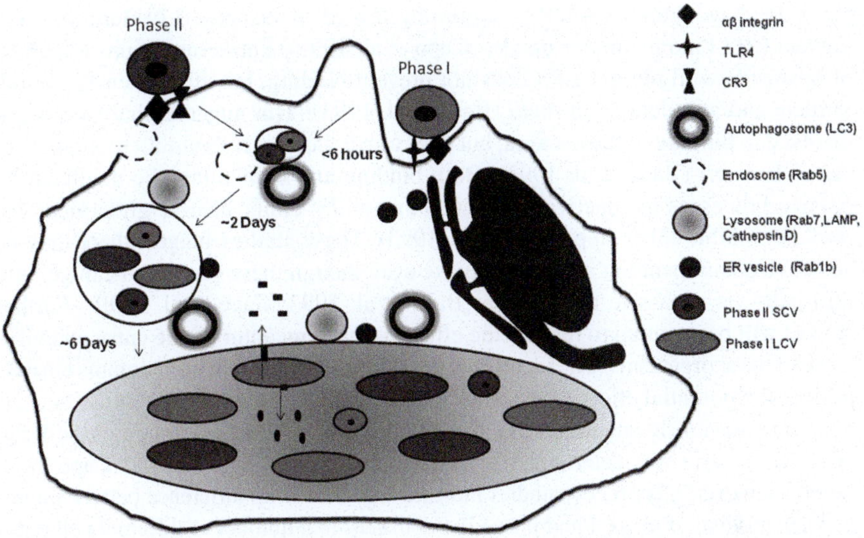

Fig. 2.1 Model of *C. burnetii* invasion and the establishment of the PV. The *SCV* forms of the *phase I* and *phase II* variants both interact with $\alpha_3\beta_v$ integrin and *CR3* or *TLR4* respectively. Adherence of *phase I* variants induces membrane ruffling through the activation of Rho GTPase. Subsequent internalization of both *phase I* and *phase II* into tight fitting phagosomes that interact with endosomal and autophagic vesicles result in recruitment of *Rab5* and *LC3* to the phagosomal membrane. Protein production by *C. burnetii* is required for interaction with the autophagy pathway and for stalling the interaction with lysosomal compartments. After 2 h the phagosome interacts with lysosomal vesicles to acquire *Lamp1*, *Rab7*, and acid hydrolases. The morphological change from *SCV* to *LCV* begin, which coincides with bacterial replication. The mature PV requires continuous protein synthesis by *C. burnetii* to maintain its spacious architecture. The PV retains its fusogenic capabilities and continues to interact with lysosomes, autophagosomes and vesicles from the ER. Although secretion of type IV effectors into the cytoplasm and type II effectors into the lumen of the PV is only depicted in the mature PV, these systems are both presumed to be active and participate in the maturation of the *C. burnetii* containing phagosome into the mature PV

result in actin re-organization and invasion (Boyle and Finlay 2003). In addition to $\alpha_v\beta_3$ integrin attachment, avirulent phase II *C. burnetii* also engage the CR3 receptor, which was speculated to account for the increased binding and internalization of this passage isolate (Fig. 2.1; Capo et al. 1999). Antibodies directed against CR3 inhibited uptake of phase II *C. burnetii,* albeit with a reduced impact compared with antibodies directed to $\alpha_v\beta_3$ integrin suggesting that $\alpha_v\beta_3$ integrin is the dominant receptor for *C. burnetii* and CR3 is a secondary receptor for the uptake of avirulent phase II bacteria (Capo et al. 1999). The ability of phase II *C. burnetii* to interact with multiple receptors may account for its ability to infect tissue culture cells at a ~10–100 fold higher rate than phase I *C. burnetii* (Moos and Hackstadt 1987; Howe et al. 2010; Miller et al. 2004). An alternative hypothesis for the increased infectivity of phase II *C. burnetii* is that the increased hydrophobicity of the bacterial membrane, due to the absence of O-antigen as compared to phase I, increases the interactions with host cells. Interestingly, pre-treatment of host cells with phase I bacteria inhibits phase II phagocytosis via CR3, suggesting that virulent phase I bacteria actively prevent CR3 binding and/or uptake (Capo et al. 1999). Furthermore, pre-treatment of host cells with phase I LPS does not prevent binding, but inhibits entry of both virulent and avirulent *C. burnetii* (Baca et al. 1993). This suggests that LPS is not involved in binding of the bacteria, but does affect the host cells ability to internalize the *C. burnetii* (Baca et al. 1993). Both binding and uptake may be mediated by passive processes as inactivation of *C. burnetii* by either heat or glutaraldehyde does not inhibit either step (Baca et al. 1993). These studies suggest there may be an active component to *C. burnetii* uptake even though these bacterium can get into host cells via a passive mechanism (Baca et al. 1993; Capo et al. 1999). Further studies will be required to resolve the effects of each receptor on the entry and virulence of these organisms. The identity of the adhesin for $\alpha_v\beta_3$ integrin remains elusive, as does the potential effect of other uncharacterized adhesins and receptors.

It was previously suggested that the difference in virulence between phase I and phase II bacteria was determined by intracellular growth and survival (Capo et al. 1999). However, more recent studies indicate that there is no difference between intracellular survival of phase I and phase II *C. burnetii* in a number of different cell types (Howe et al. 2010). Therefore, it has become increasingly important to determine how the different interactions between phase I and phase II *C. burnetii* and host cells elicit differential responses in order to elucidate the differences in virulence between these two isolates. [The different interactions between phase I and phase II with host cells also mediate of the differential pathogenesis of *C. burnetii* variants]. Consequently, future studies will be required to identify receptors and ligands on *C. burnetii* passage variants and host cells to clarify the molecular interactions involved in the establishment of an infection. In addition, many of these studies have been completed using non-clonal passage variants of *C. burnetii*, not the genetically characterized RSA439 phase II clone four, adding uncertainty to interpretation and comparison of the results.

2.3.4 Adherence of *C. burnetii* Induces Cellular Responses

The interactions of *C. burnetii* with the host cell surface elicit several significant responses. The interactions between virulent phase I and avirulent phase II have been shown to cause differential responses, which may impact disease outcome and/or pathogenesis. Pre-incubation of host cells with virulent phase I *C. burnetii* prevented subsequent uptake of avirulent phase II bacteria via CR3 (Capo et al. 1999). Phase I *C. burnetii* excluded CR3 from pseudopodal extensions, therefore, preventing interactions with CR3 during uptake and suggesting that phase I *C. burnetii* actively inhibit interaction with CR3 (Fig. 2.1; Capo et al. 2003). The ability to exclude CR3 from the pseudopodal membrane extension may prevent the interaction of CR3 with phase I and provide a mechanism for the reduction of internalization compared to phase II (Capo et al. 2003). However, it appears more likely that the increase hydrophobicity of *C. burnetii* phase II accounts for increased binding and internalization. Confirmatory studies will be required to resolve the potential effects of $\alpha_v\beta_3$ integrin and CR3 on binding and uptake of virulent phase I (RSA493) and avirulent phase II (RSA439) *C. burnetii*, respectively. For example, these data predict that cells lacking the ability to express CR3 would ingest phase I and phase II *C. burnetii* with equal efficiency. Furthermore, it is unclear if the interaction with $\alpha_v\beta_3$ integrin is critical to the infection process. Future experiments using β_3 knock-out mice may determine if, as predicted, they are relatively resistant to infection by phase I *C. burnetii* (Hodivala-Dilke et al. 1999).

Host cell interactions with both phase I and phase II *C. burnetii* elicited TNF production (Howe et al. 2010; Dellacasagrande et al. 2000). The production of TNF resulting from phase I *C. burnetii* was dependent on adhesion to host cells and may require stimulation through an LPS receptor-mediated process (Dellacasagrande et al. 2000). One study determined that phase I *C. burnetii* induced the secretion of more TNF-α earlier during infection of macrophages, while phase II *C. burnetii* induced maximum TNF-α later in the infection (Tujulin et al. 1999). A separate study using dendritic cells found that phase II, but not phase I *C. burnetii* were able to induce the secretion of TNF (Shannon et al. 2005). These conflicting data suggest that both phase I and phase II *C. burnetii* cause the production of TNF by macrophages, but not dendritic cells. Phase I LPS may block the interaction of phase I membrane proteins with receptors on dendritic cells, which leads to the production of TNF. Further studies will be required to determine precisely how phase I and phase II bacteria illicit production of this and other key inflammatory molecules.

There are other reported discrepancies in how host cells respond to phase I and phase II *C. burnetii*. For example, when phase I *C. burnetii* bound to host cells membrane ruffling was induced, a response that was dependent on actin cytoskeleton reorganization (Meconi et al. 2001). This was a transient phenotype that did not occur when phase II bacteria were delivered to host cells, was dependent on Rho GTPase and dependent on the activation of protein tyrosine kinases since PTK inhibitors abolished the formation of these cellular protrusions (Meconi et al. 2001). Interestingly, PTK activation negatively regulated bacterial phagocytosis, which was contrary to previous reports with other bacteria such a *Shigella* that require PTK activation for internalization (Meconi et al. 2001). Interestingly, phase I LPS alone can induce the formation of filopodia via actin reorganization whereas phase II LPS has no effect on macrophages, indicating that both the membrane ruffling and TNF production were due to the presence of S-LPS in phase I *C. burnetii* (Honstettre et al. 2004). It remains to be determined why *C. burnetii* phase I would actively inhibit uptake into the host cells, especially considering that *C. burnetii* is an obligate intracellular pathogen.

Both purified phase I and phase II LPS induced the production of IL-1α, which comparable to TNFα activates the expression of resistance to intracellular infections (Tujulin et al. 1999). However, only phase I *C. burnetii* were able to induce the production of IL-1α during infection of macrophages, phase II bacteria did not cause the secretion of IL-1α even after 20 h of infection (Tujulin et al. 1999). Together these results suggest that a number of different signaling pathways are induced due to binding of *C. burnetii* to host cells. It remains to be determined if these responses are important for the establishment of a productive infection or alternatively if they are involved in the clearance of the *C. burnetii*.

2.3.5 TLR Interactions

There has been significant, although somewhat conflicting data on how *C. burnetii* pathogen-associated molecular patterns (PAMPs) engage two major toll-like-receptors (TLR), TLR2 and TLR4 (Zamboni et al. 2004; Honstettre et al. 2004; Shannon et al. 2005). Zamboni and colleagues (2004) clearly demonstrated that *C. burnetii* lipid A acts as an antagonist for TLR4 and these results are supported by structural predictions of this tetra-acylated molecule. Yet several TLR4 dependent responses have also been reported. Interactions with *C. burnetii* phase I with TLR4 were involved in bacterial uptake as internalization of bacteria is reduced by ~40% in TLR 4 –/– macrophages (Fig. 2.1; Honstettre et al. 2004). In contrast, the uptake of avirulent phase II *C. burnetii* was similar in wild type and TLR 4 –/– macrophages (Honstettre et al. 2004). This suggests that the interaction between phase I *C. burnetii* and TLR 4 induces uptake, even though it is known that *C. burnetii* LPS is weakly endotoxic and acts as a TLR4 antagonist (Zamboni et al. 2004). Interestingly, phase I LPS but not phase II LPS was capable of inducing membrane ruffling in TLR4 macrophages (Honstettre et al. 2004). This suggested that the interaction of LPS with host cells both induced and inhibited internalization (Honstettre et al. 2004; Meconi et al. 2001). As mentioned earlier, interactions between LPS and host cells caused membrane ruffling through activation of PTK and inhibition of this process caused an increase in phagocytosis of phase I *C. burnetii* (Meconi et al. 2001). However, studies using TLR4 wild-type and knock-out macrophages suggested that membrane ruffling increased phagocytosis (Honstettre et al. 2004).

Phase I and phase II *C. burnetii* differentially activated TLR 2 (Shannon et al. 2005). These studies were performed using maturation of human dendritic cells (DCs) as a read-out for TLR 2 activation. Phase I *C. burnetii* infected and grew within human DCs without activating TLR 2 or causing maturation (Shannon et al. 2005). A separate study found that phase II *C. burnetii* can interact with TLR2 on macrophages, which results in the production of TNF-α (Zamboni et al. 2004). Furthermore, it was determined that TLR2 –/– macrophages were highly permissive to infection with phase II *C. burnetii* indicating that activation of TLR2 may limit replication of the bacteria (Zamboni et al. 2004). This suggests that the inability of phase I *C. burnetii* to interact with TLR2 may account for the virulence of this isolate compared to phase II. The inability of phase I to interact with TRL2 was a passive process where the smooth LPS may shield against a TLR2 agonist (Shannon et al. 2005). The interaction of *C. burnetii* with the immune system probably dictates the outcome of infection. Therefore, production of smooth LPS by *C. burnetii* may prevent TLR ligation avoiding DC maturation, which would be considered an immune evasion strategy preventing a robust Th-1 response that occurs with *C. burnetii* phase II infection (Schromm et al. 2000). The interactions that occur at the host cell surface between *C. burnetii* and host cell receptors may result in the differences in infectivity *in vivo* between phase I and phase II bacteria as the survival and maturation of the phagolysosomes appear similar for these variants in many cell types (Howe et al. 2010).

2.3.6 *Life Cycle*

2.3.6.1 Small and Large Cell Variants

C. burnetii has a very intriguing lifecycle that, like *Chlamydia* spp., involves different cell variants (McCaul and Williams 1981). The infectious particles are referred to as small cell variants (SCV), which are typically rod shaped and 0.2–0.5 μm in length with highly condensed chromatin (Coleman et al. 2004). The SCV is the form found in the extracellular environment and is highly resistant to heat, desiccation, osmotic shock, UV light and various chemical disinfectants (Williams 1991). In addition to being extremely resilient, *C. burnetii* is metabolically inactive as the SCV (Coleman et al. 2004). Originally it was suggested that *C. burnetii* may produce spores that develop into SCV as McCaul and Williams observed a spore-like particle in large cell variants (LCV) (McCaul and Williams 1981). However, failure to stain these particles with spore stain (dipicolinic acid) does not support this hypothesis, and genome sequencing demonstrated a complete lack of homologues for the formation (Williams 1991; Seshadri et al. 2003). LCVs are typically pleomorphic, metabolically active, and can exceed 1 μm in length (Coleman et al. 2004). Upon internalization, the endosome begins maturation into a parasitophorous vacuole (PV) and the *C. burnetii* begin the shift from SCV to LCV (Fig. 2.1). This shift occurs in response to the acidification of the PV (pH 4.7–4.8) (Hackstadt and Williams 1981). At 8 h post-infection PVs can be viewed that contain both SCV and LCV by TEM, however, by 16 h post-infection the PVs only contain LCV indicating that the transition is complete (Coleman et al. 2004). This transformation from SCV to LCV occurs without replication and has been classified as the initial lag period in growth (Coleman et al. 2004).

This transition in host cells correlates with the loss of ScvA, a highly basic DNA binding protein that likely plays a role in chromatin condensation (Coleman et al. 2004). Exponential growth of LCV is observed from 2 to 6 days post-infection followed by a re-appearance of ScvA positive SCV *C. burnetii*, which completes the biphasic lifestyle (Fig. 2.1; Coleman et al. 2004). It is presumed that this intracellular morphogenesis is essential for the virulence of *C. burnetii* and therefore, proteins that are differentially expressed between these two cell variants could be potential virulence factors, such as being involved in immune evasion. Early studies indicated that the protein profiles of SCV and LCV were distinct (McCaul et al. 1991). Proteomic analysis using 2D gel electrophoresis determined that there were 48 proteins that were greater than two-fold more abundant in LCVs than SCVs (Coleman et al. 2007). Of these 48 proteins only 15 were characterized by MS and, not surprisingly, a number of these proteins were predicted to be associated with cellular division and ribosomal functions correlating well with LCVs being the more metabolically active form (Coleman et al. 2007). Further studies will be required to determine the function of the hypothetical proteins identified in this study and to characterize the other 33 differentially expressed proteins, which may include novel virulence factors. There were six proteins that were at least two-fold more abundant in SCVs and these may include factors that are involved in immune evasion of the innate surveillance prior to *C. burnetii* invasion of monocytes and macrophages (Coleman et al. 2007). Consequently, the biphasic lifestyle of *C. burnetii* can clearly be considered a virulence mechanism since it provides a stable form in the environment that transforms into the infective form upon invasion of host macrophages.

2.3.6.2 Parasitophorous Vacuole Development

After internalization into the host cell, *C. burnetii* containing endosomes undergo a series of maturation steps to evolve into a parasitophorous vacuole (PV) with lysosomal characteristics (Heinzen et al. 1996). Typically bacteria that are imported into the cell via phagocytosis are killed and degraded in the phagolysosome, however *C. burnetii* is unique in its ability to survive and replicate in this harsh environment. Normal phagosomal maturation proceeds through a series of increasingly acidic phagosomes that sequentially acquire different proteins by fusion and fission events (Kinchen and Ravichandran 2008). Early endosomal markers include EEA1 and Rab5, while late endosomal markers include Rab7, LAMPS, and cathepsin D (Kinchen and Ravichandran 2008). The parasitophorous vacuole generated by *C. burnetii* is characterized by an abundance of Rab7 (Beron et al. 2002). Furthermore, sequential acquisition of Rab5 and Rab7 indicates that *C. burnetii* transits through the regular endocytic pathway (Romano et al. 2007). Finally, the mature PV contains LAMP-1, the lysosomal hydrolase acid phosphatase, and cathepsin D, all of which are characteristic markers of phagolysosomal compartments (Fig. 2.1; Heinzen et al. 1996; Howe et al. 2010). Maturation into the parasitophorous vacuole is associated with protein synthesis by *C. burnetii* and requires ~24–48 h to transition the large replicative vacuole (Fig. 2.1; Heinzen et al. 1996; Howe et al. 2003). Formalin-fixed and metabolically active *C. burnetii* both end up in PV compartments containing the lysosomal enzyme acid phosphatase, however, the kinetics of acquisition are different. The PV containing metabolically active *C. burnetii* is stalled in its maturation as the acquisition of acid phosphatase may be delayed by up to ~2 h (Howe et al. 2003). This suggests that *C. burnetii* actively modifies this compartment and is not just simply a passive bystander in the events that occur after phagocytosis. The mechanisms that mediate these events are clearly examples of virulence mechanisms but specific proteins are only beginning to be identified. For example, a recent study demonstrated that the type IV secretion apparatus is essential for intracellular replication, suggesting that secretion of effectors through the membrane of the PV is necessary for maturation and replication of *C. burnetii* (Fig. 2.1; Carey et al. 2011). Furthermore, this compartment is not static as it is highly fusogenic and interacts with both endocytic and phagocytic pathways (Fig. 2.1; Heinzen et al. 1996). Interestingly, new data suggest that the PV interacts with the ER; the PV is decorated with Rab1b, which may allow tethering of vesicles derived from the early secretory pathway (Fig. 2.1; Campoy et al. 2010). Interaction with the ER may provide a source of membrane to produce the large PV necessary for *C. burnetii* replication, which again supports the notion that this compartment is not static (Campoy et al. 2010). The PV membrane is enriched in cholesterol, which is unusual for a lysosomal-like compartment (Howe and Heinzen 2006). Furthermore, *C. burnetii* infection causes an up-regulation in expression of the genes involved in cholesterol production during mid-log phase of growth at which point the PV occupies most of the internal host cell volume (Howe and Heinzen 2006; Coleman et al. 2004). Certain inhibitors of the cholesterol pathway completely inhibit the development of the *C. burnetii* containing PV and prevent replication (Howe and Heinzen 2006). As cholesterol synthesis occurs in the ER, it is not surprising that the PV interacts with this compartment. Further studies are necessary to determine the exact players in the development of the PV and the role of cholesterol during infection of *C. burnetii*.

2.3.6.3 Interactions Linking the PV with Autophagy and Apoptosis

It has been proposed that *C. burnetii* initiates the delay in fusion to the terminal PV to interact specifically with the autophagic pathway (Romano et al. 2007). Indeed the mature PV is decorated with L3C a specific marker of autophagic vacuoles (Fig. 2.1; Beron et al. 2002). In addition, activation of the autophagic pathway favors *C. burnetii* infection and bacterial replication (Gutierrez et al. 2005). L3C is recruited to the *C. burnetii* vacuole as early as 5 min after infection and this localization is dependent on protein production by *C. burnetii* as chloramphenicol prevents co-localization (Fig. 2.1; Romano et al. 2007). Another autophagic protein, Beclin 1, also localizes to the *C. burnetii* PV and knockdown of this protein through siRNA reduces the number and size of the generated PVs. The current model is that interactions with the autophagy pathway early in infection result in delivery of nutrients and membranes to the maturing PV (Voth and Heinzen 2007). However, further studies will be required to determine the effects of autophagy on *C. burnetii* infection and to identify the pathogen proteins involved in this recruitment. It was speculated several years ago that the type IV secretion system encoded by the *C. burnetii* plays a role in the maturation of the large parasitophorous vacuole via secretion of effectors into the cytoplasm (Zamboni et al. 2003). Very recently it was determined that *C. burnetii* induces anti-apoptotic activity potentially through the interplay between Beclin 1 and Bcl-2, an event that is also dependent on protein synthesis (Vazquez and Colombo 2009). It is tempting to conclude that the interaction between Beclin1 and Bcl-2 on the *C. burnetii* PV prevents the release of cytochrome C from the mitochondria, a phenomenon previously associated with *C. burnetii* infection and a known function of Bcl-2 (Luhrmann and Roy 2007; Yang et al. 1997). *C. burnetii* infection also causes sustained activation of Akt and Erk1/2 pro-survival proteins, which also contribute to the inhibition of apoptosis induced by *C. burnetii* (Voth and Heinzen 2009). As mentioned above, it was predicted that type IV secreted effectors were involved in the induction the anti-apoptotic state in infected cells and recently the first effector involved in this process, AnkG, was described (Lührmann et al. 2010). AnkG interacts with the host protein p32, which may contribute to its anti-apoptotic abilities (Lührmann et al. 2010). It is not uncommon for intracellular pathogens that prevent apoptosis to interact with several components of the apoptotic pathway; this situation is also observed during *Legionella pneumophila* infections (Abu-Zant et al. 2007). In addition to the anti-apoptotic ability of *C. burnetii*, there were three distinct phases of PV maturation that required protein production: the recruitment of LC3, the delay in phagosomal fusion, and the production of the large PV with fusogenic properties (Romano et al. 2007). Recently, it was discovered that the actin cytoskeleton plays a role in the production and maintenance of the PV (Aguilera et al. 2009). RhoA and Cdc42 likely participate in this process since they are recruited to the PV and chloramphenicol treatment prevented co-localization of RhoA with the PV (Aguilera et al. 2009). It is possible that of several of these processes may be regulated by effectors of the type IV secretion system and future studies will help determine the function of a host of newly identified effectors (Chen et al. 2010).

2.3.7 Ability to Subvert Oxidative Stress

To survive and replicate within the harsh environment of the PV, *C. burnetii* must avoid macrophage microbicidal mechanisms including degradative enzymes such as cathepsin D protease and reactive oxygen species (ROS) that normally traffic to this phagolysosomal-like compartment (Howe et al. 2010). For example, phagocytosis of *C. burnetii* by human neutrophils fails to produce ROS, which is in contrast to *Staphylococcus aureus* which causes a respiratory burst upon phagocytosis (Akporiaye et al. 1990). It was later determined that phagocytosis of *C. burnetii* by human neutrophils prevented assembly of NADPH oxidase on the phagolysosomal membrane, which prevents production of ROS and thus is a mechanism of immune evasion and bacterial persistence (Siemsen et al. 2009). More recent data suggests that a potentially secreted acid phosphatase prevents the assembly of NADPH on the phagolysosome preventing the production of ROS (Hill and Samuel 2010). The acid phosphatase of *C. burnetii* may prevent the phosphorylation of NADPH component necessary for assembly by a similar mechanism to the acid phosphatase of *Francisella tularensis* (Mohapatra et al. 2010). *C. burnetii* may also evade immune responses by affecting the peptide presentation by MHC class II molecules, which are loaded with peptides in lysosomal compartments such as the PV (Brodsky et al. 1999). In addition, *C. burnetii* has a DNA repair system that is up-regulated under oxidative stress and may be important in maintaining the integrity of the genome during replication in the PV (Mertens et al. 2008).

2.4 Conclusions

This chapter examined the phylogenetics, genomics, and virulence of *C. burnetii* focusing on identifying potential virulence factors. *C. burnetii* is unique in its ability to grow and divide in the PV a compartment with many similarities to lysosomal compartments, compartments that are normally associated with degradation. The ability of *C. burnetii* to grow in axenic media will fundamentally advance our knowledge about this unique pathogen and facilitate the identification of unique proteins that are involved in survival in such a harsh environment. *C. burnetii* subverts the host cell immune system through passive and active mechanisms and current model of these events will be important to define new virulence factors and their effects on the host cells.

References

Abu-Zant A, Jones S, Asare R, Suttles J, Price C, Graham J, Kwaik YA (2007) Anti-apoptotic signalling by the Dot/Icm secretion system of L. pneumophila. Cell Microbiol 9:246–264. %U http://dx.doi.org/10.1111/j.1462-5822.2006.00785.x

Aguilera M, Salinas R, Rosales E, Carminati S, Colombo MI, Beron W (2009) Actin dynamics and Rho GTPases regulate the size and formation of parasitophorous vacuoles containing *Coxiella burnetii*. Infect Immun 77:4609–4620. %U http://iai.asm.org/cgi/content/abstract/77/10/4609

Akporiaye ET, Stefanovich D, Tsosie V, Baca G (1990) *Coxiella burnetii* fails to stimulate human neutrophil superoxide anion production. Acta Virol 34:64–70

Al-Khodor S, Price CT, Kalia A, Abu Kwaik Y (2010) Functional diversity of ankyrin repeats in microbial proteins. Trends Microbiol 18:132–139. %U http://www.sciencedirect.com/science/article/B6TD0-4XVGGCC-2/2/9f35429b076db88abfd89cb482b24b77

Amano K, Williams JC, Missler SR, Reinhold VN (1987) Structure and biological relationships of *Coxiella burnetii* lipopolysaccharides. J Biol Chem 262:4740–4747

Arricau-Bouvery N, Hauck Y, Bejaoui A, Frangoulidis D, Bodier CC, Souriau A, Meyer H, Neubauer H, Rodolakis A, Vergnaud G (2006) Molecular characterization of *Coxiella burnetii* isolates by infrequent restriction site-PCR and MLVA typing. BMC Microbiol 6:38

Ayers M, Howell PL, Burrows LL (2010) Architecture of the type II secretion and type IV pilus machineries. Future Microbiol 5:1203–1218

Baca OG, Klassen DA, Aragon AS (1993) Entry of *Coxiella burnetti* into host cells. Acta Virol 37:143–155

Baker S, Hanage WP, Holt KE (2010) Navigating the future of bacterial molecular epidemiology. Curr Opin Microbiol 13:640–645. %U http://www.sciencedirect.com/science/article/B6VS2-511KRNF-1/2/3507c23bc681b509a4ffd1a78df53743

Bansal A (2005) Bioinformatics in microbial biotechnology – a mini review. Microb Cell Fact 4:19. %U http://www.microbialcellfactories.com/content/4/1/19

Beare PA, Samuel JE, Howe D, Virtaneva K, Porcella SF, Heinzen RA (2006) Genetic diversity of the Q fever agent, *Coxiella burnetii*, assessed by microarray-based whole-genome comparisons. J Bacteriol 188:2309–2324

Beare PA, Unsworth N, Andoh M, Voth DE, Omsland A, Gilk SD, Williams KP, Sobral BW, Kupko JJ, Porcella SF, Samuel JE, Heinzen RA (2009) Comparative genomics reveal extensive transposon-mediated genomic plasticity and diversity among potential effector proteins within the genus *Coxiella*. Infect Immun 77:642–656. %U http://iai.asm.org/cgi/content/abstract/77/2/642

Beare P, Sandoz K, Omsland A, Rockey D, Heinzen R (2011) Advances in genetic manipulation of obligate intracellular bacterial pathogens. Front Microbiol 2:97

Beron W, Gutierrez MG, Rabinovitch M, Colombo MI (2002) *Coxiella burnetii* localizes in a Rab7-labeled compartment with autophagic characteristics. Infect Immun 70:5816–5821. %U http://iai.asm.org/cgi/content/abstract/70/10/5816

Boyle EC, Finlay BB (2003) Bacterial pathogenesis: exploiting cellular adherence. Curr Opin Cell Biol 15:633–639. %U http://www.sciencedirect.com/science/article/B6VRW-495VP58-1/2/3044a9c7b18bdda19a8987ae516255e4

Brodsky FM, Lem L, Solache A, Bennett EM (1999) Human pathogen subversion of antigen presentation. Immunol Rev 168:199–215. %U http://dx.doi.org/10.1111/j.1600-065X.1999.tb01294.x

Campoy EM, Zoppino FCM, Colombo MI (2010) The early secretory pathway contributes to the growth of the Coxiella-replicative niche. Infect Immun. doi: 10.1128/IAI.00688-10. %U http://iai.asm.org/cgi/content/abstract/IAI.00688-10v1

Capo C, Lindberg FP, Meconi S, Zaffran Y, Tardei G, Brown EJ, Raoult D, Mege J-L (1999) Subversion of monocyte functions by *Coxiella burnetii*: impairment of the cross-talk between αvß3 integrin and CR3. J Immunol 163:6078–6085. %U http://jimmunol.org/content/163/11/6078.abstract

Capo C, Moynault A, Collette Y, Olive D, Brown EJ, Raoult D, Mege J-L (2003) *Coxiella burnetii* avoids macrophage phagocytosis by interfering with spatial distribution of complement receptor 3. J Immunol 170:4217–4225. %U http://jimmunol.org/content/170/8/4217.abstract

Carey KL, Newton HJ, Lührmann A, Roy CR (2011) The *Coxiella burnetii* Dot/Icm system delivers a unique repertoire of type IV effectors into host cells and is required for intracellular replication. PLoS Pathog 7:e1002056

Casadevall A (2008) Evolution of intracellular pathogens. Annu Rev Microbiol 62:19–33

Chakraborty S, Monfett M, Maier TM, Benach JL, Frank DW, Thanassi DG (2008) Type IV pili in *Francisella tularensis*: roles of pilF and pilT in fiber assembly, host cell adherence, and virulence. Infect Immun 76:2852–2861. %U http://iai.asm.org/cgi/content/abstract/76/7/2852

Chen C, Banga S, Mertens K, Weber MM, Gorbaslieva I, Tan Y, Luo Z-Q, Samuel JE (2010) Large-scale identification and translocation of type IV secretion substrates by *Coxiella burnetii*. Proc Natl Acad Sci USA 107:21755–21760. %U http://www.pnas.org/content/107/50/21755.abstract

Chiang P, Burrows LL (2003) Biofilm formation by hyperpiliated mutants of *Pseudomonas aeruginosa*. J Bacteriol 185:2374–2378. %U http://jb.asm.org/cgi/content/abstract/185/7/2374

Coleman SA, Fischer ER, Howe D, Mead DJ, Heinzen RA (2004) Temporal analysis of *Coxiella burnetii* morphological differentiation. J Bacteriol 186:7344–7352. %U http://jb.asm.org/cgi/content/abstract/186/21/7344

Coleman SA, Fischer ER, Cockrell DC, Voth DE, Howe D, Mead DJ, Samuel JE, Heinzen RA (2007) Proteome and antigen profiling of *Coxiella burnetii* developmental forms. Infect Immun 75:290–298. %U http://iai.asm.org/cgi/content/abstract/75/1/290

Dellacasagrande J, Ghigo E, Machergui-el S, Hammami Toman R, Raoult D, Capo C, Mege J-L (2000) Alpha vbeta 3 integrin and bacterial lipopolysaccharide are involved in *Coxiella burnetii*-stimulated production of tumor necrosis factor by human monocytes. Infect Immun 68:5673–5678. %U http://iai.asm.org/cgi/content/abstract/68/10/5673

Denison A, Thompson H, Massung R (2007) IS1111 insertion sequences of *Coxiella burnetii*: characterization and use for repetitive element PCR-based differentiation of *Coxiella burnetii* isolates. BMC Microbiol 7:91. %U http://www.biomedcentral.com/1471-2180/7/91

Forsberg Å, Guina T (2007) Type II secretion and type IV pili of Francisella. Ann NY Acad Sci 1105:187–201. %U http://dx.doi.org/10.1196/annals.1409.016

Gimenez DF (1964) Staining Rickettsiae in yolk-sac cultures. Stain Technol 39:135–140

Glazunova O, Roux V, Freylikman O, Sekeyova Z, Fournous G, Tyczka J, Tokarevich N, Kovacava E, Marrie TJ, Raoult D (2005) *Coxiella burnetii* genotyping. Emerg Infect Dis 11:1211–1217

Gutierrez MG, Vázquez CL, Munafó DB, Zoppino FCM, Berón W, Rabinovitch M, Colombo MI (2005) Autophagy induction favours the generation and maturation of the Coxiella-replicative vacuoles. Cell Microbiol 7:981–993. %U http://dx.doi.org/10.1111/j.1462-5822.2005.00527.x

Hackstadt T (1983) Estimation of the cytoplasmic pH of *Coxiella burnetii* and effect of substrate oxidation on proton motive force. J Bacteriol 154:591–597. %U http://jb.asm.org/cgi/content/abstract/154/2/591

Hackstadt T (1986) Antigenic variation in the phase I lipopolysaccharide of *Coxiella burnetii* isolates. Infect Immun 52:337–340. %U http://iai.asm.org/cgi/content/abstract/52/1/337

Hackstadt T (1988) Steric hindrance of antibody binding to surface proteins of *Coxiella burnetti* by phase I lipopolysaccharide. Infect Immun 56: 802–807. %U http://iai.asm.org/cgi/content/abstract/56/4/802

Hackstadt T, Williams JC (1981) Stability of the adenosine 5′-triphosphate pool in *Coxiella burnetii*: influence of pH and substrate. J Bacteriol 148:419–425. %U http://jb.asm.org/cgi/content/abstract/148/2/419

Hager AJ, Bolton DL, Pelletier MR, Brittnacher MJ, Gallagher LA, Kaul R, Skerrett SJ, Miller, SI, Guina T (2006) Type IV pili-mediated secretion modulates Francisella virulence. Mol Microbiol 62:227–237. %U http://dx.doi.org/10.1111/j.1365-2958.2006.05365.x

Heinzen R, Stiegler GL, Whiting LL, Schmitt SA, Mallavia LP, Frazier ME (1990) Use of pulsed field gel electrophoresis to differentiate *Coxiella burnetii* Strainsa. Ann NY Acad Sci 590:504–513. %U http://dx.doi.org/10.1111/j.1749-6632.1990.tb42260.x

Heinzen RA, Scidmore MA, Rockey DD, Hackstadt T (1996) Differential interaction with endocytic and exocytic pathways distinguish parasitophorous vacuoles of *Coxiella burnetii* and *Chlamydia trachomatis*. Infect Immun 64:796–809. %U http://iai.asm.org/cgi/content/abstract/64/3/796

Hendrix LR, Samuel JE, Mallavia LP (1991) Differentiation of *Coxiella burnetii* isolates by analysis of restriction-endonuclease-digested DNA separated by SDS-PAGE. J Gen Microbiol 137:269–276. %U http://mic.sgmjournals.org/cgi/content/abstract/137/2/269

Hill J, Samuel JE (2010) *Coxiella burnetii* acid phosphatase: inhibiting the release of reactive oxygen intermediates in polymorphonuclear leukocytes. Infect Immun. doi:10.1128/IAI.01011-10. %U http://iai.asm.org/cgi/content/abstract/IAI.01011-10v1

Hodivala-Dilke KM, Mchugh KP, Tsakiris DA, Rayburn H, Crowley D, Ullman-Culleré M, Ross FP, Coller BS, Teitelbaum S, Hynes RO (1999) β3-integrin–deficient mice are a model for Glanzmann thrombasthenia showing placental defects and reduced survival. J Clin Invest 103:229–238. %U http://www.jci.org/articles/view/5487

Honstettre A, Ghigo E, Moynault A, Capo C, Toman R, Akira S, Takeuchi O, Lepidi H, Raoult D, Mege J-L (2004) Lipopolysaccharide from *Coxiella burnetii* is involved in bacterial phagocytosis, filamentous actin reorganization, and inflammatory responses through toll-like receptor 4. J Immun 172:3695–3703. %U http://www.jimmunol.org/content/172/6/3695.abstract

Hoover TA, Culp DW, Vodkin MH, Williams JC, Thompson HA (2002) Chromosomal DNA deletions explain phenotypic characteristics of two antigenic variants, phase II and RSA 514 (Crazy), of the *Coxiella burnetii* Nine Mile strain. Infect Immun 70:6726–6733

Howe D, Heinzen RA (2006) *Coxiella burnetii* inhabits a cholesterol-rich vacuole and influences cellular cholesterol metabolism. Cell Microbiol 8:496–507

Howe D, Melnicáková J, Barák I, Heinzen RA (2003) Maturation of the *Coxiella burnetii* parasitophorous vacuole requires bacterial protein synthesis but not replication. Cell Microbiol 5:469–480. %U http://dx.doi.org/10.1046/j.1462-5822.2003.00293.x

Howe D, Shannon JG, Winfree S, Dorward DW, Heinzen RA (2010) *Coxiella burnetii* phase I and II variants replicate with similar kinetics in degradative phagolysosome-like compartments of human macrophages. Infect Immun 78:3465–3474. %U http://iai.asm.org/cgi/content/abstract/78/8/3465

Huijsmans CJJ, Schellekens JJA, Wever PC, Toman R, Savelkoul PHM, Janse I, Hermans MHA (2011) SNP-genotyping of a *Coxiella burnetii* outbreak in the Netherlands. Appl Environ Microbiol. doi:10.1128/AEM.02293-10. %U http://aem.asm.org/cgi/content/abstract/AEM.02293-10v1

Jäger C, Willems H, Thiele D, Baljer G (1998) Molecular characterization of *Coxiella burnetii* isolates. Epidemiol Infect 120:157–164. %U http://www.jstor.org/stable/3864204

Jager C, Lautenschlager S, Willems H, Baljer G (2002) *Coxiella burnetii* plasmid types QpDG and QpH1 are closely related and likely identical. Vet Microbiol 89:161–166

Kinchen JM, Ravichandran KS (2008) Phagosome maturation: going through the acid test. Nat Rev Mol Cell Biol 9:781–795. %U http://dx.doi.org/10.1038/nrm2515

Luhrmann A, Roy CR (2007) *Coxiella burnetii* inhibits activation of host cell apoptosis through a mechanism that involves preventing Cytochrome c release from mitochondria. Infect Immun 75:5282–5289. %U http://iai.asm.org/cgi/content/abstract/75/11/5282

Lührmann A, Nogueira CV, Carey KL, Roy CR (2010) Inhibition of pathogen-induced apoptosis by a *Coxiella burnetii* type IV effector protein. Proc Natl Acad Sci USA107:18997–19001. %U http://www.pnas.org/content/107/44/18997.abstract

Marrie TJ (2010) Q fever pneumonia. Infect Dis Clin North Am 24:27–41. %U http://www.sciencedirect.com/science/article/B75J9-4YDSNYG-7/2/d375c369e8de324f6cf9fe0057dea99e

Maurin M, Raoult D (1999) Q fever. Clin Microbiol Rev 12:518–553. %U http://cmr.asm.org/cgi/content/abstract/12/4/518

Mccaul TF, Williams JC (1981) Developmental cycle of *Coxiella burnetii*: structure and morphogenesis of vegetative and sporogenic differentiations. J Bacteriol 147:1063–1076. %U http://jb.asm.org/cgi/content/abstract/147/3/1063

Mccaul TF, Banerjee-Bhatnagar N, Williams JC (1991) Antigenic differences between *Coxiella burnetii* cells revealed by postembedding immunoelectron microscopy and immunoblotting. Infect Immun 59:3243–3253. %U http://iai.asm.org/cgi/content/abstract/59/9/3243

Meconi S, Capo C, Remacle-Bonnet M, Pommier G, Raoult D, Mege J-L (2001) Activation of protein tyrosine kinases by *Coxiella burnetii*: role in actin cytoskeleton reorganization and bacterial phagocytosis. Infect Immun 69:2520–2526. %U http://iai.asm.org/cgi/content/abstract/69/4/2520

Mertens K, Lantsheer L, Ennis DG, Samuel JE (2008) Constitutive SOS expression and damage-inducible addAB-mediated recombinational repair systems for *Coxiella burnetii* as potential adaptations for survival within macrophages. Mol Microbiol 69:1411–1426. %U http://dx.doi.org/10.1111/j.1365-2958.2008.06373.x

Miller JD, Curns AT, Thompson HA (2004) A growth study of *Coxiella burnetii* Nine Mile phase I and phase II in fibroblasts. FEMS Immunol Med Microbiol 42:291–297

Mohapatra NP, Soni S, Rajaram MVS, Dang PM-C, Reilly TJ, El-Benna J, Clay CD, Schlesinger LS, Gunn JS (2010) Francisella acid phosphatases inactivate the NADPH oxidase in human phagocytes. J Immunol 184:5141–5150. %U http://jimmunol.org/content/184/9/5141.abstract

Mollet C, Drancourt M, Raoult D (1998) Determination of *Coxiella burnetii* rpoB sequence and its use for phylogenetic analysis. Gene 207:97–103. %U http://www.sciencedirect.com/science/article/B6T39-3S1PY3W-F/2/db7d6374f04a8b7083a1c0c35617665f

Moos A, Hackstadt T (1987) Comparative virulence of intra- and interstrain lipopolysaccharide variants of *Coxiella burnetii* in the guinea pig model. Infect Immun 55:1144–1150. %U http://iai.asm.org/cgi/content/abstract/55/5/1144

Nguyen SV, Hirai K (1999) Differentiation of *Coxiella burnetii* isolates by sequence determination and PCR-restriction fragment length polymorphism analysis of isocitrate dehydrogenase gene. FEMS Microbiol Lett 180:249–254. %U http://www.sciencedirect.com/science/article/B6T2W-3XT6BRP-M/2/a24324a06b91710f585b988e1b67adc9

Omsland A, Cockrell DC, Fischer ER, Heinzen RA (2008) Sustained axenic metabolic activity by the obligate intracellular bacterium *Coxiella burnetii*. J Bacteriol 190:3203–3212. %U http://jb.asm.org/cgi/content/abstract/190/9/3203

Omsland A, Cockrell DC, Howe D, Fischer ER, Virtaneva K, Sturdevant DE, Porcella SF, Heinzen RA (2009) Host cell-free growth of the Q fever bacterium *Coxiella burnetii*. Proc Natl Acad Sci USA 106:4430–4434. %U http://www.pnas.org/content/106/11/4430.abstract

Omsland A, Beare PA, Hill J, Cockrell DC, Howe D, Hansen B, Samuel JE, Heinzen RA (2011a). Isolation from animal tissue and genetic transformation of *Coxiella burnetii* are facilitated by an improved axenic growth medium. Appl Environ Microbiol 77:3720–3725. %U http://aem.asm.org/cgi/content/abstract/77/11/3720

Omsland A, Beare PA, Hill J, Cockrell DC, Howe D, Hansen B, Samuel JE, Heinzen RA (2011b) Isolation from animal tissue and genetic transformation of *Coxiella burnetii* are facilitated by an improved axenic growth medium. Appl Environ Microbiol 77:3720–3725

Ormsbee RA, Peacock MG (1964) Metabolic activity in *Coxiella burnetii*. J Bacteriol 88:1205–1210. %U http://jb.asm.org/cgi/content/abstract/88/5/1205

Peacock MG, Philip RN, Williams JC, Faulkner RS (1983) Serological evaluation of O fever in humans: enhanced phase I titers of immunoglobulins G and A are diagnostic for Q fever endocarditis. Infect Immun 41:1089–1098.%U http://iai.asm.org/cgi/content/abstract/41/3/1089

Raghavan R, Hicks LD, Minnick, MF (2008) Toxic introns and parasitic intein in *Coxiella burnetii*: legacies of a promiscuous past. J Bacteriol 190:5934–5943. %U http://jb.asm.org/cgi/content/abstract/190/17/5934

Rohmer L, Fong C, Abmayr S, Wasnick M, Larson Freeman T, Radey M, Guina T, Svensson K, Hayden H, Jacobs, M, Gallagher L, Manoil C, Ernst R, Drees B, Buckley D, Haugen E, Bovee D, Zhou Y, Chang J, Levy R, Lim R, Gillett W, Guenthener D, Kang A, Shaffer S, Taylor G, Chen J, Gallis B, D'argenio D, Forsman M, Olson M, Goodlett D, Kaul R, Miller S, Brittnacher M (2007) Comparison of *Francisella tularensis* genomes reveals evolutionary events associated with the emergence of human pathogenic strains Genome Biol 8:R102. %U http://genomebiology.com/2007/8/6/R102

Romano PS, Gutierrez MG, Berón W, Rabinovitch M, Colombo MI (2007) The autophagic pathway is actively modulated by phase II *Coxiella burnetii* to efficiently replicate in the host cell. Cell Microbiol 9:891–909. %U http://dx.doi.org/10.1111/j.1462-5822.2006.00838.x

Roux V, Bergoin M, Lamaze N, Raoult D (1997) Reassessment of the taxonomic position of Rickettsiella grylli. Int J Sys Bacteriol 47:1255–1257. %U http://ijs.sgmjournals.org/cgi/content/abstract/47/4/1255

Russell-Lodrigue KE, Andoh M, Poels MWJ, Shive HR, Weeks BR, Zhang GQ, Tersteeg C, Masegi T, Hotta A, Yamaguchi T, Fukushi H, Hirai K, Mcmurray DN, Samuel JE (2009) *Coxiella burnetii* isolates cause genogroup-specific virulence in mouse and guinea pig models of acute Q fever. Infect Immun 77:5640–5650. %U http://iai.asm.org/cgi/content/abstract/77/12/5640

Samuel JE, Frazier ME, Mallavia LP (1985) Correlation of plasmid type and disease caused by *Coxiella burnetii*. Infect Immun 49:775–779. %U http://iai.asm.org/cgi/content/abstract/49/3/775

Savinelli EA, Mallavia LP (1990) Comparison of *Coxiella burnetii* plasmids to homologous chromosomal sequences present in a plasmidless endocarditis-causing isolate. Ann NY Acad Sci 590:523–533

Schromm A, Brandenburg K, Loppnow H, Moran AP, Koch MHJ, Rietschel E, Seydel U (2000) Biological activities of lipopolysaccharides are determined by the shape of their lipid A portion. Eur J Biochem 267:2008–2013

Sekeyová Z, Roux V, Raoult D (1999) Intraspecies diversity of *Coxiella burnetii* as revealed by com1 and mucZ sequence comparison. FEMS Microbiol Lett 180:61–67

Seshadri R, Samuel J (2005) Genome analysis of *Coxiella burnetii* species: insights into pathogenesis and evolution and implications for biodefense. Ann NY Acad Sci 1063:442–450. %U http://dx.doi.org/10.1196/annals.1355.063

Seshadri R, Paulsen IT, Eisen JA, Read TD, Nelson KE, Nelson WC, Ward NL, Tettelin H, Davidsen TM, Beanan MJ, Deboy RT, Daugherty SC, Brinkac LM, Madupu R, Dodson RJ, Khouri HM, Lee KH, Carty HA, Scanlan D, Heinzen RA, Thompson HA, Samuel JE, Fraser CM, Heidelberg JF (2003) Complete genome sequence of the Q-fever pathogen *Coxiella burnetii*. ProcNatl Acad Sci USA 100:5455–5460. %U http://www.pnas.org/content/100/9/5455.abstract

Shannon JG, Howe D, Heinzen RA (2005) Virulent *Coxiella burnetii* does not activate human dendritic cells: role of lipopolysaccharide as a shielding molecule. Proc Natl Acad Sci USA 102:8722–8727. %U http://www.pnas.org/content/102/24/8722.abstract

Siemsen DW, Kirpotina LN, Jutila MA, Quinn MT (2009) Inhibition of the human neutrophil NADPH oxidase by *Coxiella burnetii*. Microbes Infect 11:671–679

Skultety L, Toman R, Patoprsty V (1998) A comparative study of lipopolysaccharides from two *Coxiella burnetii* strains considered to be associated with acute and chronic Q fever. Carbohydr Polym 35:189–194

Stein A, Saunders NA, Taylor AG, Raoult D (1993) Phylogenic homogeneity of *Coxiella burnetii* strains as determinated by 16S ribosomal RNA sequencing. FEMS Microbiol Lett 113:339–344. %U http://www.sciencedirect.com/science/article/B6T2W-476HN4R-15/2/5f67777b44bc4d8ddfc89c8a15f7ad10

Svraka S, Toman R, Skultety L, Slaba K, Homan WL (2006) Establishment of a genotyping scheme for *Coxiella burnetii*. FEMS Microbiol Lett 254:268–274. %U http://dx.doi.org/10.1111/j.1574-6968.2005.00036.x

Thiele D, Willems H (1994) Is plasmid based differentiation of *Coxiella burnetii* in 'Acute' and 'Chronic' isolates still valid? Eur J Epidemiol 10:427–434. %U http://www.jstor.org/stable/3520970

Thompson HA, Hoover TA, Vodkin MH, Shaw EI (2003) Do chromosomal deletions in the lipopolysaccharide biosynthetic regions explain all cases of phase variation in *Coxiella burnetii* strains? Ann NY Acad Sci 990:664–670. %U http://dx.doi.org/10.1111/j.1749-6632.2003.tb07441.x

Toman R, Kazar J (1991) Evidence for the structural heterogeneity of the polysaccharide component of *Coxiella burnetii* strain Nine Mile lipopolysaccharide. Acta Virol 35:531–537

Tujulin E, Lilliehöök B, Macellaro A, Sjöstedt A, Norlander L (1999) Early cytokine induction in mouse P388D1 macrophages infected by *Coxiella burnetii*. Vet Immunol Immunopathol 68:159–168. %U http://www.sciencedirect.com/science/article/pii/S0165242799000239

Vazquez CL, Colombo MI (2009) *Coxiella burnetii* modulates Beclin 1 and Bcl-2, preventing host cell apoptosis to generate a persistent bacterial infection. Cell Death Differ 17:421–438. %U http://dx.doi.org/10.1038/cdd.2009.129

Vishwanath S, Hackstadt T (1988) Lipopolysaccharide phase variation determines the complement-mediated serum susceptibility of *Coxiella burnetii*. Infect Immun 56:40–44. %U http://iai.asm.org/cgi/content/abstract/56/1/40

Vodkin MH, Williams JC, Stephenson EH (1986) Genetic heterogeneity among isolates of *Coxiella burnetii*. J Gen Microbiol 132:455–463

Voth DE, Heinzen RA (2007) Lounging in a lysosome: the intracellular lifestyle of *Coxiella burnetii*. Cell Microbiol 9:829–840. %U http://dx.doi.org/10.1111/j.1462-5822.2007.00901.x

Voth DE, Heinzen RA (2009) Sustained activation of Akt and Erk1/2 is required for *Coxiella burnetii* antiapoptotic activity. Infect Immun 77:205–213. %U http://iai.asm.org/cgi/content/abstract/77/1/205

Voth DE, Howe D, Beare PA, Vogel JP, Unsworth N, Samuel JE, Heinzen RA (2009) The *Coxiella burnetii* ankyrin repeat domain-containing protein family is heterogeneous, with C-terminal truncations that influence Dot/Icm-mediated secretion. J Bacteriol 191:4232–4242. %U http://jb.asm.org/cgi/content/abstract/191/13/4232

Weisburg WG, Dobson ME, Samuel JE, Dasch GA, Mallavia LP, Baca O, Mandelco L, Sechrest JE, Weiss E, Woese CR (1989) Phylogenetic diversity of the Rickettsiae. J Bacteriol 171:4202–4206. %U http://jb.asm.org/cgi/content/abstract/171/8/4202

Willems H, Ritter M, Jager C, Thiele D (1997) Plasmid-homologous sequences in the chromosome of plasmidless *Coxiella burnetii* Scurry Q217. J Bacteriol 179:3293–3297

Williams JC (1991) Infectivity, virulence, and pathogenicity of *Coxiella burnetti* for various hosts. In: Williams JC, Thompson HA (eds) Q fever: the biology of *Coxiella burnetii*. CRC Press, Boca Raton

Yang J, Liu X, Bhalla K, Kim CN, Ibrado AM, Cai J, Peng T-I, Jones DP, Wang X (1997) Prevention of apoptosis by Bcl-2: release of cytochrome c from mitochondria blocked. Science 275:1129–1132. %U http://www.sciencemag.org/content/275/5303/1129.abstract

Zamboni DS, Mortara RA, Freymuller E, Rabinovitch M (2002) Mouse resident peritoneal macrophages partially control in vitro infection with *Coxiella burnetii* phase II. Microbes Infect 4:591–598. %U http://www.sciencedirect.com/science/article/B6VPN-45JPMBJ-2/2/e343d26d0e629d99e221099e0e20e1f8

Zamboni DS, Mcgrath S, Rabinovitch M, Roy CR (2003) *Coxiella burnetii* express type IV secretion system proteins that function similarly to components of the *Legionella pneumophila* Dot/Icm system. Mol Microbiol 49:965–976. %U http://dx.doi.org/10.1046/j.1365-2958.2003.03626.x

Zamboni DS, Campos MA, Torrecilhas ACT, Kiss K, Samuel JE, Golenbock DT, Lauw FN, Roy CR, Almeida IC, Gazzinelli RT (2004) Stimulation of toll-like receptor 2 by *Coxiella burnetii* is required for macrophage production of pro-inflammatory cytokines and resistance to infection. J Biol Chem 279:54405–54415. %U http://www.jbc.org/content/279/52/54405.abstract

Zhang GQ, To H, Yamaguchi T, Fukushi H, Hirai K (1997) Differentiation of *Coxiella burnetii* by sequence analysis of the gene (com1) encoding a 27-kDa outer membrane protein. Microbiol Immunol 41:871–877

Zogaj X, Chakraborty S, Liu J, Thanassi D G, Klose KE (2008) Characterization of the *Francisella tularensis* subsp. novicida type IV pilus. Microbiology 154:2139–2150. %U http://mic.sgmjournals.org/cgi/content/abstract/154/7/2139

Chapter 3
Defense Mechanisms Against Oxidative Stress in *Coxiella burnetii*: Adaptation to a Unique Intracellular Niche

Katja Mertens and James E. Samuel

Contents

Abstract Survival of intracellular pathogenic bacteria depends on the ability to resist host-mediated degradation and to generate a replicative niche within the host. Usually, after internalization by professional phagocytic cells, the bacteria containing vacuole or phagosome traffics through the endocytic pathway, progressively acidifies

K. Mertens
Friedrich-Loeffler-Institute, Federal Research Institute for Animal Health,
Naumburger Str. 960, 07743 Jena, Germany

J.E. Samuel (✉)
Department of Microbial and Molecular Pathogenesis, College of Medicine, Texas A&M Health Science Center, 3112 Medical Research and Education Building, Bryan, TX 77807-3260, USA
e-mail: jsamuel@medicine.tamhsc.edu

R. Toman et al. (eds.), *Coxiella burnetii: Recent Advances and New Perspectives in Research of the Q Fever Bacterium*, Advances in Experimental Medicine and Biology 984,
DOI 10.1007/978-94-007-4315-1_3,

and develops into a degradative mature phagolysosome. In this environment bacteria are exposed to a wide variety of anti-microbial agents, such as defensins, proteases, and reactive oxygen species (ROS) and reactive nitrogen species (RNS). Most parasitizing bacteria have evolved strategies to interfere with this maturation process and to direct the development of an environment that supports survival and replication. *C. burnetii* also follows this paradigm, but directs the biogenesis of a unique parasitophorous vacuole (PV), which resembles, yet is distinct from a terminal phagolysosome. Within the environment of the PV, *C. burnetii* is exposed to varying levels of ROS and RNS, which represent the primary defense mechanism of the host cell against this invading microorganism. Major mediators for ROS and RNS are superoxide (O_2^-) and nitric oxide (NO^*), generated by the cellular NADPH oxidase (phox) and inducible nitric oxide synthase (iNOS), respectively. *C. burnetii* employs several strategies to evade oxidative stress; on the host side (i) delaying phagolysosome fusion and (ii) inhibiting cellular NADPH oxidase. On the bacterial side, maintaining genome stability by (iii) evolving a preference for a low iron environment, (iv) expressing a minimal and likely crucial set of DNA repair genes and (v) detoxifying the PV by ROS and RNS degrading enzymes. Overall defense mechanisms in *C. burnetii* against oxidative and nitrosative stress and the regulation thereof are not fully defined and our knowledge is mainly based on genome sequence information. Comparison with *E. coli* as a model bacterium reveals that defense strategies of *C. burnetii* differ significantly and emphasize a highly adaptive evolution to this harsh and unique niche.

Keywords Oxidative stress • ROS and RNS scavenging enzymes • OxyR regulon • Fur regulon • SOS response

3.1 Oxidative Stress

3.1.1 Sources of ROS and RNS

Phagocytic cells, as main mediators of the innate immune response, carry two specialized and inducible systems for production of ROS and RNS; the NADPH phagocyte oxidase (Phox) and inducible nitric oxide synthase (iNOS). Phox and iNOS are expressed in mononuclear and polymorphonuclear phagocytes, with considerably higher ROS production in neutrophils and higher RNS production in macrophages (Nathan and Shiloh 2000). The phagocyte oxidase is a multimeric complex that generates superoxide radicals (O_2^-) by electron transfer to oxygen. Components of this multimeric complex are pre-synthesized in phagocytic cells with two membrane-bound components, $p22^{phox}$ and $gp91^{phox}$, and three cytosolic components, $p40^{phox}$, $p47^{phox}$ and $p67^{phox}$. The membrane-bound subunits form a heterodimeric flavohemeprotein and represent the catalytic center of Phox, designated as flavocytochrome b_{558} (Bylund et al. 2010). Upon activation by phagocytosis of particles the cytosolic Phox components translocate to the phagosome or cytoplasmic membranes, assemble with flavocytochrome b_{558} and form the functional

NADPH phagocyte oxidase complex (Badwey et al. 1980; Ohno et al. 1982). Superoxide radicals are then generated by electron transfer from NADPH, bound to the flavocytochrome b_{558} component, and transferred to oxygen across the membrane. O_2^- spontaneously dismutates to hydrogen peroxide (H_2O_2) and generates more reactive oxygen intermediates, such as the highly toxic hydroxyl radicals (*OH) via the Fenton reaction or hypochlorous acid (HOCl) in presence of the cellular myeloperoxidase (MPO) (Imlay 2008; Tlili et al. 2011; Imlay et al. 1988).

Activity of iNOS is regulated on the transcriptional level and requires *de novo* synthesis upon induction by pro-inflammatory cytokines, such as IFN-γ and TNF-α or TLR ligands (LPS) (Vila-del Sol et al. 2007). The active iNOS functions as a dimer with each monomer containing an N-terminal oxygenase domain and a C-terminal reductase domain. Dimerization occurs in presence of the cofactors H_4biopterin (folic acid), heme and the substrate L-arginine. The mature and active enzyme then traffics in an actin-dependent manner to the cytoplasmic side of the phagosome and NO* radicals are produced in a two-step reaction involving both monomer subunits (Baek et al. 1993; Albakri and Stuehr 1996; Davis et al. 2007). One monomer transfers electrons from cytosolic NADPH to the heme group of the adjacent monomer via FAD and FMN, which then converts L-arginine and oxygen to NO* and L-citrulline (Mayer and Hemmens 1997). Unlike superoxide, NO* is a gaseous radical and can diffuse freely across membranes to reach its intra-phagosomal targets (Webb et al. 2001). In this environment NO* radicals can encounter and react with ROS and catalyze the production of various toxic intermediates, such as nitrite ($^*NO_2$) and nitrogen dioxide (NO_2).

ROS and RNS act synergistic to exert their toxic effects on phagocytosed bacteria and sequentially contribute to infection control. Activation of Phox during the early events contributes to control of bacterial replication, whereas iNOS expression and RNS production synergize with ROS at a later stage of infection and prolongs the ROS-mediated effects (Lindgren et al. 2004; Vazquez-Torres et al. 2000). Diverse studies using animal models with defects in either Phox or iNOS or both emphasize the importance of these effectors of the innate immune response (Brennan et al. 2004; Cirillo et al. 2009; Qiu et al. 2009). Mice deficient for Phox, modeling the human chronic granulomatous disease, or iNOS are highly susceptible to infections with several pathogens and display a more severe course of infection than wild-type mice (Pollock et al. 1995; Lindgren et al. 2004). However bacterial latency is closely associated with iNOS expression and endogenous reactivation occurs when cell-mediated immunity becomes suppressed (Hisert et al. 2004; Stenger et al. 1996). Overall ROS and RNS are beneficial for the host, but can also lead to oxidative tissue damage, if the production of these radicals is unregulated (Callahan et al. 2001; Thiemermann and Vane 1990).

3.1.2 Mechanisms of Cellular Damage

Oxidative stress is defined as the presence of elevated levels of reactive oxygen species that cause damage to the molecular processes of an organism (Rosner and Storz 1997). The highly reactive products of Phox and iNOS mediated reactions,

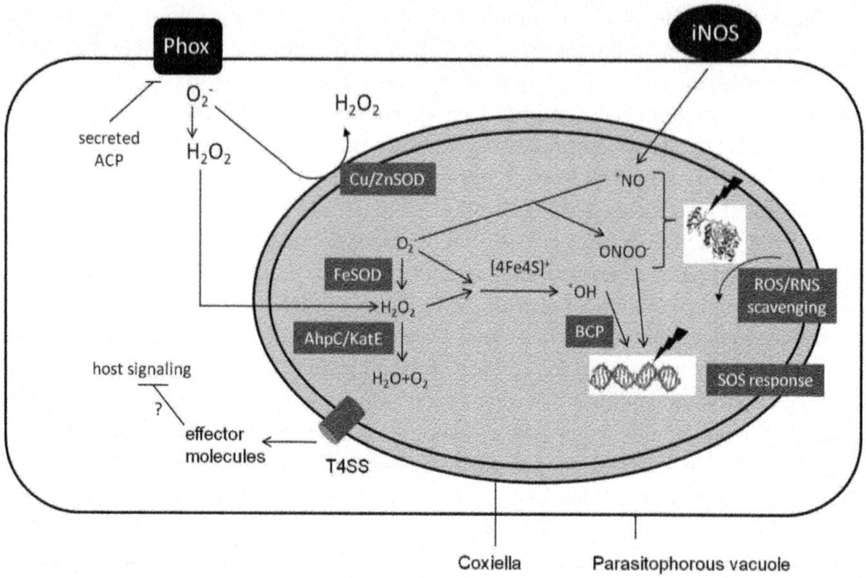

Fig. 3.1 Defense mechanisms against ROS and RNS mediated damage in *C. burnetii*. ROS and RNS generated by the host cell targets bacterial proteins by direct oxidation or oxidative release of Fe^{3+} from $[4Fe4S]^{+}$cluster. The released iron can generate $^{*}OH$ radicals via the Fenton reaction and causes oxidative damage of the DNA. *C. burnetii* expresses several detoxification enzymes, such as a cytoplasmic FeSOD, a periplasmic Cu/ZnSOD, KatE or the two alkyl hydroperoxide reductases, AhpC1 and AhpC2D, for direct inactivation of ROS. ROS and RNS scavenging systems reverse directly oxidative damage of proteins or function as radical sink. Oxidative damage of the DNA is prevented by a DNA-binding peroxidase, BCP, or by constitutive expression of the SOS response

O_2^- and NO^*, catalyze the formation of a variety of toxic intermediates with different reactivity and molecular targets (Fig. 3.1). Main targets for ROS are $[4Fe\text{-}4S]^{2+}$ cluster-containing enzymes, such as dehydratases of the Krebs cycle or DNA. Both, O_2^- and H_2O_2 can freely penetrate enzymes and catalyze the release of ferric iron (Fe^{3+}) from iron-sulfur centers. This inactivates the enzymatic function and can cause a metabolic block (Imlay 2008). Additionally, H_2O_2 directly oxidizes free unincorporated or DNA associated ferrous iron (Fe^{2+}), which results in formation of hydroxyl radicals and Fe^{3+} via the Fenton reaction. Hydroxyl radicals are potent oxidants which react at nearly diffusion-limited rates with most organic substances (Imlay and Linn 1988). $^{*}OH$ mediated oxidative attack of the DNA backbone either at the base or sugar residue causes base modifications or sugar fragmentation that ultimately results in mutations or single strand breaks, respectively. Toxic effects of hydrogen peroxide are described as 'two-mode killing'; low concentrations of H_2O_2 (1–3 mM H_2O_2) are more deleterious to DNA, whereas high concentrations (>20 mM) cause uncharacterized damage to diverse cellular components and eventually death (Imlay et al. 1988; Imlay and Linn 1988).

RNS mediated killing of intracellular bacteria is more complex and depends on the local redox environment. NO^* can directly react with DNA, prosthetic groups of proteins, proteins themselves and lipids. Many of these molecules are involved in the cellular metabolism and information processing and hence, damage has detrimental effects on the bacterial survival. Oxidative attack of the DNA by NO^* causes deamination or formation of base adducts which eventually result in abasic sites or DNA strand breaks (Juedes and Wogan 1996; Wink et al. 1991). Further DNA synthesis and replication is affected due to inhibition of the ribonucleotide reductase, likely by *S*-nitrosylation of critical amino acid residues, and mobilization of zinc from metalloproteins (Fujikura et al. 2009). *S*-nitrosylation, ADP-ribosylation, tyrosine nitration and NO^* oxidation of $[4Fe\text{-}4S]^{2+}$ cluster and heme groups are the main mechanisms of RNS damage and these inhibit many bacterial processes, particularly sensitive are bacterial respiration and enzymes of central metabolic pathways (Bogdan 2001; Ouellet et al. 2002).

3.1.3 *C. burnetii*; the Organism and Its Intracellular Niche

C. burnetii is a Gram-negative pleomorphic coccobacillus and belongs to the γ subdivision of the *Proteobacteria*, with *Legionella pneumophila* as its closest pathogen relative (Stein et al. 1993; Weisburg et al. 1989). Infections in humans with *C. burnetii* cause Q (query) fever, which manifests as an acute flu-like and often self-limiting illness or less common as chronic disease in form of endocarditis or hepatitis (Maurin and Raoult 1999). The genus *Coxiella* consists of a single species and isolates are categorized based upon genome sequence information and plasmid content into six genomic groups (Hendrix et al. 1991; Samuel et al. 1985). These groups correlate with disease outcome and isolates belonging to genomic groups I, II and III are closely associated with an acute infection, whereas isolates belonging to group IV and V are mainly associated with chronic Q fever (Beare et al. 2009; Sekeyova et al. 1999). Group VI isolates are of low virulence and were originally isolated from rodents (Stoenner and Lackman 1960).

C. burnetii is a naturally obligate intracellular bacterium and can be propagated in diverse tissue cell culture types (Baca and Paretsky 1983). The bacteria replicate to high numbers within the parasitophorous vacuole (PV), which can occupy nearly the whole cytoplasm of the host cell without affecting viability. The PV is characterized by its high fusogenicity with other *Coxiella*-containing vacuoles or vesicles from the endosomal cascade. The mature PV is slightly acidic (pH ~ 4.5) and acquires late endosomal and lysosomal markers, such as acid phosphatase, cathepsin D, Rab7 and the lysosomal glycoproteins LAMP-1 and LAMP-2 and resembles a mature phagolysosome (Howe and Mallavia 2000; Beron et al. 2002; Heinzen et al. 1996; Sauer et al. 2005). Presence of the vacuolar H^+ ATPase promotes acidification and generation of a H^+ gradient across the phagosome membrane, that normally impairs bacterial metabolism, favors the activity of hydrolytic enzymes and promotes nutrient exclusion from the phagosome (Beyenbach and Wieczorek 2006). *C. burnetii* thrives in this harsh environment and the acidic pH within the PV is an absolute requirement for metabolic activation, an as 'biochemical stratagem' termed process (Hackstadt and Williams 1981).

Active modification of the phagolysosome maturation process by secretion of effector molecules in a type IV secretion system (T4SS)-dependent manner has been described for several pathogens, such as *L. pneumophila* and *Brucella abortus* (de Jong et al. 2008; Hubber and Roy 2010). *C. burnetii* contains nearly all of the T4SS encoding *icm/dot* genes and may alter the phagosome-lysosome fusion in a novel way (Sexton and Vogel 2002; Chen et al. 2010.). Interception with the autophagic pathway for sustained nutrient delivery has been suggested and supports the idea of PV alteration by *C. burnetii* (Gutierrez et al. 2005).

Recently, axenic cultivation of *C. burnetii* has been developed and requires a complex nutrient-rich medium and microaerophilic conditions with 2.5% oxygen and 5% carbon dioxide (Omsland et al. 2008, 2009). Microaerophily is not well understood in *C. burnetii*, but the PV is considered as a low-oxygen environment and extended survival under atmospheric oxygen is attributed to the metabolic quiescent and inert small cell variant (Via et al. 2008; McCaul and Williams 1981). *C. burnetii* and several other intracellular pathogenic bacteria encode for a potential high oxygen affinity type oxidase (*cydAB*), which, for instance, maintains a transmembrane potential in *E. coli* for ATP synthesis in a low oxygen environment (Jones et al. 2007). This respiratory switch might represent one of many other uncharacterized adaptations of *C. burnetii* to its unique niche (Kana et al. 2001; Omsland et al. 2009).

It is debatable to what extent *C. burnetii* is exposed to ROS and RNS within the PV. Early reports indicated that infection of monocytes with *C. burnetii* did not elicit a detectable oxidative burst and therefore, in absence of O_2^-, no $ONOO^-$ is produced (Akporiaye et al. 1990; Howe et al. 2002). Other studies showed that IFN-γ activated macrophages indeed produced ROS and RNS and restricted replication of *C. burnetii* (Brennan et al. 2004). Comparison of growth rates in iNOS or Phox deficient IFN-γ activated macrophages indicated that both, ROS and RNS, contributed to replication control. Further, in an acidified environment like the PV, O_2^- penetrates lipid membranes and results in production of toxic ROS and RNS intermediates (Korshunov and Imlay 2002). However, with adaptation to a low-oxygen environment, the stepwise loss of oxidative stress defenses would be expected. Genome comparison of *C. burnetii* isolates indicates differences in the repertoire of antioxidant genes among isolates from different genomic groups (Table 3.1). For instance, most isolates associated with acute Q fever, belonging to genomic groups I and II, encode for a severely truncated catalase gene (*katE*) whereas the low-virulent isolate Dugway, belonging to genomic group VI, carries a frame-shifted superoxide dismutase gene (*sodC*) (Beare et al. 2009). The loss of these important oxidative stress defense mechanisms might indicates that adaptation to an obligate intracellular and microaerophilic lifestyle is an ongoing process in *C. burnetii*. Noticeable is that the loss of a functional catalase in these isolates is not associated with an apparent loss of virulence.

Table 3.1 Oxidative defense mechanisms of *C. burnetii*

Gene	CBU No.	Activity	Function in oxidative stress	Reference
Primary defense mechanisms direct detoxification				
sodB	1708	Cytoplasmic FeSOD	Dismutation of O_2^-	Heinzen et al. (1990, 1992)
sodC	1822	Periplasmic Cu/ZnSOD		R.E. Brennan, personal communication
katE	0281	Cytoplasmic catalase	Removal of H_2O_2, truncated in group I, II and V isolates	Beare et al. (2009)
ahpC1	1706	Alkyl alkyl hydroperoxide reductases	Detoxification of H_2O_2, ROOH and $ONOO^-$	LeBlanc et al. (2006); Baker et al. (2001); Hillas et al. (2000)
ahpC2D	1477/1478	Alkyl alkyl hydroperoxide reductases		
oxyR	1476	Transcription factor	Peroxide-sensing transcription factor	ERGO™ Integrated Genomics
trxB/trx	1193/2087	Thioredoxin reductase/thioredoxin	Regeneration of peroxiredoxins	ERGO™ Integrated Genomics
gshB	1875	Glutathione synthase	Repair of oxidized proteins, radical sink	ERGO™ Integrated Genomics
grxC	1520	Glutaredoxin		ERGO™ Integrated Genomics
msrA	1306	Methionine sulphoxide reductase		ERGO™ Integrated Genomics
Iron homeostasis				
fur	1301	Fur-like protein	Ferric uptake regulator	Briggs et al. (2008)
feoB		Ferrous iron transport protein B	Ferrous iron uptake	
Frg1		Fur-regulated gene 1	Putative iron-binding protein	
DNA repair				
bcp	0963	Bacterioferritin comigratory protein	DNA binding peroxidase	Hicks et al. (2010)
nth	1697	Endonuclease III, glycosylase	Base excision repair, repair of oxidative DNA damage	Wang et al. (2005)
mutT	0148	8-oxo-dGTPase		ERGO™ Integrated Genomics
mutY	0940	A/G-specific adenine DNAglycosylase		ERGO™ Integrated Genomics
tag	0383	DNA-3-methyladenine glycosylase		Grzesiuk et al. (2001)
mpg	0930	3-methylpurine DNA glycosylase		Aamodt et al. (2004)
mutL	1083	DNA mismatch repair protein MutL	Methyl mismatch repair, repair of oxidative DNA damage, suppression of DNA recombination	Wang et al. (2005)
mutS	1056	DNA mismatch repair protein MutS		

(continued)

Table 3.1 (continued)

Gene	CBU No.	Activity	Function in oxidative stress	Reference
uvrA	0274	Excinuclease complex	Nucleotide excision repair, repair of oxidative DNA damage	Cabusora et al. (2005); Park et al. (2010)
uvrB	0518			
uvrC	1185			
uvrD	2054	DNA helicase II		
recA	1054	ssDNA binding, co-protease	SOS response regulator	Mertens et al. (2008)
recA	1054	ssDNA binding, co-protease	Recombinational DNA repair	Mertens et al. (2008)
ssb	0271	ssDNA-binding protein		
addAB	1229/1230	Helicase-exonuclease type V family protein		
recFOR	0003/1501/0657	DNA replication, recombination and repair		
ruvABC	1568/1570/1567	Holiday junction helicase and nuclease		
recN	1297	DNA repair protein		
recG	0305	DNA helicase		
radA	1422	DNA repair protein		
Secondary defense mechanisms				
acp	0335	Acid phosphatase	Inhibition of oxidative burst	Hill and Samuel (2010)
icm/dot	1548–1578	Type IV secretion system	Delay phagolysosome fusion and interfering with host signaling	Chen et al. (2010), Voth and Heinzen (2009)

3.2 Defense Mechanisms Against Oxidative Stress in *C. burnetii*

Pathogenic bacteria employ several strategies to resist the toxic effects of ROS and RNS, which are categorized in evasion or suppression mechanisms and direct enzymatic degradation or scavenging of radicals. Many of these defense mechanisms function against both mediators, since cellular targets of ROS and RNS overlap. Strategies employed by *C. burnetii* to resist oxidative stress are crucial for intracellular survival and can be categorized into (i) suppression of the oxidative burst by secretion of an acid phosphatase; (ii) enzymatic degradation of radicals by expression of superoxide dismutases (SODs), catalase and peroxiredoxins as well as the expression of small scavenging molecules as radical sink or for direct repair of modified proteins; (iii) preference for a low-iron environment to avoid additional oxidative stress via the Fenton reaction and (iv) expression of DNA repair mechanisms to ensure genome stability and integrity. These adaptive defense mechanisms are usually tightly regulated by peroxide-sensing transcription factors and are well defined for *E. coli* (Storz and Imlay 1999). Following the enterobacterial paradigm, defenses in *C. burnetii* might be responsive to hydrogen peroxide exposure (OxyR) or to iron availability (Fur), but the regulatory network is usually much more complex and includes alternative sigma factors and other signaling pathways to ensure survival within the changing environments during the infection process.

3.2.1 Suppression of ROS Production by Secretion of an Acid Phosphatase

Suppression of ROS production in polymorphonuclear monocytes (PMNs) or macrophages represents an important evasion strategy employed by *C. burnetii* and depends on the secretion of an acid phosphatase (ACP, CBU0335) (Hill and Samuel 2010; Siemsen et al. 2009). Early studies noted that infection of human neutrophils with *C. burnetii* does not elicit a strong oxidative burst and this failure was later correlated with an acid phosphatase isolated from *C. burnetii* high-speed supernatants (Akporiaye et al. 1990; Baca et al. 1993). This ACP is located within the periplasmic space and is encoded in several different *C. burnetii* isolates associated with acute (group I isolate Nine Mile) or chronic Q fever (group IV isolates S and P, group V isolate K) (Baca et al. 1993). Several pathogens express ACPs during infection and suppress Phox assembly by hydrolyzation of second messengers, such as phosphatidylinositol 4,5-bisphosphate (PIP2) or *myo*-inositol 1,4,5 triphosphate (IP3) involved in macrophage activation (Saha et al. 1985, 1988). *C. burnetii* ACP did indeed hydrolyze several relevant substrates but failed to hydrolyze inositol phosphates and thus is not capable of cleaving secondary messengers to inhibit Phox assembly as previously reported (Li et al. 1996). The detected protein-tyrosine phosphatase (PTPase) activity might indicate that *C. burnetii* ACP targets host cell proteins that contain phosphotyrosine residues. For instance, *Yersinia pestis* secretes a PTPase, YopH, which dephosphorylates tyrosine kinases and signal transduction molecules to completely inhibit macrophage signaling (Cantwell et al. 2010; Tautz et al. 2005). These findings correlate with the observation that pretreatment of PMNs with *C. burnetii* ACP followed by phorbol-12-myristate-13-acetate (PMA) stimulation prevents Phox assembly and that ACP acts as agonist to extracellular signals (Hill and Samuel 2010).

3.2.2 ROS Scavenging Systems

3.2.2.1 SOD

Superoxide dismutases (SODs) are the major defense mechanism of bacteria against the toxic effects of O_2^-. SODs are ubiquitous and found in nearly all aerobic and some anaerobic microorganisms. The enzyme catalyzes the dismutation of O_2^- to H_2O_2, which is subsequently removed by catalases or peroxidases.

$$2O_2^- + H^+ \rightarrow H_2O_2 + O_2$$

SODs are categorized in three main classes depending on the metal ion cofactor; cytoplasmic MnSODs (*sodA*) or FeSODs (*sodB*) and periplasmic Cu/ZnSODs (*sodC*). *E. coli* typically expresses both cytoplasmic SODs and the periplasmic SOD as defense against endogenous and exogenous ROS. Cytoplasmic SODs are generally synthesized in abundance to protect cellular components from endogenous generated superoxide. This protection is a very fine balanced mechanism where enzyme abundance and catalytic proficiency is used to counteract the endogenous generation of superoxide (Gort and Imlay 1998). Functionality of periplasmic Cu/ZnSODs are not fully understood, but it is generally assumed that these enzymes mediate protection from exogenous superoxide, particularly when pathogenic bacteria encounter the oxidative defense of immune cells (Battistoni 2003). *C. burnetii* encodes for two SODs, a cytoplasmic FeSOD (CBU1708) and a periplasmic Cu/ZnSOD (CBU1822, Robert Brennan, personal communication, 2006) (Heinzen et al. 1990, 1992). Functional characterization in inhibition assays or complementation assays of an *E. coli sodA sodC* double mutant suggested that *C. burnetii* indeed encodes for a cytoplasmic iron-cofactored SOD and supported earlier reports of SOD activity detected in bacterial lysates (Akporiaye and Baca 1983; Heinzen et al. 1992). SODs are, besides their detoxifying function, also established virulence factors and expressed by several human pathogens. Results of mutational analysis of cytoplasmic or periplasmic SODs are contradictory, some pathogens become highly sensitive to oxidative stress encountered within the macrophage host and are readily cleared, whereas other show no growth deficits (Hassett and Cohen 1989; Franzon et al. 1990; Battistoni 2003). These conflicting results could be explained by differences in robustness of the oxidative burst induced by these pathogens or differences in the repertoire of antioxidant genes and their regulation. The only known difference between the repertoires of oxidative defense genes of the low virulent isolate Dugway and other *Coxiella* isolates is that this isolate encodes for a functional catalase but an inactive Cu/ZnSOD. This might imply that the periplasmic Cu/ZnSOD is essential for protection against exogenous ROS, especially is an acidic environment where superoxide freely passes through membranes. (Korshunov and Imlay 2002). Contrary, in *L. pneumophila*, as phylogenetic relative, inactivation of the periplasmic Cu/ZnSOD has a very limited growth effect, whereas inactivation of the cytoplasmic FeSOD is lethal (Sadosky et al. 1994, St John and Steinman 1996). This correlates with the different intracellular compartments occupied by these two pathogens; *L. pneumophila* is not likely exposed to exogenous ROS within the ER-like compartment and function of the periplasmic SOD might be dispensable, whereas *C. burnetii* resides within a phagolysosome-like vacuole and is exposed to exogenous ROS.

3.2.2.2 Catalase

Catalases or hydroperoxidases are integral components of the bacterial response to oxidative stress. Together with alkyl hydroperoxidases they limit the accumulation of H_2O_2 within the bacterial cell and catalyze the decomposition of H_2O_2 into H_2O and O_2.

$$2H_2O_2 \rightarrow 2H_2O + O_2$$

There are three different classes of catalases; monofunctional catalases, bifunctional catalases and manganese-containing catalases (Chelikani et al. 2004). Most microorganism express more than one type of catalase under the tight regulation of redox sensors (OxyR) or in dependence to the growth cycle (RpoS) (Schellhorn 1995). Only a few *C. burnetii* isolates encode for a single functional catalase, KatE (CBU0281), which is normally expressed under the control of RpoS during the onset of the stationary phase in *E. coli* (Beare et al. 2009; Tanaka et al. 1997). Isolates with a functional KatE are the group IV isolate K (Q154), which is associated with chronic Q fever and was originally isolated from the heart valve of a patient with endocarditis and the low-virulence isolate Dugway of group VI (Hackstadt 1986; Samuel et al. 1985). *C. burnetii* isolates Nine Mile and Henzerling (group I and II) as well as isolate G (group V) encode for a severely truncated KatE and are associated with acute or chronic Q fever, respectively. There are no reports available for group III isolates. It is questionable if loss of *katE* represents an adaptation to a niche where protection against exogenous oxidative stress is of greater importance (mediated by expression of the periplasmic Cu/ZnSOD) or if this loss affects persistence and contributes to the development of acute Q fever. *L. pneumophila* for instance encodes for two catalases, a periplasmic KatA and cytosolic KatB. These catalases are differentially regulated and expressed during exponential (*katB*) or stationary (*katA*) phase (Bandyopadhyay and Steinman 1998, 2000). Inactivation of either one or both catalases in *L. pneumophila* has only a slight effect on aerobic growth and results in delayed infectivity of macrophages. It is likely that expression of two catalases in dependence of the growth phase in *L. pneumophila* is essential for aerobic growth and a prerequisite for its biphasic life cycle. Indeed expression of a functional catalase in *C. burnetii* does enhance survival in vitro and in vivo, but leads to a negative growth phenotype when over-expressed. This observed growth phenotype might indicate that expression of catalase or other antioxidative genes are part of a finely balanced regulatory network and that the loss of a single functional *katE* in some *C. burnetii* isolates is negligible with adaptation to a microaerophilic lifestyle.

3.2.2.3 Peroxiredoxins

Peroxiredoxins (Prxs) are a group of non-heme proteins, which degrade H_2O_2, organic hydroperoxides (ROOH) and peroxynitrite into H_2O and their corresponding alcohols.

$$ROOH + 2e^- \rightarrow ROH + H_2O$$

Bacterial alkyl hydroperoxide reductases (AhpC) contain two conserved cysteine residues which mediate degradation of hydroperoxides in two consecutive steps; oxidation of a redox-active cysteine residue to sulfonic acid followed by formation of a disulfide bound with the second conserved, resolving cysteine residue. Most bacteria express flavoproteins in combination with peroxiredoxin, which function as disulfide reductases for regeneration of the conserved cysteine residues in the peroxiredoxin, such as the well characterized AhpC/AhpF system of *E. coli* or *S. typhimurium* (Poole 2005). Regeneration of AhpC in AhpF deficient bacteria can also be mediated by other reductase systems, such as the thioredoxin reductase (TrxR) and thioredoxin (Trx) system. *C. burnetii* encodes, like *L. pneumophila*, two peroxide scavenging alkyl hydoperoxide systems, AhpC1 (CBU1706) and AhpC2D (CBU1477/1478), with similarity to the NADPH-dependent TrxR/Trx activated AhpC of *Helicobacter pylori* or the two alkyl hydroperoxidases AhpC/AhpD of *M. tuberculosis*, respectively (Baker et al. 2001; Hillas et al. 2000). The close relation to *L. pneumophila* and high homology found for the AhpC systems suggests that these two pathogens share defense mechanisms against ROS (LeBlanc et al. 2006). AhpCs are the primary peroxide scavengers in *L. pneumophila* and are differentially regulated in a growth phase-dependent manner, as previously described for *katA* and *katB*. Mutations in either *ahpC1* or *ahpC2* results in increased sensitivity to exogenous peroxides, but no growth defects are observed under normal culture conditions or in vivo (Leblanc et al. 2006, 2008). The regenerating flavoproteins are unknown, but *L. pneumophila* as well as *C. burnetii* encode for the TrxR/Trx system (CBU1193/CBU2087, ERGO™ Integrated Genomics). The AhpC/AhpD system in *M. tuberculosis* is co-transcribed and both genes encode for two independent peroxiredoxins. The regenerating flavoprotein is unknown, but activity of both peroxiredoxins seems to be indispensable for protection against oxidative and nitrosative stress in *M. tuberculosis* (Hillas et al. 2000). Interestingly, genomic organization of the *ahpC2D* operon in *C. burnetii* is similar to the one found in *B. abortus*, *B. melintensis*, *M. bovis* and *M. leprae*. In these pathogenic organisms the oxidative stress response regulator encoding gene *oxyR* is located directly upstream of *ahpC2D* on the opposite strand (ERGO™ Integrated Genomics). It has been reported that OxyR is non-functional in virulent *M. tuberculosis*. Recent findings support that in *L. pneumophila* this regulator of oxidative stress has also lost its sensing function and has become a cell cycle regulator (Hillas et al. 2000; LeBlanc et al. 2008). It is rather unlikely that *L. pneumophila* is exposed to oxidative stress within the ER-like compartment and it has been proposed that growth stage-dependent expression of both AhpC systems and catalases might be required during transition from an extracellular to an intracellular replicative stage or during the biphasic lifecycle. Interestingly, AhpC in *M. tuberculosis* also undergoes a rather unusually regulatory cycle and is only expressed by static bacilli (Master et al. 2002). It is logical to speculate that expression of antioxidant genes in late exponential or stationary phase is associated with dormancy in *M. tuberculosis* or latency in other pathogens. However, *ahpC2* in *C. burnetii* seems to be up regulated in response to oxidative stress, but contribution to protection against oxidative stress within the PV or during different growth stages is unclear (Mertens et al. 2008).

3.2.2.4 ROS and RNS Scavenging

Bacteria employ several strategies for scavenging of ROS and RNS. This includes degradation of ROS and RNS to less toxic species or the direct reversion of oxidative damage of proteins. Low molecular weight thiols are one class of small proteins, which promote degradation of ROS and RNS and mediate the repair of oxidative and nitrosative modified proteins. One candidate is the small peptide glutathione, which reduces disulfide often in conjunction with glutaredoxin. Glutathione, L-γ-glutamyl-L-cysteinylglycine, is synthesized in two consecutive steps by γ-glutamylcysteine synthetase (*gshA*) and glutathione synthase (*gshB*). Glutathione (GSH) alone or in conjunction with glutaredoxin (*grxC*) catalyze the reduction of disulfide bounds, formed by oxidative attack of proteins from ROS and RNS. The resulting glutathione disulfide (GSSH) is then regenerated by glutaredoxin (*grxC*) in presence of NADPH (Carmel-Harel and Storz 2000). Genome annotation information indicates that *C. burnetii* encodes for both necessary genes to synthesize glutathione, a glutamate-cysteine ligase (CBU1874) and *gshB* (CBU1875) and the corresponding glutaredoxin (*grxC*, CBU1520).

Another small molecule used as ROS or RNS scavenger is methionine, either as a residue in proteins or as a free amino acid. Its oxidation to methionine sulphoxide is reversible and repaired by the methionine sulphoxide reductase in conjunction with TrxR/Trx and NADPH. *C. burnetii* encodes for the methionine sulphoxide reductase, MsrA (CBU1306), which is active on both, protein bound and free methionine sulfoxide (Boschi-Muller et al. 2005). Enzymes that mediate detoxification of RNS are bacterial hemoglobins, NO* reductases, S-nitrosoglutathione reductases or peroxynitrite reductases, but if these are employed by *C. burnetii* and what their role in pathogenesis might be is undetermined.

3.2.3 *Iron Homeostasis and Oxidative Stress*

In an acidic environment, such as the PV, superoxide radicals favor the production of hydroxyl radicals via the Fenton/Haber-Weiss reaction cycle by release of ferric iron from $[4Fe\text{-}4S]^{2+}$ clusters. This potentiates the toxic effects of ROS and a strict regulation of iron metabolism and its coupling with oxidative stress defenses is essential for bacterial survival.

$$
\begin{array}{c}
{}^{*}OH + H_2O_2 \longrightarrow HO_2{}^{*} + H_2O \\
\underline{HO_2{}^{*} + Fe^{3+} \longrightarrow O_2 + Fe^{2+} + H^{+}} \\
2H_2O_2 \xrightarrow{Fe} O_2 + 2H_2O
\end{array}
$$

Iron is the most important metal for most bacteria and any organism where it serves as cofactor for many enzymes catalyzing central metabolic pathways such as the Krebs cycle, respiration, DNA and protein synthesis. Due to its accelerating side effect on ROS toxicity, iron assimilation is tightly regulated in bacteria by a small dimeric protein, Fur (**f**erric **u**ptake **r**egulator) and Fur-like proteins. This transcription regulator coordinates the expression of iron assimilation genes with genes encoding defense mechanisms against oxidative stress, acid stress and the general metabolism. Fur is loaded with one Fe^{2+} per subunit in its active state and binds to a palindromic sequence motif, the so-called "Fur box", within the promoter region of negatively controlled genes and represses their transcription. In an iron-limited environment, Fur loses its iron cofactors and changes to an inactive state, thereby releasing the promoter regions and transcription of genes belonging to the Fur regulon to become derepressed (Touati 2000). An alternative regulatory effect of Fur is mediated by repression of a small regulatory sRNA (*ryhB*), which facilitates degradation of mRNAs from Fur positively regulated genes (Masse et al. 2007).

With the onset of infection, the host cells increasingly synthesize iron-chelating proteins, such as lactoferrin to limit iron accessibility. It is essential for most pathogenic bacteria to overcome this limitation by expression of iron acquisition or storage systems under the tight control of Fur. Inactivation of this controlling element would lead to severe growth deficits due to metabolic blocks, increased oxidative stress and higher mutation frequency. It has been reported that expression of the transferrin receptor for iron depletion is upregulated in macrophages during the onset of infection with *C. burnetii* (Howe and Mallavia 1999). Inhibition of bacterial replication in the presence of iron chelators suggested an absolute requirement for iron and led to the assumption that *C. burnetii*, similar to *L. pneumophila*, regulates iron assimilation via the Fur regulon. An intensive screen revealed that the Fur-regulon in *C. burnetii* consists of a Fur-like protein (CBU1301) and only two Fur-regulated genes; the ferrous iron transporter FeoB (CBU1766) and the putative iron-binding protein Frg1 (CBU0970) (Briggs et al. 2008). The regulatory activity of *C. burnetii* Fur was reported to be relatively low compared to *E. coli* and depends directly on iron availability. Absence of a Fur box and lack of transcriptional induction in dependence to iron supports that expression of *C. burnetii* Fur is differently regulated.

As shown for other bacteria, Fur can be regulated by either OxyR, as stress response regulator or RpoS as growth cycle regulator. However, with only two Fur-regulated genes and the absence of other potential iron acquisitions or storage encoding genes it is likely, that iron plays a rather limited role in pathogenesis of *C. burnetii*. Recent reports demonstrated the expression of a thiol-specific peroxidase (CBU0963) in *C. burnetii* that belongs to the atypical 2-cysteine subfamily of peroxiredoxins, also designated as bacterioferritin comigratory proteins (BCPs). The DNA binding activity and peroxidase activity of this BCP implies that this protein might protect DNA from the Fenton reaction mediated oxidative damage and might represent a novel adaptation of *C. burnetii* (Hicks et al. 2010).

Contrary to earlier studies, preference for a low iron environment is supported by the observation that replication of *C. burnetii* is enhanced in macrophages in the presence of iron chelators or in a mouse infection model in mice on a low iron diet. Comparison to *L. pneumophila*, a phylogenetic relative, revealed that *C. burnetii* encodes for rarely any known iron acquisition or storage proteins, besides some iron dependent pathways, such as the heme biosynthesis pathway and proteins like SodB. There is only one other example for iron dispensability during infection; *Borrelia burgdorferi*, the causative agent of Lyme disease, completely bypassed iron dependency and contains only a few iron-dependent proteins, such as SodB. This is one example of an organism which evolved to an obligate parasite by eliminating most biosynthetic pathways and the requirement for iron (Posey and Gherardini 2000). Recent studies suggested that *C. burnetii* is at an early stage of genome reduction, a common characteristic adaptation mechanism for intracellular parasitic bacteria, and might indicate that this organism also adapted to a low iron environment. This would have several advances for survival; iron restriction as host defense mechanism would become ineffective and DNA damage via the Fenton reaction would be limited (Briggs et al. 2008).

3.2.4 Regulation of Oxidative Stress Defenses

For intracellular survival and to establish a persistent infection, pathogenic bacteria depend on the ability to adapt to the changing environmental conditions within the host and to coordinate the expression of virulence factors, among them oxidative stress defense mechanisms. In *E. coli* over approximately 100 genes involved in antioxidant mechanisms are identified and regulated in response to changing levels of superoxide (SoxRS), hydrogen peroxide (OxyR, PerR, OhrR) or iron (Fur). Generally, these redox-sensing transcription factors become activated upon a conformational change due to oxidation of $[4Fe\text{-}4S]^{2+}$ cluster or active cysteine residues. A number of antioxidant enzymes and repair activities are controlled in this way. SoxRS activates SodA (MnSOD) or oxygen resistant isoenzymes of central metabolic pathways to increase the reducing capabilities of the cell. OxyR, a classic and allosteric oxidative stress response regulator, activates expression of catalases, peroxiredoxins and reductase systems as well as several DNA repair genes.

Both transcription factors, OxyR and SoxR, regulate expression of Fur to limit the amount of free iron within the cell (Mongkolsuk and Helmann 2002; Storz and Imlay 1999).

C. burnetii appears to encode only OxyR (CBU1476) as response regulator since other oxidative stress associated transcription factors have not been annotated. OxyR is a tetrameric protein and exists in two forms, reduced and oxidized, but only the oxidized form is active. Conformational change is the result of a disulfide bound formation upon H_2O_2 mediated oxidation and is responsible for OxyR activation. This mechanism allows the bacterial cell to sense oxidative stress and to directly respond to environmental changes with adaptive gene expression (Storz and Imlay 1999). However, there are several aberrations to the classical, enterobacterial OxyR regulon; in some pathogenic bacteria the genomic orientation of *oxyR* and the adjacent *ahpC*/*ahpD* genes is highly conserved (see above) and *oxyR* shares the overlapping promoter region with the adjacent gene and represses expression in an inactive, DNA-binding state (Schell 1993). In slow-growing Mycobacteria species, *ahpC* is linked to the *oxyR* locus and expression depends upon OxyR activation under oxidative stress (Pagan-Ramos et al. 2006). On the other hand in virulent *M. tuberculosis*, OxyR is non-functional and *ahpC* expression only detectable in static bacilli. It is likely that in these bacteria, a unknown OxyR-like protein or constitutive expression of stress pathways complement for the partially dysfunctional oxidative stress response (Gupta and Chatterji 2005). In contrast, oxidative stress response in *L. pneumophila* is altered and expression of several oxidative stress genes, such as *katA*, *katB* and *sodB* are not responsive to oxidative stress and are expressed during specific growth stages. Indeed, OxyR has lost its redox-sensing activity and regulates expression of *ahpC1* and *ahpC2D* in dependence of the growth cycle. This implies that in *L. pneumophila* the oxidative stress response has evolved to a growth stage regulated antioxidant defense consistent with a dimorphic growth cycle (LeBlanc et al. 2008). Despite all similarities and the close relation to *L. pneumophila, oxyR* in *C. burnetii* is likely involved in the oxidative stress induced expression of *ahpC2* (personal observation), but regulatory mechanisms or other genes belonging to the OxyR regulon are unknown.

There are several other transcription factors involved in regulation of antioxidant gene expression, illustrating the connectivity and complexity of the regulatory networks. The alternative sigma factor *rpoS* or σ^S as RNA polymerase subunit for instance directs the expression of several genes under different stresses, such as starvation, osmotic or acid stress and upon entry into stationary phase. Bacteria become increasingly resistant to stress, including oxidative stress with the onset of the stationary phase and RpoS has been shown do direct expression of *katE* and *sodC* during this growth phase. *S. typhimurium* possesses two periplasmically localized Cu/ZnSODs; SodCI is reported to be RpoS-independent and maximally expressed during exponential growth, whereas SodCII is upregulated during stationary phase in a RpoS-dependent manner (Fang et al. 1999). Stationary phase or RpoS regulation of Cu/ZnSODs has been demonstrated in several bacteria and *C. burnetii sodC* (CBU1669) is predicted to possess a putative RpoS consensus promoter recognition sequence, which suggests that this enzyme plays a role in stationary phase survival

(Robert Brennan, personal communication, 2004). As a naturally obligate intracellular pathogen and microaerophilic bacterium, the oxidative response is likely crucial for survival of *C. burnetii*. Despite the limited amount of information available, it is likely that OxyR functions at least partially as a transcription activator under oxidative stress and, together with RpoS, ensures protection within the PV.

3.3 Oxidative DNA Damage and Repair

Within the harsh environment of the PV, *C. burnetii* is constantly exposed to potentially DNA damaging agents generated by the host. Oxidative attack of the DNA by ROS and RNS causes base modifications, sugar fragmentation and eventually DNA strand breaks (Demple and Harrison 1994). Therefore, DNA repair and mechanisms to effectively counteract these detrimental effects are essential for survival within the PV. Repair mechanisms in *C. burnetii* are only partly characterized and most assumptions are based on genome sequence information and comparison to other microorganisms. Genes encoding for enzymes of direct repair, base excision repair (BER) or nucleotide excision repair (NER) are annotated and likely functional. Components of the SOS response are partly present and recombinational DNA repair is likely directed by the AddAB complex, a functional analog to the *E. coli* RecBCD complex (Mertens et al. 2008).

3.3.1 Base Excision Repair (BER)

Oxidative attack of DNA bases, especially pyrimidine bases lead to a wide variety of oxidized base products, including 8-oxo-7, 8-dihydroguanine (8-oxoG). 8-oxoG has a strong pro-mutagenic effect and mispairs during DNA replication with adenine, leading to G→T or C→A transversions. In *E. coli* 8-oxoG is removed by the DNA glycosylase MutM or putative orthologs, such as *nth*, an endonuclease III with glycosylase activity. For instance, removal of lethal or mutagenic pyrimidine lesions in *H. pylori* is mediated by *nth* and mutations in this locus lead to an increased susceptibility to oxidative stress and reduced survival in vitro and in vivo (Wang et al. 2006). *C. burnetii* does not encode for MutM but carries a putative *nth* gene (CBU1697), which might fulfill a similar role in repair of 8-oxo-G. Other components of the GO system are MutT and MutY. MutT, a 8-oxo-dGTPase scavenges the nucleotide pool for oxidized GTPs and MutY removes with 8-oxoG mispaired adenine residues from the DNA. Both genes are annotated for *C. burnetii* (CBU0148 and CBU0940) and together with *nth* represent a complete GO system.

NO* or other RNS catalyze the formation of nitrous anhydrite (N_2O_3) in presence of oxygen radicals which forms potent DNA alkylating agents. These base adducts are repaired in *E. coli* by two different glycosylases, TagA and AlkA (Grzesiuk et al. 2001). Another glycosylase as component of the BER system is *mpg*.

This 3-methylpurine DNA glycosylase has a wide range of substrates and might play a role in repair of deaminated bases, the primary product in the presence of N_2O_3 (Aamodt et al. 2004). The *tagA* and *mpg* genes are annotated for *C. burnetii* (CBU0383, CBU0930) and might mediate protection against RNS generated DNA damage.

The methyl mismatch repair system (MMR) in *C. burnetii* is only partial present with MutL (CBU1083) and MutS (CBU1056), but lack of MutH. In *H. pylori* the MMR component MutS binds to DNA containing 8-oxoG or DNA structures resembling recombination intermediates, indicating that MutS plays a role in oxidative DNA damage repair (Wang et al. 2005).

3.3.2 Nucleotide Excision Repair (NER)

Short stretches of DNA flanking a damaged region are removed via NER. This repair system recognizes a wider range of damages than BER using endonucleases with lower substrate specificity. UvrA, UvrB and UvrC form the exonuclease complex UvrABC, which mediates removal of damaged DNA together with the DNA helicase II, UvrD. Expression profiling of *M. tuberculosis* during infection of macrophages suggests that the *uvr* system plays a critical role for intracellular survival. Indeed *uvrB* is upregulated when bacteria are exposed to NO^* or H_2O_2 (Cabusora et al. 2005). The *uvrABC* (CBU0274, CBU0518, CBU1185) and *uvrD* (CBU2054) genes are annotated for *C. burnetii* and recent reports indicated that the *uvr* system might contribute to DNA repair under oxidative stress (Park et al. 2010).

3.3.3 SOS Response and Recombinational DNA Repair

Bacterial response to DNA damage is mainly regulated through RecA, which recognizes single-stranded DNA (ssDNA) regions caused by DNA damage. RecA forms a nucleoprotein filament on ssDNA and mediates strand exchange during recombinational DNA repair by the RecBCD or RecFOR pathways in *E. coli* (Kowalczykowski et al. 1994). In conjunction with the repressor protein LexA, the expression of approximately 50 genes involved in DNA repair and cell division, designated as the SOS response, are regulated by RecA. The repressor LexA binds to the so-called "SOS box" within the operator regions of these genes and represses their expression. Upon activation, RecA functions as co-protease and promotes the autocatalytic cleavage of LexA and release of the promoter regions of negatively controlled genes. This regulation allows the bacteria to sense DNA damage and to directly respond by expression of DNA repair genes, such as NER, error-prone DNA replication, delay of cell division and recombinational repair (Ennis et al. 1993; Friedberg et al. 2006; Kowalczykowski et al. 1994).

When compared to other microorganisms, such as *E. coli* or *L. pneumophila*, *C. burnetii* expresses only a minimal set of DNA repair genes to maintain DNA stability and integrity. Major components of the error-prone driven DNA replication (*umuCD*, *dinB* or *polB*) are absent and indicate that the SOS response is partially dysfunctional. Lack of the repressor LexA suggests that the regulatory element of RecA/LexA is non-functional and might result in a constitutive expression of genes, such as *recA*, *ssb*, *recFOR*, *recN*, *recG* and *ruvABC* for recombinational DNA repair. Other bacteria, such as *Rickettsia prowazekii* and *L. pneumophila* also appear to lack LexA and maintain only a reduced set of DNA repair genes. In these organism a constitutive expression of repair genes, as it was implied for *B. abortus*, might become necessary to overcome DNA damage within their intracellular niche (Roux et al. 2006). Another abbreviation to the *E. coli* model is the presence of an AddAB complex (CBU1229 and CBU1230), which is functional homolog to *recBCD* and usually found in Gram-positive bacteria. This complex is inducible under oxidative stress and mediates DNA repair via homologous recombination. This inducibility of recombinational repair in response to oxidative stress might represent a novel adaptation of *C. burnetii* to its intracellular niche (Mertens et al. 2008).

3.4 Summary and Prospects

Antioxidant defense mechanisms in *C. burnetii* are diverse and include some well-characterized proteins comparable to many other pathogenic bacteria, but also some novel mechanisms to combat oxidative stress (Table 3.1). The knowledge summarized in this review is mainly based on a few reports on enzymatic activities (SOD and catalase activity in bacterial lysates or ACP activity in bacterial supernatants) or basic genetic analysis (Fur regulon and SOS response), but most information presented is based on genome sequence information and comparison to other intracellular pathogenic bacteria. The functional characterization of these defense mechanisms is of continuing importance since Q fever is an emerging zoonotic disease with a worldwide distribution (Enserink 2010; Anderson et al. 2009). However, a detailed investigation would depend on the ability to generate specific mutations and requires large improvements of genetic manipulation techniques for *C. burnetii*.

C. burnetii has evolved complex and novel mechanisms to avoid or resist oxidative stress but remains surprisingly oxygen sensitive (microaerophilic). It is clear that these defense mechanisms against ROS and RNS are effective at different cellular levels (Fig. 3.1); Primary defenses include the expression of cytoplasmic and periplasmic SODs (FeSOD, Cu/ZnSOD), catalase (KatE) or alkyl hydroperoxide reductases (AhpC1, AhpC2D) and OxyR as potential regulator. Differences among strains in the repertoire of antioxidant genes, a severely truncated *katE* in isolates belonging to genomic groups I, II and V or the frame-shifted *sodC* in isolates of group VI, might suggest a correlation with differences in virulence or persistence, but testing this speculation would require detailed future work. The presence of a rudimentary Fur regulon and absence of iron acquisition or storage systems in *C. burnetii* suggests that iron plays only a limited role in pathogenesis. This novel adaptation, similar to the described iron dispensability of *B. burgdorferi*, has the advantage that iron limitation as a host defense mechanism and DNA damage via the Fenton reaction becomes ineffective. Compared to other microorganisms, *C. burnetii* expresses a limited set of DNA repair genes and absence of LexA, one of the major regulators of the SOS response, implies that a constitutive expression of DNA repair genes is necessary to overcome oxidative DNA damage. Inducibility of recombinational DNA repair (AddAB complex) and expression of a DNA-binding peroxidase (BCP) represent novel adaptations and ensure maintenance of DNA stability and integrity under oxidative stress. Secondary defense mechanisms of *C. burnetii* include the delay of phagolysosome fusion, likely in a T4SS-dependent manner, and allow metabolic adaptation to the environment of the PV. Suppression of the oxidative burst is mediated by secretion of an ACP that acts as agonist to extracellular signals. Taken together *C. burnetii* is protected against oxidative stress on the cellular level by expression of enzymes for direct detoxification or reversion of ROS and RNS mediated toxic effects (primary defense mechanisms) and interferes with host signaling pathways to evade oxidative stress (secondary defense mechanisms).

References

Aamodt RM, Falnes PO, Johansen RF, Seeberg E, Bjoras M (2004) The Bacillus subtilis counterpart of the mammalian 3-methyladenine DNA glycosylase has hypoxanthine and 1,N6-ethenoadenine as preferred substrates. J Biol Chem 279:13601–13606

Akporiaye ET, Baca OG (1983) Superoxide anion production and superoxide dismutase and catalase activities in *Coxiella burnetii*. J Bacteriol 154:520–523

Akporiaye ET, Stefanovich D, Tsosie V, Baca G (1990) *Coxiella burnetii* fails to stimulate human neutrophil superoxide anion production. Acta Virol 34:64–70

Albakri QA, Stuehr DJ (1996) Intracellular assembly of inducible NO synthase is limited by nitric oxide-mediated changes in heme insertion and availability. J Biol Chem 271:5414–5421

Anderson AD, Kruszon-Moran D, Loftis AD, McQuillan G, Nicholson WL, Priestley RA, Candee AJ, Patterson NE, Massung RF (2009) Seroprevalence of Q fever in the United States, 2003–2004. Am J Trop Med Hyg 81:691–694

Baca OG, Paretsky D (1983) Q fever and *Coxiella burnetii*: a model for host-parasite interactions. Microbiol Rev 47:127–149

Baca OG, Roman MJ, Glew RH, Christner RF, Buhler JE, Aragon AS (1993) Acid phosphatase activity in *Coxiella burnetii*: a possible virulence factor. Infect Immun 61:4232–4239

Badwey JA, Curnutte JT, Robinson JM, Lazdins JK, Briggs RT, Karnovsky MJ, Karnovsky ML (1980) Comparative aspects of oxidative metabolism of neutrophils from human blood and guinea pig peritonea: magnitude of the respiratory burst, dependence upon stimulating agents, and localization of the oxidases. J Cell Physiol 105:541–545

Baek KJ, Thiel BA, Lucas S, Stuehr DJ (1993) Macrophage nitric oxide synthase subunits. Purification, characterization, and role of prosthetic groups and substrate in regulating their association into a dimeric enzyme. J Biol Chem 268:21120–21129

Baker LM, Raudonikiene A, Hoffman PS, Poole LB (2001) Essential thioredoxin-dependent peroxiredoxin system from *Helicobacter pylori*: genetic and kinetic characterization. J Bacteriol 183:1961–1973

Bandyopadhyay P, Steinman HM (1998) *Legionella pneumophila* catalase-peroxidases: cloning of the *katB* gene and studies of KatB function. J Bacteriol 180:5369–5374

Bandyopadhyay P, Steinman HM (2000) Catalase-peroxidases of *Legionella pneumophila*: cloning of the *katA* gene and studies of KatA function. J Bacteriol 182:6679–6686

Battistoni A (2003) Role of prokaryotic Cu, Zn superoxide dismutase in pathogenesis. Biochem Soc Trans 31:1326–1329

Beare PA, Unsworth N, Andoh M, Voth DE, Omsland A, Gilk SD, Williams KP, Sobral BW, Kupko JJ 3rd, Porcella SF, Samuel JE, Heinzen RA (2009) Comparative genomics reveal extensive transposon-mediated genomic plasticity and diversity among potential effector proteins within the genus *Coxiella*. Infect Immun 77:642–656

Beron W, Gutierrez MG, Rabinovitch M, Colombo MI (2002) *Coxiella burnetii* localizes in a Rab7-labeled compartment with autophagic characteristics. Infect Immun 70:5816–5821

Beyenbach KW, Wieczorek H (2006) The V-type H+ATPase: molecular structure and function, physiological roles and regulation. J Exp Biol 209:577–589

Bogdan C (2001) Nitric oxide and the immune response. Nat Immunol 2:907–916

Boschi-Muller S, Olry A, Antoine M, Branlant G (2005) The enzymology and biochemistry of methionine sulfoxide reductases. Biochim Biophys Acta 1703:231–238

Brennan RE, Russell K, Zhang G, Samuel JE (2004) Both inducible nitric oxide synthase and NADPH oxidase contribute to the control of virulent phase I *Coxiella burnetii* infections. Infect Immun 72:6666–6675

Briggs HL, Pul N, Seshadri R, Wilson MJ, Tersteeg C, Russell-Lodrigue KE, Andoh M, Baumler AJ, Samuel JE (2008) A limited role for iron regulation in *Coxiella burnetii* pathogenesis. Infect Immun 76(5):2189–2201

Bylund J, Brown KL, Movitz C, Dahlgren C, Karlsson A (2010) Intracellular generation of superoxide by the phagocyte NADPH oxidase: how, where, and what for? Free Radic Biol Med 49(12):1834–1845

Cabusora L, Sutton E, Fulmer A, Forst CV (2005) Differential network expression during drug and stress response. Bioinformatics 21:2898–2905

Callahan LA, Nethery D, Stofan D, Dimarco A, Supinski G (2001) Free radical-induced contractile protein dysfunction in endotoxin-induced sepsis. Am J Respir Cell Mol Biol 24:210–217

Cantwell AM, Bubeck SS, Dube PH (2010) YopH inhibits early pro-inflammatory cytokine responses during plague pneumonia. BMC Immunol 11:29

Carmel-Harel O, Storz G (2000) Roles of the glutathione- and thioredoxin-dependent reduction systems in the *Escherichia coli* and *Saccharomyces cerevisiae* responses to oxidative stress. Annu Rev Microbiol 54:439–461

Chelikani P, Fita I, Loewen PC (2004) Diversity of structures and properties among catalases. Cell Mol Life Sci 61:192–208

Chen C, Banga S, Mertens K, Weber MM, Gorbaslieva I, Tan Y, Luo ZQ, Samuel JE (2010) Large-scale identification and translocation of type IV secretion substrates by *Coxiella burnetii*. Proc Natl Acad Sci USA 107(50):21755–21760

Cirillo SL, Subbian S, Chen B, Weisbrod TR, Jacobs WR Jr, Cirillo JD (2009) Protection of *Mycobacterium tuberculosis* from reactive oxygen species conferred by the *mel2* locus impacts persistence and dissemination. Infect Immun 77:2557–2567

Davis AS, Vergne I, Master SS, Kyei GB, Chua J, Deretic V (2007) Mechanism of inducible nitric oxide synthase exclusion from mycobacterial phagosomes. PLoS Pathog 3:e186

de Jong MF, Sun YH, den Hartigh AB, van Dijl JM, Tsolis RM (2008) Identification of VceA and VceC, two members of the VjbR regulon that are translocated into macrophages by the *Brucella* type IV secretion system. Mol Microbiol 70:1378–1396

Demple B, Harrison L (1994) Repair of oxidative damage to DNA: enzymology and biology. Annu Rev Biochem 63:915–948

Ennis DG, Little JW, Mount DW (1993) Novel mechanism for UV sensitivity and apparent UV nonmutability of *recA432* mutants: persistent LexA cleavage following SOS induction. J Bacteriol 175:7373–7382

Enserink M (2010) Infectious diseases. Questions abound in Q-fever explosion in the Netherlands. Science 327:266–267

Fang FC, Degroote MA, Foster JW, Baumler AJ, Ochsner U, Testerman T, Bearson S, Giard J-C, Xu Y, Campbell G, Laessig T (1999) Virulent Salmonella typhimurium has two periplasmic Cu, Zn-superoxide dismutases. Proc Natl Acad Sci 96:7502–7507

Franzon VL, Arondel J, Sansonetti PJ (1990) Contribution of superoxide dismutase and catalase activities to *Shigella flexneri* pathogenesis. Infect Immun 58:529–535

Friedberg EC, Walker GC, Siede W, Wood RD, Schultz RA, Ellenberger T (2006) DNA repair and mutagenesis. American Society of Microbiology Press, Washington, DC

Fujikura Y, Kudlackova P, Vokurka M, Krijt J, Melkova Z (2009) The effect of nitric oxide on vaccinia virus-encoded ribonucleotide reductase. Nitric Oxide 20:114–121

Gort AS, Imlay JA (1998) Balance between endogenous superoxide stress and antioxidant defenses. J Bacteriol 180:1402–1410

Grzesiuk E, Gozdek A, Tudek B (2001) Contribution of E. coli AlkA, TagA glycosylases and UvrABC-excinuclease in MMS mutagenesis. Mutat Res 480–481:77–84

Gupta S, Chatterji D (2005) Stress responses in mycobacteria. IUBMB Life 57:149–159

Gutierrez MG, Vazquez CL, Munafo DB, Zoppino FC, Beron W, Rabinovitch M, Colombo MI (2005) Autophagy induction favours the generation and maturation of the *Coxiella*-replicative vacuoles. Cell Microbiol 7:981–993

Hackstadt T (1986) Antigenic variation in the phase I lipopolysaccharide of *Coxiella burnetii* isolates. Infect Immun 52:337–340

Hackstadt T, Williams JC (1981) Biochemical stratagem for obligate parasitism of eukaryotic cells by *Coxiella burnetii*. Proc Natl Acad Sci USA 78:3240–3244

Hassett DJ, Cohen MS (1989) Bacterial adaptation to oxidative stress: implications for pathogenesis and interaction with phagocytic cells. FASEB J 3:2574–2582

Heinzen RA, Frazier ME, Mallavia LP (1990) Nucleotide sequence of *Coxiella burnetii* superoxide dismutase. Nucleic Acids Res 18:6437

Heinzen RA, Frazier ME, Mallavia LP (1992) *Coxiella burnetii* superoxide dismutase gene: cloning, sequencing, and expression in *Escherichia coli*. Infect Immun 60:3814–3823

Heinzen RA, Scidmore MA, Rockey DD, Hackstadt T (1996) Differential interaction with endocytic and exocytic pathways distinguish parasitophorous vacuoles of *Coxiella burnetii* and *Chlamydia trachomatis*. Infect Immun 64:796–809

Hendrix LR, Samuel JE, Mallavia LP (1991) Differentiation of *Coxiella burnetii* isolates by analysis of restriction-endonuclease-digested DNA separated by SDS-PAGE. J Gen Microbiol 137:269–276

Hicks LD, Raghavan R, Battisti JM, Minnick MF (2010) A DNA-binding peroxiredoxin of *Coxiella burnetii* is involved in countering oxidative stress during exponential-phase growth. J Bacteriol 192:2077–2084

Hill J, Samuel JE (2010) *Coxiella burnetii* acid phosphatase: inhibiting the release of reactive oxygen intermediates in polymorphonuclear leukocytes. Infect Immun 79(1):414–420

Hillas PJ, del Alba FS, Oyarzabal J, Wilks A, Ortiz De Montellano PR (2000) The AhpC and AhpD antioxidant defense system of *Mycobacterium tuberculosis*. J Biol Chem 275:18801–18809

Hisert KB, Kirksey MA, Gomez JE, Sousa AO, Cox JS, Jacobs WR Jr, Nathan CF, McKinney JD (2004) Identification of *Mycobacterium tuberculosis* counterimmune (cim) mutants in immunodeficient mice by differential screening. Infect Immun 72:5315–5321

Howe D, Mallavia LP (1999) *Coxiella burnetii* infection increases transferrin receptors on J774A. 1 cells. Infect Immun 67:3236–3241

Howe D, Mallavia LP (2000) *Coxiella burnetii* exhibits morphological change and delays phagolysosomal fusion after internalization by J774A.1 cells. Infect Immun 68:3815–3821

Howe D, Barrows LF, Lindstrom NM, Heinzen RA (2002) Nitric oxide inhibits *Coxiella burnetii* replication and parasitophorous vacuole maturation. Infect Immun 70:5140–5147

Hubber A, Roy CR (2010) Modulation of host cell function by *Legionella pneumophila* type IV effectors. Annu Rev Cell Dev Biol 26:261–283

Imlay JA (2008) Cellular defenses against superoxide and hydrogen peroxide. Annu Rev Biochem 77:755–776

Imlay JA, Linn S (1988) DNA damage and oxygen radical toxicity. Science 240:1302–1309

Imlay JA, Chin SM, Linn S (1988) Toxic DNA damage by hydrogen peroxide through the Fenton reaction in vivo and in vitro. Science 240:640–642

Jones SA, Chowdhury FZ, Fabich AJ, Anderson A, Schreiner DM, House AL, Autieri SM, Leatham MP, Lins JJ, Jorgensen M, Cohen PS, Conway T (2007) Respiration of *Escherichia coli* in the mouse intestine. Infect Immun 75:4891–4899

Juedes MJ, Wogan GN (1996) Peroxynitrite-induced mutation spectra of pSP189 following replication in bacteria and in human cells. Mutat Res 349:51–61

Kana BD, Weinstein EA, Avarbock D, Dawes SS, Rubin H, Mizrahi V (2001) Characterization of the *cydAB*-encoded cytochrome bd oxidase from *Mycobacterium smegmatis*. J Bacteriol 183:7076–7086

Korshunov SS, Imlay JA (2002) A potential role for periplasmic superoxide dismutase in blocking the penetration of external superoxide into the cytosol of Gram-negative bacteria. Mol Microbiol 43:95–106

Kowalczykowski SC, Dixon DA, Eggleston AK, Lauder SD, Rehrauer WM (1994) Biochemistry of homologous recombination in *Escherichia coli*. Microbiol Rev 58:401–465

Leblanc JJ, Davidson RJ, Hoffman PS (2006) Compensatory functions of two alkyl hydroperoxide reductases in the oxidative defense system of *Legionella pneumophila*. J Bacteriol 188:6235–6244

Leblanc JJ, Brassinga AK, Ewann F, Davidson RJ, Hoffman PS (2008) An ortholog of OxyR in *Legionella pneumophila* is expressed postexponentially and negatively regulates the alkyl hydroperoxide reductase (*ahpC2D*) operon. J Bacteriol 190:3444–3455

Li YP, Curley G, Lopez M, Chavez M, Glew R, Aragon A, Kumar H, Baca OG (1996) Protein-tyrosine phosphatase activity of *Coxiella burnetii* that inhibits human neutrophils. Acta Virol 40:263–272

Lindgren H, Stenmark S, Chen W, Tarnvik A, Sjostedt A (2004) Distinct roles of reactive nitrogen and oxygen species to control infection with the facultative intracellular bacterium *Francisella tularensis*. Infect Immun 72:7172–7182

Masse E, Salvail H, Desnoyers G, Arguin M (2007) Small RNAs controlling iron metabolism. Curr Opin Microbiol 10:140–145

Master SS, Springer B, Sander P, Boettger EC, Deretic V, Timmins GS (2002) Oxidative stress response genes in *Mycobacterium tuberculosis*: role of *ahpC* in resistance to peroxynitrite and stage-specific survival in macrophages. Microbiology 148:3139–3144

Maurin M, Raoult D (1999) Q fever. Clin Microbiol Rev 12:518–553

Mayer B, Hemmens B (1997) Biosynthesis and action of nitric oxide in mammalian cells. Trends Biochem Sci 22:477–481

McCaul TF, Williams JC (1981) Developmental cycle of *Coxiella burnetii*: structure and morphogenesis of vegetative and sporogenic differentiations. J Bacteriol 147:1063–1076

Mertens K, Lantsheer L, Ennis DG, Samuel JE (2008) Constitutive SOS expression and damage-inducible AddAB-mediated recombinational repair systems for *Coxiella burnetii* as potential adaptations for survival within macrophages. Mol Microbiol 69:1411–1426

Mongkolsuk S, Helmann JD (2002) Regulation of inducible peroxide stress responses. Mol Microbiol 45:9–15

Nathan C, Shiloh MU (2000) Reactive oxygen and nitrogen intermediates in the relationship between mammalian hosts and microbial pathogens. Proc Natl Acad Sci USA 97:8841–8848

Ohno Y, Hirai K, Kanoh T, Uchino H, Ogawa K (1982) Subcellular localization of H2O2 production in human neutrophils stimulated with particles and an effect of cytochalasin-B on the cells. Blood 60:253–260

Omsland A, Cockrell DC, Fischer ER, Heinzen RA (2008) Sustained axenic metabolic activity by the obligate intracellular bacterium *Coxiella burnetii*. J Bacteriol 190:3203–3212

Omsland A, Cockrell DC, Howe D, Fischer ER, Virtaneva K, Sturdevant DE, Porcella SF, Heinzen RA (2009) Host cell-free growth of the Q fever bacterium *Coxiella burnetii*. Proc Natl Acad Sci USA 106:4430–4434

Ouellet H, Ouellet Y, Richard C, Labarre M, Wittenberg B, Wittenberg J, Guertin M (2002) Truncated hemoglobin HbN protects *Mycobacterium bovis* from nitric oxide. Proc Natl Acad Sci USA 99:5902–5907

Pagan-Ramos E, Master SS, Pritchett CL, Reimschuessel R, Trucksis M, Timmins GS, Deretic V (2006) Molecular and physiological effects of mycobacterial oxyR inactivation. J Bacteriol 188:2674–2680

Park SH, Lee HW, Cao W (2010) Screening of nitrosative stress resistance genes in *Coxiella burnetii*: involvement of nucleotide excision repair. Microb Pathog 49:323–329

Pollock JD, Williams DA, Gifford MA, Li LL, Du X, Fisherman J, Orkin SH, Doerschuk CM, Dinauer MC (1995) Mouse model of X-linked chronic granulomatous disease, an inherited defect in phagocyte superoxide production. Nat Genet 9:202–209

Poole LB (2005) Bacterial defenses against oxidants: mechanistic features of cysteine-based peroxidases and their flavoprotein reductases. Arch Biochem Biophys 433:240–254

Posey JE, Gherardini FC (2000) Lack of a role for iron in the Lyme disease pathogen. Science 288:1651–1653

Qiu H, Kuolee R, Harris G, Chen W (2009) Role of NADPH phagocyte oxidase in host defense against acute respiratory *Acinetobacter baumannii* infection in mice. Infect Immun 77:1015–1021

Rosner JL, Storz G (1997) Regulation of bacterial responses to oxidative stress. Curr Top Cell Regul 35:163–177

Roux CM, Booth NJ, Bellaire BH, Gee JM, Roop RM 2nd, Kovach ME, Tsolis RM, Elzer PH, Ennis DG (2006) RecA and RadA proteins of *Brucella abortus* do not perform overlapping protective DNA repair functions following oxidative burst. J Bacteriol 188:5187–5195

Sadosky AB, Wilson JW, Steinman HM, Shuman HA (1994) The iron superoxide dismutase of *Legionella pneumophila* is essential for viability. J Bacteriol 176:3790–3799

Saha AK, Dowling JN, Lamarco KL, Das S, Remaley AT, Olomu N, Pope MT, Glew RH (1985) Properties of an acid phosphatase from *Legionella micdadei* which blocks superoxide anion production by human neutrophils. Arch Biochem Biophys 243:150–160

Saha AK, Dowling JN, Pasculle AW, Glew RH (1988) *Legionella micdadei* phosphatase catalyzes the hydrolysis of phosphatidylinositol 4,5-bisphosphate in human neutrophils. Arch Biochem Biophys 265:94–104

Samuel JE, Frazier ME, Mallavia LP (1985) Correlation of plasmid type and disease caused by *Coxiella burnetii*. Infect Immun 49:775–779

Sauer JD, Shannon JG, Howe D, Hayes SF, Swanson MS, Heinzen RA (2005) Specificity of *Legionella pneumophila* and *Coxiella burnetii* vacuoles and versatility of *Legionella pneumophila* revealed by coinfection. Infect Immun 73:4494–4504

Schell MA (1993) Molecular biology of the LysR family of transcriptional regulators. Annu Rev Microbiol 47:597–626

Schellhorn HE (1995) Regulation of hydroperoxidase (catalase) expression in *Escherichia coli*. FEMS Microbiol Lett 131:113–119

Sekeyova Z, Roux V, Raoult D (1999) Intraspecies diversity of *Coxiella burnetii* as revealed by *com1* and *mucZ* sequence comparison. FEMS Microbiol Lett 180:61–67

Sexton JA, Vogel JP (2002) Type IVB secretion by intracellular pathogens. Traffic 3:178–185

Siemsen DW, Kirpotina LN, Jutila MA, Quinn MT (2009) Inhibition of the human neutrophil NADPH oxidase by *Coxiella burnetii*. Microbes Infect 11:671–679

St John G, Steinman HM (1996) Periplasmic copper-zinc superoxide dismutase of *Legionella pneumophila*: role in stationary-phase survival. J Bacteriol 178:1578–1584

Stein A, Saunders NA, Taylor AG, Raoult D (1993) Phylogenic homogeneity of *Coxiella burnetii* strains as determinated by 16S ribosomal RNA sequencing. FEMS Microbiol Lett 113: 339–344

Stenger S, Donhauser N, Thuring H, Rollinghoff M, Bogdan C (1996) Reactivation of latent leishmaniasis by inhibition of inducible nitric oxide synthase. J Exp Med 183:1501–1514

Stoenner HG, Lackman DB (1960) The biologic properties of *Coxiella burnetii* isolated from rodents collected in Utah. Am J Hyg 71:45–51

Storz G, Imlay JA (1999) Oxidative stress. Curr Opin Microbiol 2:188–194

Tanaka K, Handel K, Loewen PC, Takahashi H (1997) Identification and analysis of the *rpoS*-dependent promoter of *katE*, encoding catalase HPII in *Escherichia coli*. Biochim Biophys Acta 1352:161–166

Tautz L, Bruckner S, Sareth S, Alonso A, Bogetz J, Bottini N, Pellecchia M, Mustelin T (2005) Inhibition of *Yersinia* tyrosine phosphatase by furanyl salicylate compounds. J Biol Chem 280:9400–9408

Thiemermann C, Vane J (1990) Inhibition of nitric oxide synthesis reduces the hypotension induced by bacterial lipopolysaccharides in the rat in vivo. Eur J Pharmacol 182:591–595

Tlili A, Dupre-Crochet S, Erard M, Nubetae O (2011) Kinetic analysis of phagosomal production of reactive oxygen species. Free Radic Biol Med 50(3):438–447

Touati D (2000) Iron and oxidative stress in bacteria. Arch Biochem Biophys 373:1–6

Vazquez-Torres A, Xu Y, Jones-Carson J, Holden DW, Lucia SM, Dinauer MC, Mastroeni P, Fang FC (2000) *Salmonella* pathogenicity island 2-dependent evasion of the phagocyte NADPH oxidase. Science 287:1655–1658

Via LE, Lin PL, Ray SM, Carrillo J, Allen SS, Eum SY, Taylor K, Klein E, Manjunatha U, Gonzales J, Lee EG, Park SK, Raleigh JA, Cho SN, McMurray DN, Flynn JL, Barry CE 3rd (2008) Tuberculous granulomas are hypoxic in guinea pigs, rabbits, and nonhuman primates. Infect Immun 76:2333–2340

Vila-Del Sol V, Diaz-Munoz MD, Fresno M (2007) Requirement of tumor necrosis factor alpha and nuclear factor-kappaB in the induction by IFN-gamma of inducible nitric oxide synthase in macrophages. J Leukoc Biol 81:272–283

Voth DE, Heinzen RA (2009) *Coxiella* type IV secretion and cellular microbiology. Curr Opin Microbiol 12(1):74–80

Wang G, Alamuri P, Humayun MZ, Taylor DE, Maier RJ (2005) The Helicobacter pylori MutS protein confers protection from oxidative DNA damage. Mol Microbiol 58:166–176

Wang G, Alamuri P, Maier RJ (2006) The diverse antioxidant systems of *Helicobacter pylori*. Mol Microbiol 61:847–860

Webb JL, Harvey MW, Holden DW, Evans TJ (2001) Macrophage nitric oxide synthase associates with cortical actin but is not recruited to phagosomes. Infect Immun 69:6391–6400

Weisburg WG, Dobson ME, Samuel JE, Dasch GA, Mallavia LP, Baca O, Mandelco L, Sechrest JE, Weiss E, Woese CR (1989) Phylogenetic diversity of the *Rickettsiae*. J Bacteriol 171:4202–4206

Wink DA, Kasprzak KS, Maragos CM, Elespuru RK, Misra M, Dunams TM, Cebula TA, Koch WH, Andrews AW, Allen JS et al (1991) DNA deaminating ability and genotoxicity of nitric oxide and its progenitors. Science 254:1001–1003

Chapter 4
Lipopolysaccharide of *Coxiella burnetii*

Craig T. Narasaki and Rudolf Toman

Contents

Abstract A lipopolysaccharide (LPS) is considered to be one of the major determinants of virulence expression and infection of virulent *Coxiella burnetii*. The LPSs from virulent phase I (LPS I) and from avirulent phase II (LPS II) bacteria were investigated for their chemical composition, structure and biological properties. LPS II is of rough (R) type in contrast to LPS I, which is phenotypically smooth (S) and

C.T. Narasaki
Center Department of Microbial and Molecular Pathogenesis,
Texas A&M University Health Science , College Station, TX 77843, USA

R. Toman (✉)
Laboratory for Diagnosis and Prevention of Rickettsial and Chlamydial Infections, Institute of Virology, Slovak Academy of Sciences, Dubravska cesta 9, 845 05 Bratislava, Slovakia
e-mail: Rudolf.Toman@savba.sk

R. Toman et al. (eds.), *Coxiella burnetii: Recent Advances and New Perspectives in Research of the Q Fever Bacterium*, Advances in Experimental Medicine and Biology 984,
DOI 10.1007/978-94-007-4315-1_4,

contains a noticeable amount of two sugars virenose (Vir) and dihydrohydroxystreptose (Strep), which have not been found in other LPSs and can be considered as unique biomarkers of the bacterium. Both sugars were suggested to be located mostly in terminal positions of the O-specific chain of LPS I (O-PS I) and to be involved in the immunobiology of Q fever. There is a need to establish a more detailed chemical structure of LPS I in connection with prospective, deeper studies on mechanisms of pathogenesis and immunity of Q fever, its early and reliable diagnosis, and effective prophylaxis against the disease. This will also help to better understanding of host-pathogen interactions and contribute to improved modulation of pathological reactions which in turn are prerequisite for research and development of vaccines of new type. A fundamental understanding of *C. burnetii* LPS biosynthesis is still lacking. The intracellular nature of the bacterium, lack of genetic tools and its status as a selected agent have made elucidating basic physiological mechanisms challenging. The GDP-β-D-Vir biosynthetic pathway proposed most recently is an important initial step in this endeavour. The current advanced technologies providing the genetic tools necessary to screen *C. burnetii* mutants and propagate isogenic mutants might speed the discovery process.

Keywords *Coxiella burnetii* • Biosynthesis • Function • Lipopolysaccharide • Q fever • Structure

4.1 Introduction

Upon serial passage in an immunoincompetent host, virulent *C. burnetii* undergoes a modification of its LPS what is traditionally referred to as phase variation (Stoker and Fiset 1956; Ftacek et al. 2000). *C. burnetii* phase variation includes the non-reversible switch from the virulent phase I (S) LPS I, which has a full length O-PS I chain (Mayer et al. 1988; Skultety et al. 1998) to the avirulent phase II (R) LPS II, which is missing the O-PS I chain and the saccharides located in the LPS I outer core (Toman and Skultety 1996; Toman et al. 2009). Several studies have indicated differential protein expression between the small cell variant (SCV) and large cell variant (LCV) cells of the bacterium (McCaul et al. 1991; Heinzen and Hackstadt 1996; Seshadri et al. 1999; Coleman et al. 2004, 2007) but the structural studies of the LPSs isolated from these variants have not been performed thus far. Although the structure of lipid A proximal core region of LPS I, represented by LPS II, was well established (Toman and Skultety 1996), both composition and structural features of the LPS I outer core and O-PS I chain have not been fully defined thus far (Toman et al. 2009).

LPS I is of particular biological, immunological and medical significance. It plays an essential role in interaction between the microbe and host including pathogenicity and immunogenicity of the agent. It is capable of inducing antibody response and is considered to be a protective immunogen (Williams and Waag 1991; Gajdosova et al. 1994; Hussein et al. 2001). Thus, it has been reported recently that O-PS I and its epitopes play a critical role in the antibody response to *C. burnetii* (Slaba et al. 2003; Vadovic et al. 2005), are involved in both signal transduction pathways (Dellacasagrande et al. 2000a, b) and LPS trafficking (Pretat et al. 2009), and may help to provide protection when included in a subunit vaccine (Zhang et al. 2007). Furthermore, research that advances the objective of rapid and specific detection of the infectious pathogen and differentiation of the individual isolates is centered on the development of monoclonal antibodies (mAbs) directed against the unique sugars present in the O-PS I chain (Palkovicova et al. 2009). However, both LPS I and LPS II were shown to be poor endotoxins. Investigations of their endotoxin activities showed that they were up to 1,000-fold less endotoxic than LPSs of *Escherichia coli* and *Salmonella typhimurium* (Amano et al. 1987). In this chapter, the structure, genetics, biosynthesis, and biological properties of the *C. burnetii* LPS will be reviewed.

4.2 Lipopolysaccharide Overview

LPSs are complex macromolecules found in the Gram-negative bacteria cell walls that may form up to approximately 75% of the outer leaflet of their outer membranes. They are essential for the physical organization and function of these membranes, and thus, for bacterial growth and multiplication. They are targets for bacteriophages, possess binding sites for antibodies and non-immunoglobulin serum factors, and in this way, they are involved in specific recognition and elimination of bacteria by the host defense system (Alexander and Rietschel 2001). LPSs may prevent the activation of complement and uptake of bacteria by phagocytes, and thus, they play an important role in the bacterial virulence. It has been presented in numerous studies (Alexander and Rietschel 2001; Raetz and Whitfield 2002) that LPSs display a broad spectrum of biological activities such as lethal toxicity and pyrogenicity, up- and down- regulation of several biological processes and in this manner they can contribute to the pathogenic potential of many Gram-negative bacteria. Moreover, LPSs activate strongly B-cells, granulocytes and mononuclear cells being considered as potent immunostimulators. In addition, there has been ever increasing evidence on an extensive involvement of LPSs in certain signal transduction pathways (Takeda and Akira 2005; Lu et al. 2008).

LPSs of Gram-negative bacteria share the common structural features and consist usually of three structural domains: (1) lipid A, which is attached to the outer bacterial membrane of the cell wall *via* acyl tails and is responsible for the endotoxic properties of LPS, (2) a short non-repeating inner and outer core, which is attached to lipid A and extends outwardly, and (3) O-PS chain, which is usually composed of repeating saccharide units attached to the lipid A – core and is responsible for serological heterogeneity among bacterial isolates (Alexander and Rietschel 2001). Minimally for survival, Gram-negative bacteria require lipid A and one 3-deoxy-D-*manno*-oct-2-ulosonic acid (Kdo). Bacteria with only these two subunits exhibit a deep (R) LPS phenotype (Raetz and Whitfield 2002) while bacteria with lipid

A – Kdo plus additional saccharide units but not O-PS are generally described as (R) LPS bacteria. In contrast, (S) LPS bacteria are characterized by a lipid A – core attached to a fully developed O-PS chain. In addition to numerous functions of (S) LPS mentioned above, it may serve also as a permeability barrier against external agents such as hydrophobic antibiotics and to help to maintain the structural integrity of the Gram-negative cell wall (Raetz and Whitfield 2002). Moreover, (S) LPS has been a defined virulence factor in many clinically relevant pathogens. For example, (R) LPSs of *Vibrio cholerae* mutants were impaired in the intestinal epithelia colonization in the mouse model, which reflected their increased sensitivity to complement and cationic peptides (Nesper et al. 2001). Studies with uropathogenic *E. coli* O75:K5 showed that the chemistry and length of wild type O-PS degraded the complement system by inhibiting the assembly of the membrane attack complex (Joiner 1988; Burns and Hull 1998). The increased O-PS length also decreased the sensitivity of *E. coli* to neutrophil bactericidal/permeability-increasing protein (Weiss et al. 1986). The (S) LPS of *Shigella flexneri* influenced the polar localization of its invasion protein and was essential for invasiveness and subsequent inter- and intracellular spread (Sandlin et al. 1995, 1996; Van den Bosch et al. 1997). Finally, experiments with the burned mouse model and corneal mouse model showed that *Pseudomonas aeruginosa* (S) LPS is a critical virulence factor required for colonization (Rocchetta et al. 1999).

The importance of LPS as a virulence factor for the intracellular pathogens is less clear. It was shown in *Legionella spp.* that a virulence-associated LPS epitope caused most community acquired legionellosis (Helbig et al. 1995) and a phase-variable expression of LPS contributed to virulence by mediating serum complement resistance (Luneberg et al. 1998). However, the O-acetyl groups that modify the *Legionella* O-PS, hypothesized to be critical for virulence, were shown to have no impact on serum resistance (Luck et al. 2001). *Francisella tularensis* blue strain (S) LPS was more virulent than gray strain (R) LPS, but the mechanism of action is still unknown (Gunn and Ernst 2007). *Brucella* can perturb bactericidal activity in macrophages by influencing the host cell response to its advantage through its LPS (Jimenez de Bagues et al. 2005). The *C. burnetii* phase II attenuation in the immunocompetent hosts may be explained by several observations. The phase II organisms are more susceptible to complement mediated lysis relative to the phase I organisms (Vishwanath and Hackstadt 1988). It has been shown that phase I *C. burnetii* can infect and propagate in human dendritic cells (DC) while inhibiting their ability to induce maturation or inflammatory cytokine production (Shannon et al. 2005). In the same study, phase II bacteria significantly induced DC maturation and a robust inflammatory response.

4.2.1 Lipopolysaccharide Phase Variation

The generally accepted definition of phase variation is the reversible on-off switching mechanism of cell surface structures under various environmental conditions (Van der Woude and Baumler 2004). Although most O-PS operons are expressed constitutively (Raetz and Whitfield 2002), mechanisms for the LPS phase variation have been described for several bacteria. For example, the O-PS of *Yersinia enterocolitica* O3 is regulated by a repressor gene located outside of the O-PS gene cluster which is expressed at 37°C but not at 25°C (Al-Hendy et al. 1991; Skurnik and Toivanen 1993). The *Legionella pneumophila* O-PS is modulated by an unstable 30 kb genetic element that when integrated into the chromosome produces full length O-PS, but when incorporated in its native plasmid express an (R) LPS phenotype (Luneberg et al. 2001). Although *Haemophilus influenzae* lacks O-PS, its LPS structure is a critical virulence factor and is modulated by slipped-stranded mispairing (Weiser et al. 1989). Additionally, growth phase dependent expression of RfaH, a transcriptional anti-termination factor, was shown to modulate O-PS in *S. enterica* serovar Typhi (Rojas et al. 2001; Nagy et al. 2008).

In contrast, *C. burnetii* phase variation is historically defined as a non-reversible switch from virulent phase I cells containing LPS I to avirulent phase II cells with largely modified LPS II (Stoker and Fiset 1956; Williams and Waag 1991). It has been proposed that the observed LPS modification proceeds through gradual reduction of O-PS I during the phase variation yielding the LPS II which is the only LPS present in the *C. burnetii* phase II cells (Diaz and Lukacova 1998). In contrast, a completely different insight into this complex phenomenon was brought by Ftacek et al. (2000) based on their study of both composition and structure of LPSs isolated during serial passage of *C. burnetii* in embryonated hen eggs. It is a well-known fact that LPS molecules isolated from most (S) strains/isolates of Gram-negative bacteria are heterogeneous in size. They may contain some (R) LPSs, and in several cases variously truncated LPS molecules (Keenleyside and Whitfield 1999). Therefore, it can be assumed that strains/isolates of Gram-negative bacteria contain cell populations that may express multiple LPS structures. As the similar observation was also made by Ftacek et al. (2000), it has been suggested that a redistribution of the existing LPS populations takes place during the *C. burnetii* phase variation due to an increasing prevalence of those cells in the whole cell population that express LPS molecules with truncated O-PS I and those being of (R) type. It appears that in an attempt to optimize the balance between the expenditure of energy for the synthesis of complete LPS I molecules and the need to resist the microbicidal activities of the phagolysosome of the host immune system, *C. burnetii* has evolved genetic mechanisms of the transition from an energy-demanding (phase I) to a lesser energy-demanding (phase II) states. Thus, the mutants or strains/isolates that prefer biosynthesis of the incomplete LPS I subclasses might have a growth advantage.

Investigations of the O-PS biosynthesis region in the phase II Nine Mile strain (NM II, RSA 439) revealed that a large group of the genes was deleted (Hoover et al. 2002; Denison et al. 2007). Nevertheless, some phase II isolates contained no apparent deletions (Denison et al. 2007). To our best knowledge, the phase variation mechanisms have not been elucidated by bioinformatics tools in *C. burnetii* thus far and similarly, no R$\rightarrow$S transition has ever been observed in the *C. burnetii* phase II strains/isolates. Thus, molecular mechanisms influencing the LPS modifications during the *C. burnetii* phase variation remain unclear at present.

4.3 Structural Studies of a Lipopolysaccharide from *C. burnetii*

4.3.1 Composition and Structure of LPS II

The first, more detailed chemical study of LPS II reported that it had similar sugar composition to LPS I although with quantitative differences (Baca et al. 1980). The following studies showed that LPS II and LPS I differed in the composition of constituent sugars (Schramek and Mayer 1982; Amano and Williams 1984; Mayer et al. 1988). It has been suggested, however, that LPS II contains a "Kdo-like substance" in the lipid A proximal region that is different from the enteric Kdo typical for most bacterial LPSs. Moreover, the sugar sequence in the inner core region of LPS II was not fully characterized. In their initial study, Toman et al. (1993) confirmed the presence of enteric Kdo in LPS II. Later, methylation-linkage analysis of LPS II demonstrated the presence of three Kdo residues with a structural arrangement similar to that of many enterobacterial LPSs (Toman and Skultety 1994). From other sugar residues, two terminal D-mannoses (Man), 2- and 3,4-linked D-*glycero*-D-*manno*-heptoses (D,D-Hep) were established (Toman and Skultety 1996). The presence of the latter sugar was quite surprising as most bacterial LPSs contain L-*glycero*-D-*manno*-heptose as the constituent sugar of the inner core region. Fast atom bombardment (FAB)- and electrospray ionization (ESI)-mass spectrometries (MS) of the lipid A deprived LPS II established the sequence of its sugar residues (Toman and Skultety 1996). The chemical structure of lipid A proximal region of LPS II is shown in Fig. 4.1. The subsequent matrix assisted laser desorption ionization (MALDI) – MS measurements confirmed the previous results and established the molecular mass of LPS II to 2841.58 (Toman et al. 2003b).

In the last decade, many (R) LPSs have been found in pathogenic wild type bacteria including those colonizing especially the mucosal surfaces of the respiratory and urogenital tracts, such as *Neisseria meningitidis*, *N. gonorrhoeae*, *H. influenzae*, *Bordetella pertussis* or *Chlamydiaceae* (Holst 1999; Caroff et al. 2000). At present, it is not known with certainty whether these bacteria are not able to biosynthetize (S) LPSs or whether the genes encoding synthesis of O-specific chains are suppressed or deleted by regulatory mechanisms induced by environmental cues (Burns and Hull 1998; Heinrichs et al. 1999; Holst 1999). However, occurrence of such (R) type LPSs in the pathogenic wild-type bacteria indicates that the O-chain cannot be considered as the only major factor of pathogenicity in the respective bacterium and that it is dispensable without deleterious effects for it.

4.3.2 Composition and Structure of LPS I

As mentioned earlier, LPS I plays an important role in the interaction of *C. burnetii* with host, its pathogenicity and immunogenicity and is considered to be a protective immunogen (Amano and Williams 1984; Williams and Waag 1991; Gajdosova et al. 1994;

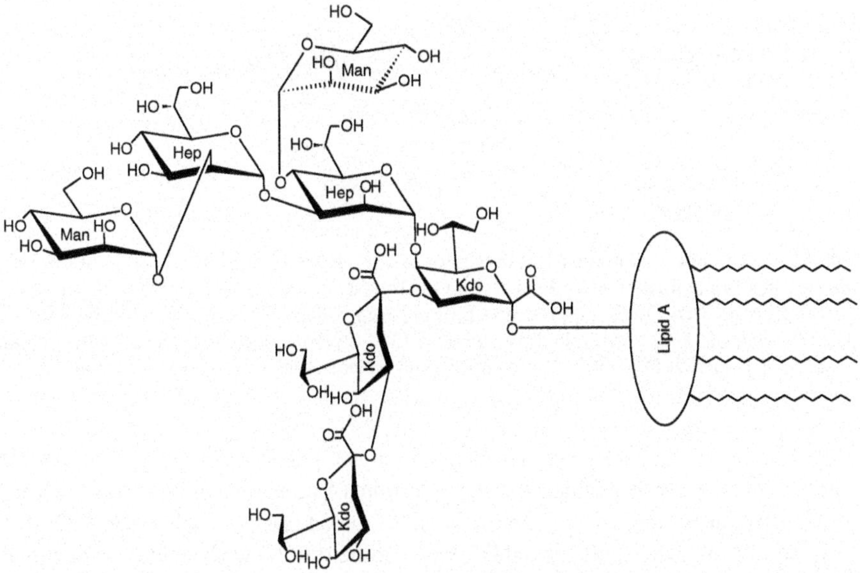

Fig. 4.1 Chemical structure of lipid A proximal region of LPS II from *Coxiella burnetii* in avirulent phase II. *Hep* D-*glycero*-D-*manno*-heptose, *Kdo* 3-deoxy-D-*manno*-oct-2-ulosonic acid, *Man* D-mannose

Hussein et al. 2001). Several structural studies of this polymer were reported in the past but the data obtained brought only limited information on its structural features (Schramek et al. 1985; Amano et al. 1987; Toman and Kazar 1991; Skultety et al. 1998). LPS I contains, in addition to the sugar residues found in LPS II and some frequently occuring sugars, two unusual sugar units in its O-PS I, namely virenose (Vir, 6-deoxy-3-C-methylgulose) and dihydrohydroxystreptose [Strep, 3-C-(hydroxymethyl)lyxose]. Both sugars have not been found in other bacterial LPSs and can be considered unique biomarkers of *C. burnetii*. The enantiomeric forms and ring conformations of both saccharides were established from the optical rotation and nuclear magnetic resonance (NMR) data (Toman et al. 1998). Vir was found to be the D-gulo enantiomer with the $^4C_1$ ring conformation and Strep was shown to be the L-lyxo enantiomer also with the $^4C_1$ conformation. However, a structural analysis of LPS I showed that Strep was present in the parent LPS I in a furanose form (Toman 1991). Therefore, it was suggested that a furanose to pyranose tautomerization took place in the course of the isolation procedure (Toman et al. 1998). Recently, candidate genes involved in the synthesis of Vir and Strep have been suggested (Hoover et al. 2002; Thompson et al. 2003) and their protein products NDP-hexose 3-C-methyltransferase TylCIII (CBU0691) and methyltransferase FkbM family (CBU0683) were identified by the proteomic analysis (Skultety et al. 2005). However, the presence of galactosaminuronyl-α-(1→6)-glucosamine structural motif reported to occur in O-PS I (Amano et al. 1987) could not be

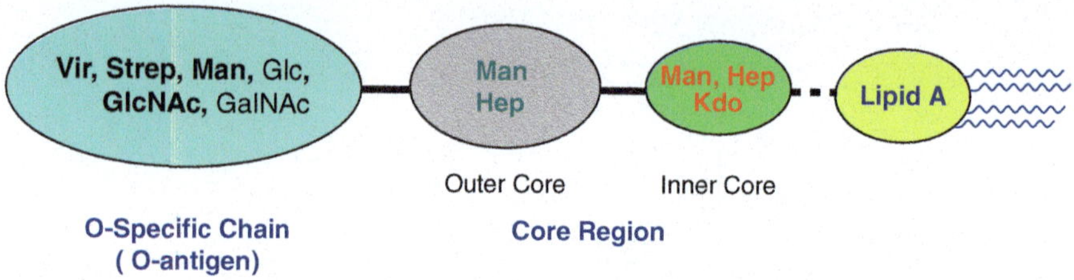

Fig. 4.2 Preliminary structural arrangement of the sugar residues of LPS I from *Coxiella burnetii* in virulent phase I. *GalNAc* N-acetyl-D-galactosamine, *GlcNAc* N-acetyl-D-glucosamine, *Glc* D-glucose, *Hep* D-*glycero*-D-*manno*-heptose, *Kdo* 3-deoxy-D-*manno*-oct-2-ulosonic acid, *Man* D-mannose, *Strep* dihydrohydroxystreptose [3-C-(hydroxymethyl)-L-lyxose], *Vir* virenose (6-deoxy-3-C-methyl-D-gulose). The sugars in bold are prevailing in the O-specific chain

confirmed in our laboratory very recently. Methylation analyses of two polysaccharide fractions of LPS I revealed the presence of terminal Vir, Strep, and Man, 4-substituted Vir, 4-substituted Man, and 2,3- and 3,4-disubstituted D,D-Hep (Vadovic et al. 2005). From amino sugars, 4-substituted D-glucosamine (GlcN) was detected. The methylation data demonstrated the pyranose form of Vir, Man, D,D-Hep and GlcN, and the furanose form of Strep. It could be anticipated from previous investigations (Toman and Skultety 1996) that 2,3- and 3,4-disubstituted D,D-Hep, and two terminal Man were from the core region of LPS I and thus, other terminal and substituted sugars should be located in the O-PS I chain. It was suggested in earlier studies (Schramek et al. 1985; Amano et al. 1987; Toman 1991; Skultety et al. 1998) that Vir and Strep were located almost exclusively in terminal positions. However, the recent findings have indicated that this might be true only for Strep as Vir is also $(1\rightarrow4)$-linked (Vadovic et al. 2005). Similarly, Man is present in terminal positions but about 23% of it could also be involved in $1\rightarrow4$ linkages. A faster progress in a more detailed structural characterization of O-PS I is hampered by the presence of several O-PS I populations differing one from another in size, shape, and chemical composition. Therefore, only a tentative structural arrangement of sugar residues in LPS I can be given at present as shown in Fig. 4.2.

In the past, the *C. burnetii* clonal derivative RSA 514 named “Crazy” (Cr) was isolated from the placental tissue of guinea pig infected with the phase I Nine Mile (NM I, RSA 493) strain for 343 days (Hackstadt et al. 1985). It reacted with phase I antiserum, but to a much lesser extent than the NM I strain, and displayed no reactivity with phase II antiserum. The isolated LPS Cr gave one band at about 14 kDa on SDS-PAGE in contrast to LPS I and LPS II (Toman et al. 2009). The compositional analysis revealed the presence of Man, Glc, D,D-Hep, Strep, and GlcN in a molar ratio 3.1:0.1:1.0:1.5:1.2, respectively. No Vir was found. Both composition and structure of LPS II were shown to be a result of large chromosomal DNA deletions in the phase II cells of NM I strain (Hoover et al. 2002). However, deletions in the variant RSA 514 were larger extending on both ends beyond the phase II deletion junctions. Surprisingly, structural complexity of the LPS Cr appeared to be intermediate to LPS I and LPS II. The reasons for this discrepancy are still unknown.

4.3.3 Composition and Structure of Lipid A

The lipid A portion of LPS is linked to the core oligosaccharide mostly *via* Kdo and serves as the hydrophobic anchor of LPS in the bacterial outer membrane. Lipid A is established as the endotoxic principle of bacterial LPSs and represents in terms of both chemical composition and structure the most conserved region in them (Alexander and Rietschel 2001; Raetz and Whitfield 2002). Lipid A plays a major role in the pathogenesis of bacterial infections and is important contributor to massive inflammation, sepsis, and septic shock leading to fatalities in Gram-negative bacteria infections. It also promotes the activation of the innate immune system *via* induction of inflammatory cytokines released by human cells (Alexander and Rietschel 2001). Structurally, lipid A is typically composed of a β-D-GlcN-(1→6)-α-D-GlcN disaccharide backbone carrying two phosphate groups at positions O-1 and O-4′. Both phosphates can be further substituted with groups such as ethanolamine, ethanolamine phosphate, ethanolamine diphosphate, GlcN, 4-amino-4-deoxy-L-arabinopyranose, and D-arabinofuranose (Rietschel et al. 1987; Vadovic et al. 2007). There are attached up to four acyl chains by ester or amide linkages to the GlcN disaccharide. These chains can then in turn be substituted by further fatty acids to provide the lipid A moieties with up to seven acyl substituents, which vary quite considerably among species in the nature, number, length, order, and saturation (Alexander and Rietschel 2001).

It appears that the major contributing factors to endotoxicity are the number and lengths of acyl chains present and the phosphorylation state of the disaccharide backbone. For example, the structures with only one phosphate at either O-1 or O-4′ appear in most assays to be ~1,000-fold less active than the highly endotoxic lipid A of *E. coli* as it was found in the naturally occurring monophosphorylated *Bacteroides fragilis* lipid A (Rietschel et al. 1987; Alexander and Rietschel 2001). However, phosphates by themselves do not appear to be essential as the substitution with the phosphono-oxyethyl group does not alter activity of the compound, suggesting that only correctly placed negative charges can restore activity (Ulmer et al. 1992; Erridge et al. 2002). The GlcN monosaccharide preparations phosphorylated and acylated in various positions lack activity in general, suggesting that the disaccharide backbone is also required for the optimum recognition by the humoral/cellular receptors (Aschauer et al. 1990). Nevertheless, lipids A with 2,3-diamino-2,3-dideoxy-D-glucose replacing GlcN in the backbone have similar activity than those having GlcN as seen, *e.g.* in *Campylobacter jejuni* (Moran et al. 1991; Alexander and Rietschel 2001).

Naturally, much work was concentrated on the role of the nature, number and lengths of acyl chains attached to the lipid A. For example, structures similar to that of the *E. coli* lipid A with two phosphates, but with seven or five fatty acids were less active by a factor of approximately 100 (Rietschel et al. 1993; Erridge et al. 2002). Thus, it appears that the hexaacylated lipid A species consisting of two 3-hydroxyacyl and two 3-acyloxyacyl residues limited in length of individual acid chains to 12 or 14 carbons of either *E. coli* or *N. meningitidis* like – types are optimally recognized by the mammalian receptors to express the full spectrum of endotoxic activities (Alexander and Rietschel 2001; Erridge et al. 2002).

Wollenweber et al. (1985) analyzed the fatty acid composition in the *C. burnetii* lipids A from LPS I and LPS II. The authors identified more than 50 different 3-acyloxyacyl residues to be involved in the amide linkage, and together with other fatty acids found, suggested an enormous heterogeneity of both lipid A components. In contrast, Toman et al. (2003a) could not find any 3-acyloxyacyl residues in the investigated lipids A from the LPSs I of Henzerling and S strains, and in addition, their fatty acid compositions were considerably less complex than those published by Wollenweber et al. (1985). The authors showed that both lipids A were tetraacylated with two amide-linked 3-hydroxy and two ester-linked nonhydroxylated fatty acids attached to the GlcN dissaccharide. Differences were found in the small structural details only. Further, in the LPS I of Priscilla strain, two tetraacylated lipid A species were found as the major components despite a noticeable microheterogeneity of the analyzed sample (Toman et al. 2004). They possessed the classical backbone of diphosphorylated GlcN disaccharide, in which both reducing-end GlcN I and terminal GlcN II carried amide-linked iso-branched or normal (n) 3-hydroxyhexadecanoic fatty acids. One of the species had ester-linked n-hexadecanoic acids at both GlcNs while the other had ester-linked anteiso-branched pentadecanoic instead of n-hexadecanoic acid at GlcN II. The lipid A moiety had a cubic inverted aggregate structure, and the inclination angle of the GlcN disaccharide backbone plane of the lipid A part with respect to the membrane normal was around 40°. The lipid A readily intercalated into phospholipid liposomes mediated by the LPS-binding protein. The lipid A-induced tumor necrosis factor (TNF) production in the human mononuclear cells was one order of magnitude lower than that found for lipid A of *S. minnesota.* In another study (Zamboni et al. 2004), an asymmetrical structure of the *C. burnetii* lipid A was suggested where a 3-acyloxyacyl chain, alternatively composed of 3-hydroxytetra- to hexadecanoic acids, was attached to the amide-linked 3-hydroxylated fatty acid at the N-2′ of the GlcN II unit. However, the proposed structure needs further verification.

It was reported (Alexander and Rietschel 2001) that variation of the lipid A domain of LPS serves as one strategy utilised by Gram-negative bacteria to promote survival by providing resistance to components of the innate immune system and helping to evade recognition by Toll-like receptor 4 (TLR 4). Thus, it was of interest to see if the long-term survival of *C. burnetii* in host led to modifications in its lipid A in comparison with the known structures for lipids A from the strains Henzerling, S and Priscilla (Toman et al. 2003a, 2004). The recent study of lipid A from the clonal derivative RSA 514 has revealed no substantial modifications in the basic chemical composition and structure proposed for lipid A of *C. burnetii* (Vadovic et al. 2009). It appears that lipids A from various strains/isolates of *C. burnetii* represent tetraacylated molecular species with the conserved basic structural features. Some differences are found in nature, length and possibly also in distribution of fatty acids along the GlcN disaccharide. These fatty acid variations contribute to a remarkable diversity of the *C. burnetii* lipid A although its biochemical synthesis is a highly conserved process like in other Gram-negative bacteria. Nevertheless, further studies are needed to confirm symmetrical (Toman et al. 2003a, 2004; Vadovic et al. 2009) or asymmetrical (Zamboni et al. 2004) structural model of the lipid A. The proposed structural features of *C. burnetii* lipid A are shown in Fig. 4.3.

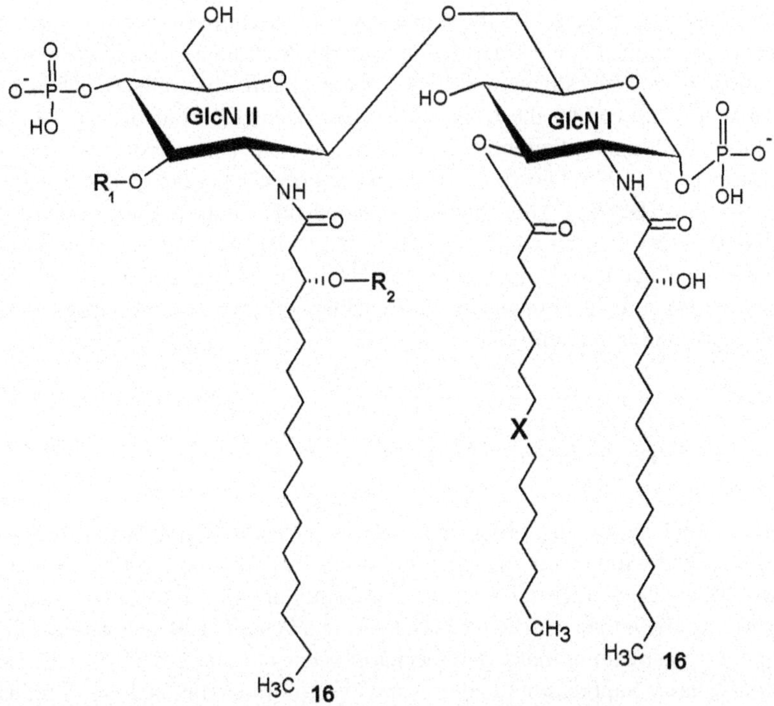

Fig. 4.3 Basic chemical structure of lipid A of *Coxiella burnetii.* Models with symmetrical $R_1 = H_3C–(CH_2)_n–CH_2–CH_2–CO$ or $H_3C–(CH_2)_n–CH(OH)–CH_2–CO$, $R_2 = H$ (Toman et al. 2003a; Toman et al. 2004; Vadovic et al. 2009) and asymmetrical $R_1 = H$, $R_2 = H_3C–(CH_2)_n–CH(OH)–CH_2–CO$ (Zamboni et al. 2004) structures have been proposed; $n = 12–14$, $X = C_{14–16}$

Distinct structural features of the *C. burnetii* lipid A might be the reason for its lower endotoxic activity as compared to the classical forms of enterobacterial lipids A, found, *e.g.* in *E. coli, N. meningitidis* or *Salmonella spp*. Nevertheless, *C. burnetii* and its LPS were shown to exhibit various immunomodulatory activities despite their low endotoxic activity (see below). This behavior could be elucidated by applying the previously presented conformational concept of endotoxicity (Seydel et al. 1999), a conical shape of the lipid A of *C. burnetii* and a sufficiently high inclination of the sugar backbone plane with respect to the membrane (Toman et al. 2004).

4.4 Biosynthesis of Lipopolysaccharides

4.4.1 Nomenclature

The complex biochemistry of LPS biosynthesis and expression causes confusion when naming and studying LPS related genes. Contributing factors that add to the confusion include different names for duplicated genes located in different loci of a bacterium, different names for the same gene in different bacteria or the same name for genes that are not related in different bacteria. Additionally, some contemporary LPS biosynthesis investigators still use nomenclature that reflects the historical assignment of a gene by the Demerec System (Demerec et al. 1966). To help to resolve the limitations inherent to the Demerec System, a new system was proposed by Reeves et al. (1996) known as the Bacterial Polysaccharide Synthesis and Gene Nomenclature (BPGN). This system expanded the number of gene possibilities from 26 (number of letters in the alphabet) to 17,576 possibilities and is organized by function. Since both Demerec and BPGN nomenclature systems are currently in use, the usage of gene names in this chapter will reflect those which are most frequently found in literature.

4.4.2 O-Polysaccharide Genetics

The structural heterogeneity observed in various O-PS chains correlates to the genetic variation found within their encoding loci (Raetz and Whitfield 2002). Despite this variation the clusters themselves tend to be highly organized (Stroeher et al. 1998; Trefzer et al. 1999; Luneberg et al. 2000; Raetz and Whitfield 2002; King et al. 2009). Present just upstream of many polysaccharide gene clusters is a 39-bp JUMPstart/Operon Polarity Suppressor site that is involved in transcriptional anti-termination shown to require RfaH (Hobbs and Reeves 1994; Wang et al. 1998). This finding is consistent with a hypothesis that the O-PS clusters are transcribed as a single locus, which may contain 7 to over 17 genes (Raetz and Whitfield 2002). In *E. coli* and *S. enterica*, the O-PS locus is usually flanked by *his* and *gnd* (Raetz and Whitfield 2002) while in *Pseudomonas spp.* is flanked by *ihfB* and *wbpM* (Rocchetta et al. 1999). It has been suggested that these flanking genes are target sites for the lateral transfer of O-PS clusters (Bisercic et al. 1991; Nelson and Selander 1994). Further supporting this hypothesis is the observation that the O-PS gene clusters typically have a GC content lower than the genome average (Samuel and Reeves 2003).

The currently accepted organizational paradigm for genes located within the O-PS locus, referred to as *wb* (BPGN) or *rfb* (Demeric), is by function: (1) pathway genes, (2) processing genes and (3) transferase genes. Pathway genes include those genes involved in the activated carbohydrate biosynthesis. These genes share a high degree of identity among a wide range of species (Trefzer et al. 1999; Raetz and Whitfield 2002; Samuel and Reeves 2003). Processing genes encode proteins that mediate O-PS transport from the cytosol to the periplasm and are usually adjacent within the *wb* locus (Raetz and Whitfield 2002). Transferase genes catalyze the sequential attachment of sugar units to the growing O-PS chain. The glycosyltransferases are often dispersed throughout the *wb* locus and have low levels of similarity with one notable exception, *wecA* (*rfe*, Demeric) (Raetz and Whitfield 2002). The *wecA* gene encodes a glycosyltransferase responsible for the transfer of the first saccharide called the primer saccharide to the hydrophobic lipid carrier molecule, undecaprenol phosphate (und-P), in *E. coli*. The initial transfer of the primer saccharide to und-P is required for the O-PS elongation in the cytosol (Raetz and Whitfield 2002). Unlike other transferases, *wecA* is located outside of the *wb* locus. Bioinformatic analysis of various WecA glycosyltransferases showed a high degree of similarity and it was once thought they might have wide substrate specificity when compared to other glycosyltransferases (Raetz and Whitfield 2002). However, recent evidence has shown that WecA is specific for only GlcNAc transfer to und-P (Rush et al. 2010).

4.4.3 O-Polysaccharide Genetic Analysis of C. burnetii

As mentioned earlier, the clonal derivative Cr had a large chromosomal deletion (25,997 bp) that overlapped the NM II deletion. The deletion should correspond to the loss of Vir in this and also the NM II strain and should explain the nonreversible phase variation observed with the NM strain (Vodkin and Williams 1986; Hoover et al. 2002). However, it does not explain the presence of LPS II in the NM II strain that lacks also Strep and the outer core saccharides. Furthermore, a recent study has shown that several phase II *C. burnetii* strains/isolates have a functional *C. burnetii wb* locus but still exhibit the (R) LPS phenotype (Denison et al. 2007). This finding might suggest that genes outside of the deleted region are required for both O-PS I and the outer core biosynthesis.

Further comparative genomic analysis revealed three regions that might contain genes necessary to express O-PS I in LPS I. Region 1 (605781–653889) includes the deleted region of the strains NM I (RSA 493) and Cr (RSA 514) defined by Hoover et al. (2002) and may contain the genes needed for GDP-β-D-Vir biosynthesis. Region 2 (775635–813578) may contain those genes needed for the expression of sugars constituting the O-PS I backbone, and region 3 (1760021–1768371) may contain the genes needed for the expression of Strep (Seshadri et al. 2003). Within the region 1, the genes that are normally associated with a *wb* operon, like genes for an activated saccharide biosynthesis, the glycosyl transferase genes and processing genes, have been identified (Hoover et al. 2002). Just outside of the NM II (RSA 439) and Cr (RSA 514) deleted regions, there are two adjacent genes that code for an ABC transporter system, O-antigen export system permease protein (CBU0703) and LPS/O-antigen export permease (CBU0704). The ABC transporter dependent pathways typically transport the linear O-PS structures *versus* branched structures characteristic of the Wzy dependent pathway (Raetz and Whitfield 2002). This evidence might support a hypothesis that the O-PS I in LPS I is essentially a linear polymer. However, this hypothesis could not be confirmed by the glycomic studies mentioned above. An exhaustive bioinformatic search has failed to identify any of the genes necessary for the Wzy dependent pathway thus far.

Although the GC content of the deleted region is slightly lower than the chromosomal GC content (43% in RSA 493 and 39% in the deleted region of RSA 439), there are no apparent target sites where the lateral transfer might occur. The bioinformatic analysis indicates no JUMPstart/ops sequence and several genes, some divergently transcribed, that may not play a role in the O-PS I biosynthesis.

4.4.4 O-Polysaccharide Biosynthesis and Export Mechanisms of C. burnetii

Three O-PS biosynthesis pathways have been described: (1) ABC-transporter dependent, (2) Wzy dependent and (3) synthase dependent. Each pathway initiates the O-PS biosynthesis by attaching a sugar-1-phosphate catalyzed by a membrane bound glycosyltransferase to und-P. The model glycosyltransferases are WecA in *E. coli* and WbaP in *Salmonella spp.*, which transfer GlcNAc and D-galactose to und-P, respectively (Raetz and Whitfield 2002). Nevertheless, other initial glycosyltransferases have been shown to initiate the O-PS biosynthesis. For example, the *Pseudomonas aeruginosa* WbpL catalyzes the transfer of N-acetyl-D-fucosamine (D-FucNAc) and N-acetyl-D-quinovosamine (D-QuiNAc) to und-P (King et al. 2009). It is likely that other initial glycosyltransferases exist but have not been characterized thus far. For example, the unique primer saccharide legionaminic acid could be transferred to und-P by an undefined enzyme in *L. pneumophila* (Knirel et al. 1994).

Structural studies indicated that GlcNAc was present in the O-PS I of LPS I from the NM I strain (Toman and Kazar 1991), and GlcNAc and GalNAc were both identified in the O-PS I of LPS I from the *C. burnetii* strain Priscilla in phase I (Hussein et al. 2001). Additionally, *C. burnetii* undecaprenyl-phosphate α-N-acetylglucosamine phosphotransferase (CBU0533) has a strong sequence homology to the characterized WecA in the model organisms. Taken together, we hypothesized that GlcNAc or GalNAc could be the initial saccharides in the O-PS I chain and CBU0533 might catalyze their attachment to und-P. To prove this hypothesis, we constructed an *E. coli* O157:H7 ZAP198 Δ*wecA* mutant by a modified protocol described by Emmerson et al. (2006). The mutant was verified genotypically by polymerase chain reaction and phenotypically by silver staining and immunoblot. The wild type *E. coli* O157:H7 LPS was restored to the ZAP198 Δ*wecA* mutant when the native *wecA* was expressed *in trans*. However, despite the bioinformatic similarity between CBU0533 to characterized WecAs, it failed to complement the ZAP198 Δ*wecA* mutant *in trans*.

Several possibilities may explain this observation: (1) CBU0533 does not catalyze the transfer of GlcNAc or GalNAc to und-P, (2) it is not active in *E. coli* O157:H7, or (3) the *E. coli* O157:H7 lipid carrier molecule is not a CBU0533 substrate. It is well known that the lipid carrier molecules are used in both eukaryotic and prokaryotic cells to transport glycans across the lipid bilayers. The most characterized prokaryotic lipid carrier molecules have 11 isoprene units, 3 in *trans* and 8 in *cis* positions for a total of 55 carbons (C_{55}). Thus, the name "undecaprenyl" phosphate derives from this finding (Raetz and Whitfield 2002). Many laboratories have shown that both length of polyisoprenol carrier molecule and nature of α-isoprene unit drive the specificity and efficiency of the initial glycosyltransferase reaction in eukaryotes (Szkopinska et al. 1992; Rush et al. 1993; D'Souza-Schorey et al. 1994; Kean et al. 1994; McLachlan and Krag 1994; Dotson et al. 1995) and in prokaryotes (Rush et al. 1997; Chen et al. 2007). An *in vitro* evidence from the Rocky Mountain

Laboratory (RML) in Hamilton, MT, has shown that CBU0533 does not transfer GlcNAc to und-P (unpublished data). In light of the bioinformatic analysis and experimental work performed in our laboratory and at RML, we have concluded that the *C. burnetii* RSA 493 CBU0533 encodes a *wecA*-like gene. Future studies on the O-PS I biosynthesis in LPS I should focus on the attachment of the first monosaccharide to the lipid carrier molecule.

After attachment of the primer saccharide to und-P, the various O-PS pathways differ in their polymerization and transport mechanisms. In the ABC transporter pathway, rather than adding O-PS units to a growing O-PS chain in the periplasm, the O-PS is completely formed at the inner leaflet of the inner membrane by a series of glycosyltransferase reactions. This pathway is found in those bacteria that generate the O-PS homopolymers while heteropolymers are usually synthesized *via* the Wzy pathway (Raetz and Whitfield 2002). The completed O-PS subsequently transits to periplasm *via* the ABC transporter system (encoded by *wzm* and *wzt*) where it is ligated to the lipid A – core by WaaL (Raetz and Whitfield 2002). How modal distribution is achieved in the ABC transporter dependent pathway is not fully understood. However, non-reducing terminal modifications have been shown to regulate the O-PS length in *E. coli* and *M. smegmatis* and may signal for its transport through its ABC transporter (Weisman and Ballou 1984; Clarke et al. 2004).

It should be mentioned here that the ABC transporters are a large class of importers and exporters observed in both eukaryotic and prokaryotic cells. Because these transporters are important in the nutrient uptake and drug resistance, they are the subject of ongoing research (Hollenstein et al. 2007). The ABC transporters that move O-PS from cytosol to periplasm are not as well characterized. However, the fundamental transport mechanism appears to be conserved. Wzm is a membrane bound protein that facilitates the O-PS transport through the inner membrane (Raetz and Whitfield 2002). The *C. burnetii* O-antigen export system permease protein (CBU0703) is adjacent to the putative O-PS I biosynthesis genes and is annotated as *wzm*. The hydropathy plots using the Kyte-Doolittle scale (Bowen 2008) of CBU0703 detected hydrophobic regions typical of the characterized Wzm proteins. The *C. burnetii* LPS/O-antigen export permease (CBU0704) is annotated as *wzt*. Bioinformatic analysis has suggested that the *C. burnetii* Wzt has a classical ATP-binding cassette motif for which the transport system is named. This enzyme is responsible for binding the und-P linked O-PS I, and providing the machinery required for transport at the expense of ATP hydrolysis.

4.4.5 *Tentative Biosynthesis Pathway of D-Virenose in C. burnetii*

A fundamental understanding of O-PS I biosynthesis in LPS I is still lacking. The intracellular nature of *C. burnetii*, lack of genetic tools and its status as a potential biowarfare agent have made elucidating of its basic physiological mechanisms challenging. Most recently, an attempt has been made to elucidate the enzymatic steps responsible for the formation of guanosine-5′-diphosphate (GDP)-β-D-Man in the bacterium, which were bioinformatically predicted as the initial steps of GDP-β-D-Vir biosynthesis (Narasaki et al. 2011). Based on both structural determination of Vir (Toman et al. 1998) and analysis of the genes deleted in the RSA 439 variant (Hoover et al. 2002), it has been hypothesized that GDP-β-D-mannose is synthesized from fructose-6-phosphate (F6P) in three successive reactions that include: (i) isomerization to D-mannose-6-phosphate (M6P) catalyzed by a mannose-6-phosphate isomerase (PMI), (ii) conversion to mannose-1-phosphate (M1P) mediated by a phosphomannomutase (PMM) and (iii) addition of GDP by a GDP-mannose pyrophosphorylase (GMP) (Fig. 4.4). GDP-β-D-mannose is then likely converted to GDP-6-deoxy-β-D-*lyxo*-hex-4-ulopyranose, a Vir intermediate, by a GDP-D-mannose dehydratase (GMD). To prove the suggested pathway in *C. burnetii*, three ORF (CBU0671, CBU0294 and CBU0689) annotated as bifunctional type II PMI, as PMM or GMD were functionally characterized by complementation of corresponding *E. coli* mutant strains and in enzymatic assays. However, CBU0671 failed to complement an *E. coli manA* (PMM) mutant strain but complementation of *manC* (GMP) mutant strain restored the capsular PS biosynthesis. CBU0294 complemented a *P. aeruginosa algC* (GMP) mutant strain and showed phosphoglucomutase activity in a *pgm E. coli* mutant strain. Despite the inability to complement a *manA* mutant, recombinant *C. burnetii* PMI protein showed PMM enzymatic activity in biochemical assays. CBU0689 showed dehydratase activity and determined kinetic parameters were in agreement with the data reported from other organisms.

GDP-6-deoxy-β-D-*lyxo*-hex-4-ulopyranose formed by GMD is the metabolic intermediate of GDP-L-fucose, GDP-colitose, GDP-perosamine, GDP-D-rhamnose and GDP-6-deoxy-D-talose (Samuel and Reeves 2003). The enzymes required to generate the final steps for GDP-perosamine (perosamine synthetase, CBU0830) and GDP-L-fucose (fucose synthetase, GDP-4-keto-6-deoxymannose epimerase/reductase, CBU0688) have been identified in the *C. burnetii* genome (Seshadri et al. 2003). However, we have not detected any perosamine or L-fucose in the compositional analyses of LPSs I obtained from various strains/isolates of *C. burnetii*. Nevertheless, the suggested biosynthesis pathway for Vir has established an initial step towards a better understanding of the O-PS I formation, which is the only known virulence factor of *C. burnetii*.

4.5 Functional Characteristics of the *C. burnetii* Lipopolysaccharide

TNF is a cytokine involved in systemic inflammation and is a member of a group of cytokines that stimulate the acute phase reaction. It is a protein with a molecular mass of 17 kDa and is produced mainly by mononuclear phagocytes following induction with an LPS (Flebbe et al. 1990). The primary role of TNF is in the regulation of immune cells. It is able to induce apoptotic cell death, to induce inflammation, and to inhibit tumorigenesis and viral replication. Dysregulation of TNF production

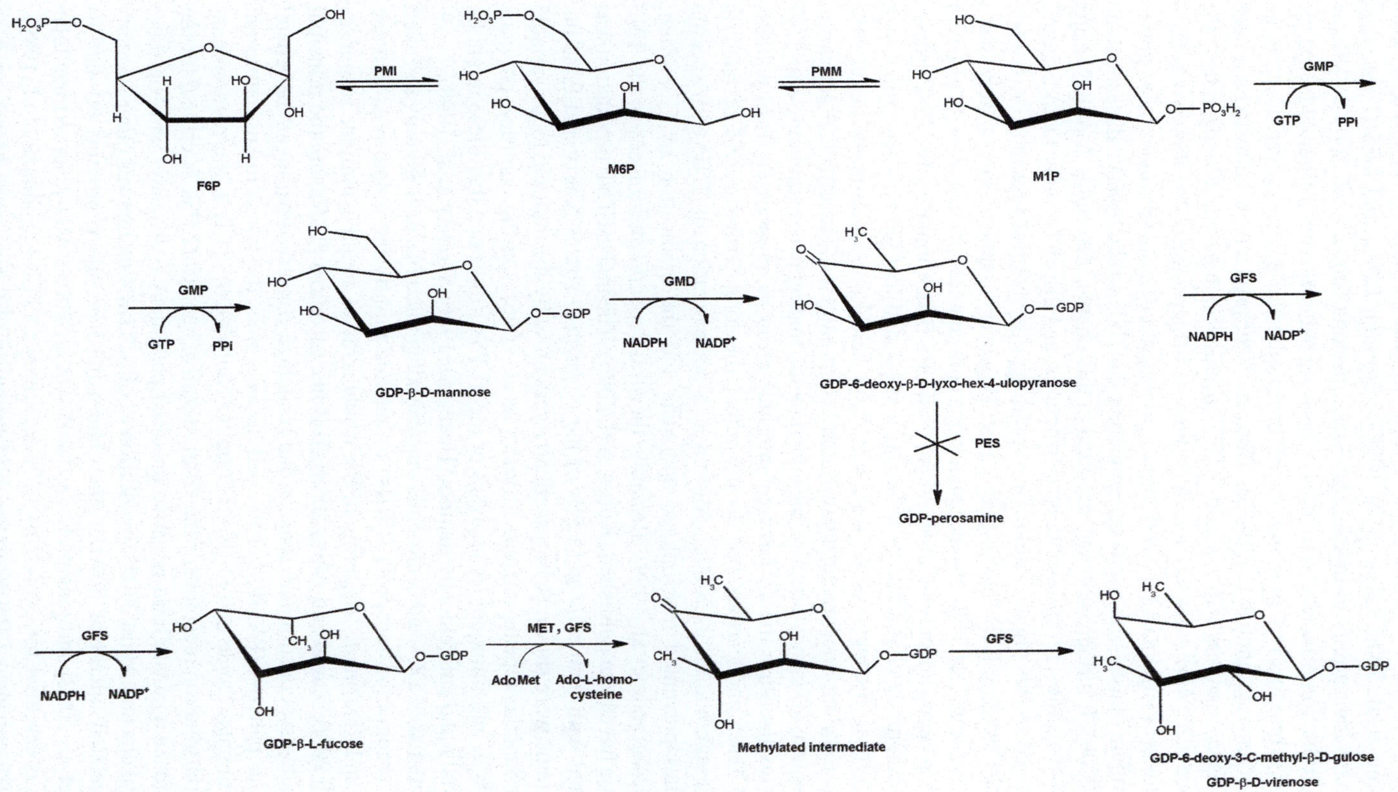

Fig. 4.4 Schematic presentation of putative GDP-β-D-virenose biosynthesis pathway. *Ado-Met* S-adenosyl methionine, *F6P* fructose-6-phosphate, *GDP* guanosine-5′-diphosphate, *GFS* fucose synthetase (GDP-4-keto-6-deoxymannose epimerase/reductase), *GMD* GDP-D-mannose dehydratase, *GMP* GDP-D-mannose pyrophosphorylase, *GTP* guanosine-5′-triphosphate, *M1P* D-mannose-1-phosphate, *M6P* D-mannose-6-phosphate, *MET* methyltransferase, *NADP* nicotinamide adenine dinucleotide phosphate, *PES* perosamine synthetase, *PMI* mannose-6-phosphate isomerase, *PMM* phosphomannomutase, PP_i inorganic pyrophosphate

has been implicated in a variety of human infectious diseases including Q fever (Locksley et al. 2001; Capo et al. 1999). Although LPS I and LPS II are weak endotoxins their ability to induce TNF was reported (Toman et al. 2004). TNF and IL-1β production is increased in monocytes from patients with Q fever endocarditis, whereas it remains low in patients with uncomplicated acute Q fever (Capo et al. 1996; Dellacasagrande et al. 2000a). Thus, the TNF production seems to be related to the disease activity. However, the TNF production does not directly reflect the virulence of *C. burnetii* since avirulent phase II bacteria were also shown to induce this cytokine (Tujulin et al. 1999). In fact, avirulent bacteria were even more potent than virulent phase I organisms at stimulating the TNF production (Dellacasagrande et al. 2000b). It has already been reported in this chapter that lipids A of LPS I and LPS II have almost identical chemical structures, and therefore, other factors should be involved in this phenomenon. One could hypothesize that presence or absence of O-PS I in these LPSs might play a role. However, it appears more likely that a limited efficiency of TNF production is not a consequence of poor activity of LPS I as this has been shown to be more potent than LPS II at stimulating the TNF release (Dellacasagrande et al. 2000b). It has been suggested that virulent bacteria, which bind poorly to monocytes, likely present fewer LPS I molecules to the target cells than do avirulent ones that efficiently bind to monocytes. The effect of activation on the uptake efficiency was also supported by the finding that neutralizing anti-TNF antibodies decrease the *C. burnetii* internalization by monocytes from patients with Q fever endocarditis but do not affect the long term survival of bacteria (Dellacasagrande et al. 2000a). Moreover, a neutralization of TNF with specific antibodies prevented macrophage apoptosis and the eradication of *C. burnetii* (Dellacasagrande et al. 1999). Recently, the strain specific differences in the TNF induction have been published (Kubes et al. 2006). The highest TNF production was monitored with both cells and LPSs of the *C. burnetii* strains Scurry and Priscilla. Both strains were shown to be genetically very close by the multiple locus variable number tandem repeats analysis (Svraka et al. 2006). On the contrary, previous SDS-PAGE and immunoblot analyses of their LPSs I indicated that these might be structurally and antigenically distinct (Hackstadt 1986; Skultety et al. 1998).

Macrophages, DCs and neutrophils sense invading microbial pathogens by so-called pattern recognition receptors (PRRs) to activate rapidly the innate host defence system and to promote triggering of the whole array of adaptive immune responses (O'Neill 2006; Kawai and Akira 2008). Toll-like receptors (TLRs) are the principal membrane-associated innate sensors that recognize conserved pathogen associated molecular patterns (PAMPs) at the cell surface or in the intracellular endosomal compartments (Doyle and O' Neill 2006; O'Neill 2006). TLR 4, the predominant signal transduction receptor for the Gram-negative bacterial LPSs, is critical for the host defence against Gram-negative bacterial pathogens (Alexander and Rietschel 2001; Takeda and Akira 2005). In addition to LPS, TLR 4 recognizes other structurally unrelated microbial PAMPs like chlamydial heat shock proteins 60 and 70 (Bulut et al. 2002) and pneumolysin (Malley et al. 2003). MD-2, an extracellular protein, is essential for conferring LPS sensitivity to TLR4 (Shimazu et al. 1999), and CD14 enhances LPS responses by facilitating LPS binding to MD-2, enables

MyD88-independent signaling pathways by LPS. TLR 2 has been shown to recognize lipoproteins and lipopeptides from Gram-positive bacteria, mycoplasma, and mycobacteria (Lien et al. 1999) in cooperation with TLR 1 or TLR 6 (Hajjar et al. 2001; O'Neill 2006). Recent results suggest, however, that TLR 4 and TLR 2 polymorphisms are associated with susceptibility to infectious diseases (Lorenz et al. 2000; Awomoyi et al. 2007).

TLR 4 was reported to be involved in the recognition of LPS I, initial activation of macrophages, and inflammatory response associated with the *C. burnetii* infection (Honstettre et al. 2004). It was shown later, however, that lipid A from *C. burnetii* antagonized activation of TLR 4 by the highly endotoxic lipids A from *E. coli* and *B. pertussis*, and by the *E. coli* LPS (Zamboni et al. 2004). A similar competitive inhibition of LPS activation of the TLR 4-dependent responses was reported for other lipids A, such as those isolated from *Rhodobacter sphaeroides* and lipid IV_A (Golenbock et al. 1991). This fact appeared to be in agreement with observations that LPSs of some bacterial species failed to activate TLR 4 and that these bacteria activated host cells through a TLR 2-dependent mechanism (Hirschfeld et al. 2001; Girard et al. 2003; Darveau et al. 2004). It has already been mentioned in this chapter that the chemical structure of *C. burnetii* lipid A differs considerably from those reported for the classical enterobacterial lipids A that activate predominantly TLR 4. The recent findings in the field have indicated that TLR 4 is involved in the cytoskeleton remodeling induced by *C. burnetii,* controls the immune response against it through granuloma formation and cytokine production, and is dispensable for the bacterial clearance *in vivo* (Honstettre et al. 2004). TLR 2 seems to be also involved in the TNF and IFN-γ production induced by avirulent variants of *C. burnetii* in macrophages from TLR 2-deficient mice (Zamboni et al. 2004). Moreover, these macrophages have been shown to be highly permissive for the intracellular growth of bacteria. Experiments *in vivo* have also indicated that TLR 2 is involved in the granuloma formation similar to that of TLR 4 (Meghari et al. 2005). Thus, a synergistic action of both TLRs has been proposed in some signal transduction pathways. However, a more detailed role of LPS I and LPS II in these events remains to be elucidated.

It has already been reported that phase I *C. burnetii* replicated within the human DCs without inducing maturation or inflammatory cytokine production (Shannon et al. 2005). In contrast, the phase II bacteria induced a dramatic maturation of the cells. A model has been proposed whereby LPS I masks TLR ligands from innate immune recognition by DCs, thereby allowing replication without significant maturation or inflammatory cytokine production. This immune evasion strategy was suggested to allow *C. burnetii* to persist in an immunocompetent host. However, these findings raised several points that are discussed in a more detail in the dedicated chapters of this book.

It has been well known for many years that during acute Q fever *C. burnetii* induces antibodies in host against its phase II (protein) antigens, while in the later stages of the disease, and especially in its persistent/chronic form, high titers of antibodies are directed against phase I (LPS I) antigen (Marrie and Raoult 1997). Initial work has already been performed towards elucidation of the interaction of phase

I antibodies with the LPS I antigen (Vadovic et al. 2005). A noticeable decrease in the serological activity of the O-antigen was monitored when Vir and Strep were selectively removed from its O-PS I chain. However, it is still not known with certainty whether the immunoreactive epitopes are located only at both sugars in terminal positions or also at those Vir residues located in the O-PS I backbone.

The presence of unique biomarkers Vir and Strep in LPS I and their favourable location in its O-PS I provides an excellent opportunity to generate mAbs against them that would enable rapid, sensitive and specific detection of the virulent form of the bacterium. So far, a mAb (IgG2b subclass) has been generated that was proved to be highly specific for the presence of Vir in the bacterium and LPS I as only Vir-containing *C. burnetii* strains/variants and their LPSs reacted with the mAb (Palkovicova et al. 2009). In addition, no cross-reaction was observed with the cells and LPSs from the selected species of the families *Rickettsiaceae, Piscirickettsiaceae, Chlamydiaceae* and *Enterobacteriaceae.*

Most recently, the intracellular trafficking of LPS I and LPS II in murine bone marrow-derived macrophages has been investigated (Pretat et al. 2009). It has been found that the intracellular trafficking of LPS I is similar to that of LPS II and that the high portions of both LPSs have not used the endosomal route. It was proposed that the LPSs transit through non-classical endocytic or uncharacterized pathways. Moreover, the results suggest that LPS I may be involved in a subverted conversion of the *C. burnetii* phagosome by interfering with the endosomal pathway.

More information on the biological and functional characteristics/properties of the *C. burnetii* LPSs can be found in the dedicated chapters of this book.

4.6 Summary

Although the *C. burnetii* lipid A and inner core structures of LPS I are well characterized, the composition and structure of its outer core and O-PS I chain are not entirely resolved. More detailed investigations are necessary in this direction, and in addition, there is a need to better characterize the role of individual LPS I components in pathogenesis and immunity of Q fever. Thus, it is important to investigate effects of these biological response modifiers on both humoral and cellular immune systems. For example, knowledge on the structure/function relationship of the LPS I antigens involved in the induction of immunity and pathological reactions during the infection and administration of the experimental vaccines is not sufficient. A better understanding of both composition and structure of the biologically active outer membrane components (LPS I and proteins) of *C. burnetii* will be of crucial importance not only in a more detailed elucidation of host-pathogen interactions but also in modulation of pathological reactions, which in turn is a prerequisite for development of vaccines of new type.

The intracellular nature of *C. burnetii*, the lack of genetic tools and its status as a selected agent have made elucidating its basic O-polysaccharide biosynthesis and export mechanisms together with other principal physiological mechanisms challenging. Future technologies that provide the genetic tools necessary to screen *C. burnetii* mutants and propagate isogenic mutants might speed the discovery process in this area.

Acknowledgement This research was supported in part by the grant 2/0026/12 from the Scientific Grant Agency of Ministry of Education of Slovak Republic and the Slovak Academy of Sciences.

References

Alexander C, Rietschel ET (2001) Bacterial lipopolysaccharides and innate immunity. J Endotoxin Res 7:167–202

Al-Hendy A, Toivanen P, Skurnik M (1991) Expression cloning of *Yersinia enterocolitica* O:3 rfb gene cluster in *Escherichia coli* K12. Microb Pathog 10:47–59

Amano K, Williams JC (1984) Chemical and immunological characterization of lipopolysaccharides from phase I and phase II *Coxiella burnetii*. J Bacteriol 160:994–1002

Amano K, Williams JC, Missler SR et al (1987) Structure and biological relationships of *Coxiella burnetii* lipopolysaccharides. J Biol Chem 262:4740–4747

Aschauer H, Grob A, Hildebrandt J et al (1990) Highly purified lipid X is devoid of immunostimulatory activity. Isolation and characterization of immunostimulating contaminants in a batch of synthetic lipid X. J Biol Chem 265:9159–9164

Awomoyi AA, Rallabhandi P, Pollin TI et al (2007) Association of TLR4 polymorphisms with asymptomatic respiratory syncytial virus infection in high-risk infants and young children. J Immunol 179:3171–3177

Baca OG, Martinez IL, Aragon AS et al (1980) Isolation and partial characterization of a lipopolysaccharide from phase II *Coxiella burnetii*. Can J Microbiol 26:819–826

Bisercic M, Feutrier JY, Reeves PR (1991) Nucleotide sequences of the gnd genes from nine natural isolates of *Escherichia coli*: evidence of intragenic recombination as a contributing factor in the evolution of the polymorphic gnd locus. J Bacteriol 173:3894–3900

Bowen RA (2008) Protein hydrophobicity plots. http://www.vivo.colostate.edu/molkit/hydropathy/index.html

Bulut Y, Faure E, Thomas L et al (2002) Chlamydial heat shock protein 60 activates macrophages and endothelial cells through Toll-like receptor 4 and MD2 in a MyD88-dependent pathway. J Immunol 168:1435–1440

Burns SM, Hull SI (1998) Comparison of loss of serum resistance by defined lipopolysaccharide mutants and an acapsular mutant of uropathogenic *Escherichia coli* O75:K5. Infect Immun 66:4244–4253

Capo C, Zugun F, Stein A et al (1996) Upregulation of tumor necrosis factor-α and interleukin-1β in Q fever endocarditis. Infect Immun 64:1638–1642

Capo C, Amirayan N, Ghigo E et al (1999) Circulating cytokine balance and activation markers of leucocytes in Q fever. Clin Exp Immunol 115:120–123

Caroff M, Brisson JR, Martin A et al (2000) Structure of the *Bordetella pertussis* 1414 endotoxin. FEBS Lett 477:8–14

Chen MM, Glover KJ, Imperiali B (2007) From peptide to protein: comparative analysis of the substrate specificity of N-linked glycosylation in *Campylobacter jejuni*. Biochemistry 46:5579–5585

Clarke BR, Cuthbertson L, Whitfield C (2004) Nonreducing terminal modifications determine the chain length of polymannose O-antigens of *Escherichia coli* and couple chain termination to polymer export via an ATP-binding cassette transporter. J Biol Chem 279:35709–35718

Coleman SA, Fischer ER, Howe D et al (2004) Temporal analysis of *Coxiella burnetii* morphological differentiation. J Bacteriol 186:7344–7352

Coleman SA, Fischer ER, Cockrell DC et al (2007) Proteome and antigen profiling of *Coxiella burnetii* developmental forms. Infect Immun 75:290–298

Darveau RP, Pham TT, Lemley K et al (2004) *Porphyromonas gingivalis* lipopolysaccharide contains multiple lipid A species that functionally interact with both toll-like receptors 2 and 4. Infect Immun 72:5041–5051

Dellacasagrande J, Capo C, Raoult D et al (1999) IFN-γ-mediated control of *Coxiella burnetii* survival in monocytes: the role of cell apoptosis and TNF. J Immunol 162:2259–2265

Dellacasagrande J, Ghigo E, Capo C et al (2000a) *Coxiella burnetii* survives in monocytes from patients with Q fever endocarditis: involvement of tumor necrosis factor. Infect Immun 68:160–164

Dellacasagrande J, Ghigo E, Machergui-El Hammami S et al (2000b) Alpha(v)beta(3) integrin and bacterial lipopolysaccharide are involved in *Coxiella burnetii*-stimulated production of tumor necrosis factor by human monocytes. Infect Immun 68:5673–5678

Demerec M, Adelberg EA, Clark AJ et al (1966) A proposal for a uniform nomenclature in bacterial genetics. Genetics 54:61–76

Denison AM, Massung RF, Thompson HA (2007) Analysis of the O-antigen biosynthesis regions of phase II isolates of *Coxiella burnetii*. FEMS Microbiol Lett 267:102–107

Diaz QM, Lukacova M (1998) Immunological consequences of *Coxiella burnetii* phase variation. Acta Virol 42:181–185

Dotson SB, Rush JS, Ricketts AD et al (1995) Mannosylphosphoryldolichol-mediated O-mannosylation of yeast glycoproteins: stereospecificity and recognition of the alpha-isoprene unit by a purified mannosyltransferase. Arch Biochem Biophys 316:773–779

Doyle SL, O' Neill LA (2006) Toll-like receptors from the discovery of NF-χB to new insights into transcriptional regulations in innate imunity. Biochem Pharmacol 72:1102–1113

D'Souza-Schorey C, McLachlan KR, Krag SS et al (1994) Mammalian glycosyltransferases prefer glycosyl phosphoryl dolichols rather than glycosyl phosphoryl polyprenols as substrates for oligosaccharyl synthesis. Arch Biochem Biophys 308:497–503

Emmerson JR, Gally DL, Roe AJ (2006) Generation of gene deletions and gene replacements in *Escherichia coli* O157:H7 using a temperature sensitive allelic exchange system. Biol Proced Online 8:153–162

Erridge C, Bennett-Guerrero E, Poxton IR (2002) Structure and function of lipopolysaccharides. Microbes Infect 4:837–851

Flebbe LM, Chapes SK, Morrison DC (1990) Activation of C3H/HeJ macrophage tumoricidal activity and cytokine release by R-chemotype lipopolysaccharide preparations. J Immunol 145:505–514

Ftacek P, Skultety L, Toman R (2000) Phase variation of *Coxiella burnetii* strain Priscilla: influence of this phenomenon on biochemical features of its lipopolysaccharide. J Endotoxin Res 6:369–376

Gajdosova E, Kovacova E, Toman R et al (1994) Immunogenicity of *Coxiella burnetii* whole cells and their outer membrane components. Acta Virol 38:339–344

Girard R, Pedron T, Uematsu S et al (2003) Lipopolysaccharides from *Legionella* and *Rhizobium* stimulate mouse bone marrow granulocytes via toll-like receptor 2. J Cell Sci 116:293–302

Golenbock DT, Hampton RY, Qureshi N et al (1991) Lipid A-like molecules that antagonize the effects of endotoxins on human monocytes. J Biol Chem 266:19490–19498

Gunn JS, Ernst RK (2007) The structure and function of *Francisella* lipopolysaccharide. Ann NY Acad Sci 1105:202–218

Hackstadt T (1986) Antigenic variation in the phase I lipopolysaccharide of *Coxiella burnetii* isolates. Infect Immun 52:337–340

Hackstadt T, Peacock MG, Hitchcock PJ et al (1985) Lipopolysaccharide variation in *Coxiella burnetii*: intrastrain heterogeneity in structure and antigenicity. Infect Immun 48:359–365

Hajjar AM, O'Mahony DS, Ozinsky A et al (2001) Cutting edge: functional interactions between toll-like receptor (TLR) 2 and TLR1 or TLR6 in response to phenol-soluble modulin. J Immunol 166:15–19

Heinrichs DE, Whitfield C, Valvano MA (1999) Biosynthesis and genetics of lipopolysaccharide core. In: Brade H, Opal SM, Vogel SN, Morrison DC (eds) Endotoxin in health and disease. Marcel Dekker, New York/Basel, pp 305–330

Heinzen RA, Hackstadt T (1996) A developmental stage-specific histone H1 homolog of *Coxiella burnetii*. J Bacteriol 178:5049–5052

Helbig JH, Luck PC, Knirel YA et al (1995) Molecular characterization of a virulence-associated epitope on the lipopolysaccharide of *Legionella pneumophila* serogroup 1. Epidemiol Infect 115:71–78

Hirschfeld M, Weiss JJ, Toshchakov V et al (2001) Signaling by toll-like receptor 2 and 4 agonists results in differential gene expression in murine macrophages. Infect Immun 69:1477–1482

Hobbs M, Reeves PR (1994) The JUMPstart sequence: a 39 bp element common to several polysaccharide gene clusters. Mol Microbiol 12:855–856

Hollenstein K, Dawson RJ, Locher KP (2007) Structure and mechanism of ABC transporter proteins. Curr Opin Struct Biol 17:412–418

Holst O (1999) Chemical structure of the core region of lipopolysaccharides. In: Brade H, Opal SM, Vogel SN, Morrison DC (eds) Endotoxin in health and disease. Marcel Dekker, New York/Basel, pp 115–154

Honstettre A, Ghigo E, Moynnault A et al (2004) Lipopolysaccharide from *Coxiella burnetii* is involved in bacterial phagocytosis, filamentous actin reorganization, and inflammatory responses through toll-like receptor 4. J Immunol 172:3695–3703

Hoover TA, Culp DW, Vodkin MH et al (2002) Chromosomal DNA deletions explain phenotypic characteristics of two antigenic variants, phase II and RSA 514 (Crazy), of the *Coxiella burnetii* nine mile strain. Infect Immun 70:6726–6733

Hussein A, Kovacova E, Toman R (2001) Isolation and evaluation of *Coxiella burnetii* O-polysaccharide antigen as immunodiagnostic reagent. Acta Virol 45:173–180

Jimenez de Bagues MP, Gross A, Terraza A et al (2005) Cellular bioterrorism: how *Brucella* corrupts macrophage physiology to promote invasion and proliferation. Clin Immunol 114:227–238

Joiner KA (1988) Complement evasion by bacteria and parasites. Ann Rev Microbiol 42:201–230

Kawai T, Akira S (2008) Toll-like receptor and RIG-I-like receptor signaling. Ann NY Acad Sci 1143:1–20

Kean EL, Rush JS, Waechter CJ (1994) Activation of GlcNAc-P-P-dolichol synthesis by mannosylphosphoryldolichol is stereospecific and requires a saturated alpha-isoprene unit. Biochemistry 33:10508–10512

Keenleyside WJ, Whitfield C (1999) Genetics and biosynthesis of lipopolysaccharide O-antigens. In: Brade H, Opal SM, Vogel SN, Morrison DC (eds) Endotoxin in health and disease. Marcel Dekker, New York/Basel, pp 331–358

King JD, Kocincova D, Westman EL et al (2009) Review: lipopolysaccharide biosynthesis in *Pseudomonas aeruginosa*. Innate Immun 15:261–312

Knirel YA, Rietschel ET, Marre R et al (1994) The structure of the O-specific chain of *Legionella pneumophila* serogroup 1 lipopolysaccharide. Eur J Biochem 221:239–245

Kubes M, Kuzmova Z, Gajdosova E et al (2006) Induction of tumor necrosis factor alpha in murine macrophages with various strains of *Coxiella burnetii* and their lipopolysaccharides. Acta Virol 50:93–99

Lien E, Sellati TJ, Yoshimura A et al (1999) Toll-like receptor 2 functions as a pattern recognition receptor for diverse bacterial products. J Biol Chem 274:33419–33425

Locksley RM, Killeen N, Lenardo MJ (2001) The TNF and TNF receptor superfamilies: integrating mammalian biology. Cell 104:487–501

Lorenz E, Mira JP, Cornish KL et al (2000) A novel polymorphism in the toll-like receptor 2 gene and its potential association with staphylococcal infection. Infect Immun 68:6398–6401

Lu YC, Yeh WC, Ohashi PS (2008) LPS/TLR4 signal transduction pathway. Cytokine 42:145–151

Luck PC, Freier T, Steudel C et al (2001) Point mutation in the active site of *Legionella pneumophila* O-acetyltransferase results in modified lipopolysaccharide but does not influence virulence. Int J Med Microbiol 291:345–352

Luneberg E, Zahringer U, Knirel YA et al (1998) Phase-variable expression of lipopolysaccharide contributes to the virulence of *Legionella pneumophila*. J Exp Med 188:49–60

Luneberg E, Zetzmann N, Alber D et al (2000) Cloning and functional characterization of a 30 kb gene locus required for lipopolysaccharide biosynthesis in *Legionella pneumophila*. Int J Med Microbiol 290:37–49

Luneberg E, Mayer B, Daryab N et al (2001) Chromosomal insertion and excision of a 30 kb unstable genetic element is responsible for phase variation of lipopolysaccharide and other virulence determinants in *Legionella pneumophila*. Mol Microbiol 39:1259–1271

Malley R, Hennecke P, Morse SC et al (2003) Recognition of pneumolysin by toll-like receptor 4 confers resistance to pneumococcal infection. Proc Natl Acad Sci USA 100:1966–1971

Marrie TJ, Raoult D (1997) Q fever – a review and issues for the next century. J Antimicrob Agents 8:145–161

Mayer H, Radziejewska-Lebrecht J, Schramek S (1988) Chemical and immunochemical studies on lipopolysaccharides of *Coxiella burnetii* phase I and phase II. Adv Exp Med Biol 228:577–591

McCaul TF, Banerjee-Bhatnagar N, Williams JC (1991) Antigenic differences between *Coxiella burnetii* cells revealed by postembedding immunoelectron microscopy and immunoblotting. Infect Immun 59:3243–3253

McLachlan KR, Krag SS (1994) Three enzymes involved in oligosaccharide-lipid assembly in Chinese hamster ovary cells differ in lipid substrate preference. J Lipid Res 35:1861–1868

Meghari S, Honstettre A, Lepidi H et al (2005) TLR2 is necessary to inflammatory response in *Coxiella burnetii* infection. Ann NY Acad Sci 1063:161–166

Moran AP, Zähringer U, Seydel U et al (1991) Structural analysis of the lipid A component of *Campylobacter jejuni* CCUG 10936 (serotype O:2) lipopolysaccharide. Description of a lipid A containing a hybrid backbone of 2-amino-2-deoxy-D-glucose and 2,3-diamino-2,3-dideoxy-D-glucose. Eur J Biochem 198:459–469

Nagy G, Palkovics T, Otto A et al (2008) "Gently rough": the vaccine potential of a *Salmonella enterica* regulatory lipopolysaccharide mutant. J Infect Dis 198:1699–1706

Narasaki CT, Mertens K, Samuel JE (2011) Characterization of the GDP-D-mannose biosynthesis pathway in *Coxiella burnetii*: the initial steps for GDP-β-D-virenose biosynthesis. PLoS One 6(10):e25514

Nelson K, Selander RK (1994) Intergeneric transfer and recombination of the 6-phosphogluconate dehydrogenase gene (gnd) in enteric bacteria. Proc Natl Acad Sci USA 91:10227–10231

Nesper J, Lauriano CM, Klose KE et al (2001) Characterization of *Vibrio cholerae* O1 El tor galU and galE mutants: influence on lipopolysaccharide structure, colonization, and biofilm formation. Infect Immun 69:435–445

O'Neill LA (2006) How toll-like receptors signal: what we know and what we don't know. Curr Opin Immunol 18:3–9

Palkovicova K, Ihnatko R, Vadovic P et al (2009) A monoclonal antibody specific for a unique biomarker virenose in a lipopolysaccharide of *Coxiella burnetii*. Clin Microbiol Infect 15:183–184

Pretat L, Toman R, Vadovic P et al (2009) Intracellular trafficking of the *Coxiella burnetii* lipopolysaccharide. Clin Microbiol Infect 15:185–187

Raetz CR, Whitfield C (2002) Lipopolysaccharide endotoxins. Ann Rev Biochem 71:635–700

Reeves PR, Hobbs M, Valvano MA et al (1996) Bacterial polysaccharide synthesis and gene nomenclature. Trends Microbiol 4:495–503

Rietschel ET, Brade H, Brade L et al (1987) Lipid A, the endotoxic centre of bacterial lipopolysaccharides: relation of chemical structure to biological activity. Prog Clin Biol Res 231:25–53

Rietschel ET, Kirikae T, Schade U et al (1993) The chemical structure of bacterial endotoxin in relation to bioactivity. Immunobiology 187:169–190

Rocchetta HL, Burrows LL, Lam JS (1999) Genetics of O-antigen biosynthesis in *Pseudomonas aeruginosa*. Microbiol Mol Biol Rev 63:523–553

Rojas G, Saldias S, Bittner M et al (2001) The rfaH gene, which affects lipopolysaccharide synthesis in *Salmonella enterica* serovar Typhi, is differentially expressed during the bacterial growth phase. FEMS Microbiol Lett 204:123–128

Rush JS, Shelling JG, Zingg NS et al (1993) Mannosylphosphoryldolichol-mediated reactions in oligosaccharide-P-P-dolichol biosynthesis. Recognition of the saturated alpha-isoprene unit of the mannosyl donor by pig brain mannosyltransferases. J Biol Chem 268:13110–13117

Rush JS, Rick PD, Waechter CJ (1997) Polyisoprenyl phosphate specificity of UDP-GlcNAc: undecaprenyl phosphate N-acetylglucosaminyl 1-P transferase from *E.coli*. Glycobiology 7:315–322

Rush JS, Alaimo C, Robbiani R et al (2010) A novel epimerase that converts GlcNAc-P-P-undecaprenol to GalNAc-P-P-undecaprenol in *Escherichia coli* O157. J Biol Chem 285:1671–1680

Samuel G, Reeves P (2003) Biosynthesis of O-antigens: genes and pathways involved in nucleotide sugar precursor synthesis and O-antigen assembly. Carbohydr Res 338:2503–2519

Sandlin RC, Lampel KA, Keasler SP et al (1995) Avirulence of rough mutants of *Shigella flexneri*: requirement of O antigen for correct unipolar localization of *IcsA* in the bacterial outer membrane. Infect Immun 63:229–237

Sandlin RC, Goldberg MB, Maurelli AT (1996) Effect of O side-chain length and composition on the virulence of *Shigella flexneri* 2a. Mol Microbiol 22:63–73

Schramek S, Mayer H (1982) Different sugar composition of lipopolysaccharides isolated from phase I and pure phase II cells of *Coxiella burnetii*. Infect Immun 38:53–57

Schramek S, Radziejewska-Lebrecht J, Mayer H (1985) 3-C-branched aldoses in lipopolysaccharide of phase I *Coxiella burnetii* and their role as immunodominant factors. Eur J Biochem 148:445–461

Seshadri R, Hendrix LR, Samuel JE (1999) Differential expression of translational elements by life cycle variants of *Coxiella burnetii*. Infect Immun 67:6026–6033

Seshadri R, Paulsen IT, Eisen JA et al (2003) Complete genome sequence of the Q-fever pathogen *Coxiella burnetii*. Proc Natl Acad Sci USA 100:5455–5460

Seydel U, Wiese A, Schromm AB et al (1999) A biophysical view on the function and activity of endotoxins. In: Brade H, Opal SM, Vogel SN, Morrison DC (eds) Endotoxin in health and disease. Marcel Dekker, New York/Basel, pp 195–219

Shannon JG, Howe D, Heinzen RA (2005) Virulent *Coxiella burnetii* does not activate human dendritic cells: role of lipopolysaccharide as a shielding molecule. Proc Natl Acad Sci USA 102:8722–8727

Shimazu R, Akashi S, Ogata H et al (1999) MD-2, a molecule that confers lipopolysaccharide responsiveness on toll-like receptor 4. J Exp Med 189:1777–1782

Skultety L, Toman R, Patoprsty V (1998) A comparative study of lipopolysaccharides from two *Coxiella burnetii* strains considered to be associated with acute and chronic Q fever. Carbohydr Polymers 35:189–194

Skultety L, Hernychova L, Toman R et al (2005) *Coxiella burnetii* whole cell lysate protein identification by mass spectrometry and tandem mass spectrometry. Ann NY Acad Sci 1063:115–122

Skurnik M, Toivanen P (1993) *Yersinia enterocolitica* lipopolysaccharide: genetics and virulence. Trends Microbiol 1:148–152

Slaba K, Hussein A, Palkovic P et al (2003) Studies on immunological role of virenose and dihydrohydroxystreptose present in the *Coxiella burnetii* phase I lipopolysaccharide. Ann NY Acad Sci 990:505–509

Stoker MG, Fiset P (1956) Phase variation of the Nine Mile and other strains of *Rickettsia burneti*. Can J Microbiol 2:310–321

Stroeher UH, Jedani KE, Manning PA (1998) Genetic organization of the regions associated with surface polysaccharide synthesis in *Vibrio cholerae* O1, O139 and *Vibrio anguillarum* O1 and O2*:* a review. Gene 223:269–282

Svraka S, Toman R, Skultety L et al (2006) Establishment of a genotyping scheme for *Coxiella burnetii*. FEMS Microbiol Lett 254:268–274

Szkopinska A, Swiezewska E, Chojnacki T (1992) On the specificity of dolichol kinase and DolPMan synthase towards isoprenoid alcohols of different chain length in rat liver microsomal membrane. Int J Biochem 24:1151–1157

Takeda K, Akira S (2005) Toll-like receptors in innate immunity. Int Immunol 17:1–14

Thompson HA, Hoover TA, Vodkin MH et al (2003) Do chromosomal deletions in the lipopolysaccharide biosynthetic regions explain all cases of phase variation in *Coxiella burnetii* strains? An update. Ann NY Acad Sci 990:664–670

Toman R (1991) Basic structural features of a lipopolysaccharide from the *Coxiella burnetii* strain Nine Mile in the virulent phase I. Acta Virol 35:224

Toman R, Kazar J (1991) Evidence for the structural heterogeneity of the polysaccharide component of *Coxiella burnetii* strain Nine Mile lipopolysaccharide. Acta Virol 35:531–537

Toman R, Skultety L (1994) Analysis of the 3-deoxy-D-manno-2-octulosonic acid region in a lipopolysaccharide isolated from *Coxiella burnetii* strain Nine Mile in phase II. Acta Virol 38:241–243

Toman R, Skultety L (1996) Structural study on a lipopolysaccharide from *Coxiella burnetii* strain Nine Mile in avirulent phase II. Carbohydr Res 283:175–185

Toman R, Skultety L, Kazar J (1993) On the determination of "Kdo-like substance" in the lipopolysaccharide from *Coxiella burnetii* strain Nine Mile in phase II. Acta Virol 37:196–198

Toman R, Skultety L, Ftacek P et al (1998) NMR study of virenose and dihydrohydroxystreptose isolated from *Coxiella burnetii* phase I lipopolysaccharide. Carbohydr Res 306:291–296

Toman R, Hussein A, Palkovic P et al (2003a) Structural properties of lipopolysaccharides from *Coxiella burnetii* strains Henzerling and S. Ann NY Acad Sci 990:563–567

Toman R, Hussein A, Slaba K et al (2003b) Further structural characteristics of the lipopolysaccharide from *Coxiella burnetii* strain Nine Mile in low virulent phase II. Acta Virol 47:129–130

Toman R, Garidel P, Andra J et al (2004) Physicochemical characterization of the endotoxins from *Coxiella burnetii* strain Priscilla in relation to their bioactivities. BMC Biochem 5:1 http://www.biomedcentral.com/1471-2091/5/1

Toman R, Skultety L, Ihnatko R (2009) *Coxiella burnetii* glycomics and proteomics – tools for linking structure to function. Ann NY Acad Sci 1166:67–78

Trefzer A, Salas JA, Bechthold A (1999) Genes and enzymes involved in deoxysugar biosynthesis in bacteria. Nat Prod Rep 16:283–299

Tujulin E, Lilliehook B, Macellaro A et al (1999) Early cytokine induction in mouse P388D1 macrophages infected by *Coxiella burnetii*. Vet Immunol Immunopathol 68:159–168

Ulmer AJ, Heine H, Feist W et al (1992) Biological activity of synthetic phosphono-oxyethyl analogues of lipid A and lipid A partial structures. Infect Immun 60:3309–3314

Vadovic P, Slaba K, Fodorova M et al (2005) Structural and functional characterization of the glycan antigens involved in immunobiology of Q fever. Ann NY Acad Sci 1063:149–153

Vadovic P, Fodorova M, Toman R (2007) Structural studies of lipid A of *Piscirickettisa salmonis*, the etiological agent of the salmonid rickettsial septicemia. Acta Virol 51:249–259

Vadovic P, Fuleova A, Ihnatko R et al (2009) Structural studies of lipid A from a lipopolysaccharide of the *Coxiella burnetii* isolate RSA 514 (Crazy). Clin Microbiol Infect 15:198–199

Van den Bosch L, Manning PA, Morona R (1997) Regulation of O-antigen chain length is required for *Shigella flexneri* virulence. Mol Microbiol 23:765–775

Van der Woude MW, Baumler AJ (2004) Phase and antigenic variation in bacteria. Clin Microbiol Rev 17:581–611

Vishwanath S, Hackstadt T (1988) Lipopolysaccharide phase variation determines the complement-mediated serum susceptibility of *Coxiella burnetii*. Infect Immun 56:40–44

Vodkin MH, Williams JC (1986) Overlapping deletion in two spontaneous phase variants of *Coxiella burnetii*. J Gen Microbiol 132:2587–2594

Wang L, Jensen S, Hallman R et al (1998) Expression of the O antigen gene cluster is regulated by RfaH through the JUMPstart sequence. FEMS Microbiol Lett 165:201–206

Weiser JN, Love JM, Moxon ER (1989) The molecular mechanism of phase variation of *Haemophilus influenzae* lipopolysaccharide. Cell 59:657–665

Weisman LS, Ballou CE (1984) Biosynthesis of the mycobacterial methylmannose polysaccharide. Identification of a 3-O-methyltransferase. J Biol Chem 259:3464–3469

Weiss J, Hutzler M, Kao L (1986) Environmental modulation of lipopolysaccharide chain length alters the sensitivity of *Escherichia coli* to the neutrophil bactericidal/permeability-increasing protein. Infect Immun 51:594–599

Williams JC, Waag DM (1991) Antigens, virulence factors, and biological response modifiers of *Coxiella burnetii*: strategies for vaccine development. In: Williams JC, Thompson HA (eds) Q fever: the biology of *Coxiella burnetii*. CRC Press, Boca Raton, pp 175–222

Wollenweber HW, Schramek S, Moll H et al (1985) Nature and linkage of fatty acids present in lipopolysaccharides of phase I and phase II *Coxiella burnetii*. Arch Microbiol 142:6–11

Zamboni DS, Campos MA, Torrecilhas ACT et al (2004) Stimulation of toll-like receptor 2 by *Coxiella burnetii* is required for macrophage production of pro-inflammatory cytokines and resistance to infection. J Biol Chem 279:54405–54415

Zhang G, Russell-Lodrigue KE, Andoh M et al (2007) Mechanisms of vaccine-induced protective immunity against *Coxiella burnetii* infection in BALB/c mice. J Immunol 179:8372–8380

Chapter 5
Components of Protective Immunity

Guoquan Zhang, Yan Zhang, and James E. Samuel

Contents

Abstract *Coxiella burnetii* is an obligate intracellular bacterium that causes a worldwide zoonotic disease, Q fever. Since *C. burnetii* infection could develop into severe chronic disease in humans, vaccination is the logical approach to prevent individuals at risk of natural and deliberate exposure. Although formalin-inactivated *C. burnetii* phase I vaccine (PIV) is effective in protecting vaccinated host against the infection in humans, widespread use of this vaccine is limited by its high incidence of adverse reactions, especially in individuals with prior immunity to the agent. Creation of a safe and effective vaccine to prevent Q fever remains an important goal for public health and international biosecurity. It is critical to clearly understand the mechanisms that involved in development of protective immunity against *C. burnetii* infection and to identify the key protective antigens for developing

G. Zhang (✉) • Y. Zhang
Department of Veterinary Pathobiology, College of Veterinary Medicine,
University of Missouri-Columbia, Columbia, MO 65211, USA
e-mail:zhangguo@missouri.edu

J.E. Samuel
Department of Microbial and Molecular Pathogenesis, College of Medicine,
Texas A&M Health Science Center, 3112 Medical Research and Education Building,
Bryan, TX 77807-3260, USA

R. Toman et al. (eds.), *Coxiella burnetii: Recent Advances and New Perspectives in Research of the Q Fever Bacterium*, Advances in Experimental Medicine and Biology 984,
DOI 10.1007/978-94-007-4315-1_5, © Springer Science+Business Media Dordrecht 2012

a safe and effective new generation vaccine against Q fever. This chapter describes new information related to the characterization of acquired immunity to *C. burnetii* vaccination and infection that will provide a fundamental understanding of the development of protective immunity against Q fever.

Keywords *Coxiella burnetii* • Protective immunity • Antibody-mediated immunity • Cell-mediated immunity • Protective antigens • Vaccine

5.1 Host Immune Responses to *C. burnetii*

Generally, the host specific immune response can be divided into antibody-mediated immunity (AMI) and cell-mediated immunity (CMI). The current immunological paradigm suggests that AMI is the principal immune response effective against extracellular bacteria while the major protective immune response against intracellular bacteria is CMI. However, recent accumulated evidence has demonstrated that antibodies (Abs) can mediate resistance to a variety of intracellular bacterial and fungal pathogens, which challenges the immunological paradigm that extracellular pathogens are controlled principally by AMI and that host protection against intracellular pathogens depends on effectors of CMI. Research on the hosts immune response to *C. burnetii* infection mainly focuses on the understanding of immune response to primary infection and vaccine-induced protective immunity against *C. burnetii* infection. Most early studies suggested that both humoral and cell-mediated immune responses are important for host defense against *Coxiella* infection, with CMI probably playing the critical role in eliminating *C. burnetii* in experimental animals, while specific Abs accelerate the process. However, one early observation that treatment of athymic mice with immune sera 24 h before challenge with *C. burnetii* had no effect on bacterial multiplication within the spleens of the T-cell-deficient animals (Humphres and Hinrichs 1981) suggested that T-cell-mediated immunity plays a critical role for the control of *C. burnetii* infection. This hypothesis was supported by a recent study which demonstrated that T cells and IFN-γ are essential for clearance of a primary *C. burnetii* infection. Since passive transfer of immune sera did not confer clear protection in early studies (Kazar et al. 1977) and careful characterization of the role of Abs in protective immunity has been lacking, Ab-mediated protective immunity has been forgotten in the development of a new generation vaccine against *C. burnetii* infection. Our recent study (Zhang et al. 2007) suggests that Abs play an important role in the protection of vaccinated host from the development of clinical disease at an early stage against *C. burnetii* challenge, while the T cell-mediated immune response is required for clearance and complete elimination of the organisms at later stages of the infection. In the following sections, we summarize our current understanding of the specific immune response to *C. burnetii* infection and vaccination with a focus on the mechanisms involved in developing protective immunity.

5.1.1 Humoral Immune Responses

Humoral, or Ab-mediated immune response is essential for host defense against bacterial pathogens. Previous studies have been shown that infection and vaccination with *C. burnetii* in animals and humans induced significant Ab responses to *C. burnetii* antigens (Behymer et al. 1975; Guigno et al. 1992; Kishimoto et al. 1977; Peacock et al. 1979, 1983) suggesting that humoral immunity plays a role in vaccine induced protection against Q fever. Behymer et al. (1975) reported there was long-term persistence of agglutinating antibodies in dairy cattle after vaccination with PIV. Worswick and Marmion (Worswick and Marmion 1985) described patients with acute Q fever that developed IgM specific Abs to phase I antigen (PI) and IgM, IgG, IgA and CF Abs to PII antigen (PII), while patients with chronic Q fever had undetectable IgM specific Ab to PI or PII antigen and induced high levels of IgG, IgA and CF Abs to both PI and PII antigens. They also indicated that seronegative volunteer vaccinations with PIV elicited dominant IgM Ab response to PI antigen and lower levels of IgM, IgG and CF Abs to PII antigen, but seropositive subjects developed IgA and IgG specific Ab response to PI antigen, and CF and IgG class Ab response to PII antigen. These studies suggested that subclass Abs may play an important role in the host immune defense and that measurement of subclass Ab responses in Q fever patients may be useful for differential diagnosis of acute and chronic Q fever. Several additional early studies demonstrated that Abs play a direct role in resistance to *C. burnetii* infections. Abinanti and Marmion (1957) first reported that mixtures of Ab and *C. burnetii* organisms were not infectious in experimental animals suggesting Ab play a role in control of *Coxiella* infection. Peacock et al. (1979) demonstrated that anti-PI IgM Ab suppressed the growth of *C. burnetii* in mouse spleen when mixed with the suspension of organisms prior to inoculation. Studies on the efficacy of formalin-killed *C. burnetii* PI and PII vaccines in humans and experimental animals demonstrated that Abs that developed against *C. burnetii* antigens were involved in the resistance to *C. burnetii* infections (Lackman et al. 1962; Ormsbee et al. 1964; Peacock et al. 1979). Several *in vitro* studies have shown that treatment of *C. burnetii* with immune serum made the organisms more susceptible to phagocytosis and to destruction by normal polymorphonuclear leukocytes or macrophages in culture (Brezina and Kazar 1965; Kazar et al. 1975; Kishimoto et al. 1976). These studies provided strong support for the hypothesis that humoral immunity was important in the development of the acquired resistance against *C. burnetii* infection. Humphres and Hinrichs (1981) found that immune serum could alter the degree of infection within the infected mice and enhance clearance of the rickettsia by macrophage population of the host. However, treatment of athymic mice with immune serum 24 h before challenge with *C. burnetii* had no effect on rickettsial multiplication within the spleens of T-cell-deficient animals. This study suggested that specific Abs were able to accelerate the initial interactions of the inductive phase of the cellular immune response and promote a more rapid development of activated macrophages to levels that could control *C. burnetii* replication.

On the other hand, this study provided evidence supporting the notion that CMI plays the critical role in controlling *C. burnetii* infection.

Recent studies have demonstrated that Ab mediate protection against intracellular pathogens through various mechanisms including direct bactericidal activity, complement activation, opsonization, cellular activation via Fc or complement receptors, and Ab-dependent cellular cytotoxicity. Ab-mediated protection has been described in host defense against *Mycobacterium tuberculosis* (Teitelbaum et al. 1998), *Listeria monocytogenes* (Edelson et al. 1999), *Candida albicans* (Han and Cutler 1995), *Histoplasma capsulatum* (Nosanchuk et al. 2003), *Cryptococcus neoformans* (Casadevall and Pirofski 2004; Mukherjee et al. 1992) and *Ehrlichia chaffeensis* (Li et al. 2001; Winslow et al. 2000). Ab-mediated direct antimicrobial activity has been demonstrated against *C. neoformans* (Rosas et al. 2001). Passive administration of Abs was recently shown to protect immunodeficient SCID mice against an obligate intracellular bacterial pathogen, *E. chaffeensis*, demonstrating that humoral immunity plays a critical role in host defense against ehrlichiosis (Li et al. 2001). This data suggests that Abs play a direct role in protection against intracellular microbial pathogens. In addition, recent findings have revealed that Abs play a significant immunoregulatory role in T cell immunity against intracellular pathogens (Moore et al. 2003). Specific Ab isotypes have been shown to modulate Th1 activation via Fc receptors (FcR) by facilitating a rapid uptake, processing and presentation of pathogen-derived antigens. This enhances the T cell response against intracellular pathogens (Casadevall and Pirofski 2004; Moore et al. 2003; Woelbing et al. 2006; Yuan et al. 1998). FcγR-mediated effector functions have been shown to play an important role in protecting against *C. neoformans* (Yuan et al. 1998). A recent study on *Leishmania major* also suggested that Ab-dependent, FcγR-mediated uptake of *L. major* is essential for optimal development of protective immunity (Woelbing et al. 2006). Furthermore, FcR-mediated Ab was able to regulate T cell immunity against chlamydial infection (Moore et al. 2003). Thus, Ab-dependent FcγRs-mediated effector functions are important for host defense against intracellular pathogens. Interestingly, despite the obligate intracellular life style of *C. burnetii,* several early studies have shown that Ab was able to confer protection against *C. burnetii* infection (Lackman et al. 1962; Ormsbee et al. 1964; Peacock et al. 1979). In addition, two recent studies (Zhang et al. 2007; Shannon et al. 2009) also demonstrated that passive transfer of immune sera from killed *C. burnetii* PI whole cell antigen immunized mice was able to provide complete protection in naive, immunocompetent mice against *C. burnetii* challenge. These studies demonstrated that AMI plays an important role in vaccine-induced protective immunity against *C. burnetii* infection. Thus, understanding the mechanisms of Ab-mediated protective immunity may provide important information for developing a safe and effective vaccine against Q fever.

One early study (Peacock et al. 1979) showed that purified human anti-PI IgM was able to suppress *C. burnetii* replication in mouse spleen when mixed with the suspension of organisms prior to inoculation of mice. This study suggested that anti-PI IgM may be able to inhibit *C. burnetii* replication via its ability to neutralize or enable bactericidal mechanisms against the organism. However, further detailed studies using both *in vitro* and *in vivo* systems will be required to clearly determine whether anti-PI IgM can neutralize or become bactericidal to *C. burnetii*. To know if Ab-mediated opsonization is involved in AMI against *C. burnetii* infection, several previous studies examined the role of Abs in opsonization of *C. burnetii*. However, there are conflicting reports on the ability of Ab opsonized *C. burnetii* to survive in phagocytes. An early study showed that immune serum opsonization was able to increase the ability of guinea pig macrophages to take up and kill *C. burnetii* (Kishimoto et al. 1977). In contrast, Hinrichs and Jerrells (1976) observed that immune serum opsonization of *C. burnetii* increased bacterial uptake by guinea pig macrophages, but had no negative effect on the intracellular replication of the organism. Likewise, Kazar et al. (1975) and Baca et al. (1984) indicated that macrophages were unable to control the growth of Ab-opsonsized *C. burnetii*. Most recently, Shannon and Heinzen (2008) showed that human monocyte-derived macrophages do not control growth of Ab-opsonized *C. burnetii in vitro*. These studies suggest that although Ab was able to increase the ability of phagocytes to uptake Ab-opsonsized *C. burnetii*, it did not affect the ability of phagocytes to control the organism replication. Since these studies were conducted using *in vitro* systems and the results may not accurately reflect what occurs *in vivo*, further studies involving histological analysis of the fate of Ab-opsonized *C. burnetii* in animal models will be necessary to determine if opsonization is an important mechanism of AMI against *C. burnetii* infection.

The complement system is an important component for AMI against many bacterial pathogens. Specific Ab bound to bacteria can activate complement, resulting in downstream effects such as lysis of bacteria via membrane attack complex (MAC), release of chemoattractants C3a and C5a, and iC3b-mediated opsonization which can synergistically increase phagocytosis. There are several mechanisms by which complement can contribute to AMI. Three major pathways have been demonstrated for activation of complement. The classical pathway, which is activated by certain isotypes of Abs bound to antigens; the alternative pathway, which is activated on microbial cell surfaces in the absence of Ab, and the lectin pathway, which is activated by a plasma lectin that binds to mannose residues on microbes. To know the roles of complement activation in AMI for *C. burnetii*, Shannon et al. (2009) recently examined if Ab provides protection in mice deficient in each complement pathways by passive transfer of immune sera from PIV-immunized mice. The results demonstrated that complement deficiency in mice did not affect the ability of immune sera to confer protection against *C. burnetii* challenge, suggesting that complement activation is not involved in AMI to *C. burnetii*. To determine if complement is involved in PIV-induced protective immunity against *C. burnetii* infection, we recently tested if complement deficiency in mice significantly affects the ability of PIV to confer protection. Our results indicated that depletion of complement in PIV-immunized mice did not affect the ability of PIV to confer protection against *C. burnetii* infection, suggesting that complement does not contribute to AMI in PIV-induced protection from *C. burnetii* infection. In summary, these data suggests that AMI to *C. burnetii* is not dependent on complement.

Since Fc receptors present on phagocytes can specifically recognize Ab-bound bacteria and thereby enhance phagocytosis and Ab-dependent FcRs-mediated effector functions are important for host defense against intracellular pathogens, FcγRs-mediated, Ab-dependent opsonization may be involved in AMI against *C. burnetii* infection. However, one recent study by Shannon et al. (2009) demonstrated that although FcR-dependent, Ab-opsonized *C. burnetii* was able to stimulate activation of mouse bone marrow-derived dendritic cells *in vitro*, FcR deficiency in mice did not affect the ability of immune sera from PIV-immunized mice to confer protection against *C. burnetii* challenge *in vivo*. These results suggest that AMI to *C. burnetii* *in vivo* is FcR-independent. In addition, to determine whether FcRs play an important role in PIV-induced protection against *C. burnetii* infection, we recently examined if Fc receptor deficiency in mice significantly affects the ability of PIV to confer protection against challenge with virulent *C. burnetii* in mice. The results indicated that Fc receptors deficiency in mice did not affect the ability of PIV to confer protection against *C. burnetii* infection (unpublished data), suggesting that FcR-mediated effector functions were not involved in AMI to *C. burnetii* vaccination. Furthermore, we examined if both complement and Fc receptor deficiency in mice significantly affects the ability of PIV to confer protection against challenge with virulent *C. burnetii*. Those results indicated that both complement and Fc receptor deficiency in mice did not affect the ability of PIV to confer protection against *C. burnetii* infection. This data suggests that complement activation and FcR-mediated effector functions may not be the essential mechanisms in vaccine induced protective immunity against *C. burnetii* infection.

In addition, it has been demonstrated that FcRs bearing effector cells were able to lyse viral or parasite-infected in the presence of specific Ab via Ab-dependent cellular cytotoxicity (ADCC) (Podleski 1976). To determine whether ADCC is a mechanism in AMI against *C. burnetii* infection, one early study (Koster et al. 1984) demonstrated that Fc receptor-bearing effector cells and immune human sera can mediate specific lysis of *C. burnetii* infected macrophages using infected murine J774 macrophage-like tumor cells as targets in a chromium-release cytotoxicity assay. This study suggested that ADCC may be involved in AMI to *C. burnetii* infection. However, the observation that FcR deficiency in mice did not affect the ability of immune sera from PIV-immunized mice to confer protection against *C. burnetii* challenge *in vivo* (Shannon et al. 2009) suggests that ADCC may not be an essential mechanism in AMI against *C. burnetii* infection.

In summary, several studies demonstrated that passive transfer of Ab was able to protect immunocompetent naive mice from *C. burnetii* challenge, but did not confer protection in controlling *C. burnetii* infection in T cell deficient mice (Humphres and Hinrichs 1981; Zhang et al. 2007). Therefore, regardless what the mechanisms enable AMI to *C. burnetii*, Ab-mediated protection is dependent on the presence of T cells. In addition, the role of Abs in protective immunity against *C. burnetii* may be more complex than what is seen with other pathogens and future studies are necessary to determine which subclass of Abs is critical and what is the mechanism of AMI against *C. burnetii* infection.

5.1.2 Cell-Mediated Immune Responses

The importance of CMI has been well documented in host defense against a number of intracellular bacterial pathogens, including *M. tuberculosis*, *L. monocytogenes*, and *Legionella pneumophila* (Neild and Roy 2004; Pamer 2004; Reece and Kaufmann 2008). The role of CMI in *C. burnetii* infection and vaccination has been studied in animals and humans. Kishimoto et al. showed that peritoneal macrophages from guinea pigs previously immunized with PI antigen were capable of killing virulent PI *C. burnetii* in the absence of homologous immune serum (Kishimoto and Walker 1976; Kishimoto et al. 1977). Subsequently, they demonstrated that both infection and vaccination with *C. burnetii* developed cell-mediated immune responses in guinea pigs as determined by the inhibition of macrophage migration and lymphocyte transformation (Kishimoto and Burger 1977; Kishimoto et al. 1978). Several studies also indicated that infection and vaccination with *C. burnetii* in humans induced long-lived ability of peripheral blood lymphocytes to proliferate when cultured with *C. burnetii* antigens (Hinrichs and Jerrells 1976; Izzo et al. 1988). These studies suggested that CMI is required for the hosts defense against *C. burnetii* infection. As in other infections with obligate intracellular pathogens, host defense in Q fever appears to be dependent on CMI in which specifically activated T cells enhance the microbicidal machinery of macrophages (Ascher et al. 1980; Kishimoto et al. 1978). Izzo et al. reported that T lymphocytes were the major contributor to the cellular immune response to *C. burnetii*, as measured by the proliferation of circulating blood lymphocytes on antigen challenge (Izzo et al. 1991). Recent studies have shown that activation of guinea pig monocytes, THP-1 monocytes, L929 murine fibroblasts, and primary mouse macrophages with IFN-γ resulted in the inhibition of *C. burnetii* replication (Dellacasagrande et al. 1999; Hinrichs and Jerrells 1976; Howe et al. 2002; Turco et al. 1984; Zamboni and Rabinovitch 2003). *In vitro* and *in vivo* studies also demonstrated that IFN-γ and TNF-α play important roles in the host defense processes leading to the elimination of *C. burnetii* (Dellacasagrande et al. 1999; Turco et al. 1984;). IFN-γ mediated killing of *C. burnetii* and death of infected monocytes were dependent on TNF-α. Thus, cell-mediated immune response is required for clearance of a *C. burnetii* infection and IFN-γ plays a key role in the control of *C. burnetii* replication. In support of this model, our recent study demonstrated that T cells and IFN-γ are essential for host defense against a primary *C. burnetii* infection (Andoh et al. 2007). In addition, our previous study (Zhang et al. 2007) showed that T cells also play a critical role in PIV-induced protective immunity against *C. burnetii* challenge in mice. The role of subset T cells in immune response to primary *C. burnetii* infection and vaccination is not well studied. Recently, Read et al. (2010) using either reconstituted immunodeficient SCID mice with $CD4^+$ T cells or $CD8^+$ T cells or depleted immunocompetent BALB/c mice of T cell subsets demonstrated that either $CD4^+$ T cells or $CD8^+$ T cells alone were able to control primary pulmonary infection with *C. burnetii*. This study suggested that both $CD4^+$ and $CD8^+$ T cells play an important role in host defense against a primary *C. burnetii* infection. However, it remains unclear how $CD4^+$ and $CD8^+$ T cells control *C. burnetii* infection and the role of $CD4^+$ and $CD8^+$ T cells in vaccine induced protective immunity against *C. burnetii* infection is not well understood.

5.2 Vaccine-Induced Protective Immunity

Several vaccines including inactivated PI or PII, attenuated PII *C. burnetii* and a chloroform: methanol residue fraction of Nine Mile phase I (NMI) have been tested in animal models and humans (Genig 1968; Kazar et al. 1982; Marmion et al. 1990; Williams et al. 1992). Early studies demonstrated that killed PI vaccine was the most effective vaccine in preventing *C. burnetii* infection in animals and humans (Izzo et al. 1988, 1991; Izzo and Marmion 1993; Kishimoto and Burger 1977; Kishimoto et al. 1978). A commercial form of PIV (Q-Vax; Commonwealth Serum Laboratories), produced from the PI Henzerling strain of *C. burnetii*, has shown an extraordinary ability to prevent human Q fever and has been licensed for use in Australia since 1989 (Marmion et al. 1984, 1990). Unfortunately, this vaccine can result in severe local or systemic adverse reactions when administered to previously infected individuals, and repeat vaccination can induce severe persistent reactions (Bell et al. 1964; Marmion et al. 1984). To overcome this problem, several different chemical extraction procedures have been applied for treatment of *C. burnetii* whole cell vaccines in attempts to develop a vaccine with good immunogenicity and reduced adverse effects (Waag et al. 1997, 2002). In addition, a few efforts have attempted to develop a sub-unit protein vaccine. Williams and coworkers demonstrated that a 29 kDa protein, P1, purified from phase I Nine Mile strain *C. burnetii*, could confer protection from a lethal challenge in mice (Williams et al. 1990). Zhang *et al.* demonstrated that a partially purified 67 kDa antigen from *C. burnetii* was able to confer full protection in both guinea pigs and mice (Zhang et al. 1994). These studies suggested that sub-unit protein vaccines were able to provide protection against *C. burnetii* infection. However, since these two proteins were not well characterized and it is unknown whether they were contaminated by other components such as LPS or other protein antigens with the same molecular weights, no single protective protein has been confirmed to deliver protection comparable to PIV.

Our studies were initially aimed at developing a new generation vaccine for Q fever that is able to confer protection against infection comparable to PIV and had the ability to be administered without prior screening for immunity. Ease of production and safety for administration in humans make recombinant antigens attractive vaccine candidates for Q fever. In an attempt to identify key protective antigens, we have engaged in a comprehensive study to identify and clone immunodominant antigens using immune sera. Twenty novel immunodominant proteins, as defined by strong reactivity with infection-derived serum, were identified and cloned (Zhang et al. 2004). However, evaluation of the protective efficacies of several recombinant proteins in a sub-lethal challenge BALB/c mice model by measuring splenomegaly indicated that these selected recombinant proteins did not individually confer significant protection against infection (Zhang and Samuel 2003). These results suggested that other unidentified antigens or multiple recombinant proteins and/or an appropriate delivery system would be required for development of protective immunity. Since current identified immunodominant antigens that are recognized by Abs failed to confer complete protection against infection, it is possible that critical T-cell antigens that are required for developing protective immunity against Q fever.

Early studies suggested that both humoral and cell-mediated immune responses are important for the development of protective immunity against *C. burnetii* infection, while CMI probably plays the critical role in eliminating the organisms. Vaccination with PIV in animals and humans induced significant antibody response against *C. burnetii* antigens, suggesting that humoral immunity plays a role in vaccine induced protective immunity against Q fever. Previous studies also demonstrated that CMI is required for vaccine induced protective immunity against *C. burnetii* infection. Similar to many intracellular pathogens, clearance of *C. burnetii* infection appears to be dependent on CMI in which specifically activated T-cells enhance the microbicidal machinery of macrophages. It is generally believed that CMI is required for elimination of *C. burnetii* from infected host and IFN-γ plays a key role in controlling *C. burnetii* replication. To provide novel approaches for development of an effective and safe vaccine against Q fever, our research focused on understanding the mechanism of vaccine-induced protective immunity against *C. burnetiii* infection. One early study showed that PIV was more protective than PIIV in a guinea pig fever model. In addition, Hackstadt et al. (1985) indicated that the LPS was structurally and antigenically distinct between PI and PII cells, whereas the protein components of the two phases were indistinguishable. The difference in LPS core polysaccharide or O side chain expression between PI and PII organisms suggested that PI-LPS may play a critical role in the development of protective immunity against *C. burnetii* infection. However, since there had been no direct evidence to demonstrate whether PI-LPS or PI unique protein antigens were involved in the development of protective immunity, it remained unclear why the protective efficacy differed between PIV and PIIV. To address this question as well as to expand our understanding the mechanisms of protective immunity against Q fever, we compared the protective efficacy of PIV and PIIV a in mouse model, measuring splenomegaly and *C. burnetii* loads in the spleen as detected by Real-time PCR. Comparing comparable doses of PIV and PIIV vaccinated mice at 14 days post-challenge demonstrated that PIV was able to elicit significant protection from the development of splenomegaly following challenge and significantly decreased *C. burnetii* loads in the spleen, while PIIV did not confer measurable protection in BALB/c mice.

To determine whether PIV and PIIV induced distinct immune responses, we compared total IgG, IgG1 and IgG2a responses to PIV and PIIV vaccination at pre-challenge and 14 days post-challenge by ELISA. Antibody responses before challenge indicated that both PIV and PIIV vaccination induced a strong antibody response, but PIV elicited a higher IgG response than PIIV, suggesting that PIV possessed stronger immunogenicity than PIIV. Interestingly, PIV elicited higher IgG2a than IgG1 at all doses, while PIIV induced higher IgG2a than IgG1 in high dose vaccinated mice, but low dose vaccinated mice generated more IgG1 than IgG2a. Comparison of antibody responses at 14 days post-challenge indicated that *C. burnetii* challenge induced dramatically higher levels of IgG2a than IgG1 in unvaccinated, PIV and PIIV vaccinated mice (data not shown). Since IgG2a is a marker of a Th1 response and IgG1 is associated with a Th2-like response in mice (Germann et al. 1995), these results suggest that both PIV and PIIV were able to induce a Th1 dominant immune response but PIV-induced stronger immune responses than PIIV at comparable doses. These data support the hypothesis that a Th1 immune response is important for host defense against *C. burnetii* infection. To further confirm this hypothesis, the abilities of PIV and PIIV in inducing cytokine responses *in vivo* were compared by Bioplex multiple cytokine detection assay with mouse sera in separate experiments. Interestingly, dramatically higher levels of IL-12p70 and IL-17, and comparatively higher levels of other Th1 cytokines, including IL-6, IFN-γ and TNF-α, were detected at similar levels in both PIV and PIIV vaccinated mice at 7 days post-vaccination and at different time points post-challenge. However, Th2 cytokines, including IL-4, IL-5 and IL-10, were undetectable or observed at very low levels (unpublished data). We also tested whether Th1- and Th2-based recall responses were different between PIV and PIIV vaccinated mice. The splenocytes from both PIV and PIIV vaccinated mice responded to mitogen (concanavalin A) stimulation, but did not demonstrate a Th1 and Th2 phenotype basis on secretion of IFN-γ and IL-4. The results support our interpretation that both PIV and PIIV were able to induce a Th1 dominant immune response but cytokine responses did not clearly correlate with the protective efficacy of PIV and PIIV. The difference between PIV and PIIV to induce antibody responses suggests anti-PI specific antibody may play an important role in protection against *C. burnetii* challenge in mice.

Since the only detectable difference between PI and PII organisms is LPS core polysaccharide O side chain expression, it suggests that PI-LPS plays a unique and critical role in the development of protective immunity. To test whether PI-LPS was able to elicit protection against *C. burnetii* infection, the protective activity of PI-LPS and PII-LPS was evaluated in BALB/c mice as measured by splenomegaly. Interestingly, PI-LPS was able to rescue the protection against challenge in PIIV vaccinated mice and vaccination with PI-LPS alone also generated a similar level of protection with PIV. However, LPS from PII organisms extracted using an isolation protocol identical to PI-LPS extraction was unable to induce protection against virulent phase I challenge. These results suggest PI-LPS is an important protective component that might be responsible for PIV-induced protection against *C. burnetii* infection. However, it remains unknown whether there are specific protective antigenic epitopes on PI-LPS.

To understand the role of humoral and cellular immunity in vaccine induced protective immunity against *C. burnetii* infection, the protective efficacy of immune sera and total splenocytes from PIV vaccinated mice was examined in BALB/c mice by adoptive immunization. The protective efficacy of immune serum and splenocytes was evaluated by measuring both splenomegaly and *C. burnetii* loads in the spleen. Our results indicate immune sera and total splenocytes from immunized mice significantly reduced splenomegaly to levels similar to PIV vaccinated mice, while unvaccinated control mice developed severe splenomegaly. In support of the splenomegaly results, significantly higher *C. burnetii com1* gene copies were detected in unvaccinated mice than PIV, IS and total splenocyte immunized mice. The results demonstrated that both passive transfer of immune sera and adoptive transfer of total splenocytes were able to transfer significant protection from PIV vaccinated donor mice to naive recipient mice against *C. burnetii* challenge. This data supports the notion that both humoral and cellular immunity are important in PIV-induced protection against *C. burnetii* infection. The finding that passive transfer of immune sera provided significant protection against *C. burnetii* infection suggests that humoral immunity plays an important role in the protective vaccination against *C. burnetii* challenge. This finding has an important implication for the design of new generation vaccines and development of therapeutic strategies against Q fever.

To assess whether Ab alone can control *C. burnetii* infection and to further understand the host components that are critical to controlling the infection, we tested whether immune sera, splenocytes, B cells and T cells isolated from PIV-vaccinated mice could protect SCID mice against a *C. burnetii* challenge. Compared to normal mouse sera and PBS control, immune sera and all immune cells protected SCID mice from the development of clinical disease and body weight loss. Interestingly, mice receiving immune sera and B cells developed severe levels of splenomegaly similar to those with normal mice sera and PBS control, while splenomegaly was significantly reduced in mice receiving immune splenocytes and T cells. In support of the splenomegaly results, similarly higher *C. burnetii* loads were detected in mice receiving PBS, normal mouse sera, immune sera, and B cells, but bacterial loads were significantly reduced in mice receiving immune splenocyte and T cells. In addition, although splenocytes from naive BALB/c protected SCID mice against body weight loss, splenomegaly and bacterial loads, the protection levels were significantly lower than with splenocytes and T cells from PIV-vaccinated mice. These results indicate that immune sera and B cells protected SCID mice against clinical disease but did not control the infection, while immune splenocytes and T cells were able to effectively protect from clinical disease and control the infection. These data suggest that immune T cells play a crucial role in PIV-induced protective immunity against *C. burnetii* infection. This data, together with earlier studies, suggest that anti-PI specific Abs play an important role in the protection of vaccinated host from the development of clinical disease at an early stage against *C. burnetii* challenge, while the T cell-mediated immune response is required for clearance and complete elimination of the organisms at the later stage of the infection. Therefore, novel vaccine approaches for Q fever should be focused on boosting both humoral and cellular immune responses.

References

Abinanti FR, Marmion BP (1957) Protective or neutralizing antibody in Q fever. Am J Hyg 66:173–195

Andoh M, Zhang G, Russell-Lodrigue KE, Shive HR, Weeks BR, Samuel JE (2007) T cells are essential for bacterial clearance, and gamma interferon, tumor necrosis factor alpha, and B cells are crucial for disease development in *Coxiella burnetii* infection in mice. Infect Immun 75:3245–3255

Ascher MS, Jahrling PB, Harrington DG, Kishimoto RA, Mcgann VG (1980) Mechanisms of protective immunogenicity of microbial vaccines: effects of cyclophosphamide pretreatment in Venezuelan encephalitis, Q fever and tularaemia. Clin Exp Immunol 41:225–236

Baca O, Akporiaye ET, Rowatt JD (1984) Possible biochemical adaptations of *Coxiella burnetii* for survival within phagocytes: effects of antibody. In: Leive L, Schlessinger D (eds) Microbiology. ASM Press, Washington DC, pp 269–272

Behymer DE, Biberstein EL, Riemann HP, Franti CE, Sawyer M (1975) Q fever (*Coxiella burnetii*) investigations in dairy cattle: persistence of antibodies after vaccination. Am J Vet Res 36:781–784

Bell EJ, Lackman DB, Meis A, Hadlow WJ (1964) Recurrent reaction at site of Q fever vaccination in a sensitized person. Milit Med 124:591–595

Brezina R, Kazar J (1965) Study of the antigenic structure of *Coxiella burnetii.* IV. Phagocytosis and opsonization in relation to the phase of *C. burnetii.* Acta Virol 9:268–274

Casadevall A, Pirofski LA (2004) New concepts in antibody-mediated immunity. Infect Immun 72:6191–6196

Dellacasagrande J, Capo C, Raoult D, Mege JL (1999) IFN-gamma-mediated control of *Coxiella burnetii* survival in monocytes: the role of cell apoptosis and TNF. J Immunol 162:2259–2265

Edelson BT, Cossart P, Unanue ER (1999) Cutting edge: paradigm revisited: antibody provides resistance to *Listeria* infection. J Immunol 163:4087–4090

Genig VA (1968) A live vaccine 1/M-44 against Q-fever for oral use. J Hyg Epid Microbiol Immunol 12:265–273

Germann T, Bongartz M, Dlugonska H, Hess H, Schmitt E, Kolbe L, Kölsch E, Podlaski FJ, Gately MK, Rüde E (1995) Interleukin-12 profoundly up-regulates the synthesis of antigen-specific complement-fixing IgG2a, IgG2b and IgG3 antibody subclasses *in vivo.* Eur J Immunol 25:823–829

Guigno D, Coupland B, Smith EG (1992) Primary humoral antibody response to *Coxiella burnetii,* the causative agent of Q fever. J Clin Microbiol 30:1958–1967

Hackstadt T, Peacock MG, Hitchcock PJ, Cole RL (1985) Lipopolysaccharide variation in *Coxiella burnetti*: intrastrain heterogeneity in structure and antigenicity. Infect Immun 48:359–365

Han Y, Cutler JE (1995) Antibody response that protects against disseminated *candidiasis.* Infect Immun 63:2714–2719

Hinrichs DJ, Jerrells TR (1976) In vitro evaluation of immunity to *Coxiella burnetii.* J Immunol 117:996–1003

Howe D, Barrows LF, Lindstrom NM, Heinzen RA (2002) Nitric oxide inhibits *Coxiella burnetii* replication and parasitophorous vacuole maturation. Infect Immun 70:5140–5147

Humphres RC, Hinrichs DJ (1981) Role of antibody in *Coxiella burnetii* infection. Infect Immun 31:641–645

Izzo AA, Marmion BP (1993) Variation in interferon-gamma responses to *Coxiella burnetii* antigens with lymphocytes from vaccinated or naturally infected subjects. Clin Exp Immunol 94:507–515

Izzo AA, Marmion BP, Worswick DA (1988) Markers of cell-mediated immunity after vaccination with an inactivated, whole-cell Q fever vaccine. J Infect Dis 157:781–789

Izzo AA, Marmion BP, Hackstadt T (1991) Analysis of the cells involved in the lymphoproliferative response to *Coxiella burnetii* antigens. Clin Exp Immunol 85:98–108

Kazar J, Skultetyova E, Brezina R (1975) Phagocytosis of *Coxiella burnetii* by macrophages. Acta Virol 19:426–431

Kazar J, El-Najdawi E, Brezina R, Schramek S (1977) Search for correlates of resistance to virulent challenge in mice immunized with *Coxiella burnetii.* Acta Virol 21:422–430

Kazar J, Brezina R, Palanova A, Tvrda B, Schramek S (1982) Immunogenicity and reactogenicity of a Q fever chemovaccine in persons professionally exposed to Q fever in Czechoslovakia. Bull WHO 60:389–394

Kishimoto RA, Burger GT (1977) Appearance of cellular and humoral immunity in guinea pigs after infection with *Coxiella burnetii* administered in small-particle aerosols. Infect Immun 16:518–521

Kishimoto RA, Walker JS (1976) Interaction between *Coxiella burnetii* and guinea pig peritoneal macrophages. Infect Immun 14:416–421

Kishimoto RA, Veltri BJ, Canonico PG, Shirey FG, Walker JS (1976) Electron microscopic study on the interaction between normal guinea pig peritoneal macrophages and *Coxiella burnetii*. Infect Immun 14:1087–1096

Kishimoto RA, Veltri BJ, Shirey FG, Canonico PG, Walker JS (1977) Fate of *Coxiella burnetii* in macrophages from immune guinea pigs. Infect Immun 15:601–607

Kishimoto RA, Johnson JW, Kenyon RH (1978) Cell-mediated immune responses of guinea pigs to an inactivated phase I *Coxiella burnetii* vaccine. Infect Immun 19:194–198

Koster FT, Kirkpatrick TL, Rowatt JD, Baca OG (1984) Antibody-dependent cellular cytotoxicity of *Coxiella burnetii*-infected J774 macrophage target cells. Infect Immun 43:253–256

Lackman DB, Frommhagen LH, Jensen FW, Lennette EH (1962) Q fever studies, 23, Antibody patterns against *Coxiella burnetii*. Am J Hyg 75:158–167

Li JS, Yager E, Reilly M, Freeman C, Reddy GR, Reilly AA, Chu FK, Winslow GM (2001) Outer membrane protein-specific monoclonal antibodies protect SCID mice from fatal infection by the obligate intracellular bacterial pathogen *Ehrlichia chaffeensis*. J Immunol 166:1855–1862

Marmion BP, Ormsbee RA, Kyrkou M, Wright J, Worswick DA, Cameron S, Esterman A, Feery B, Collins W (1984) Vaccine prophylaxis of abattoir-associated Q fever. Lancet ii:1411–1414

Marmion BP, Ormsbee RA, Kyrkou M, Wright J, Worswick DA, Izzo AA, Esterman A, Feery B, Shapiro RA (1990) Vaccine prophylaxis of abattoir-associated Q fever: eight years' experience in Australian abattoirs. Epidemiol Infect 104:275–287

Moore T, Ekworomadu CO, Eko FO, MacMillan L, Ramey K, Ananaba GA, Patrickson JW, Nagappan PR, Lyn D, Black CM, Igietseme JU (2003) Fc receptor-mediated antibody regulation of T cell immunity against intracellular pathogens. J Infect Dis 188:617–624

Mukherjee J, Scharff MD, Casadevall A (1992) Protective murine monoclonal antibodies to *Cryptococcus neoformans*. Infect Immun 60:4534–4541

Neild AL, Roy CR (2004) Immunity to vacuolar pathogens: What can we learn from *Legionella*? Cell Microbiol 6:1011–1018

Nosanchuk JD, Steenbergen JN, Shi L, Deepe GS, Casadevall A (2003) Antibodies to a cell surface histone-like protein protect against *Histoplasma capsulatum*. J Clin Invest 112:1164–1175

Ormsbee RA, Bell EJ, Lackman DB, Tallent G (1964) The influence of phase on the protective potency of Q fever vaccine. J Immunol 92:404–412

Pamer EG (2004) Immune responses to *Listeria monocytogenes*. Nat Rev Immunol 4:812–823

Peacock MG, Fiset P, Ormsbee RA, Wisseman CL (1979) Antibody response in man following a small intradermal inoculation with *Coxiella burnetii* phase I vaccine. Acta Virol 23:73–81

Peacock MG, Philip RN, Williams JC, Faulkner RS (1983) Serological evaluation of Q fever in humans: enhanced phase I titers of immunoglobulins G and A are diagnostic for Q fever endocarditis. Infect Immun 41:1089–1098

Podleski WK (1976) Cytodestructive mechanisms provoked by lymphocytes. Am J Med 61:1–8

Read AJ, Erickson S, Harmsen AG (2010) Role of CD4+ and CD8+ T cells in clearance of primary pulmonary infection with *Coxiella burnetii*. Infect Immun 78:3019–3026

Reece ST, Kaufmann SH (2008) Rational design of vaccines against tuberculosis directed by basic immunology. Int J Med Microbiol 298:143–150

Rosas AL, Nosanchuk JD, Casadevall A (2001) Passive immunization with melanin-binding monoclonal antibodies prolongs survival of mice with lethal *Cryptococcus neoformans* infection. Infect Immun 69:3410–3412

Shannon JG, Heinzen RA (2008) Infection of human monocyte-derived macrophages with *Coxiella burnetii*. Methods Mol Biol 431:189–200

Shannon JG, Cockrell DC, Takahashi K, Stahl GL, Heinzen RA (2009) Antibody-mediated immunity to the obligate intracellular bacterial pathogen *Coxiella burnetii* is Fc receptor- and complement-independent. BMC Immunol 10(26):1–8

Teitelbaum R, Glatman-Freedman A, Chen B, Robbins JB, Unanue E, Casadevall A, Bloom BR (1998) A mAb recognizing a surface antigen of *Mycobacterium tuberculosis* enhances host survival. PNAS 95:15688–15693

Turco J, Thompson HA, Winkler HH (1984) Interferon-gamma inhibits growth of *Coxiella burnetii* in mouse fibroblasts. Infect Immun 45:781–783

Waag DM, England MJ, Pitt ML (1997) Comparative efficacy of a *Coxiella burnetii* chloroform: methanol residue (CMR) vaccine and a licensed cellular vaccine (Q-Vax) in rodents challenged by aerosol. Vaccine 15:1779–1783

Waag DM, England MJ, Tammariello RF, Byrne WR, Gibbs P, Banfield CM, Pitt ML (2002) Comparative efficacy and immunogenicity of Q fever chloroform: methanol residue (CMR) and phase I cellular (Q-Vax) vaccines in cynomolgus monkeys challenged by aerosol. Vaccine 20:2623–2634

Williams JC, Hoover TA, Waag DM (1990) Antigenic structure of *Coxiella burnetii.* A comparison of lipopolysaccharide and protein antigens as vaccines against Q fever. Ann N Y Acad Sci 590:370–380

Williams JC, Peacock MG, Waag DM, Kent G, England MJ, Nelson G, Stephenson EH (1992) Vaccines against coxiellosis and Q fever Development of a chloroform:methanol residue subunit of phase I *Coxiella burnetii* for the immunization of animals. Ann N Y Acad Sci 653:88–111

Winslow GM, Yager E, Shilo K, Volk E, Reilly A, Chu FK (2000) Antibody-mediated elimination of the obligate intracellular bacterial pathogen *Ehrlichia chaffeensis* during active infection. Infect Immun 68:2187–2195

Woelbing F, Kostka SL, Moelle K, Belkaid Y, Sunderkoetter C, Verbeek S, Waisman A, Nigg AP, Knop J, Udey MC, Von Stebut E (2006) Uptake of *Leishmania major* by dendritic cells is mediated by Fcgamma receptors and facilitates acquisition of protective immunity. J Exp Med 203:177–188

Worswick D, Marmion BP (1985) Antibody responses in acute and chronic Q fever and in subjects vaccinated against Q fever. J Med Microbiol 19:281–296

Yuan R, Clynes R, Oh J, Ravetch JV, Scharff MD (1998) Antibody-mediated modulation of *Cryptococcus neoformans* infection is dependent on distinct Fc receptor functions and IgG subclasses. J Exp Med 187:641–648

Zamboni DS, Rabinovitch M (2003) Nitric oxide partially controls *Coxiella burnetii* phase II infection in mouse primary macrophages. Infect Immun 71:1225–1233

Zhang G, Samuel JE (2003) Identification and cloning protective antigens of *Coxiella burnetii* recognized by sera from mice experimentally infected with Nine Mile phase I. Ann NY Acad Sci 990:510–521

Zhang YX, Zhi N, Yu SR (1994) Protective immunity induced by 67 K outer membrane protein of phase I *Coxiella burnetii* in mice and guinea pigs. Acta Virol 38:327–332

Zhang G, Kiss K, Seshadri R, Hendrix L, Samuel JE (2004) Identification and cloning of immunodominant antigens of *Coxiella burnetii*. Infect Immun 72:844–852

Zhang G, Russell-Lodrigue KE, Andoh M, Zhang Y, Hendrix LR, Samuel JE (2007) Mechanisms of vaccine-induced protective immunity against *Coxiella burnetii* infection in BALB/c mice. J Immunol 179:8372–8380

Chapter 6
Proteome of *Coxiella burnetii*

Robert Ihnatko, Edward Shaw, and Rudolf Toman

Contents

R. Ihnatko
Department of Clinical and Experimental Medicine, Division of Cell Biology, Faculty of Health Sciences, Linköping University, S-581 85 Linköping, Sweden

E. Shaw
Department of Microbiology and Molecular Genetics, Oklahoma State University, Stillwater, OK, USA

R. Toman (✉)
Laboratory for Diagnosis and Prevention of Rickettsial and Chlamydial Infections, Institute of Virology, Slovak Academy of Sciences, Dubravska cesta 9, 845 05 Bratislava, Slovakia
e-mail: Rudolf.Toman@savba.sk

R. Toman et al. (eds.), *Coxiella burnetii: Recent Advances and New Perspectives in Research of the Q Fever Bacterium*, Advances in Experimental Medicine and Biology 984, DOI 10.1007/978-94-007-4315-1_6,

Abstract Recent proteomic studies of *C. burnetii*, the etiological agent of Q fever, have brought a deeper insight into the pathogen's physiology and offered new possibilities in investigations of inter- or intra-species relatedness. The data generated from these studies in conjunction with the current genomic sequence databases may reveal additional identities for conserved and unique *C. burnetii* biomarkers and aid in creating algorithms and/or databases that could develop into diagnostic and detection tools for the pathogen. Moreover, wide scale screening and further characterization of potential *C. burnetii* protein antigens along with a comprehensive evaluation of the humoral immune response will be of fundamental importance towards research and development of a safe and efficacious vaccine as well as improved serodiagnostic tests for rapid and sensitive detection of the Q fever pathogen. Given these advances, proteomics may make marked contributions to the improvement of human health protection against *C. burnetii* in the coming years.

Keywords *Coxiella burnetii* • Biomarker • Detection • Diagnostic • Function • Protein • Proteomics • Q fever

6.1 Introduction

The world of proteins includes a myriad of protein species that play specific tasks in almost all of the activities of living systems. These systems span the spectrum from unicellular organisms to highly sophisticated multicellular forms of life. Although the genome sequence reveals the coding capacity of an organism, the information provided about the theoretically produced proteins remains just informative. For example, it does not provide information about where, or in what quantity, the post-translationally modified proteins carry out their specific functions. The study of proteins is the main subject of proteomics. These extensive studies provide information about protein structures, their post-translationally processed forms, protein-protein interactions in a given biological compartment, at a specific time and in the defined environment. The introduction of mass spectrometry (MS) into proteomic studies has brought considerable improvement in the detection and analysis of proteins, or components of protein complexes, by its increased speed and sensitivity. Therefore, research in the field of proteomics has grown at an extremely rapid rate, providing new data from studies of many bacterial pathogens. Various proteomic techniques combined with molecular biology methods have become important tools for detection of such pathogens, elucidation of host-pathogen interactions, and unraveling the molecular mechanisms of pathogenesis for many infectious diseases. Our knowledge of proteomics is crucial to the development of new prevention and treatment strategies for infectious pathogens.

Coxiella burnetii is an extremely infectious obligate intracellular Gram-negative bacterium that causes Q fever in humans, a zoonotic disease whose numbers are thought to be underestimated (Baca and Paretsky 1983; Maurin and Raoult 1999; Arricau-Bouvery and Rodolakis 2005). In the last decade, the complete sequencing of the virulent *C. burnetii* Nine Mile phase I strain (NM I, RSA 493) genome (Seshadri et al. 2003) brought about new investigations of its proteome and an understanding of the mechanisms involved in the pathogenesis of this bacterium. This chapter attempts to provide an integrated view into the *C. burnetii* proteome based on the current knowledge from proteomic studies.

6.2 Proteomics in the Structural Analysis of *C. burnetii*

6.2.1 Proteome Analysis of C. burnetii Developmental Forms

C. burnetii undergoes an intracellular developmental cycle that generates a morphologically distinct small cell variant (SCV), a spore-like form with a remarkable ability to persist in the extracellular environment, and a large cell variant (LCV) that is the vegetative, metabolically active form (McCaul and Williams 1981). The developmental forms can be differentiated by their morphology, size, peptidoglycan, and protein content (McCaul and Williams 1981; Amano et al. 1984). Ultrastructurally, SCVs differ from LCVs by their smaller size and condensed chromatin (McCaul and Williams 1981). Analysis of total cell lysates from SCV and LCV cells using one- and two-dimensional gel electrophoresis (1- and 2-DE) revealed that the morphological differences between the cell variants correlate with unique protein expression patterns for each cell type (McCaul et al. 1991; Heinzen and Hackstadt 1996; Heinzen et al. 1996; Coleman et al. 2007). Forty-eight proteins were found twofold or more abundant in LCVs when compared to SCVs and visualized using 2-DE. Fifteen of these proteins were submitted to matrix assisted laser desorption ionization time-of-flight (MALDI-TOF) MS analysis (Coleman et al. 2007). As a result, all the identified proteins that were found more abundant in LCVs had predicted functions in general physiological processes, such as cell division, RNA and protein synthesis or processing. This is in agreement with the higher metabolic activity known to exist in LCVs compared with that of SCVs (Coleman et al. 2004). Among them, N – utilization substance protein A (CBU1433), ribosomal protein S1 (CBU0528), ribosomal protein L9 (CBU0867), elongation factor EF-Tu (CBU0236), FtsZ protein (CBU0141) – essential to the cell-division process, segregation and condensation protein B (CBU1060), chromosome partitioning protein ParB (CBU1927), 6,7-dimethyl-8-ribityllumazine synthase RibH (CBU0648), 60 kDa chaperone GroEL (CBU1718), and chaperone protein HtpG (CBU0309) were identified. However, the harsh environment in the *C. burnetii* containing parasitophorous vacuole (PV) may induce the expression of stringent starvation protein SspA (CBU1747), universal stress protein UspA (CBU1916), and thiredoxin peroxidase Bcp (CBU0963) that were found upregulated in LCVs as well (Coleman et al. 2007). On the other hand, six proteins were found to be more abundant in SCVs than in LCVs. These were: TolB protein (CBU0090), which is involved in outer membrane stability (Lazzaroni et al. 1999), GTP binding Era homolog

(CBU1502), which is involved in cell cycle control and chromosome partitioning (Caldon and March 2003), cystathionine beta-lyase MetC (CBU2025), an enzyme implicated in methionine biosynthesis that is important for bacterial protein and DNA metabolism and bacterial virulence (Ejim et al. 2004, 2007), and one uncharacterized protein (CBU2079).

In another study, several differences in outer membrane protein (OMP) composition of LCV and SCV cells have been revealed by iodination of surface antigens (Varghees et al. 2002), including a 29 kDa major OMP P1 (CBU0311), which possesses a porin activity and is present at high levels in LCVs (McCaul et al. 1991; Varghees et al. 2002), and a 34 kDa protein OMP34 that is unique to SCVs (Heinzen et al. 1999). Two other SCV-specific cytoplasmic proteins were revealed by molecular cloning (Heinzen and Hackstadt 1996; Heinzen et al. 1996), particularly a SCV protein, ScvA (CBU1267.1), and a histone H1 homolog, Hq1 (CBU0456). Both have very basic isoelectric points (pI) and are degraded in the higher developmental (LCV) stage of the bacterium. ScvA is only 30 amino acids in length with a molecular mass of 3.614 kDa. DNA-binding of ScvA suggests that the protein might be involved in DNA condensation and protection in the SCV cells (Heinzen et al. 1996). ScvA down-regulation was clearly demonstrated (Howe and Mallavia 2000) *in vivo* and *in vitro* in response to low pH, suggesting that during the transition from SCVs to LCVs, the DNA condensation and protection by ScvA is no longer required. Hq1 is 119 amino acids in length with a molecular mass of 13.1 kDa. The basic nature of this protein is conferred by its high lysine content (29%). Hq1 exhibits 34% identity with eukaryotic histone H1 and 26% identity with the histone-like protein Hc1 of *Chlamydia trachomatis* (Heinzen and Hackstadt 1996). Although the exact roles for both of these proteins are still not completely known, they are assumed to have structural roles in the formation of the condensed SCV nucleoid. This assumption is based on homology with H1, *in vitro* interactions with DNA of Hq1 and ScvA, and the labeling of chromatin by immunogold-tagged anti-ScvA antibody (Heinzen and Hackstadt 1996; Heinzen et al. 1996; Coleman et al. 2004). Binding of one or both of these proteins with genomic DNA *in vivo* could have a protective role by stabilizing the chromosome or could induce changes in DNA topology that alter gene expression (Heinzen et al. 1999). Due to the fact that ScvA is a glutamine-rich protein (23%), which is the amino acid readily converted to glutamate, the degradation of ScvA in SCV to LCV developmental transition could serve as a carbon and energy source for the bacterium early during the infectious cycle (Hackstadt and Williams 1983; Heinzen et al. 1999).

In another study, two additional developmental stage-specific proteins, particularly a 45 kDa antigen detected only in LCVs, and a 35 kDa protein that is more abundant in LCVs than in SCVs have been identified (Seshadri et al. 1999). Based on molecular cloning, both proteins were identified as homologs of the elongation factors EF-Tu and EF-Ts, respectively. The proteins are primarily involved in polypeptide chain elongation during protein biosynthesis, but they could also be involved in other important functions, e.g. in cell shape maintenance as it has been demonstrated by the colocalization of EF-Tu with the actin-like protein MreB in *Bacillus subtillis* (Defeu Soufo et al. 2010), or by the association of EF-Tu with bacterial membrane facilitating adhesion and pathogen entry into host cells and tissues (Balasubramanian et al. 2008; Barel et al. 2008).

6.2.2 Proteomic Analyses of Virulent and Avirulent Forms of *C. burnetii*

In nature, *C. burnetii* exists in a virulent (phase I) form, whereas during serial laboratory passages in embryonated hen eggs or cell culture, it undergoes a phase variation to an avirulent (phase II) form (Stoker and Fiset 1956). It is well known that this variation is accompanied with a loss of phase I antigenicity due to modifications in composition and structure of the lipopolysaccharide (LPS) and several outer membrane proteins (Krauss et al. 1977; Hackstadt et al. 1985; Ftacek et al. 2000; Toman et al. 2009). The complete sequencing of the *C. burnetii* genome revealed (Seshadri et al. 2003) more than 2,000 coding sequences for specific proteins and facilitated predictions of the bacterium's overall proteome. However, only a small percentage of the proteome has been characterized to date. One of the most common problems encountered in the proteomic analysis of bacterial cells is the dramatic difference in the relative abundance of different proteins. Protein concentrations may differ by 7–10 orders of magnitude causing low abundant proteins to easily escape detection, making it difficult to relate results of proteome profiling to the biological system. To avoid this problem, many pre-fractionation strategies have been employed to detect proteins expressed in low copy numbers within *C. burnetii*. In the initial studies (Skultety et al. 2005; Samoilis et al. 2007), whole cell lysates of *C. burnetii* NM I and NM II strains were resolved by 2-DE and liquid chromatography (LC). The proteins from 2-DE gels or collected chromatographic fractions were analyzed by peptide mass fingerprinting and various MALDI-, and electrospray ionization (ESI) tandem mass spectrometry (MS/MS) approaches (Figs. 6.1 and 6.2). However, only approximately 200 products of the *C. burnetii* open reading frames could be identified in each of the phases of the bacterium. The subcellular localization and functional classification of the identified proteins was then predicted based on the cellular process in which each protein is potentially involved. Here we will analyze the functions of some of these proteins.

Gram-negative bacteria possess a specific cell envelope that is an effective permeability barrier against most antimicrobial agents. In most cases, it consists of three layers: an outer membrane (OM) composed mainly of LPS and proteins, a lipid-protein inner membrane (IM), and a thin rigid layer of peptidoglycan located in the periplasmic space. As important elements of the cell envelope, Tol proteins of Gram-negative bacteria maintain the stability of bacterial OM and likely aid in the integration of some other OM components such as porins and LPS (Rigal et al. 1997; Lazzaroni et al. 1999; Ray et al. 2000). The Tol-Pal protein complex is formed by TolQ, TolR, and TolA proteins located in the IM, whereas the periplasmic TolB, and the Pal lipoprotein are located in the OM. The identification of the TolB precursor

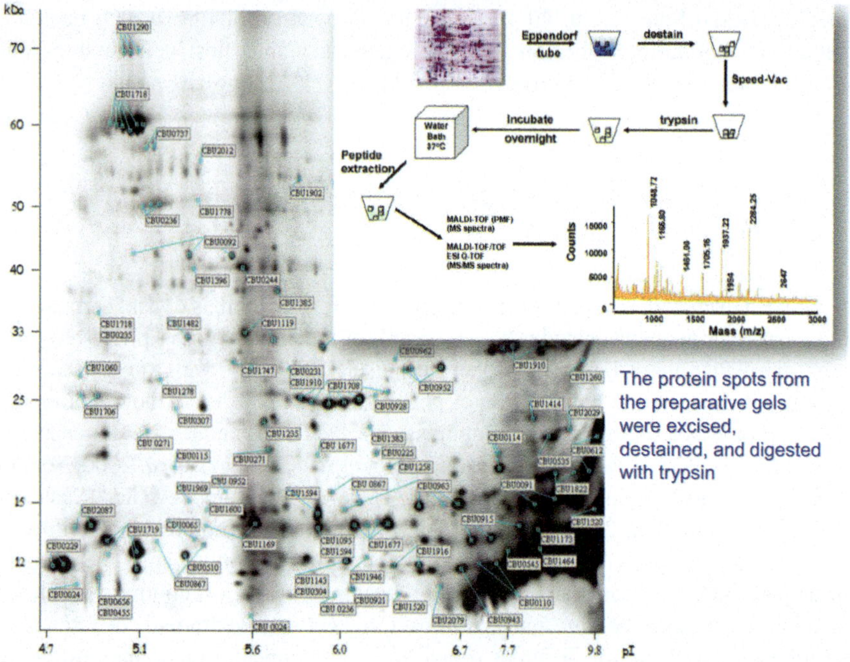

Fig. 6.1 Schematic presentation of 2-DE based proteomic analysis of the whole cell lysate of *C. burnetii* strain NM I. Assignment of the identified proteins is shown on the 2-DE gel

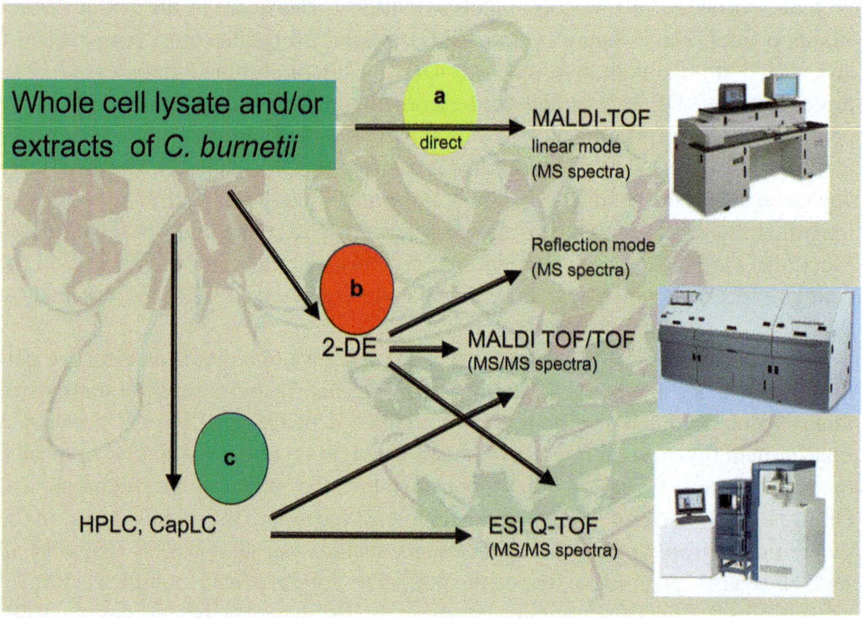

Fig. 6.2 Current MS techniques (**a**–**c**) used in analyzing the *C. burnetii* proteome. For abbreviations, see the text

(CBU0090) and the Tol system periplasmic component (CBU0092) in both phases of *C. burnetii* (Skultety et al. 2005; Samoilis et al. 2007), and the OmpA protein (CBU0307) in phase I of the microbe (Skultety et al. 2005) suggests that the function of the Tol-Pal system in the physiology and pathogenesis of *C. burnetii* could be analogous to that of the Tol-Pal system of other Gram-negative bacteria. The Tol-Pal system participates in the stabilization of the bacterial cell envelope by linking the peptidoglycan to the outer membrane, as demonstrated in *Escherichia coli* (Lloubes et al. 2001; Cascales et al. 2002). The TolB was found to be cross-linked to the peptidoglycan-associated protein OmpA and the OM lipoprotein Lpp (Clavel et al. 1998). Each of the *tol* and *pal* mutants, as well as the major lipoprotein Lpp, were found to exhibit outer membrane defects leading to bacterial hypersensitivity to harmful compounds, leakage of periplasmic proteins, and the release of outer membrane vesicles (Webster 1991; Bernadac et al. 1998; Lazzaroni et al. 1999; Cascales et al. 2000). Apart from the implication of Tol-Pal system involvement in outer membrane integrity, other studies suggest involvement of TolA in the surface expression of O-polysaccharide and synthesis of the LPS core region (Gaspar et al. 2000; Vines et al. 2005). More specifically, TolA is likely involved in the assembly of O-antigen by facilitating the transfer of O-polysaccharide to the lipid A-core molecules, or by contributing to the translocation of the O-antigen across the cytoplasmic membrane. A study with *tol* and *pal* mutants of *E. coli* K-12 strain W3110 showed a reduction in O-antigen polymerization (Vines et al. 2005) when the genes were interupted. Recently, the involvement of proteins from the Tol-Pal complex has also been described in the formation of cell envelopes in daughter cells, as mutants were found to accumulate cell envelope at the site of cell constriction (septum) separating the daughter cells (Gerding et al. 2007).

Pathogenic bacteria such as *C. burnetii* encounter extracellular and host-associated environments and have evolved highly sophisticated mechanisms for sensing the surrounding environment and adjusting gene expression appropriately (Chowdhury et al. 1996). Virulence factors control the expression of virulence genes as a consequence of the pathophysiology of infection (Chowdhury et al. 1996). In our ongoing proteomic studies of *C. burnetii*, the virulence regulator BipA (CBU0884), a highly conserved GTPase that functions to influence numerous cellular processes, has been identified (Skultety et al. 2011). In both *E. coli* and *Salmonella enterica*, BipA has been implicated in controlling bacterial motility, modulating attachment and effacement, upregulating expression of virulence genes, and mediating avoidance of host defense mechanisms (Scott et al. 2003; deLivron and Robinson 2008). The regulatory mechanisms are themselves complex and involve both transcription and translation. In this regard, we have also identified (Skultety et al. 2011) multiple, previously-unidentified *C. burnetii* transcription factors (CBU0509, CBU0780, CBU1416, CBU1417, CBU2006, and CBU1991) and a regulator of carbohydrate metabolism (CBU0744).

The outer leaflet of the OM of *C. burnetii* is composed of LPS, which is an important bacterial structural and functional determinant. It plays an important role in the interaction of *C. burnetii* with the host, its pathogenicity and immunogenicity, it is capable of inducing an antibody response, and is considered a protective immunogen (Amano and Williams 1984; Williams and Waag 1991; Gajdosova et al. 1994;

Hussein et al. 2001). Among the OM components of *C. burnetii*, the LPS was first recognized to undergo modifications of its chemical composition and structure during phase variation (Hackstadt et al. 1985; Williams and Waag 1991; Ftacek et al. 2000). In addition to the sugar residues found in *C. burnetii* phase II LPS (LPS II), (Toman and Skultety 1996), *C. burnetii* phase I LPS (LPS I) possesses two unusual sugars, namely virenose (Vir, 6-deoxy-3-C-methyl-D-gulopyranose) and dihydro-hydroxystreptose (Strep, 3-C-[hydroxymethyl]-L-lyxofuranose) located in its O-polysaccharide chain (Mayer et al. 1988; Skultety et al. 1998; Toman et al. 1998, 2009). Neither of these sugars has been found in the LPS of other enterobacterial species and are considered as unique biomarkers of *C. burnetii*. The candidate genes involved in the Vir and Strep syntheses have been suggested (Hoover et al. 2002; Thompson et al. 2003) and their protein products, NDP-hexose 3-C-methyltransferase TylCIII (CBU0691) and methyltransferase FkbM family (CBU0683), were identified by proteomic analysis (Skultety et al. 2005). Our recent proteomic search for the LPS I biosynthetic/modification enzymes (Skultety et al. 2011) resulted in the detection of isomerase (CBU0674), kinase (CBU0678), transferase (CBU1657), and three methyltransferases (CBU0682, CBU0683 and CBU0691). The phosphoheptose isomerase (CBU0674) and D-*glycero*-D-*manno*-heptose-7-phosphate 1-kinase (CBU0678) are involved in the biosynthesis of D-*glycero*-D-*manno*-heptose. The isomerase catalyzes the first step of the pathway common to L-*glycero*- and D-*glycero*-D-*manno*-heptose biosynthesis (Eidels and Osborn 1974). This enzyme is conserved in a wide range of both Gram-positive and Gram-negative bacteria (Valvano et al. 2002). By contrast, the kinase (CBU0678) is found only in the L-*glycero*-D-*manno*-heptose biosynthetic pathway (Kneidinger et al. 2002; Valvano et al. 2002). The general biosynthetic enzyme α-L-*glycero*-D-*manno*-heptose β-1,4-glucosyltransferase (CBU1657) is closely related to the D-*glycero*-D-*manno*-heptose-7-phosphate 1-kinase (CBU0678), and is required for the formation of glycosidic bonds during biosynthesis of di-, oligo-, and polysaccharides.

Methyltransferases have been considered candidate enzymes for the synthesis of Vir and Strep in LPS I (Seshadri et al. 2003; Skultety et al. 2005). It was assumed that the S-adenosyl-dependent methyl transferase (CBU0691) might be a part of the Vir biosynthetic pathway (Hoover et al. 2002). A similar enzyme was reported to participate in the mycarose biosynthetic pathway in *Streptomyces fradiae* and in the synthesis of the macrolide antibiotic tylosin (Bate et al. 2002). From a phylogenetic and evolutionary perspective, it would be of interest to find sequence homology between methyltransferases required for biosynthesis of tylosin by *S. fradiae* (Fouces et al. 2002) and the *C. burnetii* methyltransferases (CBU0682 and CBU0683). Most recently, it has been suggested (Narasaki et al. 2011) that GDP-D-mannose is a metabolic intermediate to GDP-β-Vir and the former is synthetized from fructose-6-phosphate in three successive reactions. To test the validity of the suggested pathway, three open reading frames (CBU0671, CBU0294 and CBU0689) annotated as bifunctional type II mannose-6-phosphate isomerase, as phosphomannomutase, or GDP-mannose-4,6,-dehydratase were functionally characterized by complementation of corresponding *E. coli* mutant strains and in enzymatic assays. More details are given in Chap. 4.

In addition to the proteins involved in LPS metabolism, a group of enzymes involved in low molecular mass carbohydrate metabolism was detected (Skultety et al. 2011). This group includes an epimerase (CBU1838), an esterase (CBU0419), an aldolase (CBU1778), three transferases (CBU0479, CBU1657, and CBU1887), and three kinases (CBU0846, CBU1781 and CBU2092).

Microorganisms are often exposed to toxic compounds in their environments. Thus, bacteria have evolved various active mechanisms to pump toxins out of themselves and survive. For toxin resistance, the outer membrane alone provides a relatively weak barrier. Highly efficient multidrug efflux pumps of exceptionally wide specificity interact with outer membrane channels and accessory proteins, forming the multi-subunit complexes which are capable of eliminating the toxic substances from cells (Nikaido 2001). The AcrB pump of *E. coli* can serve as an example in forming a complex with AcrA and the outer membrane channel TolC. This complex eliminates some antibiotics, detergents, and various disinfectants from the inner membrane space of bacteria (Nikaido 1998; Tsukagoshi and Aono 2000; Poole 2001; Yu et al. 2003). Similar proteins, such as the AcrB/AcrD/AcrF transporter family protein (CBU0804) and TolC (CBU0056) were found in *C. burnetii* (Skultety et al. 2005).

The harsh environment of the phagolysosome is unfavorable for most pathogens, however, this is the unique environment in which *C. burnetii* survives and proliferates. To do so, the bacterium has to express proteins with the capacity to protect it from the proteolytic attack of the host cells. Several proteins with such a function have been identified in proteomic studies (Skultety et al. 2005; Samoilis et al. 2007) to date. An important aspect of host defense against microbial pathogens is the production of reactive oxygen species within phagolysosomes. *C. burnetii* encodes multiple factors that can modulate sensitivity and expression of these toxic radicals including catalase, secreted acid phosphatase (ACP) and superoxide dismutase (SOD) enzymes which differ in their metal cofactors (Seshadri et al. 2003). Thus, the superoxide anion converting enzymes FeSOD and Cu-ZnSOD (CBU1708 and CBU1822) along with a catalase (CBU0281), and a secreted ACP (CBU0335) were identified in the high speed supernatants of ultrasonicated cells and the whole cell lysates of *C. burnetii* in phase I (Baca et al. 1993; Skultety et al. 2005). The ACP was detected in the periplasmic space of the bacterium and the level of the enzyme activity far exceeded that reported in other microorganisms such as *Legionella* and *Leishmania* spp (Baca et al. 1993). The function of the catalase in *C. burnetii* may be associated with elimination of hydrogen peroxide generated by parasite cytoplasmic SOD as well as against host-generated hydrogen peroxide (Akporiaye and Baca 1983). Other proteins that participate in the antioxidant defense of *C. burnetii* are the AhpC/Tsa family protein (CBU1706) and peroxiredoxins (as the bacterioferritin comigratory protein, CBU0963) catalyzing the reduction of hydrogen peroxide and organic hyperoxides or peroxonitrites. Additionally, the stringent starvation protein A (CBU1747), a highly conserved transcription factor among Gram-negative bacteria essential for acid tolerance (Hansen et al. 2005a, b), and two proteins (CBU1983 and CBU1916) that belong to the universal stress protein A superfamily, and are up-regulated by a wide variety of stress conditions (Kvint et al. 2003), have also been found in *C. burnetii* (Samoilis et al. 2007).

Detailed studies of the process of pathogen entry into host cells are of fundamental importance for understanding bacterial pathogenesis. *C. burnetii* encodes the enhanced entry proteins B and C (CBU1137 and CBU1136), and a precursor of macrophage infectivity potentiator (Mip) protein (CBU0630), the expression of which has also been confirmed in both phases of the bacterium (Skultety et al. 2005; Samoilis et al. 2007). The intact active site of peptidyl-prolyl cis-trans isomerase (PPIase) Mip was shown to be essential for an early establishment and initiation of intracellular infection of *L. pneumophila* (Helbig et al. 2003). The presence of the 230 amino acid gene product Mip (Cb-Mip) in the *C. burnetii* proteome has also been confirmed by molecular cloning. When the *mip* gene of *C. burnetii* was cloned into *E. coli,* the overproduced Mip protein exhibited the PPIase activity (Mo et al. 1995). Moreover, the antiserum from rabbits immunized with inactivated *C. burnetii* NM I, NM II, or Priscilla strains, as well as human serum from an individual with previous serological and clinical evidence of Q fever, showed Cb-Mip immunoreactivity (Seshu et al. 1997). The data indicate that Cb-Mip is immunogenic in both experimental and natural infections and might be an important virulence factor of *C. burnetii* analogous to that in *L. pneumophila*, where the virulence-enhancing property of Mip has been well documented (Helbig et al. 2003). Among the proteins that have been detected recently using proteomic techniques (Skultety et al. 2011) there are also three enhanced entry/infectivity proteins (CBU0085, CBU0318 and CBU1138) and four proteins involved in transport/trafficking (CBU0147, CBU0565, CBU1770 and CBU1829).

The periplasmic space of the Gram-negative bacterial cell wall is comprised of a thin, rigid layer of a polymer (peptidoglycan) consisting of repeating glycan strands of the N-acetylglucosamine-N-acetylmuramic acid disaccharide that are cross-linked by the pentapeptide side chains (Vollmer et al. 2008). The turnover of peptidoglycan in a bacterium is highly controlled and involves a complement of autolytic enzymes with specificity for either the carbohydrate or the peptide linkages of peptidoglycan. One major class of these autolysins is represented by N-acetylmuramoyl-L-alanine amidases that cleave the amide linkage between the stem peptides and the lactyl moiety of muramoyl residues. A similar role of N-acetylmuramoyl-L-alanine amidase (CBU0379) is assumed in the process of peptidoglycan turnover in both cell phases of *C. burnetii*, although it has only been found in phase II cells using proteomic techniques thus far (Samoilis et al. 2007).

6.3 Proteomics in Screening the *C. burnetii* Secretome

The investigation of mechanisms of bacterial immune modulation is crucial in understanding the pathogenesis of infection and may lead to new strategies in the treatment of bacterial infections. Proteins secreted by microorganisms into host cells facilitate the establishment of micro-environmental conditions favorable for bacterial growth and replication. *C. burnetii* is the only known intracellular bacterium which carries out its developmental cycle within the fully formed acidic phagolysosome (Voth and Heinzen 2007). Following internalization, the nascent phagolysosome ultimately develops into a large parasitophorous vacuole (PV) with lysosomal characteristics (Howe and Heinzen 2008). The biogenesis and maintenance of the PV as well as regulation of multiple events during infection requires *de novo C. burnetii* synthesis of proteins with effector functions that are actively delivered to the host cytosol by the pathogen. *C. burnetii* possesses a specialized Dot/Icm type IV secretion system (T4SS) by which bacterial proteins secreted into PV may modify the PV environment to support pathogen replication (Zamboni et al. 2003). The *C. burnetii* genome encodes a majority of the *L. pneumophila* T4SS *dot/icm* genes, with the exception of *icmR, dotJ* and *dotV* (Seshadri et al. 1999; Juhas et al. 2008). However, interaction of a protein which is functionally similar but non-homologous to IcmR of *L. pneumophila* has been identified in *C. burnetii* (Feldman et al. 2005), revealing the importance of this protein for the function of this secretion system. Moreover, similar to *L. pneumophila,* the *C. burnetii* T4SS has been found under direct control of the regulatory protein PmrA (Zusman et al. 2007). Some of the *icm/dot* genes of *C. burnetii* are capable of complementing corresponding mutations in *L. pneumophila,* suggesting that the T4SS of *C. burnetii* is active and shares similar features with its *Legionella*'s counterpart (Zusman et al. 2003). The DotB protein (CBU1645), a component of T4SS possessing the ATPase activity has also been confirmed in *C. burnetii* (Samoilis et al. 2007).

The T4SS of *C. burnetii* facilitates delivery of a large number of specific effector proteins into the host cytosol. However, these effectors are unique to *C. burnetii* and likely reflect the pathogen's unique phagolysosomal environmental niche. It has been observed that effector proteins secreted by intracellular pathogens and endosymbionts possess multiple ankyrin repeat homology domains (Ank) that serve as a platform for protein-protein interactions between bacterial and eukaryotic host cell proteins (Pan et al. 2008). The *C. burnetii* genome encodes several proteins containing Anks (Seshadri et al. 2003). This is in agreement with the expectation that the pathogen modulates the host biochemical machinery for its own purposes to assure its survival and replication. The similarity of the *Coxiella* and *Legionella* Icm/Dot systems enabled researchers to use *L. pneumophilla* as a surrogate host to measure translocation of *C. burnetii* Ank-containing proteins into host cells. Eleven *C. burnetii* proteins containing ankyrin repeats have been identified as T4SS substrates (Pan et al. 2008; Voth et al. 2009). Recently, a large-scale screening of *C. burnetii* type IV secretion substrates using a fluorescence-based β-lactamase translocation assay as well as the calmodulin-dependent adenylate cyclase assay in the surrogate *L. pneumophilla* system has been performed (Chen et al. 2010). Many candidate secretion substrates using a two-hybrid system and bioinformatic approaches were suggested. Twenty-three protein substrates identified in this study contained eukaryotic-like domains implicated in protein-protein interactions. These include coiled-coil motifs, tetratricopeptide repeats, and eukaryotic-like serine/threonine protein kinase domains (Chen et al. 2010). Moreover, three proteins (CBU0814, CBU1217 and CBUA0014) that possess an F-box motif and one protein (CBU2078) containing a filamentation induced by cyclic adenosine monophosphate (Fic) domain have also been found, suggesting that *C. burnetii* actively modulates host ubiquitination pathways and is capable of disrupting host cell processes using Fic domain-regulated AMPylation as posttranslational regulation of protein function in host cells during infection (Roy and Mukherjee 2009; Yarbrough et al. 2009). Interestingly, only 18 protein substrates were found fully conserved among *C. burnetii* strains K, G, and Dugway (Chen et al. 2010) in contrast to the highly conserved *L. pneumophilla* (Cazalet et al. 2008).

Most recently, the T4SS effector proteins of *C. burnetii* NM II secreted into the cytosol of infected Vero cells have been investigated by proteomic approaches (Samoilis et al. 2010). Proteins from purified cytoplasmic fractions were separated into 19 bands after 1-DE and the tryptic peptides obtained were analyzed with a LC-ESI-MS/MS system. The resulting data were searched against both the nonredundant NCBI protein database over all entries and the NCBI database for the proteobacteria protein species. A wide bioinformatic analysis was performed to identify eukaryotic-like motifs, repeats and domains within the protein sequences of all identified *C. burnetii* proteins with respect to their possible involvement in pathogenesis and/or survival mechanisms of the pathogen. Fifty proteins were identified that have potential as effectors secreted by *C. burnetii* into the host cell cytoplasm. The *in silico* analysis of proteins revealed that more than 50% of them possessed a characteristic motif, domain or profile of effector proteins or was somehow involved in pathogenesis or survival mechanisms. The second interesting finding was the fact that over 55% of the identified *C. burnetii* proteins were basic in nature, having a theoretical pI over 8. This is in agreement with the highly basic theoretical proteome of *C. burnetii* suggested from its genome sequence (Seshadri et al. 2003). It is believed that the basic nature of the *C. burnetii* proteome is required by the pathogen to counterbalance the phagolysosomal acidic environment. So far, only one-fifth of the identified *C. burnetii* proteins in NM I and NM II have had basic pI (Skultety et al. 2005; Samoilis et al. 2007). However, this is likely due to the fact that the techniques used for protein separations were not favorable for isolation of those proteins having a basic pI. In line with this assumption, we have recently detected 15 additional proteins with an average deduced pI value of 9.5: CBU1403, CBU1829, CBU1814, CBU1744, CBU1559, CBU1425, CBU1372, CBU1221, CBU1099, CBU1094, CBU0980, CBU0611, CBU0370, CBU0311, and CBU0085 (Skultety et al. 2011). It has been suggested (Samoilis et al. 2007) that proteins secreted into the cytoplasm of the host cell were primarily in direct contact with the acidic milieu of the phagolysosome and this might be a possible explanation for their strongly basic properties. On the other hand, the presence of effective detoxification mechanisms likely helps it to maintain its cytoplasmic pH in the physiological range. It should be mentioned that despite identifying an increasing number of potential effector molecules secreted by *C. burnetii*, information about their function and importance in pathogenesis of the bacterium requires further investigation.

6.4 Proteomics in Detection of *C. burnetii* and Q Fever Diagnosis

C. burnetii causes Q fever, a disease which ranges mostly from a flu-like illness and self-recovering mild pneumonia to severe meningoencephalitis, myocarditis or endocarditis. The pathogen is usually transmitted from animals to humans by airborne particles derived from contaminated feces and birth products. Surprisingly, only ~40% of infected humans develop clinical symptoms of the disease (Maurin and Raoult 1999). From these, an estimated 1–3% will evolve a chronic form of the disease, which can be life threatening. The most serious form of chronic Q fever is endocarditis (Raoult et al. 1986; Marrie and Raoult 1997). Various serological methods are currently used for diagnosis of Q fever but ambiguous results have been obtained in several cases (Marrie and Raoult 1997; Arricau-Bouvery and Rodolakis 2005; Slaba et al. 2005). Diagnostic methods based on polymerase chain reaction have improved detection of the infectious agent and diagnosing the disease (Arricau-Bouvery and Rodolakis 2005; Brouqui and Raoult 2006; Svraka et al. 2006; Huijsmans et al. 2011). In addition, a monoclonal antibody (mAb) that specifically recognizes Vir residues in the virulent form of *C. burnetii* has been developed (Palkovicova et al. 2009) which offers the possibility of rapid, sensitive and specific detection of the bacterium and the ability to distinguish between *C. burnetii* isolates. Implementation of mass spectrometry/proteomics may also soon contribute to progress in *C. burnetii*/Q fever diagnostic applications. The current MS techniques applied in the field are shown in Fig. 6.2. Nevertheless, both rapid and specific identification of the pathogen and discovery of antigens for early and specific serological diagnosis of the infection in environmental and clinical samples, together with the identification of protein candidates suitable for development of an efficacious and safe vaccine remain the challenge within the field.

6.4.1 Proteomics and Biomarker Identification in C. burnetii/Q Fever Diagnostic Applications

The rapid and specific identification of infectious bacteria in samples collected from patients with clinical disease is an ongoing pursuit of basic and clinical laboratories throughout the world. Studies have shown that for gravely ill patients with septicemia, time is of the absolute essence (Emonet et al. 2010). In the past 20 years, molecular biology techniques have replaced, or been used in combination with, traditional growth and phenotypic differentiation methodologies (Seng et al. 2009, 2010) in an effort to decrease the time between sample collection and laboratory confirmation. More recently, the use of MALDI-TOF MS techniques has become a viable alternate to traditional, and other molecular based, diagnostic techniques (Seng et al. 2009; Leggieri et al. 2010). Numerous studies have shown that MALDI-TOF analysis can accurately and rapidly identify bacterial pathogens from small colonies taken directly from growth plates (Seng et al. 2009; Cherkaoui et al. 2010; Emonet et al. 2010; Leggieri et al. 2010). These methods have been extended to include the identification of bacteria from blood cultures with a great deal of success (Drancourt 2010). The promise and growing use of proteomics based technology has led to the commercial development of MS instrumentation (Bruker Daltronics and Shimadzu) and data analysis software (MALDI BioTyper and Launchpad) and databases (MALDI Biotyper, SARAMIS, and an Andromas product) specifically designed for diagnostic application. These technologies are currently employed by a number of high throughput diagnostic laboratories in Europe (Emonet et al. 2010; Seng et al. 2010). Studies have shown that using this proteomics based technology significantly reduces the time and cost required to identify many pathogens from clinical samples (Seng et al. 2009; Leggieri et al. 2010).

The detection and identification of bacteria using MS techniques requires distinct spectra (or fingerprints) that can identify an organism at the genus, species, and/or strain level. Proteomic studies to identify dominant and consistent *C. burnetii* peptides using MALDI-TOF MS have revealed peptides that may be used in genus and even biotype differentiation (Shaw et al. 2004; Pierce et al. 2007; Skultety et al. 2007). The first such study used purified whole cells of *C. burnetii* NM I which had been inactivated by γ-irradiation, mixed with matrix solution either α-cyano-4-hydroxycinnamic acid (CHCA), for the 1–6 kDa range, or 3,5-dimethoxy-4-hydroxycinnamic acid (sinapinic acid, SA) for the 3–25 kDa range, mixed in 50% acitonitrile (ACN) and Milli-Q grade water containing 0.1% trifluoroacetic acid (TFA), and placed directly on stainless steel MALDI-TOF plates for analyses (Shaw et al. 2004). These studies revealed 24 biomarkers that could be consistently detected when analyzing the 1–25 kDa size range. These statistically significant biomarkers ranged from 1,829 to17,319 Da in size. Interestingly, bioinformatic efforts to identify these dominant biomarkers revealed ScvA as one of the biomarkers (Shaw et al. 2004). As mentioned earlier, ScvA is a small DNA binding peptide solely expressed in the environmentally stable SCV form of *C. burnetii* (Heinzen et al. 1996), making it an excellent biomarker for *C. burnetii* detection. A subsequent study sought to use MALDI-TOF MS analysis to determine whether biomarkers unique to virulent strains suggested to cause acute (NM I, Henzerling, and Ohio) or chronic (KAV, PAV) Q fever might express biomarkers unique from each other or avirulent (Australian QD, M44) strains (Pierce et al. 2007). Using samples prepared as stated previously and the CHCA matrix mixture, spectra in the 1–6 kDa range were analyzed for these seven strains. Biomarkers at 1,829, 2,951, and 3,612 (ScvA) Da were detected in all seven of these strains. This suggests that biomarkers consistent to all *C. burnetii* isolates might be identified. In addition, more in-depth analysis of the spectral data using Partial Least Squares-Discriminant Analysis (PLSDA) indicated that MALDI-TOF MS analysis of whole purified organism could be used to differentiate between phase I and phase II organisms as well as the strains suggested to cause acute and chronic form of the disease (Pierce et al. 2007). Another study also used MALDI-TOF MS analysis against multiple *C. burnetii* strains in an effort to define biomarkers (Skultety et al. 2007). The "acute" NM I and BUD (isolated from an acute Q fever patient in Budulov, Slovakia, 1969) strains along with the "chronic" Priscilla strain revealed numerous biomarkers shared between the "acute" strains, and between BUD and Priscilla in the 3–18 kDa range as well as numerous biomarkers unique to each strain (Skultety et al. 2007). Efforts to determine the identity of the biomarkers using Tag-Ident software and the *C. burnetii* NM I genome sequence (Seshadri et al. 2003) revealed primarily "hypothetical" peptides and did not identify the ScvA as one of the ions detected, yet clearly demonstrated the usefulness of MALDI-TOF MS in differentiating strains of *C. burnetii*.

Many of the biomarkers identified in the separate studies were unique to the study; however, some of the biomarkers identified were consistent and suggest that MALDI-TOF MS analysis represents a powerful tool to identify *C. burnetii* strains. Some of the differences are likely due to differences in sample preparation. While the first two studies used whole organism inactivated by γ-irradiation, mixed with matrix and loaded directly on the MALDI plate (Shaw et al. 2004; Pierce et al. 2007), the third study used phenol to inactivate the pathogen and 70% ACN-water containing 0.5% TFA to extract the bacteria (Skultety et al. 2007). This was followed by centrifugation of the extract, which after concentration to a small volume was prepared for MALDI-TOF in a matrix of SA (10 mg/ml) in aqueous 30% ACN with 0.5% TFA. In addition, mass spectrometers from different manufacturers were employed. Whether these differences had an effect on the ions detected remains to be seen.

MALDI-TOF MS analysis is a powerful means by which to detect and define bacteria at the genus, species, and even strain level. While still understudied, the use of MALDI-TOF MS analysis has demonstrated that unique biomarkers for *C. burnetii* can be detected and used to differentiate strains, phase variants, and possibly even disease type (acute *vs.* chronic Q fever). The high pI of the *C. burnetii* proteome lends itself to MALDI-TOF MS analysis (Arnold and Reilly 1998; Amiri-Eliasi and Fenselau 2001) as basic proteins are thought to be better characterized by this methodology. In addition, the genomes of the *C. burnetii* K, G, and Dugway strains have been sequenced and compared to the reference NM I strain (Beare et al. 2009). These sequences and annotations were not available at the time the MALDI-TOF analysis studies were performed.

Specific *C. burnetii* proteins were also identified using an ESI-MS/MS approach (Hernychova et al. 2008). Thus, proteins of the ACN extract of the bacterium were cleaved with trypsin, and the resulting peptides were separated by reversed-phase chromatography and analyzed by MS/MS. The experiment was repeated with 10 freshly prepared ACN extracts. Based on the MS/MS data, 20 distinct proteins were identified in more than three experiments. The identified proteins were mainly alkaline (75%), and almost all of them (80%) were of low (<20 kDa) molecular mass. Each protein was assigned to a functional category by comparing its amino acid sequences (obtained from NCBI) to a database of protein sequences of fully sequenced genomes and transferring the broad-level Clusters of Orthologous Groups annotations to the *C. burnetii* proteins. The proteins most likely represent the highly expressed proteins of the *C. burnetii* outer membrane that are substantially enriched for those involved in posttranslational modification (14%), cell envelope biogenesis (14%), and DNA replication, recombination, and repair (19%). Moreover, it was shown that the identified proteins represented five basic groups: (1) membrane proteins, (2) ribosomal proteins, (3) DNA-binding proteins, (4) heat shock proteins, and (5) hypothetical proteins.

In conclusion, using the spectral data generated from these studies in conjunction with the current genomic sequence data might well reveal additional identities to conserved and unique *C. burnetii* biomarkers and aid in creating algorithms and/or databases that could develop into diagnostic and detection tools for this unusual pathogen.

6.4.2 *Immunoproteomics and Protein Microarrays – Platforms for the Identification of Protein Candidates Applicable in Serological Diagnosis of Q Fever*

Recently, 2-DE immunoproteomic and protein microarray approaches have been employed for wide-scale screening of immunoreactive proteins which might be useful for Q fever serodiagnosis (Beare et al. 2008; Sekeyova et al. 2009; Vigil et al. 2010; Deringer et al. 2011). Human sera from patients with acute and chronic Q fever, or sera obtained from guinea pigs immunized with whole phase I cells of *C. burnetii*, were used in these studies. The whole cell extracts from phase I or phase II *C. burnetii* were separated by 2-DE and the resolved proteins analyzed by Western blot. Blotting membranes were probed with human or guinea pig sera and the immunoreactive proteins detected were submitted to MALDI or LC-MS/MS analyses. Twenty-nine proteins were successfully identified with both acute and chronic Q fever sera (Sekeyova et al. 2009). Moreover, two proteins, OmpH (CBU0612) and arginine repressor (CBU0480), were exclusively recognized by chronic sera suggesting they may be suitable antigenic candidates for developing Q fever endocarditis serodiagnostic assays. Nine immunoreactive proteins were identified with guinea pig immune sera employing 2-DE and MS/MS (Deringer et al. 2011). Gycerol kinase (CBU0932), malate dehydrogenase (CBU1241), succinyl-CoA synthetase alpha (CBU1396), ribonuclease PH (CBU0299), arabinose 5-phosphate isomerase (CBU0750), M20A family peptidase (CBU0103), and elongation factor EF-Ts (CBU1385) were the most immunoreactive, while glutamine dependent NAD^+ synthase (CBU0850) and cell division protein FtsA (CBU0140) displayed weak immunoreactivity.

Gene-specific transcriptionally active PCR (TAP) fragments of *C. burnetii* have been used for developing the first generation of *C. burnetii* protein microarray (Beare et al. 2008). In this work, 1988 ORFs (97.2% of *C. burnetii* coding sequences) were amplified into TAP fragments. Full-length recombinant proteins were synthesized from these by *in vitro* transcription-translation (IVTT) and 72.9% of the predicted proteome was ultimately used in the array. The protein arrays were probed with a small collection of Q fever patient sera (n = 5) and 44 seroreactive ORFs that were subsequently cloned into expression plasmids were identified.

The proteins resulting from IVTT reactions, using both plasmid and TAP templates, were printed onto small arrays enabling direct comparison of TAP and plasmid-based arrays. Both types of arrays displayed similar results with regard to antigen detection specificity. However, the sensitivity of detection was significantly higher when the plasmid-based IVTT reactions were used. In another study, an expression plasmid-based protein microarray was developed (Vigil et al. 2010) which covered 84% of the *C. burnetii* proteome. The system was tested with a large collection of patient and control sera (n=40 and 20, respectively) and identified 13 potential serodiagnostic antigens. Additionally, the seroreactive antigens were validated for their diagnostic potential using an immunostrip platform based on antigen-antibody binding similar to that of Western blot detection.

Despite a wide spectrum of immunoreactive proteins identified by anti-*C. burnetii* sera, cross-reactivity of several *C. burnetii* proteins to antigen determinants (seroreactive antigens) in other bacterial species was observed. However, some of the identified proteins do not cross-react and might serve as molecular targets in further studies to develop highly specific tests for Q fever serodiagnosis. In the following text, additional information on selected *C. burnetii* immunoreactive proteins is presented.

The CBU0092 ORF codes for a periplasmic component (YbgF) of the Tol-Pal system that is involved in outer membrane integrity. Deletion mutant studies have demonstrated (Walburger et al. 2002) that YbgF interacts with the central domain of the inner membrane protein TolA and contributes to the formation of a *trans*-envelope complex that brings the inner and outer membranes in close proximity. The YbgF homolog of *C. burnetii* was found strongly seroreactive in both 2-DE-based proteomic and protein microarray approaches (Sekeyova et al. 2009; Vigil et al. 2010; Deringer et al. 2011), yet shares only 34% homology with other bacterial YbgF proteins.

Elongation factor EF-Tu is primarily involved in polypeptide lengthening during bacterial protein biosynthesis but may function as a cytoskeleton component in bacterial cell shape maintenance (Defeu Soufo et al. 2010). Membrane association of EF-Tu has been observed in *Mycoplasma pneumonia* and *Francisella tularensis* (Balasubramanian et al. 2008; Barel et al. 2008). The elongation factors EF-Tu (CBU0236) and EF-G (CBU0235) as well as dihydrolipoamide succinyltransferase SucB (CBU1398) and ribosomal protein L7/L12 (CBU0229) of *C. burnetii* were strongly reactive with immune sera (Coleman et al. 2007; Beare et al. 2008; Sekeyova et al. 2009; Vigil et al. 2010; Deringer et al. 2011) but their potential utility in Q fever diagnosis is notably reduced due to their highly conserved nature among other bacterial species. Cross-reactivity of mAb was observed with EF-Tu of *C. burnetii* and its recombinant homolog from *C. trachomatis* (Seshadri et al. 1999). Similarly, antisera against EF-G and ribosomal L7/L12 proteins of *E. coli* cross-reacted with *C. burnetii* homologs (Baca 1978) while antibodies against *C. burnetii* cross-reacted with recombinant SucB from *Bartonella vinsonii* subsp. *berkhoffii* (Gilmore et al. 2003).

The DnaK (CBU1290) ORF encodes a prokaryotic ortholog of Hsp70 that belongs to a highly conserved family of ubiquitous molecular chaperones which are involved in myriad of biological processes. These include protein folding, transmembrane

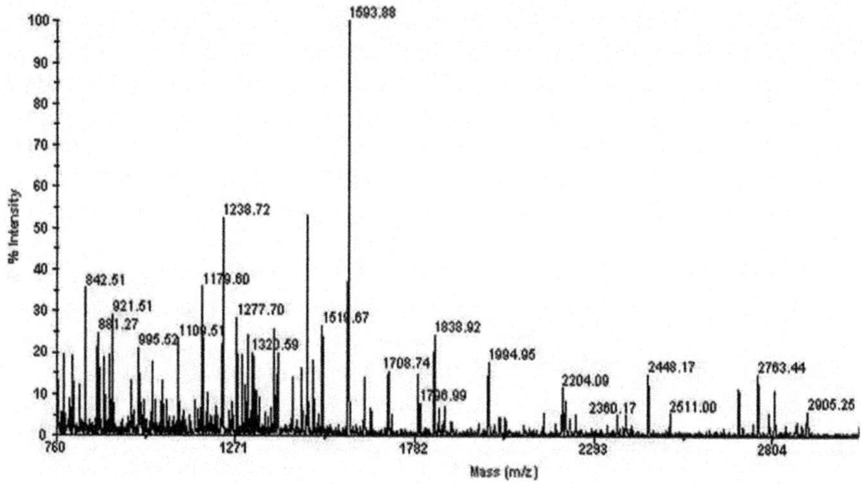

Fig. 6.3 Positive-ion MALDI-TOF mass spectrum of GroEL, heat shock protein B (GroEL/HtpB), 60 kDa, from *C. burnetii* strain NM I (RSA 493)

transport of proteins, protein degradation, protection of cells from thermal and oxidative stress, and others. The association of DnaK with the bacterial cell membrane and its ability to interact with the hosts plasminogen/plasmin system is used by several pathogens and commensals for colonization of the human gastrointestinal tract (Parkkinen and Korhonen 1989; Schaumburg et al. 2004; Lähteenmäki et al. 2005; Hurmalainen et al. 2007; Knaust et al. 2007; Xolalpa et al. 2007). Interestingly, DnaK has been found to be strongly reactive with both human and guinea pig Q fever sera in 2-DE based studies (Sekeyova et al. 2009; Deringer et al. 2011).

The role of thioredoxin peroxidase (CBU1706) in bacterial defense against oxidants has been mentioned above. The protein was found seroreactive in both 2-DE based studies (Sekeyova et al. 2009; Deringer et al. 2011). Another seroreactive protein, GroEL (CBU1718), is a 60 kDa product of the *htpB* gene that belongs to an evolutionarily conserved group of molecular chaperonins. These proteins are involved in protein folding and are present in all bacterial species as well as eukaryotic cell organelles such as mitochondria. The 60 kDa chaperonin family constitutes the major antigenic proteins within numerous pathogens and appears to participate in their pathogenesis. For example, the surface-associated HtpB/Hsp60 of *L. pneumophila* promotes attachment and invasion in HeLa cells (Garduno et al. 1998) and mediates attachment to gastric epithelial cells in case of *Helicobacter pylori* (Hoffman and Garduno 1999). Moreover, HtpB was shown to be involved in intracellular trafficking of *L. pneumophila* in CHO cells (Chong et al. 2009). Interestingly, GroEL/HtpB was identified in both cell phases of *C. burnetii* (Skultety et al. 2005; Samoilis et al. 2007) and its seroreactivity is well documented (Coleman et al. 2007; Sekeyova et al. 2009). The MALDI-TOF MS spectrum of GroEL identification is shown in Fig. 6.3. However, only a weak reactivity of GroEL was observed with guinea pig immune sera and no reactivity was observed with acute Q fever patient sera although reactivity has been reported in other studies (Williams et al. 1990; Deringer et al. 2011). Recombinant *C. burnetii* HtpB and HtpB-P1 fusion proteins (P1, CBU0311) have also been evaluated as potential candidates for developing a subunit vaccine against Q fever (Li et al. 2005; Fernandes et al. 2009). Purified native outer membrane porin P1 from *C. burnetii* was an effective inducer of a protective response against Q fever in mice (Williams et al. 1990). However in another study, recombinant P1 was unable to provide a protective effect (Zhang and Samuel 2003).

The *C. burnetii* Com1 (CBU1910) ORF encodes a 27 kDa surface protein (Muller et al. 1987; Hendrix et al. 1990, 1993) frequently detected as a reactive antigen within immune sera (Sekeyova et al. 2009, 2010; Vigil et al. 2010; Deringer et al. 2011). Recombinant Com1 has been suggested as a sensitive and specific means to detect antibodies against *C. burnetii* in patient sera (Zhang et al. 1998). However, the recombinant protein encoded by ORF CBU0937 (unknown biological function) has been shown to be more suitable for diagnosing Q fever in an ELISA assay setting (Sekeyova et al. 2010). In *C. burnetii* lysates, native CBU0937 reacted with immune sera in both 2-DE based proteomic studies (Sekeyova et al. 2009; Deringer et al. 2011). A cytosol aminopeptidase (CBU0572), ATP synthase alpha subunit (CBU1943), outer membrane OmpA-like protein (CBU0307), an uncharacterized protein (CBU1260), as well as other proteins also display immunoreactivity. Further studies of their serodiagnostic potential will prove informative.

Despite great efforts devoted to developing sensitive and reliable tests for *C. burnetii*/Q fever diagnosis which does not rely on the large-scale production of this hazardous pathogen, continued research and development is needed to develop diagnostic applications that yield unambiguous results. Likewise, there is a need for an efficient, subcellular, recombinant or LPS – conjugate vaccine against Q fever. There have been several attempts to produce a safe and effective vaccine. However, only partial success has been achieved. Four main vaccines have been used in humans: a live attenuated strain, M-44 (Freylikhman et al. 2003), a trichloroacetic acid extract of NM I strain (Kazar et al. 1982), a chloroform-methanol residue of phase I Herzenling strain (Fries et al. 1993; Waag et al. 2002), and formalin-inactivated cells of phase I Herzenling strain named Q-Vax (Marmion et al. 1990; Ackland et al. 1994). Although these vaccines showed high efficacy in establishing protection against infection, the application of them has been problematic. Vaccination often causes adverse local, or occasionally, systemic reactions in people previously sensitized to the pathogen. Therefore, skin tests, serological tests, and/or *in vitro* lymphocyte proliferation assays are prerequisites prior to vaccination. This makes vaccination time-consuming and costly for large-scale applications (Oyston and Davies 2011). The most thoroughly tested vaccine for humans is Q-Vax (Marmion et al. 1990; Ackland et al. 1994). This cellular vaccine is only produced and licensed for use in Australia. However, it induces side-effects in a significant portion of recipients (Oyston and Davies 2011).

Native and recombinant proteins have been reported (Williams et al. 1990; Zhang et al. 1994; Varghees et al. 2002; Zhang and Samuel 2003; Li et al. 2005) as candidates for development of a new-generation of Q-fever subunit vaccines. Recombinant protein subunit vaccines where the proteins were expressed in *E. coli* have shown limited efficacy. However, immunization with the same proteins in their native form following isolation directly from *C. burnetii* does result in reduced microbial colonization in non-lethal animal models. Two reasons for this discrepancy have been suggested: either post-translational modification in *C. burnetii* results in different antigenicity when compared to proteins expression in *E. coli*, or the proteins were contaminated with *C. burnetii* LPS (Shannon and Heinzen 2009) increasing their immune stimulation. Thus, wide scale screening and continued evaluation of potential *C. burnetii* antigens along with a comprehensive evaluation of the humoral immune response are crucial to the development of a safe and efficacious vaccine as well as improved diagnostic assays for rapid and sensitive detection of the Q fever pathogen.

6.5 Summary

In the recent years, there has been an immense improvement in our understanding of both the biology and pathogenicity of *C. burnetii*. These advances have been aided due to the application of high-throughput proteomic approaches. The potential of proteomics lies in the fact that it can provide a more global and detailed view of *C. burnetii* physiology allowing the elucidation of functions at the molecular level in ways that are not predictable from genomic sequence information alone.

Acknowledgement This research was supported in part by the grant 2/0026/12 from the Scientific Grant Agency of Ministry of Education of Slovak Republic and the Slovak Academy of Sciences.

References

Ackland JR, Worswick DA, Marmion BP (1994) Vaccine prophylaxis of Q fever – a follow-up study of the efficacy of Q-Vax (CSL) 1985–1990. Med J Aust 160:704–708

Akporiaye ET, Baca OG (1983) Superoxide anion production and superoxide dismutase and catalase activities in *Coxiella burnetii*. J Bacteriol 154:520–523

Amano K, Williams JC (1984) Chemical and immunological characterization of lipopolysaccharides from phase I and phase II *Coxiella burnetii*. J Bacteriol 160:994–1002

Amano K, Williams JC, McCaul TF et al (1984) Biochemical and immunological properties of *Coxiella burnetii* cell wall and peptidoglycan-protein complex fractions. J Bacteriol 160:982–988

Amiri-Eliasi B, Fenselau C (2001) Characterization of protein biomarkers desorbed by MALDI from whole fungal cells. Anal Chem 73:5228–5231

Arnold RJ, Reilly JP (1998) Fingerprint matching of *E. coli* strains with matrix-assisted laser desorption/ionization time-of-flight mass spectrometry of whole cells using a modified correlation approach. Rapid Commun Mass Spectrom 12:630–636

Arricau-Bouvery N, Rodolakis A (2005) Is Q fever an emerging or re-emerging zoonosis? Vet Res 36:327–349

Baca OG (1978) Comparison of ribosomes from *Coxiella burnetii* and *Escherichia coli* by gel electrophoresis, protein synthesis, and immunological techniques. J Bacteriol 136:429–432

Baca OG, Paretsky D (1983) Q fever and *Coxiella burnetii*: a model for host-parasite interactions. Microbiol Rev 47:127–149

Baca OG, Roman MJ, Glew RH et al (1993) Acid phosphatase activity in *Coxiella burnetii*: a possible virulence factor. Infect Immun 61:4232–4239

Balasubramanian S, Kannan TR, Baseman JB (2008) The surface-exposed carboxyl region of *Mycoplasma pneumoniae* elongation factor Tu interacts with fibronectin. Infect Immun 76:3116–3123

Barel M, Hovanessian AG, Meibom K et al (2008) A novel receptor – ligand pathway for entry of *Francisella tularensis* in monocyte-like THP-1 cells: interaction between surface nucleolin and bacterial elongation factor Tu. BMC Microbiol 8:145

Bate N, Butler AR, Smith IP et al (2002) The mycarose-biosynthetic genes of *Streptomyces fradiae*, producer of tylosin. Microbiology 146:139–146

Beare PA, Chen C, Bouman T et al (2008) Candidate antigens for Q fever serodiagnosis revealed by immunoscreening of a *Coxiella burnetii* protein microarray. Clin Vaccine Immunol 15:1771–1779

Beare PA, Unsworth N, Andoh M et al (2009) Comparative genomics reveal extensive transposon-mediated genomic plasticity and diversity among potential effector proteins within the genus *Coxiella*. Infect Immun 77:642–656

Bernadac A, Gavioli M, Lazzaroni JC et al (1998) *Escherichia coli* tol-pal mutants form outer membrane vesicles. J Bacteriol 180:4872–4878

Brouqui P, Raoult D (2006) New insight into the diagnosis of fastidious bacterial endocarditis. FEMS Immunol Med Microbiol 47:1–13

Caldon CE, March PE (2003) Function of the universally conserved bacterial GTPases. Curr Opin Microbiol 6:135–139

Cascales E, Gavioli M, Sturgis JN et al (2000) Proton motive force drives the interaction of the inner membrane TolA and outer membrane pal proteins in *Escherichia coli*. Mol Microbiol 38:904–915

Cascales E, Bernadac A, Gavioli M (2002) Pal lipoprotein of *Escherichia coli* plays a major role in outer membrane integrity. J Bacteriol 184:754–759

Cazalet C, Jarraud S, Ghavi-Helm Y et al (2008) Multigenome analysis identifies a worldwide distributed epidemic *Legionella pneumophila* clone that emerged within a highly diverse species. Genome Res 18:431–441

Chen C, Banga S, Mertens K et al (2010) Large-scale identification and translocation of type IV secretion substrates by *Coxiella burnetii*. Proc Natl Acad Sci USA 107:21755–21760

Cherkaoui A, Hibbs J, Emonet S et al (2010) Comparison of two matrix-assisted laser desorption ionization-time of flight mass spectrometry methods with conventional phenotypic identification for routine identification of bacteria to the species level. J Clin Microbiol 48: 1169–1175

Chong A, Lima CA, Allan DS et al (2009) The purified and recombinant *Legionella pneumophila* chaperonin alters mitochondrial trafficking and microfilament organization. Infect Immun 77: 4724–4739

Chowdhury R, Sahu GK, Das J (1996) Stress response in pathogenic bacteria. J Biosci 21:149–160

Clavel T, Germon P, Vianney A et al (1998) TolB protein of *Escherichia coli* K-12 interacts with the outer membrane peptidoglycan-associated proteins Pal, Lpp and OmpA. Mol Microbiol 29:359–367

Coleman SA, Fischer ER, Howe D et al (2004) Temporal analysis of *Coxiella burnetii* morphological differentiation. J Bacteriol 186:7344–7352

Coleman SA, Fischer ER, Cockrell DC et al (2007) Proteome and antigen profiling of *Coxiella burnetii* developmental forms. Infect Immun 75:290–298

Defeu Soufo HJ, Reimold C, Linne U et al (2010) Bacterial translation elongation factor EF-Tu interacts and colocalizes with actin-like MreB protein. Proc Natl Acad Sci USA 107:3163–3168 deLivron MA, Robinson VL (2008) *Salmonella enterica* serovar typhimurium BipA exhibits two distinct ribosome binding modes. J Bacteriol 190:5944–5952

Deringer JR, Chen C, Samuel JE et al (2011) Immunoreactive *Coxiella burnetii* Nine Mile proteins separated by 2D electrophoresis and identified by tandem mass spectrometry. Microbiology 157:526–542

Drancourt M (2010) Detection of microorganisms in blood specimens using matrix-assisted laser desorption ionization time-of-flight mass spectrometry: a review. Clin Microbiol Infect 16:1620–1625

Eidels L, Osborn MJ (1974) Phosphoheptose isomerase, first enzyme in the biosynthesis of aldoheptose in *Salmonella typhimurium*. J Biol Chem 249:5642–5648

Ejim LJ, D'Costa VM, Elowe NH et al (2004) Cystathionine beta-lyase is important for virulence of *Salmonella enterica* serovar Typhimurium. Infect Immun 72:3310–3314

Ejim LJ, Blanchard JE, Koteva KP et al (2007) Inhibitors of bacterial cystathionine beta-lyase: leads for new antimicrobial agents and probes of enzyme structure and function. J Med Chem 50:755–764

Emonet S, Shah HN, Cherkaoui A et al (2010) Application and use of various mass spectrometry methods in clinical microbiology. Clin Microbiol Infect 16:1604–1613

Feldman M, Zusman T, Hagag S et al (2005) Coevolution between nonhomologous but functionally similar proteins and their conserved partners in the *Legionella* pathogenesis system. Proc Natl Acad Sci USA 102:12206–12211

Fernandes I, Rousset E, Dufour P et al (2009) Evaluation of the recombinant Heat shock protein B (HspB) of *Coxiella burnetii* as a potential antigen for immunodiagnostic of Q fever in goats. Vet Microbiol 134:300–304

Fouces R, Mellado E, Diez B et al (2002) The tylosin biosynthetic cluster from *Streptomyces fradiae*: genetic organization of the left region. Microbiology 145:855–868

Freylikhman O, Tokarevich N, Suvorov A et al (2003) *Coxiella burnetti* persistence in three generations of mice after application of live attenuated human M-44 vaccine against Q fever. Ann NY Acad Sci 990:496–499

Fries LF, Waag DM, Williams JC (1993) Safety and immunogenicity in human volunteers of a chloroform-methanol residue vaccine for Q fever. Infect Immun 61:1251–1258

Ftacek P, Skultety L, Toman R (2000) Phase variation of *Coxiella burnetii* strain Priscilla: influence of this phenomenon on biochemical features of its lipopolysaccharide. J Endotoxin Res 6:369–376

Gajdosova E, Kovacova E, Toman R et al (1994) Immunogenicity of *Coxiella burnetii* whole cells and their outer membrane components. Acta Virol 38:339–344

Garduno RA, Garduno E, Hoffman PS (1998) Surface-associated Hsp60 chaperonin of *Legionella pneumophila* mediates invasion in a HeLa cell model. Infect Immun 66:4602–4610

Gaspar JA, Thomas JA, Marolda CL (2000) Surface expression of O-specific lipopolysaccharide in *Escherichia coli* requires the function of the TolA protein. Mol Microbiol 38:262–275

Gerding MA, Ogata Y, Pecora ND et al (2007) The trans-envelope Tol-Pal complex is part of the cell division machinery and required for proper outer-membrane invagination during cell constriction in *E. coli*. Mol Microbiol 63:1008–1025

Gilmore RD Jr, Carpio AM, Kosoy MY (2003) Molecular characterization of the sucB gene encoding the immunogenic dihydrolipoamide succinyltransferase protein of *Bartonella vinsonii* subsp. *berkhoffii* and *Bartonella quintana*. Infect Immun 71:4818–4822

Hackstadt T, Williams JC (1983) pH dependence of the *Coxiella burnetii* glutamate transport system. J Bacteriol 154:598–603

Hackstadt T, Peacock MG, Hitchcock PJ et al (1985) Lipopolysaccharide variation in *Coxiella burnetii:* intrastrain heterogeneity in structure and antigenicity. Infect Immun 48:359–365

Hansen AM, Gu Y, Li M et al (2005a) Structural basis for the function of stringent starvation protein A as a transcription factor. J Biol Chem 280:17380–17391

Hansen AM, Qiu Y, Yeh N et al (2005b) SspA is required for acid resistance in stationary phase by downregulation of H-NS in *Escherichia coli*. Mol Microbiol 56:719–734

Heinzen RA, Hackstadt T (1996) A developmental stage-specific histone H1 homolog of *Coxiella burnetii*. J Bacteriol 178:5049–5052

Heinzen RA, Howe D, Mallavia LP et al (1996) Developmentally regulated synthesis of an unusually small, basic peptide by *Coxiella burnetii*. Mol Microbiol 22:9–19
Heinzen RA, Hackstadt T, Samuel JE (1999) Developmental biology of *Coxiella burnettii*. Trends Microbiol 7:149–154
Helbig JH, Konig B, Knospe H et al (2003) The PPIase active site of *Legionella pneumophila* Mip protein is involved in the infection of eukaryotic host cells. Biol Chem 384:125–137
Hendrix LR, Samuel JE, Mallavia LP (1990) Identification and cloning of a 27-kDa *Coxiella burnetii* immunoreactive protein. Ann NY Acad Sci 590:534–540
Hendrix LR, Mallavia LP, Samuel JE (1993) Cloning and sequencing of *Coxiella burnetii* outer membrane protein gene com1. Infect Immun 61:470–477
Hernychova L, Toman R, Ciampor F et al (2008) Detection and identification of *Coxiella burnetii* based on the mass spectrometric analyses of the extracted proteins. Anal Chem 80:7097–7104
Hoffman PS, Garduno RA (1999) Surface-associated heat shock proteins of *Legionella pneumophila* and *Helicobacter pylori*: roles in pathogenesis and immunity. Infect Dis Obstet Gynecol 7:58–63
Hoover TA, Culp DW, Vodkin MH et al (2002) Chromosomal DNA deletions explain phenotypic characteristics of two antigenic variants, phase II and RSA 514 (Crazy), of the *Coxiella burnetii* Nine Mile strain. Infect Immun 70:6726–6733
Howe D, Heinzen RA (2008) Fractionation of the *Coxiella burnetii* parasitophorous vacuole. Method Mol Biol 445:389–406
Howe D, Mallavia LP (2000) *Coxiella burnetii* exhibits morphological change and delays phagolysosomal fusion after internalization by J774A.1 cells. Infect Immun 68:3815–3821
Huijsmans CJJ, Schellekens JJA, Wever PC et al (2011) Single-nucleotide-polymorphism genotyping of *Coxiella burnetii* during a Q fever outbreak in the Netherlands. Appl Environ Microbiol 77:2051–2057
Hurmalainen V, Edelman S, Antikainen J et al (2007) Extracellular proteins of *Lactobacillus crispatus* enhance activation of human plasminogen. Microbiology 153:1112–1122
Hussein A, Kovacova E, Toman R (2001) Isolation and evaluation of *Coxiella burnetii* O-polysaccharide antigen as immunodiagnostic reagent. Acta Virol 45:173–180
Juhas M, Crook DW, Hood DW (2008) Type IV secretion systems: tools of bacterial horizontal gene transfer and virulence. Cell Microbiol 10:2377–2386
Kazar J, Brezina R, Palanova A et al (1982) Immunogenicity and reactogenicity of a Q fever chemovaccine in persons professionally exposed to Q fever in Czechoslovakia. Bull World Health Organ 60:389–394
Knaust A, Weber MVR, Hammerschmidt S et al (2007) Cytosolic proteins contribute to surface plasminogen recruitment of *Neisseria meningitidis*. J Bacteriol 189:3246–3255
Kneidinger B, Marolda C, Graninger M et al (2002) Biosynthesis pathway of ADP-L-glycero-β-D-manno-heptose in *Escherichia coli*. J Bacteriol 184:363–369
Krauss H, Schiefer HG, Schmatz HD (1977) Ultrastructural investigations on surface structures involved in *Coxiella burnetii* phase variation. Infect Immun 15:890–896
Kvint K, Nachin L, Diez A (2003) The bacterial universal stress protein: function and regulation. Curr Opin Microbiol 6:140–145
Lähteenmäki K, Edelman S, Korhonen TK (2005) Bacterial metastasis: the host plasminogen system in bacterial invasion. Trends Microbiol 13:79–85
Lazzaroni JC, Germon P, Ray MC et al (1999) The Tol proteins of *Escherichia coli* and their involvement in the uptake of biomolecules and outer membrane stability. FEMS Microbiol Lett 177:191–197
Leggieri N, Rida A, Francois P et al (2010) Molecular diagnosis of bloodstream infections: planning to (physically) reach the bedside. Curr Opin Infect Dis 23:311–319
Li Q, Niu D, Wen B et al (2005) Protective immunity against Q fever induced with a recombinant P1 antigen fused with HspB of *Coxiella burnetii*. Ann NY Acad Sci 1063:130–142
Lloubes R, Cascales E, Walburger A et al (2001) The Tol-Pal proteins of the *Escherichia coli* cell envelope: an energized system required for outer membrane integrity? Res Microbiol 152:523–529

Marmion BP, Ormsbee RA, Kyrkou M et al (1990) Vaccine prophylaxis of abattoir-associated Q fever: 8 years' experience in Australian abattoirs. Epidemiol Infect 104:275–287

Marrie TJ, Raoult D (1997) Q fever – a review and issues for the next century. Int J Antimicrob Agents 8:145–161

Maurin M, Raoult D (1999) Q Fever. Clin Microbiol Rev 12:518–553

Mayer H, Radziejewska-Lebrecht J, Schramek S (1988) Chemical and immunochemical studies on lipopolysaccharides of *Coxiella burnetii* phase I and phase II. Adv Exp Med Biol 228:577–591

McCaul TF, Williams JC (1981) Developmental cycle of *Coxiella burnetii*: structure and morphogenesis of vegetative and sporogenic differentiations. J Bacteriol 147:1063–1076

McCaul TF, Banerjee-Bhatnagar N, Williams JC (1991) Antigenic differences between *Coxiella burnetii* cells revealed by postembedding immunoelectron microscopy and immunoblotting. Infect Immun 59:3243–3253

Mo YY, Cianciotto NP, Mallavia LP (1995) Molecular cloning of a *Coxiella burnetii* gene encoding a macrophage infectivity potentiator (Mip) analogue. Microbiology 141:2861–2871

Muller HP, Schmeer N, Rantamäki L et al (1987) Isolation of a protein antigen from *Coxiella burnetii*. Zentralbl Bakteriol Mikrobiol Hyg A 265:277–289

Narasaki CT, Mertens K, Samuel JE (2011) Characterization of the GDP-D-mannose biosynthesis pathway in *Coxiella burnetii*: the initial steps for GDP-β-D-virenose biosynthesis. PLoS ONE 6(10):e25514

Nikaido H (1998) Multiple antibiotic resistance and efflux. Curr Opin Microbiol 1:516–523

Nikaido H (2001) Preventing drug access to targets: cell surface permeability barriers and active efflux in bacteria. Semin Cell Dev Biol 12:215–223

Oyston PCF, Davies C (2011) Q fever: the neglected biothreat agent. J Medical Microbiol 60:9–21

Palkovicova K, Ihnatko R, Vadovic P et al (2009) A monoclonal antibody specific for a unique biomarker, virenose, in a lipopolysaccharide of *Coxiella burnetii*. Clin Microbiol Infect 15(2):183–184

Pan X, Luhrmann A, Satoh A et al (2008) Ankyrin repeat proteins comprise a diverse family of bacterial type IV effectors. Science 320:1651–1654

Parkkinen J, Korhonen TK (1989) Binding of plasminogen to *Escherichia coli* adhesion proteins. FEBS Lett 250:437–440

Pierce CY, Barr JR, Woolfitt AR et al (2007) Strain and phase identification of the U.S. category B agent *Coxiella burnetii* by matrix assisted laser desorption/ionization time-of-flight mass spectrometry and multivariate pattern recognition. Anal Chim Acta 583:23–31

Poole K (2001) Multidrug resistance in Gram-negative bacteria. Curr Opin Microbiol 4:500–508

Raoult D, Piquet P, Gallais H et al (1986) *Coxiella burnetii* infection of a vascular prosthesis. N Engl J Med 315:1358–1359

Ray MC, Germon P, Vianney A et al (2000) Identification by genetic suppression of *Escherichia coli* TolB residues important for TolB-Pal interaction. J Bacteriol 182:821–824

Rigal A, Bouveret E, Lloubes R et al (1997) The TolB protein interacts with the porins of *Escherichia coli*. J Bacteriol 179:7274–7279

Roy CR, Mukherjee S (2009) Bacterial Fic proteins AMP up infection. Sci Signal 2:14

Samoilis G, Psaroulaki A, Vougas K et al (2007) Analysis of whole cell lysate from the intercellular bacterium *Coxiella burnetii* using two gel-based protein separation techniques. J Proteome Res 6:3032–3041

Samoilis G, Aivaliotis M, Vranakis I et al (2010) Proteomic screening for possible effector molecules secreted by the obligate intracellular pathogen *Coxiella burnetii*. J Proteome Res 9:1619–1626

Schaumburg J, Diekmann O, Hagendorff P et al (2004) The cell wall subproteome of *Listeria monocytogenes*. Proteomics 4:2991–3006

Scott K, Diggle MA, Clarke SC (2003) TypA is a virulence regulator and is present in many pathogenic bacteria. Brit J Biomed Sci 60:168–170

Sekeyova Z, Kowalczewska M, Decloquement P et al (2009) Identification of protein candidates for the serodiagnosis of Q fever endocarditis by an immunoproteomic approach. Eur J Clin Microbiol Infect Dis 28:287–295

Sekeyova Z, Kowalczewska M, Vincentelli R et al (2010) Characterization of antigens for Q fever serodiagnostics. Acta Virol 54:173–180

Seng P, Drancourt M, Gouriet F et al (2009) Ongoing revolution in bacteriology: routine identification of bacteria by matrix-assisted laser desorption ionization time-of-flight mass spectrometry. Clin Infect Dis 49:543–551

Seng P, Rolain JM, Fournier PE et al (2010) MALDI-TOF-mass spectrometry applications in clinical microbiology. Future Microbiol 5:1733–1754

Seshadri R, Hendrix LR, Samuel JE (1999) Differential expression of translational elements by life cycle variants of *Coxiella burnetii*. Infect Immun 67:6026–6033

Seshadri R, Paulsen IT, Eisen JA et al (2003) Complete genome sequence of the Q-fever pathogen *Coxiella burnetii*. Proc Natl Acad Sci USA 100:5455–5460

Seshu J, McIvor KL, Mallavia LP (1997) Antibodies are generated during infection to *Coxiella burnetii* macrophage infectivity potentiator protein (Cb-Mip). Microbiol Immunol 41:371–376

Shannon JG, Heinzen RA (2009) Adaptive immunity to the obligate intracellular pathogen *Coxiella burnetii*. Immunol Res 43:138–148

Shaw EI, Moura H, Woolfitt AR et al (2004) Identification of biomarkers of whole *Coxiella burnetii* phase I by MALDI-TOF mass spectrometry. Anal Chem 76:4017–4022

Skultety L, Toman R, Patoprsty V (1998) A comparative study of lipopolysaccharides from two *Coxiella burnetii* strains considered to be associated with acute and chronic Q fever. Carbohydr Polym 35:189–194

Skultety L, Hernychova L, Toman R et al (2005) *Coxiella burnetii* whole cell lysate protein identification by mass spectrometry and tandem mass spectrometry. Ann NY Acad Sci 1063:115–122

Skultety L, Hernychova L, Bereghazyova E et al (2007) Detection of specific spectral markers of *Coxiella burnetii* isolates by MALDI-TOF mass spectrometry. Acta Virol 51:55–58

Skultety L, Hajduch M, Flores-Ramirez G et al (2011) Proteomic comparison of virulent phase I and avirulent phase II of *Coxiella burnetii*, the causative agent of Q fever. J Proteomics 74:1974–1984

Slaba K, Skultety L, Toman R (2005) Efficiency of various serological techniques for diagnosing *Coxiella burnetii* infection. Acta Virol 49:123–127

Stoker MG, Fiset P (1956) Phase variation of the Nine Mile and other strains of *Rickettsia burneti*. Can J Microbiol 2:310–321

Svraka S, Toman R, Skultety L et al (2006) Establishment of a genotyping scheme for *Coxiella burnetii*. FEMS Microbiol Lett 254:268–274

Thompson HA, Hoover TA, Vodkin MH et al (2003) Do chromosomal deletions in the lipopolysaccharide biosynthetic regions explain all cases of phase variation in *Coxiella burnetii* strains? An update. Ann NY Acad Sci 990:664–670

Toman R, Skultety L (1996) Structural study on a lipopolysaccharide from *Coxiella burnetii* strain Nine Mile in avirulent phase II. Carbohydr Res 283:175–185

Toman R, Skultety L, Ftacek P et al (1998) NMR study of virenose and dihydrohydroxystreptose isolated from *Coxiella burnetii* phase I lipopolysaccharide. Carbohydr Res 306:291–296

Toman R, Skultety L, Ihnatko R (2009) *Coxiella burnetii* glycomics and proteomics – tools for linking structure to function. Ann NY Acad Sci 1166:67–78

Tsukagoshi N, Aono R (2000) Entry into and release of solvents by *Escherichia coli* in an organic-aqueous two-liquid-phase system and substrate specificity of the AcrAB-TolC solvent-extruding pump. J Bacteriol 182:4803–4810

Valvano MA, Messner P, Kosma P (2002) Novel pathways for biosynthesis of nucleotide-activated glycero-manno-heptose precursors of bacterial glycoproteins and cell surface polysaccharides. Microbiology 148:1979–1989

Varghees S, Kiss K, Frans G et al (2002) Cloning and porin activity of the major outer membrane protein P1 from *Coxiella burnetii*. Infect Immun 70:6741–6750

Vigil A, Ortega R, Nakajima-Sasaki R et al (2010) Genome-wide profiling of humoral immune response to *Coxiella burnetii* infection by protein microarray. Proteomics 10:2259–2269

Vines ED, Marolda CL, Balachandran A et al (2005) Defective O-antigen polymerization in tolA and pal mutants of *Escherichia coli* in response to extracytoplasmic stress. J Bacteriol 187:3359–3368

Vollmer W, Blanot D, De Pedro MA (2008) Peptidoglycan structure and architecture. FEMS Microbiol Rev 32:149–167

Voth DE, Heinzen RA (2007) Lounging in a lysosome: the intracellular lifestyle of *Coxiella burnetii*. Cell Microbiol 9:829–840

Voth DE, Howe D, Beare PA et al (2009) The *Coxiella burnetii* ankyrin repeat domain-containing protein family is heterogeneous, with C-terminal truncations that influence Dot/Icm-mediated secretion. J Bacteriol 191:4232–4242

Waag DM, England MJ, Tammariello RF et al (2002) Comparative efficacy and immunogenicity of Q fever chloroform:methanol residue (CMR) and phase I cellular (Q-Vax) vaccines in cynomolgus monkeys challenged by aerosol. Vaccine 20:2623–2634

Walburger A, Lazdunski C, Corda Y (2002) The Tol/Pal system function requires an interaction between the C-terminal domain of TolA and the N-terminal domain of TolB. Mol Microbiol 44:695–708

Webster RE (1991) The tol gene products and the import of macromolecules into *Escherichia coli*. Mol Microbiol 5:1005–1011

Williams JC, Waag DM (1991) Antigens, virulence factors, and biological response modifiers of *Coxiella burnetii:* strategies for vaccine development. In: Williams JC, Thompson HA (eds) Q Fever: the biology of *Coxiella burnetii*. CRC Press, Boca Raton FL, pp 175–223

Williams JC, Hoover TA, Waag DM et al (1990) Antigenic structure of *Coxiella burnetii*. A comparison of lipopolysaccharide and protein antigens as vaccines against Q fever. Ann NY Acad Sci 590:370–380

Xolalpa W, Vallecillo AJ, Lara M et al (2007) Identification of novel bacterial plasminogen-binding proteins in the human pathogen *Mycobacterium tuberculosis*. Proteomics 7:3332–3341

Yarbrough ML, Li Y, Kinch LN et al (2009) AMPylation of Rho GTPases by *Vibrio* VopS disrupts effector binding and downstream signaling. Science 323:269–272

Yu EW, Aires JR, Nikaido H (2003) AcrB multidrug efflux pump of *Escherichia coli*: composite substrate-binding cavity of exceptional flexibility generates its extremely wide substrate specificity. J Bacteriol 185:5657–5664

Zamboni DS, McGrath S, Rabinovitch M et al (2003) *Coxiella burnetii* expresses type IV secretion system proteins that function similarly to components of the *Legionella pneumophila* Dot/Icm system. Mol Microbiol 49:965–976

Zhang GQ, Samuel JE (2003) Identification and cloning potentially protective antigens of *Coxiella burnetii* using sera from mice experimentally infected with Nine Mile phase I. Ann NY Acad Sci 990:510–520

Zhang YX, Zhi N, Yu SR et al (1994) Protective immunity induced by 67 K outer membrane protein of phase I *Coxiella burnetii* in mice and guinea pigs. Acta Virol 38:327–332

Zhang GQ, Hotta A, Ho T et al (1998) Evaluation of a recombinant 27-kDa outer membrane protein of *Coxiella burnetii* as an immunodiagnostic reagent. Microbiol Immunol 42:423–428

Zusman T, Yerushalmi G, Segal G (2003) Functional similarities between the icm/dot pathogenesis systems of *Coxiella burnetii* and *Legionella pneumophila*. Infect Immun 71:3714–3723

Zusman T, Aloni G, Halperin E et al (2007) The response regulator PmrA is a major regulator of the icm/dot type IV secretion system in *Legionella pneumophila* and *Coxiella burnetii*. Mol Microbiol 63:1508–1523

Chapter 7
Coxiella Subversion of Intracellular Host Signaling

S. Kauser Hussain and Daniel E. Voth

Contents

Abstract *Coxiella burnetii* is a highly infectious bacterial pathogen that replicates in a specialized vacuole inside eukaryotic cells. Due to a prolonged growth cycle, *Coxiella* continuously manipulates cellular processes to generate this parasitophorous vacuole (PV) and promote host cell viability. Here, we discuss recent findings that indicate *Coxiella* modulates several host signaling pathways to influence survival and ensure intracellular replication. The pathogen actively inhibits apoptotic cell death and activates the pro-survival kinases Akt and Erk1/2 to promote host viability. *Coxiella's* anti-apoptotic activity also involves the interface between autophagy and apoptosis, which is regulated by the interaction of autophagy-related Beclin-1 and anti-apoptotic Bcl-2. Additionally, *Coxiella* requires host kinase activity for PV biogenesis and maintenance. Thus, signaling modulation by *Coxiella* is critical for multiple aspects of host cell parasitism. Collectively, recent signaling studies

S.K. Hussain • D.E. Voth (✉)
Department of Microbiology and Immunology, University of Arkansas for Medical Sciences, 4301 W. Markham Street, 72205 Little Rock, AR, USA
e-mail: dvoth@uams.edu

R. Toman et al. (eds.), *Coxiella burnetii: Recent Advances and New Perspectives in Research of the Q Fever Bacterium*, Advances in Experimental Medicine and Biology 984, DOI 10.1007/978-94-007-4315-1_7,

have enhanced our understanding of the unique *Coxiella*-host cell interaction. Identification of bacterial factors that regulate signaling events will further our ability to model this intriguing infectious process.

Keywords *Coxiella* • Vacuole • Apoptosis • Signaling • Autophagy

7.1 Introduction

Bacterial pathogens have evolved a battery of mechanisms to modulate the hostile host environment encountered during infection and cause disease. Intracellular bacterial parasites comprise a group of intriguing organisms that rely on the internal eukaryotic cell environment for nutrient acquisition and formation of a protected replication niche. Intracellular pathogens have adapted to their unique lifestyle by counteracting antibacterial responses such as cytokine production, immune detection, lysosomal degradation, and cell death. Manipulation of these processes intricately controls the infectious process and ultimately ensures an efficient infection. Conversely, combating these activities often results in clearance of the pathogen and resolution of disease. Thus, regulation of intracellular host processes has become a major focus of this growing field of infectious disease.

Coxiella burnetii is the intracellular bacterial agent of the zoonosis human Q fever, which can present as acute or chronic disease. Humans are typically exposed to *Coxiella* via inhalation of contaminated aerosols, resulting in deposition of infectious organisms in the alveolar spaces (Raoult et al. 2005). Hence, alveolar phagocytic cells are considered the initial reservoir wherein *Coxiella* promotes formation of a parasitophorous vacuole (PV) in which to replicate (Voth and Heinzen 2007). The early *Coxiella*-containing phagosome matures through interactions with host autophagosomes as a potential source of nutrients and membrane for the expanding vacuole (Romano et al. 2006). As the vacuole enlarges through continual fusion with lysosomes, autophagosomes, and fluid phase endosomes, *Coxiella* converts from a small cell variant morphological form into a replication-proficient large cell variant form that divides by binary fission throughout a prolonged growth cycle (Coleman et al. 2004). The mature PV is unique among intracellular pathogen compartments in possessing an acidic (pH ~ 5.0), phagolysosome-like nature (Akporiaye et al. 1983). Indeed, the PV contains active acid hydrolases and retains degradative activity against other bacterial cells (Howe et al. 2010). Strikingly, *Coxiella* replicates to high numbers and thrives in this hostile environment designed to dispose of invading pathogens. Therefore, PV maintenance is of utmost importance to *Coxiella* and the bacterium actively promotes vacuole biogenesis and maintenance to ensure a successful infection.

Coxiella protein synthesis is required for PV biogenesis and antibiotic treatment results in accumulation of tight-fitting phagolysosomes containing individual organisms (Howe et al. 2003). These small vacuoles are unable to fuse with other PV or host vesicles and expansion and replication is stalled. To direct PV biogenesis,

Coxiella likely employs its Dot/Icm type IV secretion system (T4SS), which delivers bacterial proteins, termed substrates or effectors, to the host cytosol where they interact with eukaryotic proteins to regulate infection events. Recent evidence indicates that *Coxiella* encodes numerous Dot/Icm substrates predicted to control a vast array of host processes. For example, *Coxiella* isolates collectively encode 11 Dot/Icm substrates with ankyrin repeat domains (Voth et al. 2009; Pan et al. 2008), which are eukaryotic motifs that mediate protein-protein interactions. These and other Dot/Icm-translocated proteins are predicted to be major virulence factors used by *Coxiella* to control infection events, including subversion of signaling pathways. In this chapter, recent discoveries will be discussed regarding intracellular host signaling pathways manipulated by *Coxiella* to control host viability and PV formation.

7.2 *Coxiella* Exhibits Potent Anti-apoptotic Activity

A costly detriment to an intracellular parasite is the demise of its preferred host cell. Many intracellular pathogens, including *Coxiella* (Coleman et al. 2004), display slow replication rates, necessitating a viable host cell throughout a prolonged growth cycle. Thus, these organisms have developed methods to promote host cell survival until replication is complete and cellular release can ensue. A common target of these organisms is apoptosis (Labbe and Saleh 2008), a form of ordered, non-inflammatory eukaryotic cell death broadly consisting of two pathways: extrinsic and intrinsic (Fig. 7.1). Extrinsic apoptosis is mediated by ligation of cell surface death receptors that activate intracellular proteolytic caspase cascades (Jin and El-Deiry 2005). Subsequent activation of downstream, or effector, caspases such as caspase-3 ultimately results in DNA fragmentation and cell death. Intrinsic apoptosis is regulated by intracellular events that trigger mitochondrial release of cytochrome *c* (Jin and El-Deiry 2005). Cytochrome *c* release is regulated by a panel of interacting anti- and pro-apoptotic mitochondrial surface proteins containing BH3 homology domains. In the cytosol, cytochrome *c* forms a complex with Apaf-1 and caspase-9 termed the apoptosome, which activates effector caspases, triggering DNA damage and death. It is increasingly clear that intracellular pathogens can completely halt cell death.

Anti-apoptotic activities have been described for multiple intracellular pathogens including *Rickettsia*, *Salmonella*, *Mycobacterium*, *Legionella*, and *Chlamydia*. *Salmonella typhimurium* secretes SopB, a type III effector protein that activates the host pro-survival kinase Akt during early stages of epithelial cell infection, resulting in decreased caspase activation (Knodler et al. 2005). *Mycobacterium tuberculosis* inhibits mitochondrial-dependent death by directly altering activity of the BH3 domain protein Bad (Maiti et al. 2001). *Legionella pneumophila* also directly engages mitochondrial proteins by secreting effector proteins that target pro-apoptotic BNIP3 and Bcl-rambo (Banga et al. 2007). *Chlamydia spp*. are perhaps the most prolific anti-apoptotic organisms described to date. *Chlamydia spp.* use a multi-faceted approach to inhibit apoptosis including activation of Akt

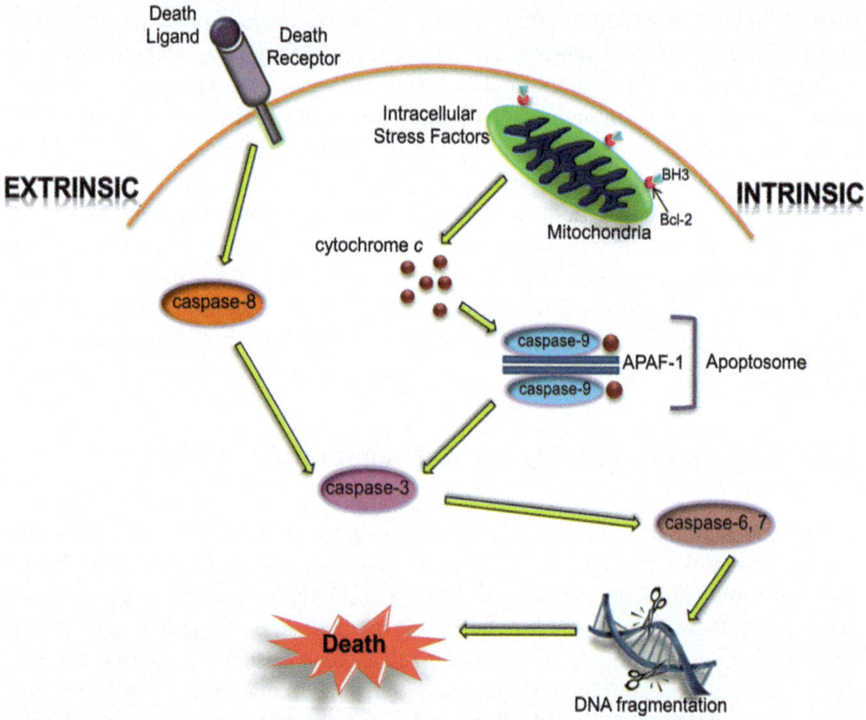

Fig. 7.1 Overview of extrinsic and intrinsic apoptosis. Extrinsic apoptosis ensues following ligation of cell surface death receptors that trigger activation of proteolytic initiator caspase proteins, such as caspase-8. Intrinsic apoptosis is initiated by stress factors that allow mitochondrial release of cytochrome *c* through the interaction of pro-apoptotic BH3 domain proteins and anti-apoptotic proteins such as Bcl-2. Cytochrome *c* release results in formation of a multi-protein complex, termed the apoptosome, that mediates caspase-9 activation with the assistance of apoptotic protease activating factor-1 (Apaf-1). Following activation of initiator caspases, both apoptotic pathways converge on effector caspases, such as caspase-3 and -7, which trigger downstream DNA fragmentation and cell death

(Verbeke et al. 2006), inhibition of cytochrome *c* release (Fan et al. 1998), and degradation of pro-apoptotic BH3 domain mitochondrial proteins (Fischer et al. 2004; Dong et al. 2005; Ying et al. 2005), an activity mediated by the chlamydial protease CPAF (Pirbhai et al. 2006). Finally, many intracellular pathogens promote a pro-survival host gene expression program that relies on the transcription factor NF-κB (Clifton et al. 1998; Losick and Isberg 2006; Abu-Zant et al. 2006; Wahl et al. 2003; Dhiman et al. 2007).

Coxiella has taken a page from the pathogens above and actively promotes host cell survival by altering multiple steps of the apoptotic pathway. The organism antagonizes caspase-3, caspase-9, and poly (ADP-ribose) polymerase (PARP) processing following either staurosporine (intrinsic apoptosis) or TNF-α (extrinsic apoptosis) treatment of THP-1 macrophage-like cells (Voth et al. 2007b). *Coxiella* also inhibits caspase-3 activation in primary primate alveolar macrophages, which represent a *Coxiella* target cell, implicating the *in vivo* importance of promoting cell survival. *Coxiella* also exhibits anti-apoptotic activity in HeLa (human epithelial) and CHO (Chinese hamster ovary) cells via inhibition of mitochondrial cytochrome *c* release (Luhrmann and Roy 2007). However, unlike *Chlamydia spp.*, mitochondrial BH3 domain proteins are not degraded during *Coxiella* infection (Luhrmann and Roy 2007; Voth et al. 2007b). Importantly, virulent *Coxiella* also inhibits caspase-3 and PARP processing (Voth et al. 2007b), indicating anti-apoptotic activity is important during natural infection by disease-causing organisms. Furthermore, bacterial protein synthesis is required for the organism's anti-apoptotic effects, indicating active regulation of host cell survival. It is reasonable to predict that one or more Dot/Icm substrates are responsible for this potent anti-apoptotic activity. Indeed, closely related *Legionella pneumophila* secretes the Dot/Icm substrates SidF (Banga et al. 2007), SdhA (Laguna et al. 2006), and LegK1 (Ge et al. 2009) to antagonize mitochondrial-mediated apoptosis and activate NF-κB. However, *Coxiella* does not encode homologs of these proteins, suggesting a pathogen-specific repertoire of anti-apoptotic effectors.

Coxiella also regulates host apoptosis at the transcriptional level by altering expression of 30 survival-related genes (Voth et al. 2007b). For example, infected THP-1 cells show increased expression of *cIAP2*, *a1/bfl-1*, and *bag1*, which promote survival, and decreased expression of pro-apoptotic *bax*, *bim*, *bik*, *casp2*, and *casp6*, contributing to an overall anti-apoptotic state. cIAP2 and A1/Bfl-1 protein production also dramatically increases during infection. Previous studies demonstrated that *cIAP2* and *a1/bfl-1* are regulated by NF-κB transcriptional activity (Zong et al. 1999; Chu et al. 1997), which is important for the anti-apoptotic potential of other intracellular pathogens (see above). Interestingly, NF-κB translocates to the host nucleus during early stages of *Coxiella* infection and substantial nuclear levels persist throughout intracellular growth (Voth et al. 2007a), suggesting *Coxiella* promotes sustained activation of this versatile transcription factor.

7.3 A Link Between Autophagy and *Coxiella* Anti-apoptotic Activity

Coxiella engages host autophagosomes throughout infection as a potential source of nutrients and membrane for the maturing PV. In fact, the PV decorates with the autophagy marker LC3 as early as 5 min post-infection (Gutierrez et al. 2005; Romano et al. 2006). Additionally, activation of autophagy by amino acid depravation or exogenous treatment with rapamycin stimulates infection and enhances PV formation. Recent evidence indicates a potential link between *Coxiella* interactions with autophagosomes and the pathogen's ability to antagonize apoptosis. During intracellular growth, the PV decorates with the autophagy-related protein Beclin-1 and the anti-apoptotic mitochondrial protein Bcl-2 (Vazquez and Colombo 2010). The interaction between these two proteins is critical for both PV biogenesis and inhibition of apoptosis as evidenced by increased death of cells expressing low levels of Beclin-1 or a Beclin-1 mutant deficient for Bcl-2 binding. These intriguing new studies underscore the importance of crosstalk between multiple pathways for proper PV formation and sustenance of host viability.

7.4 *Coxiella* Activates Host Pro-survival Signaling Proteins

It is clear that intracellular pathogens control apoptosis at both transcriptional and post-translational levels to ensure a viable niche throughout their growth cycle. As discussed above, *Coxiella* inhibits mitochondrial cytochrome *c* release and caspase activation and induces a pro-survival transcriptional program (Luhrmann and Roy 2007; Voth et al. 2007b). A recent study also demonstrates the role of host kinase-directed signaling in the *Coxiella* anti-apoptotic response (Voth and Heinzen 2009). During infection of THP-1 cells, two kinase activation events are observed. First, c-Jun, Hsp27, JNK, and p38 MAPK are phosphorylated at 2 hours post-infection (hpi), then de-phosphorylated at 12–24 hpi, suggesting these proteins are activated as an initial host response to phagocytosis of *Coxiella*. Conversely, the pro-survival kinases Akt and Erk1/2 are phosphorylated at 6 hpi and remain phosphorylated throughout infection, indicating prolonged activation of both proteins and regulation of downstream targets by *Coxiella*. The pathogen actively promotes increased Akt and Erk1/2 activation as elevated phosphorylation levels are not observed in cells infected in the presence of bacterial protein synthesis inhibitors. Akt and Erk1/2 are heavily involved in maintaining a viable eukaryotic cell via phosphorylation. Akt regulates numerous pathways that promote survival including activation of the FOXO family of transcription factors and direct regulation of caspase-9 activity (Manning and Cantley 2007), while Erk1/2 activates survival-related transcription factors such as Elk-1 and cAMP response binding element, or CREB (McCubrey et al. 2006). Thus, Akt and Erk1/2 are prime targets for apoptosis intervention by *Coxiella*. Furthermore, treatment of infected cells with Akt and Erk1/2 pathway inhibitors negates *Coxiella's* anti-apoptotic effect, suggesting these pathways are critical for the pathogen's ability to promote survival. However, the upstream and downstream components of these cascades that are modulated during infection, and the bacterial proteins responsible, are currently unknown.

7.5 A Role for Host Signaling Cascades in PV Development

Mammalian cells also use phosphorylation-based signaling to control a diverse array of responses not directly related to apoptosis. Kinase-based cascades, such as Akt and Erk1/2, are efficient scaffolds that control sequential phosphorylation events to regulate downstream substrate activity (McCubrey et al. 2006). Eukaryotic kinases are also controlled by their own phosphorylation, providing for tightly regulated responses. Not surprisingly, bacterial pathogens have evolved mechanisms to subvert distinct signaling pathways and influence infection events for the benefit of the pathogen (Bhavsar et al. 2007). Aside from apoptosis, intracellular pathogens utilize eukaryotic signaling for cellular entry, modulation of cytokine production, and alteration of the host cell cycle (Knodler et al. 2001; Bhavsar et al. 2007). However, the role of mammalian kinase signaling in formation of bacterial replication vacuoles has not been elucidated.

Recent studies in our laboratory have uncovered a role for host phosphorylation signaling in PV biogenesis and maintenance (Hussain et al. 2010). Thirteen signaling proteins including protein kinase C (PKC), cAMP-dependent protein kinase, and calmodulin kinase II are involved in PV formation, and inhibition of these molecules adversely affects *Coxiella* growth. These results suggest the organism must engage host signaling that is not directly involved in survival to promote PV formation and maintenance. Additionally, several kinases are differentially phosphorylated throughout intracellular growth. Furthermore, virulent *Coxiella* isolates activate PKC during infection, suggesting PKC signaling is regulated during natural infection. However, the ultimate effects of pathogen-modulated kinase activity are unknown.

7.6 Conclusions and Future Perspectives

Intracellular pathogens are adept at directing host pathways during intracellular growth to ensure a viable, sustainable niche in which to replicate. As discussed in this chapter, *Coxiella* actively regulates host apoptosis, autophagy, and phosphorylation cascades to provide a proper replication environment (Fig. 7.2). However, the bacterial proteins that regulate signaling events have not been identified and represent a substantial void in our understanding of *Coxiella*-host cell interactions. As mentioned above, *Coxiella* produces a specialized Dot/Icm T4SS during infection that delivers bacterial proteins directly to the host cytoplasm. It is intriguing to predict that a subset of effector proteins controls the pathogen's anti-apoptotic activity through direct interactions with host proteins. Indeed, exciting new evidence demonstrates *Coxiella* Dot/Icm-translocated AnkG binds to host p32, inhibiting the pro-apoptotic properties of this protein and promoting cell survival (Luhrmann et al. 2010). In addition to identifying novel Dot/Icm substrates using heterologous models such as closely-related *L. pneumophila*, recent advances in *Coxiella* host cell-free growth (Omsland et al. 2009) and genetic manipulation (Beare et al. 2009) will enhance our ability to identify and functionally characterize pathogen virulence factors. Furthermore, recent success using transposon mutagenesis (Beare et al. 2009) will allow screening of *Coxiella* mutant libraries to identify strains defective for anti-apoptotic activity and kinase activation. These anticipated studies will shed important insight into *Coxiella* virulence factors and pathogenic mechanisms.

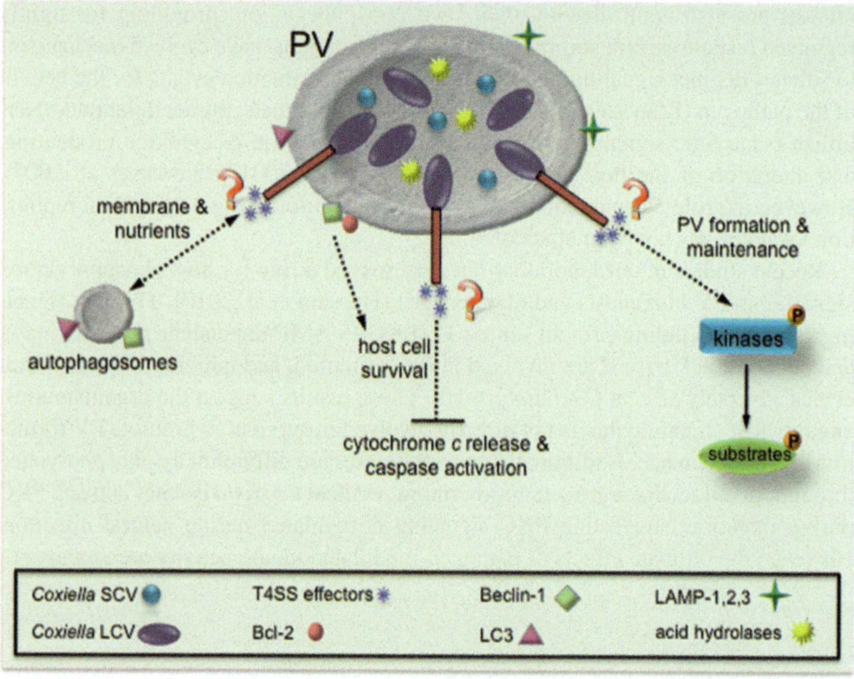

Fig. 7.2 Manipulation of intracellular host signaling by *Coxiella*. In susceptible host cells, *Coxiella* generates a membrane bound vacuolar compartment with features of phagolysosomes, including acidic pH and the presence of acid hydrolases and lysosomal membrane markers (*e.g.*, LAMP-1, -2, -3). PV formation is aided by interactions with autophagosomes (denoted by labeling with LC3) that provide membrane for the expanding vacuole and may provide nutrients to replicating organisms. PV biogenesis relies on functional host kinase activity as evidenced by phosphorylation (p) of downstream substrates. During a lengthy infectious cycle, *Coxiella* actively antagonizes apoptotic cell death through inhibition of cytochrome *c* release and caspase processing, up-regulation of an anti-apoptotic transcriptional program, and activation of pro-survival kinases. Additionally, the autophagy-related protein Beclin-1 interacts with anti-apoptotic Bcl-2 to promote cell survival. These processes are likely controlled by the activity of numerous *Coxiella* proteins, such as anti-apoptotic AnkG, that are delivered to the host cytosol by the Dot/Icm T4SS

Acknowledgements Research in the Voth laboratory is supported by funding to D. E. V. from NIH/NIAID (K22AI081753 and R01AI087669), the American Heart Association (BGIA3080001), and the Arkansas Biosciences Institute.

References

Abu-Zant A, Jones S, Asare R, Suttles J, Price C, Graham J, Kwaik YA (2006) Anti-apoptotic signalling by the Dot/Icm secretion system of *L. pneumophila*. Cell Microbiol 9:246–264

Akporiaye ET, Rowatt JD, Aragon AA, Baca OG (1983) Lysosomal response of a murine macrophage-like cell line persistently infected with *Coxiella burnetii*. Infect Immun 40:1155–1162

Banga S, Gao P, Shen X, Fiscus V, Zong WX, Chen L, Luo ZQ (2007) *Legionella pneumophila* inhibits macrophage apoptosis by targeting pro-death members of the Bcl2 protein family. Proc Natl Acad Sci U S A 104:5121–5126

Beare PA, Howe D, Cockrell DC, Omsland A, Hansen B, Heinzen RA (2009) Characterization of a *Coxiella burnetii ftsZ* mutant generated by *Himar1* transposon mutagenesis. J Bacteriol 191:1369–1381

Bhavsar AP, Guttman JA, Finlay BB (2007) Manipulation of host-cell pathways by bacterial pathogens. Nature 449:827–834

Chu ZL, Mckinsey TA, Liu L, Gentry JJ, Malim MH, Ballard DW (1997) Suppression of tumor necrosis factor-induced cell death by inhibitor of apoptosis c-IAP2 is under NF-kappaB control. Proc Natl Acad Sci U S A 94:10057–10062

Clifton DR, Goss RA, Sahni SK, Van Antwerp D, Baggs RB, Marder VJ, Silverman DJ, Sporn LA (1998) NF-kappa B-dependent inhibition of apoptosis is essential for host cellsurvival during *Rickettsia rickettsii* infection. Proc Natl Acad Sci U S A 95:4646–4651

Coleman SA, Fischer ER, Howe D, Mead DJ, Heinzen RA (2004) Temporal analysis of *Coxiella burnetii* morphological differentiation. J Bacteriol 186:7344–7352

Dhiman R, Raje M, Majumdar S (2007) Differential expression of NF-kappaB in mycobacteria infected THP-1 affects apoptosis. Biochim Biophys Acta 1770:649–658

Dong F, Pirbhai M, Xiao Y, Zhong Y, Wu Y, Zhong G (2005) Degradation of the proapoptotic proteins Bik, Puma, and Bim with Bcl-2 domain 3 homology in *Chlamydia trachomatis*-infected cells. Infect Immun 73:1861–1864

Fan T, Lu H, Hu H, Shi L, Mcclarty GA, Nance DM, Greenberg AH, Zhong G (1998) Inhibition of apoptosis in chlamydia-infected cells: blockade of mitochondrial cytochrome c release and caspase activation. J Exp Med 187:487–496

Fischer SF, Vier J, Kirschnek S, Klos A, Hess S, Ying S, Hacker G (2004) *Chlamydia* inhibit host cell apoptosis by degradation of proapoptotic BH3-only proteins. J Exp Med 200:905–916

Ge J, Xu H, Li T, Zhou Y, Zhang Z, Li S, Liu L, Shao F (2009) A *Legionella* type IV effector activates the NF-kappaB pathway by phosphorylating the IkappaB family of inhibitors. Proc Natl Acad Sci U S A 106:13725–13730

Gutierrez MG, Vazquez CL, Munafo DB, Zoppino FC, Beron W, Rabinovitch M, Colombo MI (2005) Autophagy induction favours the generation and maturation of the *Coxiella*-replicative vacuoles. Cell Microbiol 7:981–993

Howe D, Melnicakova J, Barak I, Heinzen RA (2003) Maturation of the *Coxiella burnetii* parasitophorous vacuole requires bacterial protein synthesis but not replication. Cell Microbiol 5:469–480

Howe D, Shannon JG, Winfree S, Dorward DW, Heinzen RA (2010) *Coxiella burnetii* phase I and II variants replicate with similar kinetics in degradative phagolysosome-like compartments of human macrophages. Infect Immun 78:3465–3474

Hussain SK, Broederdorf LJ, Sharma UM, Voth DE (2010) Host kinase activity is required for *Coxiella burnetii* parasitophorous vacuole formation. Front Microbiol 1:137. doi:10.3389/fmicb.2010.00137

Jin Z, El-Deiry WS (2005) Overview of cell death signaling pathways. Cancer Biol Ther 4:139–163

Knodler LA, Celli J, Finlay BB (2001) Pathogenic trickery: deception of host cell processes. Nat Rev Mol Cell Biol 2:578–588

Knodler LA, Finlay BB, Steele-Mortimer O (2005) The *Salmonella* effector protein SopB protects epithelial cells from apoptosis by sustained activation of Akt. J Biol Chem 280:9058–9064

Labbe K, Saleh M (2008) Cell death in the host response to infection. Cell Death Differ 15:1339–1349

Laguna RK, Creasey EA, Li Z, Valtz N, Isberg RR (2006) A *Legionella pneumophila*-translocated substrate that is required for growth within macrophages and protection from host cell death. Proc Natl Acad Sci U S A 103:18745–18750

Losick VP, Isberg RR (2006) NF-kappaB translocation prevents host cell death after low-dose challenge by *Legionella pneumophila*. J Exp Med 203:2177–2189

Luhrmann A, Roy CR (2007) *Coxiella burnetii* inhibits activation of host cell apoptosis through a mechanism that involves preventing cytochrome c release from mitochondria. Infect Immun 75:5282–5289

Luhrmann A, Nogueira CV, Carey KL, Roy CR (2010) Inhibition of pathogen-induced apoptosis by a *Coxiella burnetii* type IV effector protein. Proc Natl Acad Sci USA 107:18997–19001

Maiti D, Bhattacharyya A, Basu J (2001) Lipoarabinomannan from *Mycobacterium tuberculosis* promotes macrophage survival by phosphorylating bad through a phosphatidylinositol 3-kinase/Akt pathway. J Biol Chem 276:329–333

Manning BD, Cantley LC (2007) AKT/PKB signaling: navigating downstream. Cell 129:1261–1274

Mccubrey JA, Steelman LS, Abrams SL, Lee JT, Chang F, Bertrand FE, Navolanic PM, Terrian DM, Franklin RA, D'assoro AB, Salisbury JL, Mazzarino MC, Stivala F, Libra M (2006) Roles of the RAF/MEK/ERK and PI3K/PTEN/AKT pathways in malignant transformation and drug resistance. Adv Enzyme Regul 46:249–279

Omsland A, Cockrell DC, Howe D, Fischer ER, Virtaneva K, Sturdevant DE, Porcella SF, Heinzen RA (2009) Host cell-free growth of the Q fever bacterium *Coxiella burnetii*. Proc Natl Acad Sci U S A 106:4430–4434

Pan X, Luhrmann A, Satoh A, Laskowski-Arce MA, Roy CR (2008) Ankyrin repeat proteins comprise a diverse family of bacterial type IV effectors. Science 320:1651–1654

Pirbhai M, Dong F, Zhong Y, Pan KZ, Zhong G (2006) The secreted protease factor CPAF is responsible for degrading pro-apoptotic BH3-only proteins in *Chlamydia trachomatis*-infected cells. J Biol Chem 281:31495–31501

Raoult D, Marrie T, Mege J (2005) Natural history and pathophysiology of Q fever. Lancet Infect Dis 5:219–226

Romano PS, Gutierrez MG, Beron W, Rabinovitch M, Colombo MI (2006) The autophagic pathway is actively modulated by phase II *Coxiella burnetii* to efficiently replicate in the host cell. Cell Microbiol 9:891–909

Vazquez CL, Colombo MI (2010) *Coxiella burnetii* modulates Beclin 1 and Bcl-2, preventing host cell apoptosis to generate a persistent bacterial infection. Cell Death Differ 17:421–438

Verbeke P, Welter-Stahl L, Ying S, Hansen J, Hacker G, Darville T, Ojcius DM (2006) Recruitment of BAD by the *Chlamydia trachomatis* vacuole correlates with host-cell survival. PLoS Pathog 2:e45

Voth DE, Heinzen RA (2007) Lounging in a lysosome: the intracellular lifestyle of *Coxiella burnetii*. Cell Microbiol 9:829–840

Voth DE, Heinzen RA (2009) Sustained activation of Akt and Erk1/2 is required for *Coxiella burnetii* antiapoptotic activity. Infect Immun 77:205–213

Voth DE, Howe D, Heinzen RA (2007a) *Coxiella burnetii* inhibits apoptosis and modulates cell survival signaling in macrophages *FASEB Microbial Pathogenesis Meeting*

Voth DE, Howe D, Heinzen RA (2007b) *Coxiella burnetii* inhibits apoptosis in human THP-1 cells and monkey primary alveolar macrophages. Infect Immun 75:4263–4271

Voth DE, Howe D, Beare PA, Vogel JP, Unsworth N, Samuel JE, Heinzen RA (2009) The *Coxiella burnetii* ankyrin repeat domain-containing protein family is heterogeneous, with C-terminal truncations that influence Dot/Icm-mediated secretion. J Bacteriol 191:4232–4242

Wahl C, Maier S, Marre R, Essig A (2003) *Chlamydia pneumoniae* induces the expression of inhibitor of apoptosis 2 (c-IAP2) in a human monocytic cell line by an NF-kappaB-dependent pathway. Int J Med Microbiol 293:377–381

Ying S, Seiffert BM, Hacker G, Fischer SF (2005) Broad degradation of proapoptotic proteins with the conserved Bcl-2 homology domain 3 during infection with *Chlamydia trachomatis*. Infect Immun 73:1399–1403

Zong WX, Edelstein LC, Chen C, Bash J, Gelinas C (1999) The prosurvival Bcl-2 homolog Bfl-1/A1 is a direct transcriptional target of NF-kappaB that blocks TNFalpha-induced apoptosis. Genes Dev 13:382–387

Chapter 8
The *Coxiella burnetii* Parasitophorous Vacuole

Eric Ghigo, María I. Colombo, and Robert A. Heinzen

Contents

E. Ghigo
URMITE CNRS UMR 6236 – IRD 3R198, Institut Fédératif de Recherche 48, Université de la Méditerranée , Marseille, France

M.I. Colombo
Laboratorio de Biología Celular y Molecular, Instituto de Histología y Embriología (IHEM)-CONICET, Facultad de Ciencias Médicas, Universidad Nacional de Cuyo , Mendoza, Argentina

R.A. Heinzen (✉)
Coxiella Pathogenesis Section, Laboratory of Intracellular Parasites, Rocky Mountain Laboratories, National Institute of Allergy and Infectious Diseases, National Institutes of Health, 903 South 4th street, Hamilton, MT 59840, USA
e-mail: rheinzen@niaid.nih.gov

R. Toman et al. (eds.), *Coxiella burnetii: Recent Advances and New Perspectives in Research of the Q Fever Bacterium*, Advances in Experimental Medicine and Biology 984, DOI 10.1007/978-94-007-4315-1_8,

Abstract *Coxiella burnetii* is a bacterial intracellular parasite of eucaryotic cells that replicates within a membrane-bound compartment, or "parasitophorous vacuole" (PV). With the exception of human macrophages/monocytes, the consensus model of PV trafficking in host cells invokes endolysosomal maturation culminating in lysosome fusion. *C. burnetii* resists the degradative functions of the vacuole while at the same time exploiting the acidic pH for metabolic activation. While at first glance the mature PV resembles a large phagolysosome, an increasing body of evidence indicates the vacuole is in fact a specialized compartment that is actively modified by the pathogen. Adding to the complexity of PV biogenesis is new data showing vacuole engagement with autophagic and early secretory pathways. In this chapter, we review current knowledge of PV nature and development, and discuss disparate data related to the ultimate maturation state of PV harboring virulent or avirulent *C. burnetii* lipopolysaccharide phase variants in human mononuclear phagocytes.

Keywords Macrophage • Parasitophorous vacuole • Endosome • Lysosome • Autophagy • Integrin • Lipopolysaccharide • Monocyte • Phase variation • Secretory pathway

8.1 Introduction

Soon after identification of *Coxiella burnetii* as the cause of human Q fever, the organism was classified as a rickettsial agent due to its obligate intracellular nature and staining properties (Cox 1939). The original assignment of *C. burnetii* to the α-proteobacterial order *Rickettsiales* (Weiss and Moulder 1984) was changed during the molecular era with genetic information indicating appropriate placement within the γ-proteobacteria order *Legionellales*, which includes the facultative intracellular pathogen *Legionella pneumophila* (Weisburg et al. 1989). Indeed, *C. burnetii* and *L. pneumophila* share several traits including aerosol transmission, a tropism for alveolar macrophages during natural infection, developmental forms adapted to intracellular survival and environmental transmission, and subversion of host cell functions by effector proteins secreted by a specialized Dot/Icm type IV secretion system (T4SS) (Voth and Heinzen 2007; Newton et al. 2010). Although macrophage invasion and survival are central to pathogenesis by both *C. burnetii* and *L. pneumophila*, the organisms occupy unique intracellular niches and utilize different subversion strategies once internalized by their macrophage hosts.

Natural isolates of *C. burnetii* undergo a virulent to avirulent transition upon serial passage in embryonated eggs or tissue culture. This transition can be serologically defined using post-vaccination antisera and is referred to as "phase variation". Lipopolysaccharide (LPS) *O*-antigen of virulent organisms is the primary surface antigen recognized by phase I antiserum. Avirulent *C. burnetii,* recognized by phase II, but not phase I antiserum, produce a severely truncated LPS lacking *O*-antigen and some core sugars (Hackstadt et al. 1985). Some, but not all, *C. burnetii* in phase

II contain a large chromosomal deletion enriched in genes involved in *O*-antigen biosynthesis (Denison et al. 2007; Beare et al. 2006). The virulence function of *C. burnetii* LPS is unrelated to lipid A, as both phase I and phase II organisms of the Nine Mile reference strain have the same tetra-acylated structure that weakly interacts with Toll-like receptor (TLR) 4 (Zamboni et al. 2004). Instead, virulence appears related to *O*-antigen masking of the organism's outer surface that inhibits complement deposition (Vishwanath and Hackstadt 1988) and antibody access to surface proteins (Hackstadt 1988). Masking is also proposed to restrict TLR receptor interactions of immune cells with non-LPS ligands (Shannon et al. 2005). Interestingly, despite weak lipid A-TLR4 interactions, full-length phase I LPS stimulates dramatic F (filamentous)-actin rearrangement in human macrophages in a TLR4-dependent manner (Capo et al. 1999, 2003; Honstettre et al. 2004). Consequently, *C. burnetii* protein interactions with complement receptor 3 (CR3) are inhibited, an effect speculated to result in different trafficking of phase I and phase II organisms in human mononuclear phagocytes (discussed in more detail below) (Barry et al. 2011).

LPS *O*-antigen does not appear to act directly as a ligand in *C. burnetii* uptake as phase II organisms are internalized 10–100 times more efficiently then phase I organisms by cultured professional phagocytes, epithelial cells and fibroblasts (Moos and Hackstadt 1987). The lower carbohydrate content of phase II LPS is thought to make the organism hydrophobic, thereby facilitating non-specific host plasma membrane interactions (Williams et al. 1981). The more accessible surface protein ligands of phase II *C. burneii* may further enhance uptake (Hackstadt 1988).

By default, the phagocytic process results in material being sequestered within a nascent phagosome that traffics through the endocytic pathway (Haas 2007). The first step of phagosome maturation is the intermingled fusion/fission events with early endosomes. Phagosomes acquire several markers, such as early endosome antigen-1 (EEA 1) and the small GTPase Rab5. Early phagosomes progressively transform into compartments that present features of late endosomes. Markers of late endosomes, such as the small GTPase Rab7 and lysosomal membrane-associated protein-1 (LAMP-1), gradually replace early endosomal markers (Scott et al. 2003; Henry et al. 2004). The pH of early phagosomes is around 6.0, with acquisition of the vacuolar proton pump ATPase (V-H^+-ATPase) by late phagosomes leading to an intraphagosomal pH of 4.5–5.5 (Scott et al. 2003). Finally, late phagosomes fuse with lysosomes that contain hydrolytic enzymes such as cathepsin D, thus leading to the formation of phagolysosomes in which bacteria can be destroyed (Scott et al. 2003). Consequently, numerous bacterial pathogens have developed specific strategies to avoid this intracellular fate, thereby enhancing their survival within host cells (Scott et al. 2003). Bacteria such as *Listeria*, *Shigella* and *Rickettsia* escape from nascent phagosomes to the cytosol to avoid destruction in phagolysosomes (Cossart and Sansonetti 2004). A different strategy used by several pathogens involves interference with normal phagolysosome biogenesis that leads to formation of vacuoles supporting replication (Alonso and Garcia-del Portillo 2004). For example, *B. abortus* vacuoles interact with the endoplasmic reticulum but not with the classical endocytic network (Meresse et al. 1999b). Furthermore,

Mycobacterium phagosomes exchange material such as transferrin with early endosomes but are unable to fuse with late endosomes (Scott et al. 2003; Rohde et al. 2007). Finally, *Salmonella* resides in atypical phagosomes that are neither early nor late phagosomes. Membranes of the trans-Golgi network surround *Salmonella*-containing phagosomes, suggesting interactions with endocytic and biosynthetic pathways (Holden 2002).

In contrast with other intracellular bacteria, the *C. burnetii* parasitophorous vacuole (PV) has extensive interactions with the endolysosomal pathway. With the exception of human mononuclear phagocytes, studies using several cell types have consistently demonstrated PV containing virulent phase I strains of *C. burnetii*, or the avirulent Nine Mile phase II (NMII) RSA439 strain, fully mature through the endolysosomal cascade to resemble a large phagolysosome (Voth and Heinzen 2007). However, disparate models have been published on trafficking of vacuoles harboring virulent Nine Mile phase I (NMI) RSA493 organisms or NMII in human mononuclear phagocytes (Ghigo et al. 2002; Howe et al. 2010). NMI is considered a reference strain with NMII being a high passage isogenic variant of NMI. NMII has been extensively characterized and harbors a large chromosomal (~26 kb) deletion of *O*-antigen biosynthesis genes (Moos and Hackstadt 1987; Hoover et al. 2002). The clonality and large chromosomal deletion of NMII render it non-revertable to full virulence. For the purpose of discussion, we will refer to the opposing models of NMI and NMII trafficking as the differential trafficking (DT) and similar trafficking (ST) models. In the DT model, maturation of PV containing NMI stalls at a late endosomal stage, resulting in a vacuole permissive for pathogen survival and growth (Ghigo et al. 2002). Conversely, PV sheltering NMII fully mature into a phagolysosomal-like compartment that contains active lysosomal hydrolases and is bacteriocidal (Ghigo et al. 2002). Thus, in the DT model, virulence of phase variants is attributed to the ultimate maturation state of their respective PVs. In the ST model, NMI and NMII both replicate in phenotypically-indistinguishable PV that fully mature through the endolysosomal pathway to fuse with lysosomes. In this phagolysosome-like vacuole, phase variants resist degradation and replicate at equal rates (Howe et al. 2010).

Here, we review early and recent work on *C. burnetii* PV nature and development. Furthermore, we discuss disparate data related to the ultimate maturation state of PV harboring *C. burnetii* NMI or NMII phase variants in human mononuclear phagocytes.

8.2 Early Studies of the *Coxiella* Vacuole

In 1937, Edward Derrick of Brisbane, Australia, published his careful and extensive description of a new clinical entity in humans called Q (query) fever (Derrick 1937). He successfully isolated the infectious agent – now known as *Coxiella burnetii* – in guinea pigs but was unable to visualize the pathogen in infected tissues. Using inocula provided by Derrick, Macfarlane Burnet and Mavis Freeman

(Burnet and Freeman 1937) infected mice and successfully stained organisms in spleen sections that they stated "occur as intracellular microcolonies of close-packed individuals, nearly always sharply circumscribed within an oval or circular outline." Although not specifically mentioned, it is clear from their elegant sketches that organisms were restrained in an intracellular compartment. Interestingly, unpublished notes of Derrick showed strikingly similar drawings (Cooke 2008). A year later, Harold Cox at the Rocky Mountain Laboratories in Hamilton, Montana, USA, successfully cultivated the newly isolated NMI strain of *C. burnetii* in tissue cultures of minced chick embryos and alluded to their vacuolar nature (Cox 1938). Subsequent electron microscopy clearly demonstrated intracellular NMI surrounded by a limiting membrane (Handley et al. 1967).

The biological nature of the *C. burnetii* PV began to unravel with the seminal findings of Burton and co-workers (Burton et al. 1971, 1978). Based on cytochemical localization of the lysosomal enzymes acid phosphatase and 5′-nucleotidase in NMI-infected mouse L929 cells, they suggested *C. burnetii* resides in a secondary lysosome, i. e., the product of primary lysosome fusion with a phagosome. They proposed both significant replication and degradation of *C. burnetii* in the vacuole (Burton et al. 1971). A similar fusion event in spleen reticular cells from mice infected with the Luga strain of *C. burnetii* was described by Ariel et al. (1973) who used the term "phagolysosome" to describe the *C. burnetii* vacuole. Burton et al. (1978) went on to show acid phosphatase activity in NMI and NMII vacuoles of persistently infected (6–10 months) mouse L929 cells and African green monkey kidney (Vero) cells, with Vero cell vacuoles also containing abundant membrane whorls, or "myelin configurations" (Fig. 8.1).

By definition, a phagolysosome is a degradative organelle containing acid activated hydrolases. The proton pumping V-H^+-ATPase is responsible for lowering the vacuole's lumenal pH to approximately 5.0 to achieve optimal hydrolytic enzyme activity (Luzio et al. 2007). Consistent with the presence of two active lysosomal enzymes (Burton et al. 1971), Akporiaye et al. (1983) subsequently demonstrated that NMI PV in persistently-infected J774 murine macrophages acquire thorium dioxide from primary lysosomes and acidify to pH 5.2, thus strengthening the emerging model of PV-lysosome fusion.

8.2.1 Acid Activation of Metabolism

Shortly after the original clinical isolation of *C. burnetii,* robust growth of the pathogen was achieved in embryonated hen's eggs and cultured cell lines (Ormsbee 1952; Cox 1938). Thus, it was perplexing in early metabolic studies why intact host cell-free bacteria, unlike bacterial lysates, displayed little metabolic activity in different physiologic buffers (Weiss 1973). This puzzle was solved by Hackstadt and Williams (1981) who astutely recognized the implications Burton et al.'s (1975, 1978) findings. In a landmark study published in 1981, and before the actual determination of PV pH, they showed that transport, catabolism and incorporation

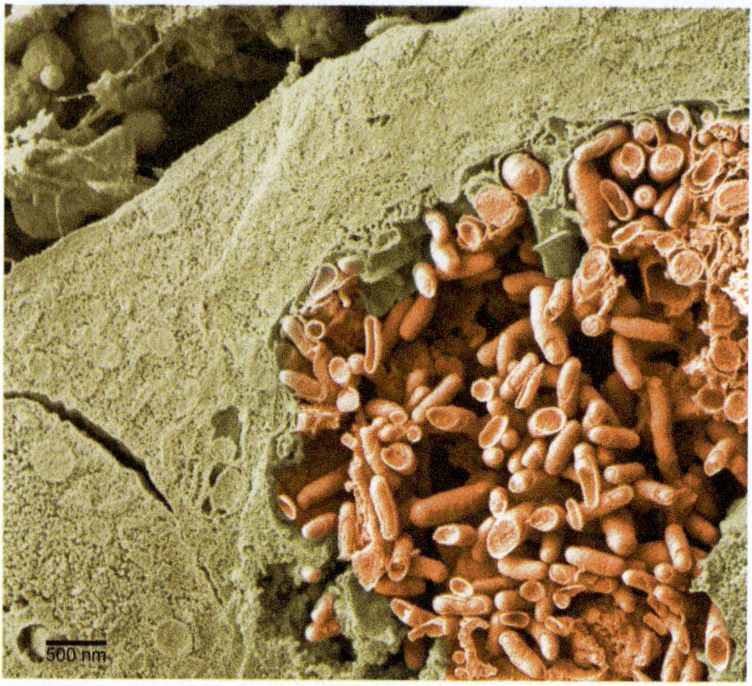

Fig. 8.1 The *C. burnetii* parasitophorous vacuole (PV). Scanning electron micrograph showing a cryo-prepared Vero cell infected with *C. burnetii* Nine Mile (phase II) for 4 days. The host cell cytoplasm (tan) contains a large PV filled with *C. burnetii* (*orange*) (Image courtesy of Beth Fischer, Rocky Mountain Laboratories)

of both glucose and glutamate by host cell-free *C. burnetii* NMI are highly stimulated under acidic conditions mimicking that of a phagolysosome (Hackstadt and Williams 1981). Importantly, they also demonstrated an intracellular growth requirement for vacuolar acidity by showing cessation of NMI growth following neutralization of the PV with lysosomotropic amines (Hackstadt and Williams 1981). The discovery of "acid activated" *C. burnetii* metabolism led to a series of studies that identified optimal conditions for transport and utilization of nutrients, maintenance of the ATP pool, and membrane energization (Hackstadt and Williams 1984).

8.3 The *Coxiella* Parasitophorous (PV) Vacuole

A resurgence in *C. burnetii* PV characterization occurred in the mid-1990s. Enabled by new reagents, such as vital stains and organelle-specific antibodies, and enhanced imaging capabilities, such as laser scanning confocal microscopy, researchers could now more precisely monitor phagosome maturation. These tools, coupled with a variety of cultured cell model systems, allowed several groups to dramatically expand our knowledge of *C. burnetii* PV biogenesis. With the exception of NMI organisms in human mononuclear phagocytes (discussed in detail below), cumulative evidence supports a model whereby *C. burnetii* PV fully mature through the endolysosomal cascade to acquire characteristics of a phagolysosome. However, this is an overly simplified model as *C. burnetii* modifies the vacuole to create a specialized compartment with unusual fusogenicity with other endolysosomal compartments and interactions with autophagy/secretory pathways.

8.3.1 PV Maturation/Biology in Animal Cells and Human Epithelial Cells/Trophoblasts

Although the primary targets of *C. burnetii* during natural infection are monocytes/ macrophages (Stein et al. 2005; Khavkin and Tabibzadeh 1988), the organism has an impressive ability to infect a wide variety of cultured epithelial cells, fibroblasts, and macrophage-like cells (Voth and Heinzen 2007). In this section, we discuss *C. burnetii* interactions with animal cells (both professional and non-professional phagocytes) and human epithelial cells/trophoblasts.

In mouse L929 cells and P388D1 macrophages, live or dead NMI or NMII *C. burnetii* are internalized by phagocytosis at equal rates, indicating *C. burnetii* plays a passive role in cellular uptake (Baca et al. 1993a; Tujulin et al. 1998). Rearrangement of host cell F (filamentous)- actin is required for pathogen uptake as internalization is inhibited by treating cells with actin depolymerizing agents such as cytochalasin D (Baca et al. 1993a). The bacterial ligand(s) mediating attachment are unknown; however, it is likely proteinaceous as treatment of *C. burnetii* with proteases hinders uptake (Baca et al. 1993a). Pretreatment of NMI with the cationic peptide CAP37 dramatically enhances infection of L929 cells without a deleterious effect on the pathogen (Aragon et al. 1995). An opsonon-like activity of the peptide is speculated to promote invasiveness (Aragon et al. 1995).

In addition to a role in pathogen uptake, F-actin rearrangements regulate the size and formation of the *C. burnetii* PV (Aguilera et al. 2009). In HeLa cells, a web of F-actin surrounds mature PV harboring NMII. Treatment with latrunculin B or cytochalasin D following uptake results in smaller PV with reduced fusogenicity. PV decorate with both wild type and constitutively active forms of the Rho family GTPases Cdc42 and RhoA, noted regulators of actin dynamics. Actin accumulation may benefit PV expansion by providing tracks for vesicular fusion and/or enhancing structural integrity. A similar actin web surrounds the vacuole (inclusion) occupied by the intracellular bacterium *Chlamydia trachomatis,* and is thought to provide a structural scaffold (Kumar and Valdivia 2008).

In animal cells and human epithelial cells/trophoblasts, studies assessing pH, enzymatic activities, and lysosomal protein markers agree that nascent NMI or NMII-containing phagosomes fully mature through the default endocytic pathway to acquire characteristics of a phagolysosome (Heinzen et al. 1996; Howe and Mallavia 2000;

Burton et al. 1971, 1978; Akporiaye et al. 1983; Howe et al. 2003; Romano et al. 2007; Campoy et al. 2011; Aguilera et al. 2009; Grieshaber et al. 2002). Ratiometric pH-sensitive probes consistently show PV with a phagolysosomal-like pH (~5) in J774 mouse macrophages infected with NMI (Akporiaye et al. 1983), Vero cells infected with NMII (Grieshaber et al. 2002), and P388D1 and L929 cells infected with NMII or the virulent Priscilla and Q212 strains of *C. burnetii* (Maurin et al. 1992a, b). A cursory examination of the kinetics of PV maturation has been conducted in several cell lines. In CHO cells, the early endosomal marker Rab5 is found on PV containing NMII as early as 5 min after bacterial uptake, with maximum association at 20 min (Romano et al. 2007). The late endosomal/lysosomal marker Rab7 has maximal PV association between 40 and 60 min post-infection, when Rab5 levels are decreasing (Romano et al. 2007). Phagosome recruitment of lysosomal hydrolases is a defining marker of lysosome fusion; however, trafficking of the lysosomal enzymes acid phosphatase and cathepsin D to the *C. burnetii* PV appears delayed in NMI-infected J774 macrophages and NMII-infected CHO cells respectively, taking approximately 1–2 h to reach high levels (Howe and Mallavia 2000; Romano et al. 2007). These maturation kinetics are substantially slower than those of phagosomes containing inert particles such as latex beads, which acquire lysosomal markers by 15 min after uptake (Oh and Swanson 1996). Cathepsin D localizes to mature PV harboring replicating NMII in both HeLa and CHO cells (Heinzen et al. 1996; Aguilera et al. 2009; Romano et al. 2007) and PV containing replicating NMI in human BeWo trophoblasts (Ben Amara et al. 2010). General protease activity has also been demonstrated for mature NMII PV in HeLa cells (Campoy et al. 2011). Other late endosome/lysosomal markers that decorate mature PV in animal cells and human epithelial cells/ trophoblasts include the V-H^+-ATPase (Heinzen et al. 1996), the lysosomal glycoproteins LAMP-1, LAMP-2 and LAMP-3 (CD63) (Ghigo et al. 2002; Heinzen et al. 1996; Shannon et al. 2005; Sauer et al. 2005; Beare et al. 2009a), and syntaxin 8 (Fig. 8.2), a t-SNARE involved in late endosome-lysosome fusion (Luzio et al. 2009).

8.3.1.1 Autophagy/Secretory Pathway Interactions

A potential mechanism for delayed *C. burnetii* phagosome maturation involves interactions with autophagosomes (Gutierrez et al. 2005; Romano et al. 2007). As early as 5 min after NMII uptake by CHO cells, the nascent phagosome acquires the autophagosomal marker microtubule-associated protein light-chain 3 (LC3) and the marker remains associated through at least 3 days post infection (Romano et al. 2007; Beron et al. 2002; Campoy et al. 2011; Gutierrez et al. 2005). Rab24, a small GTPase involved in autophagy, also decorates mature NMII PV (Gutierrez et al. 2005). Overexpression of LC3 or starvation-induced autophagy both result in reduced recruitment of cathepsin D by the *C. burnetii* phagosome at 1 h post-infection (Romano et al. 2007). Thus, engagement of the autophagy pathway promotes delayed PV-lysosome fusion. In CHO and HeLa cells, NMII infection itself induces autophagy as evidenced by increased conversion of LC3-1 to LC3-II

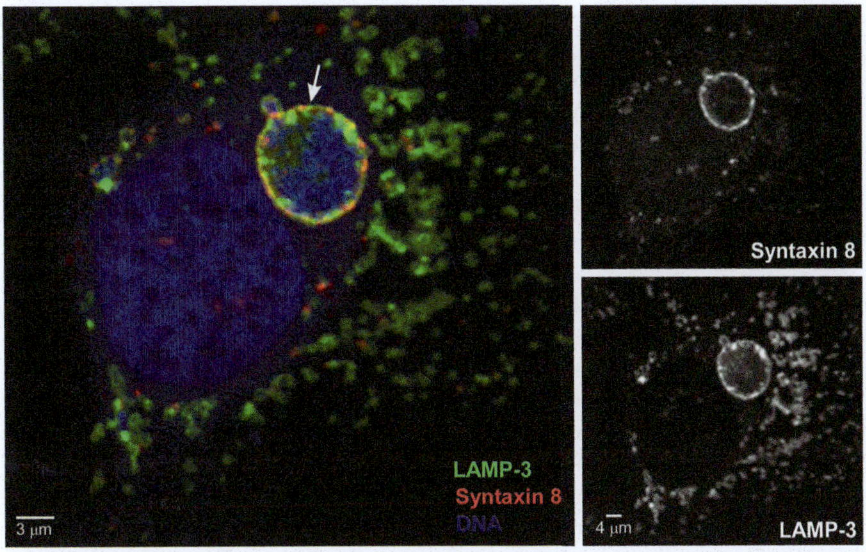

Fig. 8.2 The *C. burnetii* PV decorates with the t-SNARE syntaxin 8. HeLa cells infected with *C. burnetii* Nine Mile (phase II) for 2 days were stained by immunofluorescence for syntaxin 8 (*red*) and LAMP-3 (CD63) (*green*). Host and *C. burnetii* DNA were stained with DRAQ5 (*blue*). syntaxin 8 and LAMP-3 decorate the PV membrane (*arrow*) (Image courtesy of Dale Howe, Rocky Mountain Laboratories)

(Vazquez and Colombo 2010; Romano et al. 2007). An increasing body of evidence indicates that autophagy mediates innate immunity by delivering both intravacuolar and cytoplasmic bacterial pathogens to lysosomes for degradation (Deretic and Levine 2009). However, *C. burnetii* clearly benefits from interactions with the autophagic pathway as induction enhances PV formation and bacterial replication (Gutierrez et al. 2005).

An interplay between autophagy and apoptosis pathways in *C. burnetii* PV development has recently been described (Vazquez and Colombo 2010). In NMII-infected HeLa cells, Beclin 1, an important regulator of autophagy, is recruited to the PV membrane. Beclin 1 overexpression favors development of PV while siRNA knockdown reduces PV size. Beclin 1 has a BH2 domain that binds Bcl-2 family proteins involved in apoptosis, including anti-apoptotic Bcl-2. Interestingly, the beneficial effect of PV-localized Beclin-1 is optimized upon binding Bcl-2. Thus, *C. burnetii* infection modulates both autophagy and apoptotic pathways through Beclin 1/Bcl-2 interactions to promote successful infection.

New data indicates that *C. burnetii* benefits from PV interactions with the early secretory pathway (Campoy et al. 2011). In CHO and RAW murine macrophages cells, NMII PV recruit Rab1b, a small GTPase responsible for anterograde transport between the endoplasmic reticulum (ER) and Golgi apparatus. Overexpression of a dominant-negative form of Rab1b results in smaller PV and less *C. burnetii* replication. Disruption of the Golgi apparatus, and consequently the secretory pathway, by treating infected cells with brefeldin A or expression of a dominant-negative form of the small GTPase Sar1, also results in significantly smaller PV. Rab1b decoration is most noticeable on mature PV (>24 h post-infection), suggesting secretory pathway engagement is important during the expansion phase of PV development. Interestingly, autophagosome formation also requires Rab1 and functional ER exist sites (Zoppino et al. 2010), thereby implicating interactions between ER transport processes and autophagy in PV biogenesis.

8.3.1.2 *Coxiella* Modulation of PV Biogenesis

Collectively, the current model of *C. burnetii* PV biogenesis now includes endocytic, autophagy and secretory pathway interactions. As mentioned above, the *C. burnetii* phagosome matures with delayed kinetics. This behavior requires active *C. burnetii* protein synthesis as PV containing organisms inactivated with formalin or chloramphenicol mature with normal phagosome kinetics (Romano et al. 2007; Howe and Mallavia 2000) and fail to recruit LC3 (Romano et al. 2007). How *C. burnetii* benefits from stalled phagosome maturation is unclear. Two possibilities are (1) stalling affords the organism extra time to metabolically activate and express proteins ultimately required for survival within a phagolysosomal environment, and (2) stalling is necessary for differentiation of the non-replicating small cell variant (SCV) developmental form into the replicating large cell variant (LCV) developmental form (Voth and Heinzen 2007; Coleman et al. 2004). However, contrary data demonstrate that NMII inactivated with chloramphenicol persist in a viable state for days in lysosome-like vacuoles of Vero cells (Howe et al. 2003), and that purified NMII LCV and SCV are equally infectious (Omsland et al. 2009; Coleman et al. 2004).

In addition to PV biogenesis, *C. burnetii* protein synthesis is required for maintenance of the mature PV structure and its recruitment of lysosomes (Howe et al. 2003). Coincident with *C. burneti* entry into log phase (~2 days post-infection), mature PV containing NMII in Vero cells undergo a dramatic expansion, resulting in a large and spacious vacuole that is easily visible by light microscopy (Coleman et al. 2004). However, following a 12 h treatment with chloramphenicol, these vacuoles collapse and lose their spaciousness and phase translucent appearance (Howe et al. 2003). Moreover, fusion between mature PV and the lysosomal compartment, as evidenced by recruitment of latex bead-containing phagolysosomes, is substantially reduced in infected J774 macrophages treated with chloramphenicol (Howe et al. 2003). Although the *C. burnetii* protein effectors of PV maturation and maintenance are unknown, some are presumably delivered directly to the host cytosol via the pathogen's T4SS (Voth and Heinzen 2009a).

The PV filled with stationary phase *C. burnetii* (~6 days post-infection) can encompass nearly the entire host cell cytoplasm (Coleman et al. 2004). Interestingly, this parasitic burden does not impose obvious cytopathic effects and there is no concerted lytic event associated with *C. burnetii* host cell egress. Indeed, L929 and J774 infected with NMI or NMII continue to divide normally with the PV segregating to one of two daughter cells (Roman et al. 1986; Baca et al. 1985). Segregation of NMII PV does not apparently lead to defects in host cell cytokinesis as shown for vacuoles harboring *C. trachomatis* (Grieshaber et al. 2006).

8.3.2 PV Maturation/Biology in Human Mononuclear Phagocytes

As mentioned above, there are conflicting models regarding the ultimate maturation state of PV harboring phase I and phase II *C. burnetii* in human mononuclear phagocytes and the outcome of infection (Ghigo et al. 2002; Howe et al. 2010). These studies have exclusively used the NMI and NMII strains of *C. burnetii.*

8.3.2.1 NMI and NMII: Differential Trafficking (DT) Model

In the following sections, data supporting the DT model are presented. Selective receptor engagement by NMI and NMII that induces different cytoskeletal rearrangements and signaling is discussed, as is the involvement of TLR4 in host cell interactions. Moreover, the maturation process of PV containing NMI or NMII is described along with the modulation of trafficking by cytokines. The DT model asserts that the avirulence of NMII is associated with residence in a vacuole that fuses with lysosomes. Conversely, virulence of NMI is associated with residence in a vacuole that stalls at a late endosomal stage, thereby allowing pathogen growth (Ghigo et al. 2002).

Integrin-Dependent Phagocytosis

A critical element of the DT model is the engagement of different phagocyte receptors by NMI and NMII that ultimately affects PV outcome. Phagocytosis is an ancestral defense mechanism directed against microbial invasion. Professional mononuclear phagocytes, including circulating monocytes and tissue macrophages, ingest and degrade a wide variety of microorganisms such as bacteria, virus, fungi and protozoa. To discriminate between infectious agents, mononuclear phagocytes possess a restricted number of phagocytic receptors including Fc receptors and complement receptor (CR3, a heterodimer consisting of αM and β2 integrin proteins CD18 and CD11b) that recognize opsonized microorganisms, the mannose receptor that recognizes conserved motifs on pathogens, scavenger receptors that recognize diacyl lipids from the bacteria surface, and TLRs that recognize microbial structures such as LPS, peptidoglycan or flagellin (Taylor et al. 2005).

Numerous pathogens exploit the phagocytic process to infect host cells. In human mononuclear phagocytes, *C. burnetii* has developed a survival strategy based on subversion of receptor-mediated phagocytosis (Capo et al. 1999, 2003). In contrast to NMII, NMI organisms are poorly internalized by monocytes. Whereas the uptake of NMII is mediated by both αvβ3 integrin and CR3, the internalization of NMI involves only the engagement of αvβ3 integrin. The phagocytic efficiency of CR3 depends on its activation through αvβ3 integrin and an integrin-associated protein (CD47), a molecule physically and functionally associated with β3 integrins. Indeed, macrophages from CD47-deficient mice are unable to ingest NMII bacteria through CR3. It has been demonstrated that the inhibitory mechanism mediated by NMI does not target CD47 since they do not down-regulate the expression of CD47 (Capo et al. 1999).

The functional consequence of inhibition of αvβ3 integrin/CR3 crosstalk is survival of NMI in monocytes (Capo et al. 1999). Conversely, monocytes eliminate NMII, suggesting CR3 engagement is deleterious to *C. burnetii*. The involvement of CR3 in the microbicidal activity of myeloid cells such as monocytes and macrophages seems to depend on the nature and opsonization of pathogens (Stuart and Ezekowitz 2005). *Escherichia coli*, *Salmonella* sp. and *Pseudomonas aeruginosa* are eliminated after CR3-mediated internalization. In contrast, *Bordetella* sp. avoid killing by taking advantage of CR3 uptake (Agramonte-Hevia et al. 2002). CR3 engagement does not affect the intracellular survival of *Mycobacterium tuberculosis* (Hirsch et al. 1994) and CR3 is not involved in the development of *M. tuberculosis* infection (Hu et al. 2000).

Cytoskeleton Remodelling

C. burnetii virulence is associated in monocytes with cytoskeletal rearrangement (Meconi et al. 1998). NMI stimulates membrane protrusions accompanied with increased content in F-actin and transient and intense reorganization of F-actin. In contrast, NMII does not induce any change in cell morphology, actin polymerization and F-actin reorganization. The mechanism used by NMI to induce cytoskeletal rearrangement likely requires actin polymerization and the tension of actin-myosin filaments since F-actin colocalizes with myosin in cell protrusions. It has also been demonstrated that contact between *C. burnetii* and monocytes is necessary to induce cytoskeleton reorganization: bacterial supernatants do not stimulate F-actin reorganization and bacteria are in close apposition with F-actin protrusions.

The cytoskeletal reorganization induced by NMI is related to its inefficient uptake by human monocytes (Capo et al. 2003). Indeed, CD11b and CD18 molecules, the two components of CR3, are excluded from the protrusions induced by these bacteria, but not αvβ3 integrin, suggesting that a physical cross-talk between αvβ3 integrin and CR3 is needed to activate CR3. When CR3 is localized within protrusions induced by unrelated inducers of cytoskeletal rearrangement, including a chemoattractant such as RANTES (Regulated on Activation Normal T cell Expressed and Secreted), or THP-1 monocytes expressing Nef protein, the uptake of NMI is increased, demonstrating that the localization of CR3 in the vicinity of αvβ3 integrin facilitates *C. burnetii* uptake. In RANTES-stimulated monocytes and Nef-expressing monocytes, in which CR3 is distributed in the proximity of αvβ3 integrin, the replication of NMI is inhibited. Again, the mode of entry of *C. burnetii* into monocytes seems to govern the intracellular fate of organisms. Hence, the localization of CR3 is critical for *C. burnetii* uptake and also for the control of bacterial replication.

The cytoskeleton reorganization induced in monocytes by NMI is related to activation of protein tyrosine kinases (PTK). NMI induces early PTK activation and the tyrosine phosphorylation of several endogenous proteins, including Hck and Lyn, two Src-related kinases (Meconi et al. 2001). PTK activation reflects *C. burnetii* virulence since NMII bacteria do not stimulate PTK. Tyrosine-phosphorylated molecules colocalize with F-actin inside cell protrusions induced by NMI. PTK and Src kinase inhibitors block the formation of cell protrusions and F-actin rearrangement induced by NMI, demonstrating that tyrosine kinases are involved in the cytoskeletal reorganization induced by these bacteria. These inhibitors also increase the uptake of NMI but have no effect on the uptake of NMII, demonstrating functional links between PTK activation, cytoskeletal reorganization and *C. burnetii* uptake. It has been demonstrated that the activation of PTK also provides an uptake signal for several invasive pathogens such as *Listeria monocytogenes*, enteropathogenic *E. coli*, *Helicobacter pylori* and *Campylobacter* species (Cossart and Sansonetti 2004). Concerning the interaction of *C. burnetii* and monocytes, PTK activation may result in the formation of membrane ruffles that limit the redistribution of CR3 in contact areas between αvβ3 integrin and *C. burnetii*. Alternatively, PTK may target αvβ3 integrin, thus interfering with the cross-talk between αvβ3 integrin, CR3, and the actin cytoskeleton (Meconi et al. 1998; Patil et al. 1999). In contrast, some bacterial pathogens inhibit PTK and PTK-mediated microbicidal responses. For example, a tyrosine phosphatase of *Yersinia* spp. and *Salmonella enterica* serovar Typhimurium induces the disruption of the actin cytoskeleton and regulates bacterial uptake (Cossart and Sansonetti 2004).

Involvement of Toll-Like Receptors

Innate and adaptive immune responses are initiated by the recognition of microbial molecules through TLRs. Among TLRs, TLR4 and TLR2 are involved in the recognition of Gram-negative bacteria and Gram-positive bacteria, respectively. TLR4 specifically recognizes prototypic LPSs whereas TLR2 recognizes bacterial lipoproteins, proteoglycans, lipopeptides or LPS from *Porphyromonas gingivalis* (Underhill 2004).

TLR4 is involved in uptake of NMI by macrophages (Honstettre et al. 2004). Polymyxin B, which interferes with LPS binding, inhibits uptake of NMI, but not NMII. The uptake of NMI is also reduced in murine macrophages that do not express TLR4 whereas the uptake of NMII organisms remains unaffected. It has been also demonstrated that the uptake of NMI is independent of TLR2. Besides its role in the uptake of NMI, TLR4 is also involved in *C. burnetii*-induced F-actin reorganization, highlighting again the close relationship between *C. burnetii* uptake and cytoskeleton organization. The mechanism that connects TLR4 and the actin cytoskeleton remains hypothetical, even if the LPS-TLR complex is known to induce a transient F-actin remodeling in a p38- and ERK-dependent pathway (West et al. 2004).

TLR4 controls the immune response against *C. burnetii* through granuloma formation and cytokine production (Honstettre et al. 2004). NMI infection of wild type mice results in sustained formation of granulomas, indicative of a protective immune response, but granuloma formation is only transient in TLR4-deficient mice. The altered granuloma formation is associated with decreased levels of interferon (IFN)-γ. The levels of interleukin (IL)-10, known to impair the production of inflammatory type 1 cytokines, are also reduced in TLR4-deficient mice, suggesting that cytokines are involved in the defective formation of granulomas.

In contrast, TLR4 is dispensable for the survival of NMI in macrophages as growth occurs in TLR4-deficient macrophages (Honstettre et al. 2004). TLR4 is also dispensable for NMI infection of mice. Indeed, NMI infection of tissues and bacterial clearance are similar in wild type and TLR4-deficient mice (Honstettre et al. 2004). Similarly to TLR4, TLR2 is involved in inflammatory and immune responses to NMI but is not necessary for bacterial clearance (Meghari et al. 2005). These results may be explained by the results of Zamboni et al. (2004) who demonstrated that purified lipid A from NMI or NMII fails to activate TLR2 and TLR4. We can suppose that some functions induced by *C. burnetii* need TLR4 (and/or TLR2) whereas other functions are independent of TLR4 (and/or TLR2). It has been suggested that TLRs may regulate phagosome maturation in macrophages (Blander and Medzhitov 2004) even if this hypothesis has been debated (Yates and Russell 2005). TLR4 does not control the maturation of *C. burnetii* phagosomes since NMI colocalize with LAMP-1 but not with cathepsin D in wild type and TLR4-deficient macrophages (Honstettre et al. 2004).

Intracellular Trafficking

In human monocytes and macrophages, NMI survive and replicate whereas NMII is eliminated (Fig. 8.3a, b) (Ghigo et al. 2002, 2006). During the first hours after uptake, nascent NMI and NMII phagosomes interact with intracellular compartments related to the early endosomal network, as revealed by the presence of a marker such as EEA1. This association of *C. burnetii* with the early endocytic network is transient, demonstrating that the interaction between the early endosomal compartment and *C. burnetii* phagosomes is normal. Later, *C. burnetii* phagosomes interact with late endosomes, as demonstrated by their colocalization with LAMP-1, CD63, and mannose-6-phosphate receptor. Mature PV containing NMI and NMII also accumulate V-H^+-ATPase and acquire a pH around 4.5–5, demonstrating that *C. burnetii* virulence is not related to the acidic pH of PV. In contrast, *C. burnetii* virulence seems associated with a defective acquisition of Rab7, a small GTPase involved in phagosome maturation (Desjardins et al. 1994). Indeed, Rab7 is acquired by PV containing NMII but only in part by those containing NMI (Ghigo et al. 2002). The different levels of Rab7 acquisition may explain the defective fusion of PV containing NMI with lysosomes because the amount of Rab proteins on endocytic organelles is critical for the fusion process (Henry et al. 2004; Rink et al. 2005). The partial acquisition of Rab7 by PV containing NMI also suggests that

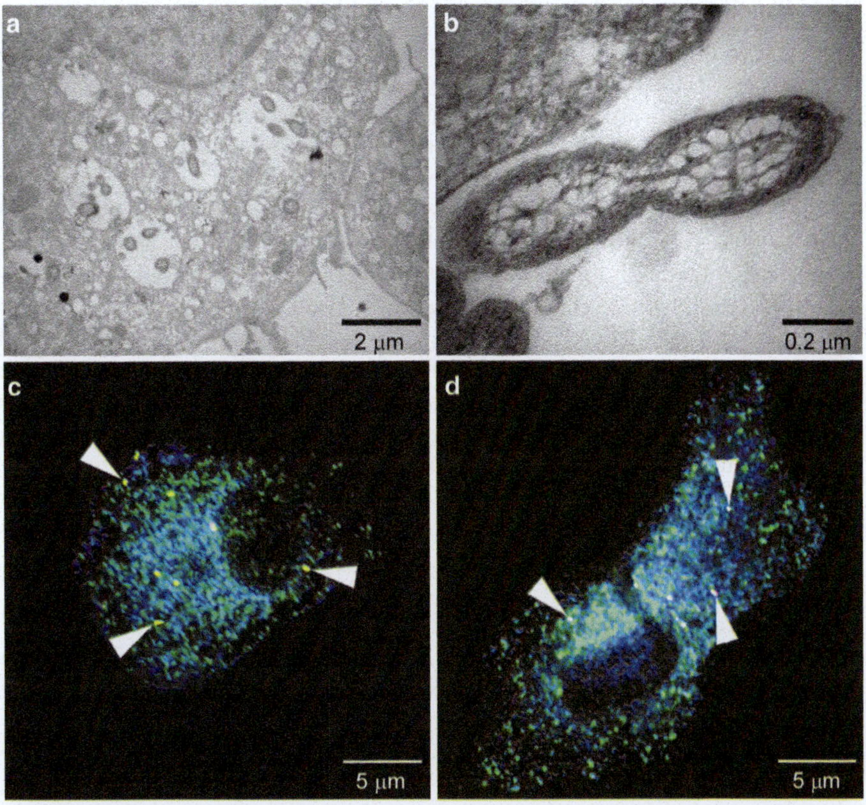

Fig. 8.3 Differential trafficking of *C. burnetii* Nine Mile phase I (NMI) and Nine Mile phase II (NMII) in human mononuclear phagocytes. (**a** and **b**) Electron micrographs of vacuoles containing NMI in human monocyte-derived macrophages. A dividing organism is observed in (**b**). Confocal microscopy of human monocyte-derived macrophages infected with NMI (**c**) or NMII (**d**) for 96 h. Cells were stained by immunofluorescence using antibodies directed against *C. burnetii* (*red*), LAMP-1 (*blue*), and cathepsin D (*green*). Cathepsin D localizes with NMII, but not NMI (*arrowheads*) (Images courtesy of Eric Ghigo, Université de la Méditerranée)

downstream maturation events are impaired. Cathepsin D, a lysosomal protease, does not accumulate in PV containing NMI. The lack of cathepsin D colocalization with NMI is due to defective fusion of PV with lysosomes, as evidenced with a lysosomotropic probe. The impaired acquisition of cathepsin D by PV containing NMI is related to bacterial virulence since NMII colocalize with cathepsin D (Fig. 8.3c, d). Taken together, these data indicate that the survival of NMI in mononuclear phagocytes is associated with an altered maturation of PV.

This hijacking strategy is reminiscent of the escape mechanism used by *S. typhimurium* (Holden 2002). However, the involved molecular mechanisms are different: while NMI PV acquire Rab7 only in part, *Salmonella*-containing vacuoles recruit LAMP-1 in a Rab7-dependent manner (Meresse et al. 1999a). It is hypothesized that Rab7 regulates vesicle traffic in late endocytosis (Bucci et al. 2000;

Zerial and McBride 2001). The fact that NMI does not modify the early acquisition of EEA1 but alters the acquisition of Rab7 by its PV strengthens this hypothesis. It is likely that the survival strategy of *C. burnetii* in mononuclear phagocytes is based on the interference of Rab7 recruitment at the surface of *C. burnetii*-containing vacuoles, thus leading to the formation of PV unable to interact with lysosomes.

The discrepancies in the literature regarding NMI and NMII trafficking may reflect the nature of engaged receptors. As discussed earlier, the route of *C. burnetii* internalization may influence the intracellular fate of the bacteria (Capo et al. 1999, 2003). The importance of pathogen routes of internalization has been clearly demonstrated for mycobacteria. Indeed, IgG-opsonised mycobacteria internalized via Fc receptors are delivered to phagolysosomes and killed whereas unopsonized mycobacteria replicate in vacuoles having characteristics of an early endosome (Scott et al. 2003). These observations suggest that the nature of the receptor engaged during bacterial entry may govern the molecular mechanisms involved in phagosome maturation (Scott et al. 2003).

Modulation of Trafficking by Cytokines

Cytokines are well known to modulate the microbicidal activity of macrophages. IFN-γ, an inflammatory Th1 cytokine, activates macrophages, leading to the control or elimination of several intracellular pathogens (Santic et al. 2005). Intracellularly, IFN-γ induces the killing of NMI by naive monocytes (Dellacasagrande et al. 1999) and the maturation of PV containing NMI as demonstrated by their acquisition of cathepsin D (Ghigo et al. 2002). IFN-γ acts through two distinct mechanisms. First, the addition of IFN-γ to monocytes infected with NMI stimulates PV-lysosome fusion without affecting PV pH. In contrast, IFN-γ decreases the pH of *Mycobacterium*-containing phagosomes (Schaible et al. 1999) and rescues phagosome maturation (Tsang et al. 2000). Second, the treatment of monocytes by IFN-γ prior to infection with NMI induces the alkalinization of *C. burnetii* PV independently of V-H^+-ATPase exclusion. IFN-β also induces the alkalinization of the trans-Golgi network by inhibiting V-H^+-ATPase activity (Sidhu et al. 1999). The mechanism of phagosome alkalinization mediated by IFN-γ is still unknown. It has been shown that IFN-γ inhibits the remodeling of *Legionella*-containing phagosomes into endoplasmic reticulum-derived vesicles (Santic et al. 2005). It is noteworthy that acute Q fever is characterized by the production of IFN-γ whereas in patients with chronic Q fever, the production of IFN-γ is defective (Koster et al. 1985; Izzo and Marmion 1993). Monocytes from patients with chronic Q fever are unable to kill NMI and exhibit a defective maturation of bacterial phagosomes (Ghigo et al. 2004). The production of IFN-γ by patients with acute Q fever may control *C. burnetii* infection through PV-lysosome fusion (Ghigo et al. 2002, 2004).

IL-10, an immunoregulatory cytokine, also modulates trafficking of NMI. IL-10 is known to support the intracellular replication of numerous bacterial pathogens (Blauer et al. 1995; Park and Skerrett 1996). Q fever is characterized by the overproduction of IL-10 (Capo et al. 1996a, b; Honstettre et al. 2003) and IL-10 stimulates

NMI replication (Ghigo et al. 2004). Monocytes from convalescent patients recovering from acute Q fever exhibit fusion of NMI PV with lysosomes and, subsequently, efficient bacterial killing. It is related to the activity of the disease since NMI killing and PV-lysosome fusion are restored in patients who have recovered from chronic Q fever. Interestingly, NMI killing by monocytes and PV maturation are modulated by IL-10. The neutralization of endogenous IL-10 overproduced by patients with ongoing chronic Q fever enhances NMI killing by monocytes to the level of cured patients and restores PV-lysosome fusion. Similarly, adding recombinant IL-10 to monocytes from patients with cured chronic Q fever prevents NMI killing and induces PV maturation blockage (Ghigo et al. 2004). These results concerning the role of IL-10 in the inhibition of phagolysosome fusion are consistent with those of Via et al. (1998) who found that maturation of mycobacterial phagosomes is improved in macrophages from IL-10 deficient mice.

8.3.2.2 NMI and NMII: Similar Trafficking (ST) Model

An opposing model by Howe et al. (2010) proposes similar trafficking (ST) of NMI and NMII in human mononuclear phagocytes, and that the virulence of *C. burnetii* is unrelated to PV maturation status. Data for the ST model is primarily derived using human THP-1 cells, differentiated into macrophage-like cells using phorbol 12-myristate 13-acetate, and primary human peripheral blood monocyte-derived macrophages (HMDM).

In both HMDM and THP-1 cells, coinfection experiments show a common PV can support growth of both NMI and NMII (Howe et al. 2010). In THP-1 cells, there is no difference in the percentage of NMI and NMII PV decorated with Rab5, Rab7, CD63, and cathepsin D at early (8 h) and late (72 h) time points post-infection. At 72 h post-infection, greater than 80% of NMI and NMII PV colocalize with the late endosomal/lysosomal markers Rab7, CD63, and cathepsin D. Mature PV containing NMI or NMII readily degrade bovine serum albumin in a cathepsin D-dependent manner, and contain proteolytically-active cathepsins B, K and L as assessed by the cleavage of cathepsin-specific fluorogenic peptides. Protease activity correlates with the ability of NMI and NMII PV to completely degrade *E. coli* within 15 min. Collectively, these data suggest NMI and NMII reside in phenotypcially-indistinguishable PV that fully mature through the endolysosomal pathway. In both HMDM and THP-1 cells, phase variants replicate with similar kinetics, achieving roughly 2–3 logs of growth before reaching stationary phase (Howe et al. 2010). Thus, in the ST model, NMI and NMII phase variants replicate in PV that retain phagolysosome degradative activities, and *C. burnetii* resistance to this environment represents a pathogenic strategy.

Using microscopy and/or genome equivalent PCR assays, several independent studies have also revealed growth of NMII in human monuclear phagocytes. NMII grows robustly in CD63-positive PV of human monocyte-derived dendritic cells (Shannon et al. 2005; Omsland et al. 2010; Sauer et al. 2005). Primary alveolar macrophages from humans (Fig. 8.4a, b) and cynomolgous monkeys (*Macaca fascicularis*) also support growth of NMII, with an approximately 3 log increase in genome equivalents at 6 days post-infection documented for the latter cell type (J. Hill and R. A. Heinzen, unpublished data). With respect to human monocyte/macrophage-like cell lines, NMII infection of THP-1 cells was used to decipher *C. burnetii* anti-apoptotic signaling (Voth et al. 2007; Voth and Heinzen 2009b), probe the roles of kinases/phosphatases in PV formation (Hussain et al. 2010), identify pathogen proteins secreted into the host cell cytoplasm (Chen et al. 2010; Voth et al. 2011), and to define host cell transcriptional responses to infection (Mahapatra et al. 2010; Ren et al. 2003). Furthermore, NMII-infected MonoMac-1 cells were employed to characterize pathogen clearance activities of TLR agonists (Lubick et al. 2007). Electron micrographs of PV harboring NMII in THP-1 cells are depicted in Fig. 8.4c, d.

8.3.2.3 Reconciliation of DT and ST Trafficking Models

Subtleties between laboratories in the microscopic scoring of PV cellular markers might partially explain the different phenotypes of vacuoles harboring NMI in the DT and ST models. The complicated post-translational processing of cathepsin D makes this marker particularly problematic (Zaidi et al. 2008). What is more difficult to reconcile is why NMII grows robustly in the ST model while showing no growth in the DT model. Studies related to both models used non-opsonized bacteria; thus, the routes of NMI and NMII entry should be similar. In fact, Fc receptor-mediated uptake of NMI by HMDM, J774 cells, or primary guinea pig peritoneal macrophages does not inhibit pathogen replication (Hinrichs and Jerrells 1976; Baca et al. 1984; Shannon et al. 2009). An intriguing possibility is that the NMII strain used in DT studies has a novel genetic mutation(s) that restricts growth in human mononuclear phagocytes, but not in continuous lines such as L929 cells (Ghigo et al. 2002). NMII is a high passage derivative of low passage NMI that was plaque-cloned after 90 egg passages (Amano and Williams 1984). The strain has subsequently undergone extensive (and undocumented) laboratory passages in various labs. Thus, it is possible that the NMII strains used in ST and DT studies are genetically different. Minor genetic differences can confer permissive or non-permissive inter-strain growth phenotypes in intracellular bacteria. For example, mutations in *C. trachomatis trpA*, encoding tryptophane synthase, are associated with growth in the presence of IFNγ and implicated in tissue tropisms (Caldwell et al. 2003). A *C. burnetii* strain that cannot adequately metabolize and grow inside a given host cell should, by default, traffic to a phagolysosomal compartment, a behavior observed with chloramphenicol-treated organisms (Howe et al. 2003). The NMII used in studies related to the ST model was derived from a single source (Rocky Mountain Laboratories, Hamilton, Montana). Recent derivation of this strain's genome sequence shows that, in addition to the 25,992 bp deletion of LPS biosynthesis genes, it has 18 single nucleotide polymorphisms relative to NMI, none of which

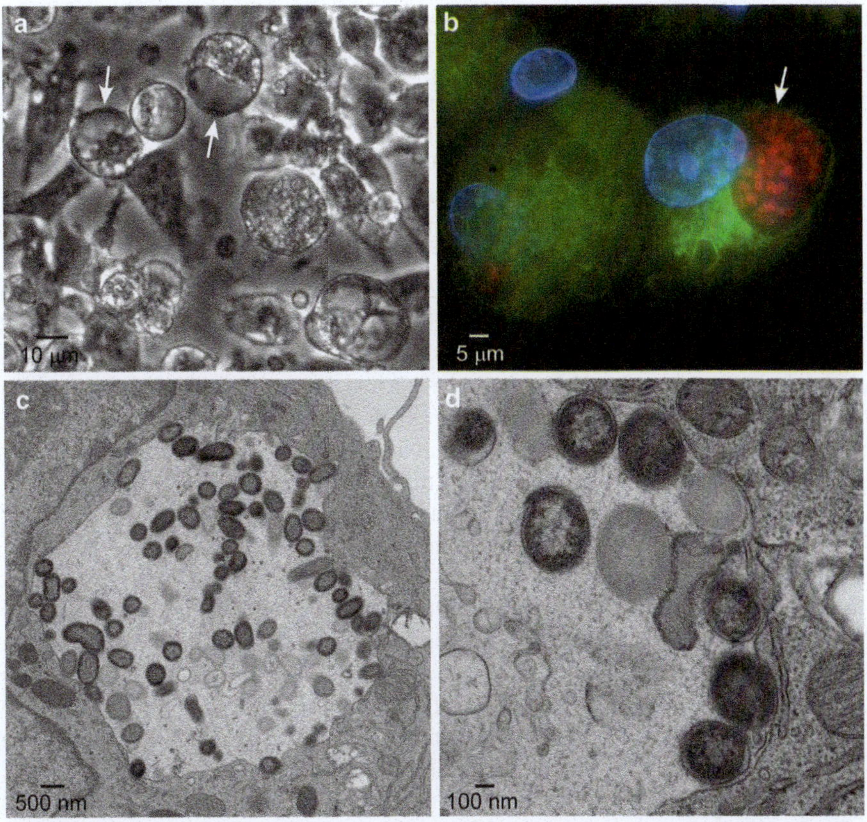

Fig. 8.4 Growth of *C. burnetii* Nine Mile (phase II) in mononuclear phagocytes. Phase contrast (**a**) and immunofluorescence (**b**) images of primary human alveolar macrophages infected with *C. burnetii* Nine Mile (phase II) for 3 days. Cells in (**b**) were stained by immunofluorescence using antibodies directed against *C. burnetii* (*red*) and the LAMP-3 (CD-63) (*green*). Host and *C. burnetii* DNA were stained with DRAQ5 (*blue*). Large PV (*arrows*) containing *C. burnetii* are seen in both micrographs. (**c**) and (**d**) Electron micrographs of THP-1 human monocyte-like cells infected with *C. burnetii* Nine Mile (phase II) for 4 days. Prior to infection, cells were differentiated into adherent, macrophage-like cells by treatment with PMA. Micrographs show large and spacious PV harboring multiple *C. burnetii*. Many organisms have intimate contact with the PV membrane that may facilitate cytosolic delivery of effector proteins required for PV biogenesis (Images in panel **a** and **b** are courtesy of Dan Voth, University of Arkansas for Medical Science. Images in panels **c** and **d** are courtesy of Dale Howe and Beth Fischer, Rocky Mountain Laboratories)

are predicted to disrupt proteins required for intracellular growth (P. A. Beare and R. A. Heinzen, unpublished data). Comparison of the genome sequence of NMII organisms associated with DT and ST models might be illuminating. Perhaps more informative would be testing of additional non-revertable phase II clones of different *C. burnetii* strains. Unfortunately, such clones are currently unavailable.

8.3.3 The Murine Macrophage Anomaly

While debate continues on the outcome of NMII infection of human mononuclear phagocytes, the strain clearly has severe growth defects relative to NMI in primary mouse macrophages (Honstettre et al. 2004; Sauer et al. 2005; Zamboni et al. 2002, 2004; Zamboni 2004; Zamboni and Rabinovitch 2003, 2004). This behaviour contrasts markedly with the proficient growth observed in multiple mouse macrophage-like cells (Baca et al. 1981; Briggs et al. 2008; Roman et al. 1986; Maurin et al. 1992b; Zamboni et al. 2002) and L929 fibroblasts (Howe et al. 2002; Baca et al. 1985; Burton et al. 1978; Roman et al. 1986; Maurin et al. 1992b; Zamboni et al. 2002). As mentioned above, surface TLR ligands of NMII, such as lipoproteins, are thought to be more accessible due the strain's lack of LPS *O*-antigen (Shannon et al. 2005). Thus, potent activation of the primary mouse macrophage pathogen recognition system by NMII may induce production of a cellular effector that limits replication. Consistent with this idea is the observation that primary mouse macrophages from TLR2 knockout mice are markedly more permissive for growth of NMII than wild type macrophages, and produce less pro-inflammatory IL-12 and TNF-α (Zamboni et al. 2004). Moreover, the corresponding knockout mouse is more susceptible to NMII infection (Ochoa-Reparaz et al. 2007). Nitric oxide (NO) induced by TNF-α may be the critical effector limiting NMII growth in primary mouse macrophages (Brennan et al. 2004; Zamboni and Rabinovitch 2003, 2004; Howe et al. 2002). NMII infection induces significant amounts of NO and treatment of macrophages with inducible NO synthase (iNOS) inhibitors, such as aminoguanidine, enhances NMII replication (Zamboni and Rabinovitch 2003). Furthermore, NMII growth is improved in macrophages from iNOS knockout mice (Zamboni and Rabinovitch 2004). TNF-α also controls mycobacteria infection of primary mouse macrophages in an iNOS-dependent manner (Bekker et al. 2001).

C. burnetii appears exquisitely sensitive to oxidative and nitrosative stress (Brennan et al. 2004) and the organism has evolved active mechanisms of avoidance. For example, infection of J774 macrophages or human neutrophils by *C. burnetii* results in little superoxide anion production (Baca et al. 1984; Akporiaye et al. 1990; Hill and Samuel 2011; Siemsen et al. 2009). In human neutrophils, this behavior involves pathogen inhibition of the host cell NADPH oxidase complex via the activity of a secreted acid phosphatase (Hill and Samuel 2011; Siemsen et al. 2009; Baca et al. 1993b, 1994; Li et al. 1996). The heightened sensitivity of *C. burnetii* to oxidants may partially explain the requirement for a microaerobic (2.5% O_2) environment to support pathogen growth in a recently established host cell-free growth system (Omsland et al. 2009).

Interestingly, relative to NMI, NMII also stimulates potent activation and maturation of human monocyte-derived dendritic cells and their subsequent release of pro-inflammatory cytokines (i. e., IL-12 and TNF-α) (Shannon et al. 2005). Despite activation, NMI grows with the same kinetics as NMII in these cells (Shannon et al. 2005). However, this behavior is predicted to result in potentiated innate and adaptive immune responses that prevents productive NMII infection of animals (Shannon et al. 2005).

8.4 Summary

The *C. burnetii* PV is a unique intracellular niche and a fascinating example of pathogen-host adaptation. The acid activated metabolism of the organism represents a biochemical strategy that ensures replication only within an acidified compartment while also conferring metabolic quiescence and stability outside the host cell (Hackstadt and Williams 1981). Lysosomal fusion with the PV is documented in many cells types (Voth and Heinzen 2007). How does *C. burnetii* resist degradation by lysosomal hydrolases and other toxic elements such as antimicrobial peptides? Resistance appears to be an inherent property of *C. burnetii*, as non-metabolizing organisms remain viable for several days in lysosome-like vacuoles of Vero cells (Howe et al. 2003). Full-length LPS is not required for resistance as NMI and NMII grow equally well in multiple cells types. An unusual cell wall containing protease-resistant peptidoglycan-associated proteins may be a resistance mechanism (Amano et al. 1984).

Early elucidation of *C. burnetii*'s metabolic requirements set the stage for the recent development of a hypoxic (2.5% O_2) host cell-free growth method using Acidified Citrate Cysteine Medium (ACCM) (Omsland et al. 2009). The precise biochemical and physiochemical character of the PV luminal environment remains an interesting question. However, one would surmise that it likely resembles ACCM in being rich in amino acids/peptides, critical carbon and energy sources of *C. burnetii*, with low oxygen content. Amino acids are likely delivered to the PV via fusion with autophagosomes and/or late endosomes/lysosomes carrying cargo expected to be rich in peptides (Gutierrez et al. 2005; Heinzen et al. 1996). A similar mechanism for nutrient delivery is proposed for *Leishmania mexicana*, a protozoan pathogen with a PV phenotypically similar to *C. burnetii*'s (Schaible et al. 1999). Furthermore, oxygen gradients across biological membranes can be associated with oxygen concentrations considerably lower than atmospheric oxygen levels (i.e., ~20%) (Hu et al. 1992).

Biogenesis of the PV begins with receptor-mediated phagocytosis. The *C. burnetii* protein adhesin(s) mediating internalization is unknown. The organism encodes multiple proteins with integrin binding RGD motifs that could potentially interact with cellular integrin receptors (Ruoslahti 1996; Seshadri et al. 2003; Beare et al. 2009b). The nascent phagosome quickly becomes fusogenic with other cellular vesicles, including autophagosomes, that contribute membrane to the enlarging PV. Bacterial pathogens utilize several strategies to manipulate the host cell membrane machinery including subversion of phosphoinosotide metabolism (Weber et al. 2009), and the functions of SNAREs and Rab family GTPases (Wesolowski and Paumet 2010; Cossart and Roy 2010). *C. burnetii* proteins predicted to modulate these and other host factors reside within a repertoire of proteins secreted into the host cytosol by the pathogen's Dot/Icm T4SS. To date, 49 *C. burnetii* Dot/Icm secretion substrates have been identified (Voth et al. 2009, 2011; Pan et al. 2008; Luhrmann et al. 2010). However, only AnkG currently has a defined function in inhibiting apoptosis via a mechanism involving binding of the pro-apoptotic protein gC1qR (p32) (Luhrmann et al. 2010). An F-actin mesh surrounds the mature

PV, but what functional role(s) does this structure play and are other cytoskeletal elements, such as microtubules, involved in PV development?

Another lingering question regards the immune function of *C. burnetii* PV. In HeLa cells, the vacuole functions as the loading compartment of the class II antigen presentation pathway (Lem et al. 1999). The class II MHC molecules HLA-DM and HLA-DR accumulate in the vacuole, which seems to affect the dynamics and repertoire of peptide loading. Consequently, defects in immune recognition of *C. burnetii* may occur. The distribution of MHC class I is unaffected by *C. burnetii* infection (Lem et al. 1999); however, whether the PV participates in antigen cross-presentation (Houde et al. 2003) is unknown.

The prior obligate intracellular nature of *C. burnetii* imposed significant experimental obstacles to unraveling the molecular details of PV biogenesis and host cell manipulation. However, novel genetic tools and host cell-free growth (Beare et al. 2011; Omsland et al. 2009) will enable novel lines of investigation that should entice new investigators into the field. Studies from additional research groups will eventually resolve discrepant data on NMI and NMII trafficking in human mononuclear phagocytes. An improved understanding of *C. burnetii* cellular microbiology will aid our ability to model mechanisms used by the pathogen to cause animal and human disease.

Acknowledgements We thank Dale Howe and Stacy Gilk for critical reading of the manuscript. This research was supported by the CNRS (E. G.), the Agencia Nacional de Promoción Científica y Tecnológica and SECTyP (Universidad Nacional de Cuyo) (M. I. C.) and the Intramural Research Program of the National Institutes of Health, National Institute of Allergy and Infectious Diseases (R. A. H.)

References

Agramonte-Hevia J, Gonzalez-Arenas A, Barrera D, Velasco-Velazquez M (2002) Gram-negative bacteria and phagocytic cell interaction mediated by complement receptor 3. FEMS Immunol Med Microbiol 34:255–266

Aguilera M, Salinas R, Rosales E, Carminati S, Colombo MI, Beron W (2009) Actin dynamics and Rho GTPases regulate the size and formation of parasitophorous vacuoles containing *Coxiella burnetii*. Infect Immun 77:4609–4620

Akporiaye ET, Rowatt JD, Aragon AA, Baca OG (1983) Lysosomal response of a murine macrophage-like cell line persistently infected with *Coxiella burnetii*. Infect Immun 40: 1155–1162

Akporiaye ET, Stefanovich D, Tsosie V, Baca G (1990) *Coxiella burnetii* fails to stimulate human neutrophil superoxide anion production. Acta Virol 34:64–70

Alonso A, Garcia-Del Portillo F (2004) Hijacking of eukaryotic functions by intracellular bacterial pathogens. Int Microbiol 7:181–191

Amano K, Williams JC (1984) Chemical and immunological characterization of lipopolysaccharides from phase I and phase II *Coxiella burnetii*. J Bacteriol 160:994–1002

Amano K, Williams JC, Mccaul TF, Peacock MG (1984) Biochemical and immunological properties of *Coxiella burnetii* cell wall and peptidoglycan-protein complex fractions. J Bacteriol 160:982–988

Aragon AS, Pereira HA, Baca OG (1995) A cationic antimicrobial peptide enhances the infectivity of *Coxiella burnetii*. Acta Virol 39:223–226

Ariel BM, Khavkin TN, Amosenkova NI (1973) Interaction between *Coxiellae burnetii* and the cells in experimental Q-rickettsiosis. Histologic and electron microscope studies. Pathol Microbiol 39:412–423

Baca OG, Akporiaye ET, Aragon AS, Martinez IL, Robles MV, Warner NL (1981) Fate of phase I and phase II *Coxiella burnetii* in several macrophage-like tumor cell lines. Infect Immun 33:258–266

Baca O, Akporiaye ET, Rowatt JD (1984) Possible biochemical adaptations of *Coxiella burnetii* for survival within phagocytes: effects of antibody. In: Leive L, Schlessinger D (eds) Microbiology 1984. ASM Press, Washington, DC

Baca OG, Scott TO, Akporiaye ET, Deblassie R, Crissman HA (1985) Cell cycle distribution patterns and generation times of L929 fibroblast cells persistently infected with *Coxiella burnetii*. Infect Immun 47:366–369

Baca OG, Klassen DA, Aragon AS (1993a) Entry of *Coxiella burnetii* into host cells. Acta Virol 37:143–155

Baca OG, Roman MJ, Glew RH, Christner RF, Buhler JE, Aragon AS (1993b) Acid phosphatase activity in *Coxiella burnetii*: a possible virulence factor. Infect Immun 61:4232–4239

Baca OG, Li YP, Kumar H (1994) Survival of the Q fever agent *Coxiella burnetii* in the phagolysosome. Trends Microbiol 2:476–480

Barry AO, Mege JL, Ghigo E (2011) Hijacked phagosomes and leukocyte activation: an intimate relationship. J Leukoc Biol 89:373–382

Beare PA, Samuel JE, Howe D, Virtaneva K, Porcella SF, Heinzen RA (2006) Genetic diversity of the Q fever agent, *Coxiella burnetii*, assessed by microarray-based whole-genome comparisons. J Bacteriol 188:2309–2324

Beare PA, Howe D, Cockrell DC, Omsland A, Hansen B, Heinzen RA (2009a) Characterization of a *Coxiella burnetii ftsZ* mutant generated by *Himar1* transposon mutagenesis. J Bacteriol 191:1369–1381

Beare PA, Unsworth N, Andoh M, Voth DE, Omsland A, Gilk SD, Williams KP, Sobral BW, Kupko JJ 3rd, Porcella SF, Samuel JE, Heinzen RA (2009b) Comparative genomics reveal extensive transposon-mediated genomic plasticity and diversity among potential effector proteins within the genus *Coxiella*. Infect Immun 77:642–656

Beare PA, Sandoz KM, Omsland A, Rockey DD, Heinzen RA (2011) Advances in genetic manipulation of obligate intracellular bacterial pathogens. Front Microbiol. 2:97

Bekker LG, Freeman S, Murray PJ, Ryffel B, Kaplan G (2001) TNF-alpha controls intracellular mycobacterial growth by both inducible nitric oxide synthase-dependent and inducible nitric oxide synthase-independent pathways. J Immunol 166:6728–6734

Ben Amara A, Ghigo E, Le Priol Y, Lepolard C, Salcedo SP, Lemichez E, Bretelle F, Capo C, Mege JL (2010) *Coxiella burnetii*, the agent of Q fever, replicates within trophoblasts and induces a unique transcriptional response. PLoS One 5:e15315

Beron W, Gutierrez MG, Rabinovitch M, Colombo MI (2002) *Coxiella burnetii* localizes in a Rab7-labeled compartment with autophagic characteristics. Infect Immun 70:5816–5821

Blander JM, Medzhitov R (2004) Regulation of phagosome maturation by signals from toll-like receptors. Science 304:1014–1018

Blauer F, Groscurth P, Schneemann M, Schoedon G, Schaffner A (1995) Modulation of the antilisterial activity of human blood-derived macrophages by activating and deactivating cytokines. J Interferon Cytokine Res 15:105–114

Brennan RE, Russell K, Zhang G, Samuel JE (2004) Both inducible nitric oxide synthase and NADPH oxidase contribute to the control of virulent phase I *Coxiella burnetii* infections. Infect Immun 72:6666–6675

Briggs HL, Pul N, Seshadri R, Wilson MJ, Tersteeg C, Russell-Lodrigue KE, Andoh M, Baumler AJ, Samuel JE (2008) Limited role for iron regulation in *Coxiella burnetii* pathogenesis. Infect Immun 76:2189–2201

Bucci C, Thomsen P, Nicoziani P, Mccarthy J, Van Deurs B (2000) Rab7: a key to lysosome biogenesis. Mol Biol Cell 11:467–480

Burnet FM, Freeman M (1937) Experimental studies on the virus of "Q" fever. Med J Aust 2:299–305

Burton PR, Kordova N, Paretsky D (1971) Electron microscopic studies of the rickettsia *Coxiella burnetii*: entry, lysosomal response, and fate of rickettsial DNA in L-cells. Can J Microbiol 17:143–150

Burton PR, Stueckemann J, Paretsky D (1975) Electron microscopy studies of the limiting layers of the rickettsia *Coxiella burneti*. J Bacteriol 122:316–324

Burton PR, Stueckemann J, Welsh RM, Paretsky D (1978) Some ultrastructural effects of persistent infections by the rickettsia *Coxiella burnetii* in mouse L cells and green monkey kidney (Vero) cells. Infect Immun 21:556–566

Caldwell HD, Wood H, Crane D, Bailey R, Jones RB, Mabey D, Maclean I, Mohammed Z, Peeling R, Roshick C, Schachter J, Solomon AW, Stamm WE, Suchland RJ, Taylor L, West SK, Quinn TC, Belland RJ, Mcclarty G (2003) Polymorphisms in *Chlamydia trachomatis* tryptophan synthase genes differentiate between genital and ocular isolates. J Clin Invest 111:1757–1769

Campoy EM, Zoppino FC, Colombo MI (2011) The early secretory pathway contributes to the growth of the *Coxiella*-replicative niche. Infect Immun 79:402–413

Capo C, Zaffran Y, Zugun F, Houpikian P, Raoult D, Mege JL (1996a) Production of interleukin-10 and transforming growth factor beta by peripheral blood mononuclear cells in Q fever endocarditis. Infect Immun 64:4143–4147

Capo C, Zugun F, Stein A, Tardei G, Lepidi H, Raoult D, Mege JL (1996b) Upregulation of tumor necrosis factor alpha and interleukin-1 beta in Q fever endocarditis. Infect Immun 64:1638–1642

Capo C, Lindberg FP, Meconi S, Zaffran Y, Tardei G, Brown EJ, Raoult D, Mege JL (1999) Subversion of monocyte functions by *Coxiella burnetii*: impairment of the cross-talk between $\alpha_v\beta_3$ integrin and CR3. J Immunol 163:6078–6085

Capo C, Moynault A, Collette Y, Olive D, Brown EJ, Raoult D, Mege JL (2003) *Coxiella burnetii* avoids macrophage phagocytosis by interfering with spatial distribution of complement receptor 3. J Immunol 170:4217–4225

Chen C, Banga S, Mertens K, Weber MM, Gorbaslieva I, Tan Y, Luo ZQ, Samuel JE (2010) Large-scale identification and translocation of type IV secretion substrates by *Coxiella burnetii*. Proc Natl Acad Sci U S A 107:21755–21760

Coleman SA, Fischer ER, Howe D, Mead DJ, Heinzen RA (2004) Temporal analysis of *Coxiella burnetii* morphological differentiation. J Bacteriol 186:7344–7352

Cooke RA (2008) Q fever. Was Edward Derrick's contribution undervalued? Med J Aust 189:660–662

Cossart P, Roy CR (2010) Manipulation of host membrane machinery by bacterial pathogens. Curr Opin Cell Biol 22:547–554

Cossart P, Sansonetti PJ (2004) Bacterial invasion: the paradigms of enteroinvasive pathogens. Science 304:242–248

Cox HR (1938) A filter-passing infectious agent isolated from ticks. III. Description of organism and cultivation experiments. Public Health Rep 53:2270–2276

Cox HR (1939) Sudies of a filter-passing infectious agent isolated from ticks. V. Further attempts to cultivate in cell-free media. Suggested classification. Public Health Rep 54:1822–1827

Dellacasagrande J, Capo C, Raoult D, Mege JL (1999) IFN-gamma-mediated control of *Coxiella burnetii* survival in monocytes: the role of cell apoptosis and TNF. J Immunol 162:2259–2265

Denison AM, Massung RF, Thompson HA (2007) Analysis of the O-antigen biosynthesis regions of phase II isolates of *Coxiella burnetii*. FEMS Microbiol Lett 267:102–107

Deretic V, Levine B (2009) Autophagy, immunity, and microbial adaptations. Cell Host Microbe 5:527–549

Derrick EH (1937) "Q" fever, a new fever entity: clinical features, diagnosis, and laboratory investigation. Med J Aust 2:281–299

Desjardins M, Celis JE, Van Meer G, Dieplinger H, Jahraus A, Griffiths G, Huber LA (1994) Molecular characterization of phagosomes. J Biol Chem 269:32194–321200

Ghigo E, Capo C, Tung CH, Raoult D, Gorvel JP, Mege JL (2002) *Coxiella burnetii* survival in THP-1 monocytes involves the impairment of phagosome maturation: IFN-gamma mediates its restoration and bacterial killing. J Immunol 169:4488–4495

Ghigo E, Honstettre A, Capo C, Gorvel JP, Raoult D, Mege JL (2004) Link between impaired maturation of phagosomes and defective *Coxiella burnetii* killing in patients with chronic Q fever. J Infect Dis 190:1767–1772

Ghigo E, Capo C, Raoult D, Mege JL (2006) Intracellular life of *Coxiella burnetii* in macrophages: insight Into Q fever. Curr Immunol Rev 2:225–232

Grieshaber S, Swanson JA, Hackstadt T (2002) Determination of the physical environment within the *Chlamydia trachomatis* inclusion using ion-selective ratiometric probes. Cell Microbiol 4:273–283

Grieshaber SS, Grieshaber NA, Miller N, Hackstadt T (2006) *Chlamydia trachomatis* causes centrosomal defects resulting in chromosomal segregation abnormalities. Traffic 7:940–949

Gutierrez MG, Vazquez CL, Munafo DB, Zoppino FC, Beron W, Rabinovitch M, Colombo MI (2005) Autophagy induction favours the generation and maturation of the *Coxiella*-replicative vacuoles. Cell Microbiol 7:981–993

Haas A (2007) The phagosome: compartment with a license to kill. Traffic 8:311–330

Hackstadt T (1988) Steric hindrance of antibody binding to surface proteins of *Coxiella burnetti* by phase I lipopolysaccharide. Infect Immun 56:802–807

Hackstadt T, Williams JC (1981) Biochemical stratagem for obligate parasitism of eukaryotic cells by *Coxiella burnetii*. Proc Natl Acad Sci U S A 78:3240–3244

Hackstadt T, Williams JC (1984) Metabolic adaptations of *Coxiella burnetii* to intraphagolysosomal growth. In: Lieve L, Schlessinger D (eds) Microbiology 1984. ASM Press, Washington, DC

Hackstadt T, Peacock MG, Hitchcock PJ, Cole RL (1985) Lipopolysaccharide variation in *Coxiella burnetii*: intrastrain heterogeneity in structure and antigenicity. Infect Immun 48:359–365

Handley J, Paretsky D, Stueckemann J (1967) Electron microscopic observations of *Coxiella burnetii* in the guinea pig. J Bacteriol 94:263–267

Heinzen RA, Scidmore MA, Rockey DD, Hackstadt T (1996) Differential interaction with endocytic and exocytic pathways distinguish parasitophorous vacuoles of *Coxiella burnetii* and *Chlamydia trachomatis*. Infect Immun 64:796–809

Henry RM, Hoppe AD, Joshi N, Swanson JA (2004) The uniformity of phagosome maturation in macrophages. J Cell Biol 164:185–194

Hill J, Samuel JE (2011) *Coxiella burnetii* acid phosphatase inhibits the release of reactive oxygen intermediates in polymorphonuclear leukocytes. Infect Immun 79:414–420

Hinrichs DJ, Jerrells TR (1976) In vitro evaluation of immunity to *Coxiella burnetii*. J Immunol 117:996–1003

Hirsch CS, Ellner JJ, Russell DG, Rich EA (1994) Complement receptor-mediated uptake and tumor necrosis factor-alpha-mediated growth inhibition of *Mycobacterium tuberculosis* by human alveolar macrophages. J Immunol 152:743–753

Holden DW (2002) Trafficking of the *Salmonella* vacuole in macrophages. Traffic 3:161–169

Honstettre A, Imbert G, Ghigo E, Gouriet F, Capo C, Raoult D, Mege JL (2003) Dysregulation of cytokines in acute Q fever: role of interleukin-10 and tumor necrosis factor in chronic evolution of Q fever. J Infect Dis 187:956–962

Honstettre A, Ghigo E, Moynault A, Capo C, Toman R, Akira S, Takeuchi O, Lepidi H, Raoult D, Mege JL (2004) Lipopolysaccharide from *Coxiella burnetii* is involved in bacterial phagocytosis, filamentous actin reorganization, and inflammatory responses through Toll-like receptor 4. J Immunol 172:3695–3703

Hoover TA, Culp DW, Vodkin MH, Williams JC, Thompson HA (2002) Chromosomal DNA deletions explain phenotypic characteristics of two antigenic variants, phase II and RSA 514 (crazy), of the *Coxiella burnetii* Nine Mile strain. Infect Immun 70:6726–6733

Houde M, Bertholet S, Gagnon E, Brunet S, Goyette G, Laplante A, Princiotta MF, Thibault P, Sacks D, Desjardins M (2003) Phagosomes are competent organelles for antigen cross-presentation. Nature 425:402–406

Howe D, Mallavia LP (2000) *Coxiella burnetii* exhibits morphological change and delays phagolysosomal fusion after internalization by J774A.1 cells. Infect Immun 68:3815–3821

Howe D, Barrows LF, Lindstrom NM, Heinzen RA (2002) Nitric oxide inhibits *Coxiella burnetii* replication and parasitophorous vacuole maturation. Infect Immun 70:5140–5147

Howe D, Melnicakova J, Barak I, Heinzen RA (2003) Maturation of the *Coxiella burnetii* parasitophorous vacuole requires bacterial protein synthesis but not replication. Cell Microbiol 5:469–480

Howe D, Shannon JG, Winfree S, Dorward DW, Heinzen RA (2010) *Coxiella burnetii* phase I and II variants replicate with similar kinetics in degradative phagolysosome-like compartments of human macrophages. Infect Immun 78:3465–3474

Hu H, Sosnovsky G, Swartz HM (1992) Simultaneous measurements of the intra- and extra-cellular oxygen concentration in viable cells. Biochim Biophys Acta 1112:161–166

Hu C, Mayadas-Norton T, Tanaka K, Chan J, Salgame P (2000) *Mycobacterium tuberculosis* infection in complement receptor 3-deficient mice. J Immunol 165:2596–2602

Hussain SK, Broederdorf LJ, Shama UM, Voth DE (2010) Host kinase activity is required for *Coxiella burnetii* parasitophorous vacuole formation. Front Microbiol 1:137

Izzo AA, Marmion BP (1993) Variation in interferon-gamma responses to *Coxiella burnetii* antigens with lymphocytes from vaccinated or naturally infected subjects. Clin Exp Immunol 94:507–515

Khavkin T, Tabibzadeh SS (1988) Histologic, immunofluorescence, and electron microscopic study of infectious process in mouse lung after intranasal challenge with *Coxiella burnetii*. Infect Immun 56:1792–1799

Koster FT, Williams JC, Goodwin JS (1985) Cellular immunity in Q fever: specific lymphocyte unresponsiveness in Q fever endocarditis. J Infect Dis 152:1283–1289

Kumar Y, Valdivia RH (2008) Actin and intermediate filaments stabilize the *Chlamydia trachomatis* vacuole by forming dynamic structural scaffolds. Cell Host Microbe 4:159–169

Lem L, Riethof DA, Scidmore-Carlson M, Griffiths GM, Hackstadt T, Brodsky FM (1999) Enhanced interaction of HLA-DM with HLA-DR in enlarged vacuoles of hereditary and infectious lysosomal diseases. J Immunol 162:523–532

Li YP, Curley G, Lopez M, Chavez M, Glew R, Aragon A, Kumar H, Baca OG (1996) Protein-tyrosine phosphatase activity of *Coxiella burnetii* that inhibits human neutrophils. Acta Virol 40:263–272

Lubick K, Radke M, Jutila M (2007) Securinine, a GABAA receptor antagonist, enhances macrophage clearance of phase II *C. burnetii*: comparison with TLR agonists. J Leukoc Biol 82:1062–1069

Luhrmann A, Nogueira CV, Carey KL, Roy CR (2010) Inhibition of pathogen-induced apoptosis by a *Coxiella burnetii* type IV effector protein. Proc Natl Acad Sci U S A 107:18997–19001

Luzio JP, Pryor PR, Bright NA (2007) Lysosomes: fusion and function. Nat Rev Mol Cell Biol 8:622–632

Luzio JP, Parkinson MD, Gray SR, Bright NA (2009) The delivery of endocytosed cargo to lysosomes. Biochem Soc Trans 37:1019–1021

Mahapatra S, Ayoubi P, Shaw EI (2010) *Coxiella burnetii* Nine Mile II proteins modulate gene expression of monocytic host cells during infection. BMC Microbiol 10:244

Maurin M, Benoliel AM, Bongrand P, Raoult D (1992a) Phagolysosomal alkalinization and the bactericidal effect of antibiotics: the *Coxiella burnetii* paradigm. J Infect Dis 166:1097–1102

Maurin M, Benoliel AM, Bongrand P, Raoult D (1992b) Phagolysosomes of *Coxiella burnetii*-infected cell lines maintain an acidic pH during persistent infection. Infect Immun 60:5013–5016

Meconi S, Jacomo V, Boquet P, Raoult D, Mege JL, Capo C (1998) *Coxiella burnetii* induces reorganization of the actin cytoskeleton in human monocytes. Infect Immun 66:5527–5533

Meconi S, Capo C, Remacle-Bonnet M, Pommier G, Raoult D, Mege JL (2001) Activation of protein tyrosine kinases by *Coxiella burnetii*: role in actin cytoskeleton reorganization and bacterial phagocytosis. Infect Immun 69:2520–2526

Meghari S, Honstettre A, Lepidi H, Ryffel B, Raoult D, Mege JL (2005) TLR2 is necessary to inflammatory response in *Coxiella burnetii* infection. Ann N Y Acad Sci 1063:161–166

Meresse S, Steele-Mortimer O, Finlay BB, Gorvel JP (1999a) The rab7 GTPase controls the maturation of *Salmonella typhimurium*-containing vacuoles in HeLa cells. EMBO J 18:4394–4403
Meresse S, Steele-Mortimer O, Moreno E, Desjardins M, Finlay B, Gorvel JP (1999b) Controlling the maturation of pathogen-containing vacuoles: a matter of life and death. Nat Cell Biol 1:183–188
Moos A, Hackstadt T (1987) Comparative virulence of intra- and interstrain lipopolysaccharide variants of *Coxiella burnetii* in the guinea pig model. Infect Immun 55:1144–1150
Newton HJ, Ang DK, Van Driel IR, Hartland EL (2010) Molecular pathogenesis of infections caused by *Legionella pneumophila*. Clin Microbiol Rev 23:274–298
Ochoa-Reparaz J, Sentissi J, Trunkle T, Riccardi C, Pascual DW (2007) Attenuated *Coxiella burnetii* phase II causes a febrile response in gamma interferon knockout and Toll-like receptor 2 knockout mice and protects against reinfection. Infect Immun 75:5845–5858
Oh YK, Swanson JA (1996) Different fates of phagocytosed particles after delivery into macrophage lysosomes. J Cell Biol 132:585–593
Omsland A, Cockrell DC, Howe D, Fischer ER, Virtaneva K, Sturdevant DE, Porcella SF, Heinzen RA (2009) Host cell-free growth of the Q fever bacterium *Coxiella burnetii*. Proc Natl Acad Sci U S A 106:4430–4434
Omsland A, Gilk SD, Shannon JG, Beare PA, Voth DE, Howe D, Cockrell DC, Heinzen RA (2010) Exploring the cause of human Q fever: recent advances in *Coxiella burnetii* research. In: Georgiev V (ed) National Institute of Allergy and Infectious Diseases: Intramural Research, vol 3. Springer, New York
Ormsbee RA (1952) The growth of *Coxiella burnetii* in embryonated eggs. J Bacteriol 63:73–86
Pan X, Luhrmann A, Satoh A, Laskowski-Arce MA, Roy CR (2008) Ankyrin repeat proteins comprise a diverse family of bacterial type IV effectors. Science 320:1651–1654
Park DR, Skerrett SJ (1996) IL-10 enhances the growth of *Legionella pneumophila* in human mononuclear phagocytes and reverses the protective effect of IFN-gamma: differential responses of blood monocytes and alveolar macrophages. J Immunol 157:2528–2538
Patil S, Jedsadayanmata A, Wencel-Drake JD, Wang W, Knezevic I, Lam SC (1999) Identification of a talin-binding site in the integrin beta(3) subunit distinct from the NPLY regulatory motif of post-ligand binding functions. The talin n-terminal head domain interacts with the membrane-proximal region of the beta(3) cytoplasmic tail. J Biol Chem 274:28575–28583
Ren Q, Robertson SJ, Howe D, Barrows LF, Heinzen RA (2003) Comparative DNA microarray analysis of host cell transcriptional responses to infection by *Coxiella burnetii* or *Chlamydia trachomatis*. Ann N Y Acad Sci 990:701–713
Rink J, Ghigo E, Kalaidzidis Y, Zerial M (2005) Rab conversion as a mechanism of progression from early to late endosomes. Cell 122:735–749
Rohde K, Yates RM, Purdy GE, Russell DG (2007) *Mycobacterium tuberculosis* and the environment within the phagosome. Immunol Rev 219:37–54
Roman MJ, Coriz PD, Baca OG (1986) A proposed model to explain persistent infection of host cells with *Coxiella burnetii*. J Gen Microbiol 132:1415–1422
Romano PS, Gutierrez MG, Beron W, Rabinovitch M, Colombo MI (2007) The autophagic pathway is actively modulated by phase II *Coxiella burnetii* to efficiently replicate in the host cell. Cell Microbiol 9:891–909
Ruoslahti E (1996) RGD and other recognition sequences for integrins. Annu Rev Cell Dev Biol 12:697–715
Santic M, Molmeret M, Abu Kwaik Y (2005) Maturation of the *Legionella pneumophila*-containing phagosome into a phagolysosome within gamma interferon-activated macrophages. Infect Immun 73:3166–3171
Sauer JD, Shannon JG, Howe D, Hayes SF, Swanson MS, Heinzen RA (2005) Specificity of *Legionella pneumophila* and *Coxiella burnetii* vacuoles and versatility of *Legionella pneumophila* revealed by coinfection. Infect Immun 73:4494–4504
Schaible UE, Schlesinger PH, Steinberg TH, Mangel WF, Kobayashi T, Russell DG (1999) Parasitophorous vacuoles of *Leishmania mexicana* acquire macromolecules from the host cell cytosol via two independent routes. J Cell Sci 112:681–693

Scott CC, Botelho RJ, Grinstein S (2003) Phagosome maturation: a few bugs in the system. J Membr Biol 193:137–152

Seshadri R, Paulsen IT, Eisen JA, Read TD, Nelson KE, Nelson WC, Ward NL, Tettelin H, Davidsen TM, Beanan MJ, Deboy RT, Daugherty SC, Brinkac LM, Madupu R, Dodson RJ, Khouri HM, Lee KH, Carty HA, Scanlan D, Heinzen RA, Thompson HA, Samuel JE, Fraser CM, Heidelberg JF (2003) Complete genome sequence of the Q-fever pathogen *Coxiella burnetii*. Proc Natl Acad Sci U S A 100:5455–5460

Shannon JG, Howe D, Heinzen RA (2005) Virulent *Coxiella burnetii* does not activate human dendritic cells: role of lipopolysaccharide as a shielding molecule. Proc Natl Acad Sci U S A 102:8722–8727

Shannon JG, Cockrell DC, Takahashi K, Stahl GL, Heinzen RA (2009) Antibody-mediated immunity to the obligate intracellular bacterial pathogen *Coxiella burnetii* is Fc receptor- and complement-independent. BMC Immunol 10:26

Sidhu GS, Singh AK, Sundarrajan RN, Sundar SV, Maheshwari RK (1999) Role of vacuolar H(+)-ATPase in interferon-induced inhibition of viral glycoprotein transport. J Interferon Cytokine Res 19:1297–1303

Siemsen DW, Kirpotina LN, Jutila MA, Quinn MT (2009) Inhibition of the human neutrophil NADPH oxidase by *Coxiella burnetii*. Microbes Infect 11:671–679

Stein A, Louveau C, Lepidi H, Ricci F, Baylac P, Davoust B, Raoult D (2005) Q fever pneumonia: virulence of *Coxiella burnetii* pathovars in a murine model of aerosol infection. Infect Immun 73:2469–2477

Stuart LM, Ezekowitz RA (2005) Phagocytosis: elegant complexity. Immunity 22:539–550

Taylor PR, Martinez-Pomares L, Stacey M, Lin HH, Brown GD, Gordon S (2005) Macrophage receptors and immune recognition. Annu Rev Immunol 23:901–944

Tsang AW, Oestergaard K, Myers JT, Swanson JA (2000) Altered membrane trafficking in activated bone marrow-derived macrophages. J Leukoc Biol 68:487–494

Tujulin E, Macellaro A, Lilliehook B, Norlander L (1998) Effect of endocytosis inhibitors on *Coxiella burnetii* interaction with host cells. Acta Virol 42:125–131

Underhill DM (2004) Toll-like receptors and microbes take aim at each other. Curr Opin Immunol 16:483–487

Vazquez CL, Colombo MI (2010) *Coxiella burnetii* modulates Beclin 1 and Bcl-2, preventing host cell apoptosis to generate a persistent bacterial infection. Cell Death Differ 17:421–438

Via LE, Fratti RA, Mcfalone M, Pagan-Ramos E, Deretic D, Deretic V (1998) Effects of cytokines on mycobacterial phagosome maturation. J Cell Sci 111:897–905

Vishwanath S, Hackstadt T (1988) Lipopolysaccharide phase variation determines the complement-mediated serum susceptibility of *Coxiella burnetii*. Infect Immun 56:40–44

Voth DE, Heinzen RA (2007) Lounging in a lysosome: the intracellular lifestyle of *Coxiella burnetii*. Cell Microbiol 9:829–840

Voth DE, Heinzen RA (2009a) *Coxiella* type IV secretion and cellular microbiology. Curr Opin Microbiol 12:74–80

Voth DE, Heinzen RA (2009b) Sustained activation of Akt and Erk1/2 is required for *Coxiella burnetii* antiapoptotic activity. Infect Immun 77:205–213

Voth DE, Howe D, Heinzen RA (2007) *Coxiella burnetii* inhibits apoptosis in human THP-1 cells and monkey primary alveolar macrophages. Infect Immun 75:4263–4271

Voth DE, Howe D, Beare PA, Vogel JP, Unsworth N, Samuel JE, Heinzen RA (2009) The *Coxiella burnetii* ankyrin repeat domain-containing protein family is heterogeneous, with C-terminal truncations that influence Dot/Icm-mediated secretion. J Bacteriol 191:4232–4242

Voth DE, Beare PA, Howe D, Sharma UM, Samoilis G, Cockrell DC, Omsland A, Heinzen RA (2011) The *Coxiella burnetii* cryptic plasmid is enriched in genes encoding type IV secretion system substrates. J Bacteriol 193:1493–1503

Weber SS, Ragaz C, Hilbi H (2009) Pathogen trafficking pathways and host phosphoinositide metabolism. Mol Microbiol 71:1341–1352

Weisburg WG, Dobson ME, Samuel JE, Dasch GA, Mallavia LP, Baca O, Mandelco L, Sechrest JE, Weiss E, Woese CR (1989) Phylogenetic diversity of the *Rickettsiae*. J Bacteriol 171:4202–4206

Weiss E (1973) Growth and physiology of rickettsiae. Bacteriol Rev 37:259–283

Weiss E, Moulder JW (1984) Order I. Rickettsiales Gieszczkiewicz 1939, 25[AL]. The Williams & Wilkens Co., Baltimore

Wesolowski J, Paumet F (2010) SNARE motif: a common motif used by pathogens to manipulate membrane fusion. Virulence 1:319–324

West MA, Wallin RP, Matthews SP, Svensson HG, Zaru R, Ljunggren HG, Prescott AR, Watts C (2004) Enhanced dendritic cell antigen capture via toll-like receptor-induced actin remodeling. Science 305:1153–1157

Williams JC, Peacock MG, Mccaul TF (1981) Immunological and biological characterization of *Coxiella burnetii*, phases I and II, separated from host components. Infect Immun 32:840–851

Yates RM, Russell DG (2005) Phagosome maturation proceeds independently of stimulation of toll-like receptors 2 and 4. Immunity 23:409–417

Zaidi N, Maurer A, Nieke S, Kalbacher H (2008) Cathepsin D: a cellular roadmap. Biochem Biophys Res Commun 376:5–9

Zamboni DS (2004) Genetic control of natural resistance of mouse macrophages to *Coxiella burnetii* Infection in vitro: macrophages from restrictive strains control parasitophorous vacuole maturation. Infect Immun 72:2395–2399

Zamboni DS, Rabinovitch M (2003) Nitric oxide partially controls *Coxiella burnetii* phase II infection in mouse primary macrophages. Infect Immun 71:1225–1233

Zamboni DS, Rabinovitch M (2004) Phagocytosis of apoptotic cells increases the susceptibility of macrophages to infection with *Coxiella burnetii* Phase II through down-modulation of nitric oxide production. Infect Immun 72:2075–2080

Zamboni DS, Mortara RA, Freymuller E, Rabinovitch M (2002) Mouse resident peritoneal macrophages partially control in vitro infection with *Coxiella burnetii* phase II. Microbes Infect 4:591–598

Zamboni DS, Campos MA, Torrecilhas AC, Kiss K, Samuel JE, Golenbock DT, Lauw FN, Roy CR, Almeida IC, Gazzinelli RT (2004) Stimulation of toll-like receptor 2 by *Coxiella burnetii* is required for macrophage production of pro-inflammatory cytokines and resistance to infection. J Biol Chem 279:54405–54415

Zerial M, Mcbride H (2001) Rab proteins as membrane organizers. Nat Rev Mol Cell Biol 2:107–117

Zoppino FC, Militello RD, Slavin I, Alvarez C, Colombo MI (2010) Autophagosome formation depends on the small GTPase Rab1 and functional ER exit sites. Traffic 11:1246–1261

Chapter 9
Coxiella burnetii Secretion Systems

Justin A. McDonough, Hayley J. Newton, and Craig R. Roy*

Contents

Abstract The ability of bacteria to transport proteins across their membranes is integral for interaction with their environment. Distinct families of secretion systems mediate bacterial protein secretion. The human pathogen, *Coxiella burnetii* encodes components of the Sec-dependent secretion pathway, an export system used for type IV pilus assembly, and a complete type IV secretion system. The type

*The first two authors contributed equally.

J.A. McDonough • H.J. Newton • C.R. Roy (✉)
Section of Microbial Pathogenesis, Yale University School of Medicine,
295 Congress Avenue, New Haven 06536, CT, USA
e-mail: craig.roy@yale.edu

R. Toman et al. (eds.), *Coxiella burnetii: Recent Advances and New Perspectives in Research of the Q Fever Bacterium*, Advances in Experimental Medicine and Biology 984,
DOI 10.1007/978-94-007-4315-1_9, © Springer Science+Business Media Dordrecht 2012

IVB secretion system in *C. burnetii* is functionally analogous to the *Legionella pneumophila* Dot/Icm secretion system. Both *L. pneumophila* and *C. burnetii* require the Dot/Icm apparatus for intracellular replication. The Dot/Icm secretion system facilitates the translocation of many bacterial effector proteins across the bacterial and vacuole membranes to enter the host cytoplasm where the effector proteins mediate their specific activities to manipulate a variety of host cell processes. Several studies have identified cohorts of *C. burnetii* Dot/Icm effector proteins that are predicted to be involved in modulation of host cell functions. This chapter focuses specifically on these secretion systems and the role they may play during *C. burnetii* replication in eukaryotic host cells.

Keywords Bacterial secretion • Dot/Icm • Effector proteins • Type IV secretion system • Virulence determinant

9.1 Introduction

Secreted and surface proteins of bacteria are essential for bacterial physiology, virulence, and recognition by host immune mechanisms. Transport of these proteins from the cytoplasm to the bacterial cell envelope or outside of the bacterial cell is dependent on the function of specific secretion systems. These systems recognize defined protein signatures and actively move these proteins across lipid membranes. Bacterial pathogens often have one or more specialized secretion systems that act on specific subsets of bacterial proteins involved in interaction with eukaryotic host cells. In these instances, numerous gene products function in concert to form a highly specialized molecular apparatus designed to deliver macromolecules into host cells. The importance of these secretion systems in promoting virulence makes them important candidates for research studies and attractive targets for new antimicrobial therapies.

Coxiella burnetii is a Gram-negative facultative intracellular pathogen and the causative agent of Q fever in humans (Madariaga et al. 2003). *C. burnetii* invades host cells and over several days directs the formation of a unique vacuolar compartment that resembles a lysosome (Voth and Heinzen 2007). Following uptake, the nascent *C. burnetii* vacuole sequentially acquires markers of early and late endosomes, eventually forming a single spacious compartment referred to as the *C. burnetii*-containing vacuole (CCV), which demonstrates robust fusion with host endocytic vesicles and expands to occupy much of the host cytoplasm. The CCV lumen has a pH of approximately 4.7 and contains lysosomal hydrolases, such as cathepsin D (Heinzen et al. 1996). The lysosomal membrane proteins LAMP-1 and LAMP-2 are present on the CCV and this compartment is rich in cholesterol (Heinzen et al. 1996; Howe and Heinzen 2006). In addition, proteins usually found on autophagosomes, such as microtubule-associated protein light-chain 3 (LC3) and Rab24, are present on the CCV membrane (Beron et al. 2002). It is predicted that bacterial proteins delivered into the lumen of this vacuole and across the vacuole membrane will be important determinants of host pathogenesis. Thus, a more complete understanding of *C. burnetii* secretion should provide important insight into the mechanisms by which this pathogen creates and maintains this specialized vacuole.

During infection, the bacteria exist as two morphologically distinct forms. A metabolically inert small cell variant is stable in the environment and capable of infecting a wide range of host cells (McCaul and Williams 1981). Once the spacious CCV is established the small cell variant morphs into a large cell variant capable of replication to high numbers within the CCV (Coleman et al. 2004). Like several other intracellular bacterial pathogens, *C. burnetii* is able to block the intrinsic death pathway and prevent the induction of host cell apoptosis in infected cells, which is important for survival within cells over long periods of time (Voth et al. 2007; Luhrmann and Roy 2007).

There is evidence to suggest that establishment of the CCV is distinct from the default transport pathway that delivers endocytic cargo to lysosomes. Because the CCV is highly fusogenic, fusing promiscuously with other endocytic vesicles and phagosomes residing in the cell, it suggests that *C. burnetii* overrides the host mechanisms that regulate lysosome biogenesis (Howe et al. 2003a). Inhibition of *C. burnetii* protein synthesis by chloramphenicol treatment impedes fusogenicity and prevents formation and maintenance of the CCV, indicating that continued synthesis of bacterial proteins is required to maintain this specialized organelle (Howe et al. 2003b), predicting a role for secreted bacterial products in this process.

The genomes of several strains of *C. burnetii* have been sequenced and reveal genes predicted to encode proteins used by *C. burnetii* to survive and replicate within host cells (Seshadri et al. 2003; Beare et al. 2009), including homologues to known secretion systems and putative substrates of these systems. The importance of these secretion systems in the ecology and evolution of pathogenesis are current areas of investigation in the field of *C. burnetii* biology.

9.2 Secretion Systems Encoded by *Coxiella burnetii*

The ability of intracellular bacteria to subvert host mechanisms and survive within mammalian cells is often dependent on the function of a subset of bacterial proteins that interact with host cell factors. To function in this manner, these virulence factors are translocated across bacterial membranes, and in many cases, the host compartmental membrane that contains the bacteria. Gram-negative pathogens use sophisticated secretion systems for this purpose and much work has been devoted to understanding how these molecular machines function (Cambronne and Roy 2006). The genome of *C. burnetii* predicts functional versions of the Sec-dependent, type I, type IV pilus, and type IV secretion systems (described below); and is missing canonical components defining the type II and type III secretion systems, type V autotransporters, and type VI secretion (Seshadri et al. 2003; Beare et al. 2009).

Table 9.1 *C. burnetii* homologs of Gram-negative inner membrane transport components

	ORF name	
	***C. burnetii* Nine Mile strain**	***C. burnetii* Dugway strain**
Sec dependent pathway		
SecY	CBU0258	CBUD1834
SecE	CBU0224	CBUD1868
SecG	CBU1449	CBUD0543
SecD	CBU1142	CBUD1240
SecF	CBU1141	CBUD1238
YajC	CBU1143	CBUD1241
YidC	CBU1920	CBUD0201
SecA	CBU0147	CBUD1958
SecB	CBU1519	CBUD0465
SecM	None	None
SRP		
Ffh	CBU0450	CBUD1624
4.5S	None	None
SRP receptor		
FtsY	CBU1903	CBUD0216
Tat (twin arginine translocation) pathway		
TatA	None	None
TatB	None	None
TatC	None	None
TatE	None	None
Signal peptidase		
SPase I (LepB-1)	CBU1099	CBUD1201
SPase I (LepB-2)	CBU1504	CBUD0480
SPase II (LspA)–lipoprotein SPase	CBU0397	CBUD1671

9.2.1 Sec-Dependent Secretion

Targeted export of proteins across the bacterial inner membrane by the Sec-dependent general secretion pathway is a ubiquitous process in eubacteria. Most of the Sec system components, including signal recognition particle (SRP)-mediated membrane insertion components, are found in the *C. burnetii* genome (Table 9.1; Seshadri et al. 2003; Lima et al. 2009; Mori and Ito 2001). Consistent with this pathway being operational in *C. burnetii* there are numerous genes encoding proteins with predicted amino (N)-terminal signal peptides that function in directing unfolded proteins to the canonical Sec pathway for transport across the bacterial inner membrane (Seshadri et al. 2003). Several of these signal sequence-containing proteins have potential roles in virulence, including phospholipase D (CBU0968), carboxypeptidase (CBU1261), and acid phosphatase (CBU0335).

Many, but not all bacteria utilize a second dedicated secretion apparatus called the Twin-arginine translocase (Tat) for delivering folded peptides across the inner membrane (De Buck et al. 2008; Lee et al. 2006). In some pathogens, such as *Pseudomonas aeruginosa* and *Legionella pneumophila*, the Tat pathway is required for the transport of specific virulence factors across the inner membrane (Voulhoux et al. 2001; Rossier and Cianciotto 2005). Once exported to the periplasm these proteins typically engage the type II secretion system for translocation across the bacterial envelope. Unlike *L. pneumophila*, the *C. burnetii* genome lacks homologues to the known functional components (*tatA*, *tatB*, *tatC*, *tatE*) of the twin-arginine translocase. Thus, the Tat pathway does not appear to be operational in *C. burnetii*.

9.2.2 Type I Secretion

The type I secretion apparatus spans the bacterial cell envelope and is similar to ATP-binding cassette (ABC) transporters. It is composed of a specific outer membrane protein (OMP), an ABC protein, and a membrane fusion protein (MFP) (Thanabalu et al. 1998). The *C. burnetii* genome has a gene *tolC* (CBU0056) that is predicted to encode a protein homologous to the type I secretion system outer membrane channel protein TolC, and is therefore potentially competent for type I secretion. Although the function of *C. burnetii* TolC is unknown, a TolC homolog of *L. pneumophila* is involved in multidrug efflux function and required for virulence, suggesting TolC may play a similar role in *C. burnetii* (Ferhat et al. 2009).

9.2.3 Type IV Pilus System

The type II secretion (T2S) system works in conjunction with the general secretion pathway to ultimately deliver proteins across the Gram-negative outer membrane (Cianciotto 2005). Proteins destined for type II translocation have an amino terminal signal sequence and are capable of Sec or Tat dependent export to the periplasm, where they then engage the type II secretion apparatus. The apparatus is composed mainly of a multimeric outer membrane complex formed from pore-forming secretin proteins. Some components of the T2S system are evolutionarily-related to Gram-negative type IV pili, which uses a modified version of type II secretion for pilus biogenesis (Hobbs and Mattick 1993; Sauvonnet et al. 2000).

The full repertoire of genes encoding members of a prototypical T2S system is lacking in the *C. burnetii* genome, including the major and minor pilin genes *pulGHIJK*; however, several *C. burnetii* genes encode proteins with homology to components of the bacterial type IV pili (T4P) system. These include the conserved

ATPase PilB, the outer membrane secretin PilQ, the prepilin peptidase T4P PilD, and the transmembrane protein PilC. These components are also conserved in most T2S systems. In addition, the genome of *C. burnetii* encodes homologs to *pilE*, *pilF*, and the major type IV prepilin gene *pilA*. Of note is the lack of a *pilT* homolog in *C. burnetii*, which encodes an ATPase essential for pilus retraction. It is interesting to speculate that *C. burnetii* may secrete a subset of signal sequence-containing proteins through the type IV pilus, as has been reported for *Francisella novicida* (Hager et al. 2006). Several signal sequence-containing proteins of *C. burnetii* are potential virulence protein candidates for T4P-related secretion based on the presence of one or more eukaryotic-like motifs within the mature protein sequence. An example is the predicted protein encoded by *enhC*, which has homology to EnhC (Enhanced entry protein C) of *L. pneumophila*, a factor that is involved in early stages of host cell infection (Liu et al. 2008). The *enhC* product has a predicted Sec-dependent signal sequence, and 21 tandemly-arranged tetratricopeptide repeats (TPRs)–a motif found in eukaryotic adaptor proteins that functions in signal transduction. Whether the *C. burnetii* EnhC protein resides on the bacterial cell surface and participates in pathogen uptake remain unanswered questions.

9.2.4 Type IV Secretion System

Type IV secretion systems of Gram-negative bacteria are related to the conjugal transfer systems that participate in exchange of broad-host range plasmids between bacterial cells (Segal et al. 1998; Christie and Vogel 2000; Vogel et al. 1998). Virulence-associated type IV secretion systems are found in several bacterial pathogens including *Agrobacterium tumefaciens*, *Bordetella pertussis*, *Helicobacter pylori*, *Bartonella henselae*, *Brucella suis*, *Rickettsia prowazekii*, *Xanthomonas* spp., *L. pneumophila*, and *C. burnetii*, and function in the secretion and delivery of bacterial effector proteins into host cells (Alvarez-Martinez and Christie 2009). Bacterial type IV secretion systems are classified as either type IVA or type IVB based on homology to the *A. tumefaciens virB/virD4* system or the *L. pneumophila* Dot/Icm system, respectively (Christie and Vogel 2000). The *L. pneumophila* Dot/Icm system is ancestrally related to the *tra/trb*-encoded Inc plasmid system and has been shown to mediate transfer of the IncQ plasmid RSF1010 between *L. pneumophila* strains, as well as to *E. coli* (Segal et al. 1998; Vogel et al. 1998). During host infection, *L. pneumophila* uses the Dot/Icm type IV secretion system to deliver a large repertoire of effector proteins into host cells that are important in establishing a niche for intracellular survival and replication of the bacteria (Hubber and Roy 2010). The *C. burnetii* genome encodes an intact type IVB secretion system highly homologous to the *L. pneumophila* Dot/Icm system (Segal and Shuman 1999; Seshadri et al. 2003). This locus contains homologues to most of the known *L. pneumophila dot/icm* genes and is organized similarly with few exceptions. No other type IV-related systems are encoded by *C. burnetii*. Because the *C. burnetii* Dot/Icm system is the only virulence determinant to date that has been shown to be essential for bacterial replication inside eukaryotic host cells, a more detailed description of this apparatus and the effectors translocated by the system are provided in the following sections.

9.3 Genomic Analysis of the *Coxiella burnetii* Dot/Icm System

L. pneumophila, the causative agent of Legionnaire's disease, is a Gram-negative intracellular pathogen that survives within and kills free-living amoebae and human macrophages (Rowbotham 1980; Horwitz and Silverstein 1980; Nash et al. 1984). The intracellular success of *L. pneumophila* is dependent on the function of a type IVB secretion system called Dot/Icm (Defect in organelle trafficking/Intracellular multiplication) (Marra et al. 1992; Berger and Isberg 1993). The Dot/Icm system is a major virulence determinant of *L. pneumophila* and is required for the creation of a vacuole that evades endocytic maturation and develops into an endoplasmic reticulum-derived organelle that supports replication of *L. pneumophila* within eukaryotic host cells (Kagan and Roy 2002; Horwitz 1983; Swanson and Isberg 1995).

The *dot/icm* genes in *L. pneumophila* are located on two chromosomal loci and encode 24 protein components that assemble to form a membrane-bound secretion apparatus that functions in the translocation of effector proteins across both the bacterial envelope and the host cell vacuolar membrane (Swanson and Isberg 1995; Vogel et al. 1998; Segal et al. 1998). Early evidence from *C. burnetii* genomic sequencing revealed putative gene products with high amino acid homology to Dot/Icm proteins of *L. pneumophila* (Seshadri et al. 2003; Sexton and Vogel 2002). *C. burnetii* is the only other animal pathogen in which the Dot/Icm secretion system has been identified; however, genome sequencing of a phytopathogen *Xanthomonas* sp. suggests the presence of a plasmid-encoded *dot/icm* locus (Thieme et al. 2005). The *C. burnetii dot/icm* genes are found on a 35 kb region of the chromosome and there is synteny with the genes in *L. pneumophila* (Fig. 9.1). Comparison of Dot/Icm proteins reveals that most of the components found in the *L. pneumophila* system are present in the *C. burnetii* system (Seshadri et al. 2003). *C. burnetii* appears to lack an *icmM* gene and instead has two genes that encode proteins with similarity to the *icmL* product. Given that there is a high degree of similarity between the *L. pneumophila icmL* and *icmM* products, the second *C. burnetii icmL* gene is likely to encode a protein that functions similarly to the *L. pneumophila icmM* product. Initially, *C. burnetii* was thought to also lack the *icmR* gene, which encoded a protein that forms a complex with the *icmQ* product in *L. pneumophila*. However, *C. burnetii* has a gene immediately upstream of *icmQ* that encodes a 49 amino acid protein called CoxigA (CBU1634a) that is functionally homologous to the *L. pneumophila* IcmR protein. CoxigA is a member of the FIR family of proteins, which are

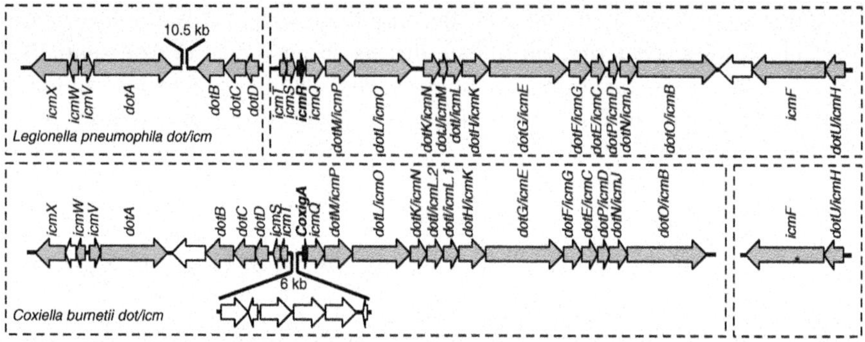

Fig. 9.1 Genomic organization of the genes encoding Dot/Icm systems. The *L. pneumophila* and *C. burnetii dot/icm* genes represented as *shaded arrows* are aligned to demonstrate synteny. Genes that are not required for Dot/Icm-dependent processes are shown as *white arrows*. Genes that are linked together on the chromosome are contained within the dashed boxes. The location and size of intervening sequences are indicated. The asterisk indicates the location of a stop codon within *C. burnetii icmF* gene

Functional Homologues of IcmR identified in other species of *Legionella*. FIR proteins are encoded by a family of hypervariable genes that are located in a similar genomic position upstream of *icmQ* (Feldman and Segal 2004; Feldman et al. 2005).

9.4 Transcriptional Regulation of the *dot/icm* Genes

Analysis of RNA transcribed from a subset of *C. burnetii dot/icm* genes has demonstrated transcription of three linkage groups, which would correspond to three operons consisting of *icmW-CBU1651-icmX*, *icmV-dotA-CBU1647* and *icmT-icmS-icmD-icmC-icmB-CBU1646* (Morgan et al. 2010b) (see Fig. 9.1). Examination of the relative transcript levels showed a general trend towards increased transcription of these *dot* and *icm* genes during the initial 16 h of *C. burnetii* infection of a eukaryotic host cell, and a subsequent drop in transcript levels by 24 h post-infection (Coleman et al. 2004; Morgan et al. 2010b). Changes in the RNA levels for *icmT* were subtle, yet corresponded to IcmT protein levels increasing two to three fold during the initial 24 h of infection and then remaining constant (Morgan et al. 2010b). These data suggest that the *dot/icm* locus is upregulated in response to infection and that this may facilitate the establishment of the replicative vacuole and a productive infection.

Presumably, a number of different regulatory factors work in concert to coordinate the expression of both components of the Dot/Icm apparatus and the Dot/Icm effectors that are delivered into host cells by the type IV system. A cohort of regulatory proteins has been shown to tightly control the expression of virulence-related genes in *L. pneumophila*. Most virulence genes seem to be upregulated as *L. pneumophila* enter stationary phase, including genes encoding the Dot/Icm system. The two-component response regulators PmrA, CpxR and LetA, the stationary phase sigma factor RpoS and the ppGpp synthetase RelA, have all been shown to play an important role in regulating virulence in *L. pneumophila* (Zusman et al. 2007; Altman and Segal 2008; Gal-Mor and Segal 2003; Lynch et al. 2003; Shi et al. 2006; Rasis and Segal 2009; Bachman and Swanson 2001). Homologues of each of these Dot/Icm regulators are present in *C. burnetii*. The stationary phase sigma factor RpoS of *C. burnetii* shares 50% identity with *L. pneumophila* RpoS and can functionally complement an *Escherichia coli rpoS* null mutant (Seshadri and Samuel 2001). Interestingly, the *C. burnetii* RpoS protein is abundant in the replicating large cell variants of *C. burnetii* and undetectable in the stationary small cell variants, suggesting that the genes regulated by RpoS in *C. burnetii* may expressed at a higher level during exponential phase and down regulated in stationary phase, which would be consistent with the pattern of *dot/icm* gene expression observed. A consensus promoter sequence for RpoS, however, has not been clearly defined, and the role of this sigma factor in Dot/Icm regulation in *C. burnetii* remains to be determined (Melnicakova et al. 2003).

There is evidence that regulation of the *C. burnetii* Dot/Icm system is mediated directly by the response regulator PmrA. The *C. burnetii pmrA* gene (CBU1227) encodes a protein in which the third alpha helix in the predicted structure is identical to that predicted for the *L. pneumophila* PmrA protein. Because this region is a determinant of the helix-loop-helix motif that mediates DNA recognition it was anticipated that these two response regulators would recognize a similar consensus sequence in the promoters of PmrA-regulated genes (Zusman et al. 2007). Bioinformatical analysis confirmed the presence of PmrA regulatory sequences upstream of five putative promoter regions controlling the transcription of operons that express 23 *dot/icm* genes. Gel mobility shift assays confirmed that the *C. burnetii* PmrA protein bound to the regulatory sequence upstream of these genes. Furthermore, a PmrA-regulated transcriptional fusion to *lacZ* was introduced into a *L. pneumophila pmrA* mutant strain to show that the *C. burnetii* PmrA protein could restore transcriptional activity (Zusman et al. 2007).

In *L. pneumophila*, PmrA also regulates the transcription of a large number of Dot/Icm effector proteins, and the presence of the PmrA consensus sequence has been used as a signature to identify Dot/Icm effectors (Zusman et al. 2007; Chen et al. 2010). Analysis of the *C. burnetii* genome identified 68 genes containing this upstream consensus regulatory element. It has been suggested that *C. burnetii* genes that have been called *cig* for coregulated with i*cm* genes, would be attractive candidates for encoding Dot/Icm effectors (Zusman et al. 2007). As detailed in the following sections, subsequent independent studies have now confirmed that several of the *cig* genes encode Dot/Icm effectors.

9.5 Dot/Icm Apparatus Assembly

The high degree of similarity between the *L. pneumophila* Dot/Icm components and those encoded by *C. burnetii* predicts that the two systems produce structurally analogous secretion systems. There is a great deal of interest in understanding how the individual Dot/Icm components assemble to build the machinery that translocates effector proteins into host cells. Partial modeling of the *L. pneumophila* Dot/Icm secretion apparatus has been achieved using both biochemical and genetic approaches. Protein-protein interactions, identification of the subcellular localization of individual Dot/Icm proteins and examination of the stability of each component of the system in systematic knockout backgrounds have provided some insight into the relationships between individual key components of the type IVB secretion system (Vincent et al. 2006). A core subcomplex, consisting of the inner membrane proteins DotG and DotF with the outer membrane protein DotH and the outer membrane lipoproteins DotC and DotD, combines to span both bacterial membranes (Vincent et al. 2006). It is predicted that DotC and DotD facilitate the outer membrane insertion of DotH to form the outer membrane pore and DotF and DotG can form both homo- and heterodimers in the inner membrane to interact with the DotC-DotD-DotH outer membrane complex. Given that the *C. burnetii* homologues share approximately 50% similarity to the *L. pneumophila* core subcomplex it is expected that this structure is also a key component of the *C. burnetii* Dot/Icm apparatus.

Surprisingly, despite being evolutionarily distinct and possessing more than double the number of proteins, the Dot/Icm type IVB apparatus produces a core structure that appears to be related to the VirB apparatus in *A. tumefaciens*, which is a type IVA system (Vincent et al. 2006). The VirB apparatus consists or a core complex containing the inner membrane protein VirB8 interacting with the membrane spanning protein VirB10 and the outer membrane proteins VirB7 and VirB9 (Krall et al. 2002). Most recently, the crystal structure of *L. pneumophila* DotD has been reported, revealing a structure surprisingly distinct from VirB7, the presumed type IVA counterpart (Nakano et al. 2010). Rather, the C-terminus of *L. pneumophila* DotD displays similarity to the N-terminal subdomain of secretins, the outer membrane components of type II and type III secretion systems (Nakano et al. 2010). This suggests DotD may create a channel in the bacterial outer membrane through which effectors potentially travel.

Proteins of the *C. burnetii* Dot/Icm secretion system have been visualized by immunofluorescence and electron microscopy. Antibodies to IcmT, IcmV and DotH clearly demonstrated polar localization of these proteins, particularly in the large cell variant of *C. burnetii* (Morgan et al. 2010a). Electron micrographs indicate that the poles of *C. burnetii* can interact with the CCV membrane, which is where protein translocation by the Dot/Icm system should be occurring; however, the functional relevance of these observations has yet to be determined. It is unclear whether polar localization of the Dot/Icm apparatus is important for the efficient translocation of substrates; however, this is not a unique observation as a range of bacterial pathogens express virulence factors and secretion system components preferentially at the bacterial poles (Morgan et al. 2010a; Jaumouille et al. 2008; Carlsson et al. 2009; Jain et al. 2006).

9.6 Functional Analysis of the *Coxiella burnetii* Dot/Icm System

Several lines of evidence indicate that the *C. burnetii* Dot/Icm system is functional and has similar features to the *L. pneumophila* Dot/Icm secretion system. Reverse transcription-PCR indicates that *dot/icm* genes are expressed during *C. burnetii* infection of Vero cells, including the genes *dotB*, *icmQ*, *icmS*, and *icmW* (Zamboni et al. 2003). These genes are of particular interest because they represent type IV secretion components that are not present in the conjugal DNA transfer system and likely have specific roles in translocation of virulence factors in *L. pneumophila* and *C. burnetii*. In addition, genetic complementation was used to test functional similarity by asking whether expression of individual *C. burnetii dot/icm* genes in a corresponding *L. pneumophila* mutant having the gene of interest deleted could restore *L. pneumophila* growth in eukaryotic cells. It was found that the *C. burnetii* genes encoding *dotB*, *icmS*, *icmW*, and *icmT* restored intracellular replication when expressed in a corresponding *L. pneumophila* deletion strain (Zamboni et al. 2003; Zusman et al. 2003). In contrast, *L. pneumophila icmB*, *icmJ*, *icmO*, *icmP*, *icmQ* mutants could not be complemented by the corresponding *C. burnetii* gene product (Zamboni et al. 2003; Zusman et al. 2003).

Genetic complementation studies have been further verified by studies examining protein-protein interactions. The IcmS and IcmW proteins are small proteins that appear to form a stable complex in the *L. pneumophila* cytosol (Zuckman et al. 1999; Ninio et al. 2005). These proteins interact directly with many different effectors and are believed to function as chaperones that facilitate the recognition of translocation signals in the effectors that enable these proteins to be recognized as substrates by the secretion apparatus (Cambronne and Roy 2007; Ninio et al. 2005). As was shown for the *L. pneumophila* proteins, it was found that the *C. burnetii* IcmW and IcmS proteins interact with each other (Zusman et al. 2003). Additionally, cross species interactions between the *C. burnetii* and *L. pneumophila* IcmW and IcmS proteins supported the observation that the *C. burnetii* genes encoding either component could complement function in a corresponding *L. pneumophila* mutant (Zusman et al. 2003). The cytoplasmic IcmQ protein is essential for Dot/Icm function and through homotypic interactions is capable of forming pores in lipid membranes *in vitro* (Dumenil et al. 2004; Raychaudhury et al. 2009). The activity of IcmQ is regulated by the IcmR protein, which is capable of forming a protein complex with IcmQ (Coers et al. 2000; Raychaudhury et al. 2009). The *C. burnetii* IcmQ protein did not interact with the *L. pneumophila* IcmR protein, which explains why the *C. burnetii icmQ* gene did not complement the *L. pneumophila icmQ* deletion mutant (Zamboni et al. 2003; Zusman et al. 2003). In summary, these studies indicate the two systems are closely related and that some of the Dot/Icm components can function in both systems. The observation that the proteins that comprise the membrane-bound translocation machine are not interchangeable between the two systems suggests that divergent evolution has resulted in structural constraints that prevent these *C. burnetii* components from interacting correctly with components of the *L. pneumophila* Dot/Icm apparatus.

The host cell processes targeted by *C. burnetii* and the specific function of *C. burnetii* effectors are anticipated to be quite unique. Unlike other intracellular bacteria, *C. burnetii* does not mediate phagosome arrest or escape, and therefore does not require effector proteins that function in this regard. Study of the *C. burnetii* intracellular lifestyle and interactions with host cells can provide some clues as to functions that may be encoded by Dot/Icm effectors. Specific traits of the developing CCV are dependent on *C. burnetii* protein synthesis. Blocking *C. burnetii* protein synthesis, with the application of chloramphenicol, obstructs the capacity of the CCV to promote fusion of endocytic vesicles and lysosomes, and prevents formation of the characteristic large spacious vacuole in which the bacteria reside (Howe et al. 2003a, b). Chloramphenicol treatment also significantly reduces the capacity of *C. burnetii* to block the intrinsic death pathway and inhibit host cell apoptosis (Luhrmann and Roy 2007; Voth et al. 2007). The ability of *C. burnetii* to actively perturb vesicular fusion and apoptosis signaling are examples of putative functions for specific Dot/Icm effector proteins.

To determine whether the *C. burnetii* Dot/Icm system plays an important role in host cell infection, a mutant deficient in this apparatus was isolated. Using random transposon mutagenesis, an insertion in the *dotI/icmL.1* gene of *C. burnetii* was isolated (Carey et al. 2011). The *icmL::Tn* mutant grows similarly to the isogenic parental strain in axenic medium, indicating that the Dot/Icm system is not required for replication in defined medium. When the *icmL::Tn* mutant was used to infect mammalian host cells, a complete defect in *C. burnetii* intracellular growth was observed. The *icmL::Tn* mutant phenotype is similar to the effect chloramphenicol treatment has on wild type *C. burnetii* replication in cultured host cells (Carey et al. 2011). These findings indicate that the Dot/Icm secretion system is essential for successful infection of host cells by *C. burnetii* and indicates that the effectors translocated by this secretion system must play an important role in creating a host cell compartment that permits *C. burnetii* replication.

9.7 Effectors of the Dot/Icm Secretion System

In *L. pneumophila* there have been over 200 different proteins identified that are translocated into host cells by the Dot/Icm secretion system (Burstein et al. 2009; Huang et al. 2011). Most of these Dot/Icm effectors are dispensable for intracellular replication of *L. pneumophila,* highlighting genetic and functional redundancy among the effector repertoire. Despite this, several *L. pneumophila* Dot/Icm effectors have been shown to have a profound influence on specific processes that are subverted during host cell infection, including subversion of vesicular trafficking, modulation of host GTPase function and localization, and promoting the ubiquitination of host and bacterial proteins (see review articles (Newton et al. 2010; Ensminger and Isberg 2009; Shin and Roy 2008)).

Using strategies that were successful in identification of *L. pneumophila* Dot/Icm effectors, several studies have reported the identification of *C. burnetii* effectors that are translocated into host cells by the Dot/Icm system (Pan et al. 2008; Voth et al. 2009; Chen et al. 2010) (Table 9.2). Protein fusions have been the primary method used to detect protein translocation into host cells by the Dot/Icm system. The basis of this detection scheme is to fuse a putative effector protein to an enzyme with a novel activity that can indicate delivery of the hybrid into the host cytosol. The calmodulin-activated adenylate cyclase, CyaA, from *Bordetella pertussis* is a commonly used reporter to examine bacterial translocation of effector proteins into host cells (Sory and Cornelis 1994; Chen et al. 2010). After being transported to the host cytosol, the Cya protein binds to calmodulin and then efficiently converts ATP into cAMP. Thus, the production of cAMP is measured after infection to determine the efficiency by which the effector hybrid has been translocated into the host cytosol by the Dot/Icm system. Another reporter system that provides both visual and quantitative means to measure protein translocation is the TEM-1 β-lactamase fluorescence-based system (BlaM) (Charpentier and Oswald 2004). The BlaM-based system takes advantage of a fluorescent substrate, CCF2-AM, composed of a coumarin and fluorescein moiety. Excitation of the coumarin moiety at 409 nm results in FRET to the fluorescein moiety and subsequent fluorescence emission at 520 nm. Translocated BlaM fusion proteins cleave the β-lactam ring linking coumarin and fluorescein disrupting the FRET reaction, which shifts fluorescence emission to 447 nm. This change in the fluorescence signal in the cytosol of infected host cells indicates translocation of the BlaM-effector protein fusion by the Dot/Icm system (Charpentier and Oswald 2004).

Both Cya and BlaM reporter assays have been used to demonstrate Dot/Icm-dependent translocation of *C. burnetii* effector proteins expressed in *L. pneumophila* (Chen et al. 2004; de Felipe et al. 2008). Recent advances in *C. burnetii* genetics have also provided means to express reporter fusion proteins in *C. burnetii,* which has confirmed that effectors translocated by the *L. pneumophila* Dot/Icm system are also translocated by *C. burnetii* during host cell infection (Chen et al. 2010). In addition, translocation of the endogenously-produced effector proteins AnkG and CBU0077 has been demonstrated during *C. burnetii* infection of host cells by immunoblot analysis of fractionated host cells that were infected with *C. burnetii* (Pan et al. 2008; Carey et al. 2011). Lastly, the *C. burnetii icmL:Tn* mutant does not show any translocation of a plasmid-expressed BlaM-CBU0077 fusion protein after infection of host cells, demonstrating clearly that the Dot/Icm system is essential for translocation of effector proteins (Carey et al. 2011).

Table 9.2 ***C. burnetii* effectors.** Shown are genes in the four sequenced strains of *C. burnetii* that or *C. burnetii*

Nine Mile RSA493	Size kDa (bp)	Dugway 5 J108-111	Size kDa (bp)	K Q154	Size kDa (bp)
CBUA0014 (CoxU3)	44.5 (1,254)	DNA Absent		DNA Absent	
CBU0041 (CoxCC1)	82.7 (2,133)	CBUD2061	82.7 (2,133)	CbuK1986	82.7 (2,133)
CBU0071	36.4 (975)	CBUD2035 (*ankP*)	76.7 (2,043)	**CbuK1982**	76.0 (2,013)
CBU0072 (*ankA*)	44.3 (1,164)	CBUD2034	53.0 (1,407)	**CbuK1982**	44.3 (1,164)
CBU0077	29.6 (792)	CBUD2029	29.6 (792)	CbuK1977	29.6 (792)
CBU0080	18.7 (480)	**CBUD2028**	23.1 (600)	CbuK1976	38.1 (987)
CBU0129 (CoxCC2)	28.5 (348)	**CBUD1978**	24.4 (630)	**Frameshift–no ORF**	
CBU0144/0145 (*ankB*)	41.9 (1,140)	**CBUD1960**	41.0 (1,116)	CbuK1907	39.9 (1,080)
CBU0175 (CoxK1)	28.9 (741)	CBUD1925	28.9 (741)	CbuK0362	28.9 (741)
CBU0295	56.8 (1,467)	CBUD1787	56.8 (1,467)	**CbuK0462**[a]	34.2 (888)
CBU0329	23.2 (597)	CBUD1750	68.4 (1,764)	**CbuK0525**	43.0 (1,107)
CBU0410 (CoxCC3)	68.4 (1,737)	CBUD1664	68.4 (1,737)	**CbuK1657**	53.1 (1,467)
CBU0414 (CoxH1)	30.4 (786)	CBUD1656	48.4 (1,248)	**CbuK1651**	34.6 (894)
CBU0425	51.9 (1,365)	CBUD1648[a]	54.0 (1,248)	**CbuK1643**	14.5 (378)
CBU0447 (*ankF*)	21.3 (555)	CBUD1627	21.3 (555)	CbuK1411	21.3 (555)
CBU0635	55.3 (1,488)	CBUD0646	55.3 (1,488)	**CbuK1621**	38.4 (1,029)
CBU0781 (*ankG*)	38.6 (1,017)	CBUD0829	38.6 (1,017)	CbuK0651	25.7 (669)
CBU0794 (CoxCC4)	53.1 (1,395)	**CBUD0861**[a]	46.0 (1,200)	CbuK0662[a]	39.5 (1,029)
CBU0801 (*rimL*, CoxH2)	16.9 (438)	CBUD0868	16.9 (438)	CbuK0670	16.9 (438)
CBU0814 (CoxU1)	12.6 (321)	**CBUD0882**	47.9 (1,266)	CbuK0684	70.2 (1,854)

encode effectors of the Dot/Icm system based on translocation assays in either *L. pneumophila*

G Q212	Size kDa (bp)	Features of Protein	Identification Method	Reference
DNA Absent		F-box, coiled-coil motif, encoded on the QpH1 plasmid	Putative PmrA binding motif Homology to Dot/Icm substrate	Chen et al. (2010)
CbuG1970	82.7 (2,133)	2 coiled-coil motifs	Putative PmrA binding motif	Chen et al. (2010)
DNA absent		6 Ank repeats	Ankyrin repeat domain containing protein	Voth et al. (2009)
DNA Absent		4 Ank repeats	Ankyrin repeat domain containing protein	Pan et al. (2008); Voth et al. (2009)
CbuG1938	29.6 (792)	4 TM domains	Proximity to CBU0080	Carey et al. (2011)
CbuG1937[a]	34.3 (888)	Helix-Loop-Helix motif in CbuK1976 and CbuG1937.	Screen for translocation signals	Carey et al. (2011)
CbuG1886	51 (1,311)	coiled-coil motif	Putative PmrA binding motif	Chen et al. (2010)
CbuG1870	36.0 (984)	7 Ank repeats	Ankyrin repeat domain containing protein	Pan et al. (2008); Voth et al. (2009)
CbuG1837	28.9 (741)	Predicted serine/threonine kinase	Homology to Dot/Icm substrate	Chen et al. (2010)
CbuG1711	51.3 (1323)	3 SLRs in CBU0295, CBUD1787 and CbuG1711.	Screen for translocation signals	Carey et al. (2011)
CbuG1677	738 (28.9)		Proximity to CBU0328	Carey et al. (2011)
CbuG1603	68.4 (1,737)	coiled-coil motif	Putative PmrA binding motif	Chen et al. (2010)
CbuG1597[a]	45.1 (1161)		Homology to Dot/Icm substrate	Chen et al. (2010)
CbuG1587	56.2 (1,482)		Screen for translocation signals	Carey et al. (2011)
CbuG1564	21.3 (555)	3 Ank repeats	Ankyrin repeat domain containing protein	Pan et al. (2008); Voth et al. (2009)
CbuG1369	55.3 (1,488)	6 TM domains	Screen for translocation signals	Carey et al. (2011)
CbuG1220	38.6 (1,017)	2 Ank repeats, coiled-coil motif	Ankyrin repeat domain containing protein	Pan et al. (2008); Voth et al. (2009)
CbuG1207	55.4 (672)	coiled-coil motif, CbuG1207 has a C-terminal repeat	Putative PmrA binding motif	Chen et al. (2010)
CbuG1199	14.5 (372)	Ribosomal protein S18 alanine acetyltransferase	Homology to Dot/Icm substrate	Chen et al. (2010)
CbuG1185[a]	31.6 (834)	F-box motif, RCC1/BLIPII domain, 1 TM domain	Putative PmrA binding motif Homology to Dot/Icm substrate	Chen et al. (2010)

(continued)

Table 9.2 (continued)

Nine Mile RSA493	Size kDa (bp)	Dugway 5 J108-111	Size kDa (bp)	K Q154	Size kDa (bp)
CBU0881 (CoxCC5)	26.3 (666)	DNA Absent		DNA Absent	
CBU0937 (CoxDFB1)	53.4 (1,452)	CBUD1137	53.4 (1,452)	CbuK0900	53.4 (1,452)
CBU1025	88.8 (2,361)	CBUD1019 (*ankH*)	96.6 (2,574)	**CbuK0815**	64.2 (1,713)
CBU1045 (CoxDFB2)	42.7 (825)	CBUD0997	65.5 (1,749)	**CbuK0793**	17.4 (444)
CBU1107	35.5 (918)	CBUD1210	122.3 (3,177)	**CbuK0977**	18.1 (468)
CBU1108	25.8 (657)	As above		**As above**	
CBU1213[a]-	74.9 (1,974)	CBUD1298 (*ankI*)	82.8 (2,181)	DNA Absent	
CBU1217 (CoxU2)	57.8 (1,494)	**CBUD1301**	44.0 (1,025)	CbuK1077	55.9 (1,440)
CBU1253/1254	66.6 (1,752)	**CBUD1338 (*ankJ*)**	75.6 (1,986)	**CbuK1113**	35.3 (915)
CBU1314 (CoxCC6)	24.4 (645)	CBUD1402	24.4 (645)	CbuK1177	24.4 (645)
CBU1379 (CoxK2)	78.0 (2,031)	**CBUD1462**	59.2 (1,344)	CbuK1237	76.1 (1,980)
CBU1406[a] (CoxDFB3)	17.6 (468)	CBUD0588	32.1 (837)	CbuK1480	32.1 (834)
CBU1425 (CoxDFB4)	15.9 (453)	CBUD0572	15.9 (453)	CbuK1498	15.9 (453)
CBU1457 (CoxTPR1)	78.3 (2,031)	CBUD0496[a]	71.0 (1,848)	CbuK1685[a]	74.4 (1,932)
CBU1460 (CoxCC7)	29.6 (786)	**CBUD0501**	27.3 (714)	**CbuK1689**	19.7 (519)
CBU1461 (CoxCC8)	71.4 (1,899)	**CBUD0503**	79.6 (2,103)	CbuK1690	86.0 (2,271)
CBU1524 (CoxCC9)	25.1 (660)	CBUD0462	36.8 (996)	CbuK1751	24.3 (642)
CBU1525	43.9 (1,146)	CBUD0461	115.4 (3,033)	**CbuK1752**	52.5 (1,752)
CBU1532	16.5 (438)	CBUD0454	49.7 (1,332)	**CbuK1760[a]**	28.1 (762)
CBU1543 (CoxCC10)	22.3 (567)	CBUD0444	23.3 (594)	CbuK1769	23.3 (594)

G Q212	Size kDa (bp)	Features of Protein	Identification Method	Reference
CbuG1121	26.3 (666)	coiled-coil motif	Putative PmrA binding motif	Chen et al. (2010)
CbuG1068	53.4 (1,452)	coiled-coil motif	Interaction with DotF	Chen et al. (2010)
CbuG0983[a]	14.9 (390)	14 Ank repeats	Ankyrin repeat domain containing protein	Voth et al. (2009)
CBUG0959	65.5 (1,749)		Interaction with DotF	Chen et al. (2010)
DNA Absent			Proximity to CBU1108	Carey et al. (2011)
DNA Absent			Homology to CBU1525	Carey et al. (2011)
CbuG0789	82.8 (2,181)	4 Ank repeats, *cig*	Ankyrin repeat domain containing protein	Voth et al. (2009); Zusman et al. (2007)
CbuG0795	33.8 (864)	2 internal repeat domains	Homology to Dot/Icm substrate	Chen et al. (2010)
CbuG0758[a]	25.8 (669)	2 Ank repeats	Ankyrin repeat domain containing protein	Voth et al. (2009)
CbuG0696	27.2 (717)	coiled-coil motif	Putative PmrA binding motif	Chen et al. (2010)
CbuG0633	72.2 (1,860)	Putative serine/threonine protein kinase	Homology to Dot/Icm substrate	Chen et al. (2010)
CbuG0608[a]	27.0 (693)		Interaction with DotF	Chen et al. (2010)
CbuG0591	15.9 (453)	N-terminal signal sequence and domain common to Rickettsial 17 kDa surface antigen proteins	Interaction with DotF	Chen et al. (2010)
CbuG0554[a]	77.6 (2,013)	3 SLRs and coiled-coil motif.	Putative PmrA binding motif	Chen et al. (2010)
CbuG0550	14.9 (387)	coiled-coil motif	Putative PmrA binding motif	Chen et al. (2010)
CbuG0549	61.4 (1,617)	1 TM domain, coiled-coil motif, 1 Ank repeat,	Putative PmrA binding motif	Chen et al. (2010)
CbuG0468	24.3 (642)	coiled-coil motif	Proximity to CBU1525 Putative PmrA binding motif	Carey et al. (2011), Chen et al. (2010)
CbuG0485[a]	36.1 (948)		Screen for translocation signals	Carey et al. (2011)
CbuG0477	49.7 (1,332)	5 TM domains, coiled-coil motif	Proximity to CBU1525	Carey et al. (2011)
CbuG0467	20.5 (519)	1 TM domain	Putative PmrA binding motif	Chen et al. (2010)

(continued)

Table 9.2 (continued)

Nine Mile RSA493	Size kDa (bp)	Dugway 5 J108-111	Size kDa (bp)	K Q154	Size kDa (bp)
CBU1556 (CoxCC11)	64.6 (1,704)	CBUD0430	64.6 (1,704)	CbuK1785	64.6 (1,704)
CBU1569 (CoxCC12)	63.1 (1,644)	CBUD0419	63.1 (1,644)	**CbuK1796**	45.6 (897)
CBU1599 (CoxCC13)	67.9 (1,803)	**CBUD0392**	49.8 (1,326)	**CbuK1827**	66.6 (1,773)
CBU1636 (CoxCC14)	45.0 (1,176)	**CBUD0362**	20.8 (555)	**CbuK1859**	29.8 (795)
CBU1751 (CoxDFB5)	48.8 (1,263)	CBUD0250	48.8 (1,263)	CbuK0255	48.8 (1,263)
CBU1757 (*ankM*)	73.4 (1,938)	CBUD0245	96.8 (2,556)	**CbuK0249**	34.2 (900)
CBU1769 (CoxH3)	22.8 (618)	CBUD0236	22.8 (618)	CbuK0236	22.8 (618)
CBU1776	23.7 (609)	CBUD0231	36.2 (924)	CbuK0231	36.2 (924)
CBU1823 (CoxH4)	40.3 (1,050)	CBUD0053	40.3 (1,050)	**CbuK0119**	28.5 (747)
CBU1825 (CoxDFB5)	13.3 (345)	CBUD0054	16.1 (420)	CbuK0120	16.2 (420)
CBU1963	72.9 (1,899)	CBUD2063	81.4 (2,115)	CbuK2014[a]	76.8 (1,995)
CBU2052 (CoxCC15)	34.0 (903)	CBUD2147	33.7 (897)	CbuK2099	33.7 (897)
CBU2056	25.4 (663)	CBUD2151	25.6 (669)	CbuK2102	25.6 (669)
CBU2059	52.6 (1,389)	CBUD2154	61.0 (1,593)	**CbuK2105**	22.7 (600)
CBU2078 (CoxFIC1)	39.9 (1,047)	CBUD2174	39.9 (1,047)	CbuK2125	39.9 (1,047)
DNA absent		**CBUD1552**	23.1 (630)	CbuK1330 (*ankN*)	58.0 (1,593)
DNA absent		CBUD1108 (*ankO*)	83.9 (2,223)	DNA absent	

Putative pseudogenes are indicated in *bold type*, and *gray shading* represents a gene that is
TM predicted transmembrane domain, *SLR* Sel1-like repeat
[a]ORF has 3′ truncation, therefore the resulting protein, lacking the C-terminus, is possibly not

G Q212	Size kDa (bp)	Features of Protein	Identification Method	Reference
CbuG0454	64.7 (1,707)	4 TM domains, coiled-coil motif	Putative PmrA binding motif	Chen et al. (2010)
CbuG0443	38.2 (1,002)	2 coiled-coil motifs	Putative PmrA binding motif	Chen et al. (2010)
CbuG0414	59.2 (1,566)	1 coiled-coil motif	Putative PmrA binding motif	Chen et al. (2010)
CbuG0386	15.9 (459)	1 coiled-coil motif (only in CBUD0362 and CbuK1859)	Putative PmrA binding motif	Chen et al. (2010)
CbuG0133	48.8 (1,263)	2 coiled-coil motifs	Interaction with DotF	Chen et al. (2010)
CbuG0139	90.9 (2,397)	8 Ank repeats, *cig*	Ankyrin repeat domain containing protein	Voth et al. (2009)
CbuG0151	22.8 (618)	Putative alpha/beta hydrolase	Homology to Dot/Icm substrate	Chen et al. (2010)
DNA absent			Proximity to CBU1780	Carey et al. (2011)
CbuG0196	26.2 (690)	*cig*	Homology to CBU1963Homology to Dot/Icm substrate Putative PmrA binding motif	Carey et al. (2011), Zusman et al. (2007); Chen et al. (2010)
CbuG0197	16.3 (420)		Homology to CBU1823 Interaction with DotF	Carey et al. (2011), Chen et al. (2010)
CbuG1971[a]	24.0 (612)		Downstream of CBU1957	Carey et al. (2011)
CbuG2057[a]	24.8 (645)	Coiled-coil motif	Proximity to CBU2056 and CBU2059 Putative PmrA binding motif	Carey et al. (2011), Chen et al. (2010)
CbuG2060[a]	13.5 (348)		Screen for translocation signals	Carey et al. (2011)
CbuG2063	22.7 (600)		Screen for translocation signals	Carey et al. (2011)
CbuG2079	39.9 (1047)	Fic domain	Fic domain containing protein	Chen et al. (2010)
CbuG1487	58.2 (1,590)	7 Ank repeats	Ankyrin repeat domain containing protein	Voth et al. (2009)
DNA absent		8 Ank repeats	Ankyrin repeat domain containing protein	Voth et al. (2009)

conserved in all four genomes

translocated

9.7.1 *Coxiella burnetii* Effectors with Ankyrin Repeats

Genome sequencing of *L. pneumophila* and *C. burnetii* has revealed a number of bacterial genes encoding proteins with motifs that are common in eukaryotic organisms but rare in bacteria. Analysis of the *L. pneumophila* genome identified 30 predicted proteins with significant similarity to eukaryotic proteins and a further 32 proteins with eukaryotic domains or motifs (Chien et al. 2004; Cazalet et al. 2004). Given the GC bias of these genes and the eukaryotic homology, it is believed the genes encoding these eukaryotic-like proteins have been acquired by interdomain horizontal transfer (de Felipe et al. 2005). Many of these *L. pneumophila* eukaryotic-like proteins are translocated by the Dot/Icm secretion system and are presumed to have effector functions.

Similarly, the *C. burnetii* genome encodes a wide variety of genes predicted to produce proteins with eukaryotic-like motifs. The gene products can be categorized into two groups. The first are proteins with motifs or domains that are generally considered to function as scaffolds that promote specific protein-protein interactions. These include scaffolds created by ankyrin repeats, tetratricopeptide repeats, coiled-coil motifs and leucine rich repeats. The second category includes proteins predicted to have enzymatic functions that can affect eukaryotic signal transduction processes. Examples include *C. burnetii* proteins with GTPase domains, motifs that promote protein ubiquitination such as the F-box domain, and kinases and phosphatases that have the potential to modify proteins and lipids.

Within the category of proteins having eukaryotic-like scaffolding domains are the ankyrin repeat-containing proteins called Anks (Pan et al. 2008; Voth et al. 2009). An ankyrin repeat is a motif of 33 amino acids that forms antiparallel α-helices with connecting loop regions. The tertiary structure of this pattern enables the loop regions to be available for specific protein-protein interactions, and eukaryotic proteins with ankyrin repeats are abundant and influence a wide range of cellular processes from transcription regulation to cytoskeletal motility (Mosavi et al. 2004). Interestingly, several other intracellular bacterial pathogens with virulence-associated secretion systems also encode proteins with ankyrin repeats, including *Anaplasma phagocytophilum*, *Erlichia chaffeensis* and *Rickettsia* spp. (Rikihisa and Lin 2010; Ogata et al. 2005). Thus, the ankyrin repeat domain has likely been subverted by these pathogens to create novel effectors that can interact with eukaryotic proteins.

The genome sequence of *Coxiella burnetii* Nine Mile RSA 493 revealed 13 genes encoding Ank proteins (Seshadri et al. 2003). These *C. burnetii* Ank proteins were tested for Dot/Icm-dependent translocation into host cells using *L. pneumophila* as a surrogate host (Pan et al. 2008). Using the Cya reporter, this study showed Dot/Icm-dependent translocation of AnkA, AnkB, AnkF and AnkG (Pan et al. 2008)(Table 9.2). The proteins AnkE, AnkH, AnkI and AnkM showed less efficient translocation, and no detectable translocation of AnkC, AnkD, AnkJ, AnkK and AnkL was observed (Pan et al. 2008). To validate that a positive signal obtained using Cya reporters expressed in *L. pneumophila* represented a *bone fide* translocation event that would occur during infection, translocation of the endogenous AnkF protein during *C. burnetii* infection was demonstrated by immunoblot analysis.

After the genomes of the *C. burnetii* Dugway strain and the human endocarditis isolates K and G were sequenced it was observed that there is a high degree of variability in the predicted sequences for the Ank proteins (Beare et al. 2009). The Dugway isolate encodes 11 Ank proteins (AnkA, AnkC, AnkD, AnkF, AnkG, AnkH, AnkI, AnkK, AnkM, AnkO and AnkP) that are longer than the Ank proteins encoded by the Nine Mile isolate, with AnkO being unique to Dugway. With the exception of five Ank proteins encoded by the Nine Mile isolate (AnkA, AnkC, AnkF, AnkG and AnkK) most of the *ank* genes possessed mutations that would either result in frameshifts or deletions that would make these proteins over 20% shorter than the corresponding version of the protein encoded by the other sequenced strains of *C. burnetii*. This suggests that pseudogenization may be occurring in the Nine Mile strain, reducing the number of functional Ank proteins that are produced. It is interesting to note that most of the strains produce truncated versions of AnkB, AnkJ and AnkL, suggesting the function of these Ank proteins may not be important for survival in specific hosts, and there are four Ank proteins (AnkC, AnkF, AnkG and AnkK) that are conserved among the sequenced genomes, which could play important roles in host persistence of *C. burnetii* in both pathogenic and non-pathogenic conditions (Beare et al. 2009).

When the Ank proteins produced by the other sequenced strains of *C. burnetii* were tested for translocation by the *L. pneumophila* Dot/Icm secretion system using the CyaA reporter assay, it was found that the only Ank proteins where translocation could not be detected were AnkC, AnkD, AnkK and AnkL (Voth et al. 2009) (Table 9.2). Additionally, it was shown that efficient Dot/Icm-dependent translocation of AnkA, AnkF, AnkG and AnkP required the chaperone protein IcmS (Voth et al. 2009). This is consistent with findings that the *C. burnetii* IcmS protein can restore function to a *L. pneumophila icmS* mutant, and that a subset of the *C. burnetii* effectors require the IcmS-IcmW chaperone complex for efficient recognition by the Dot/Icm system (Zamboni et al. 2003; Ninio et al. 2005; Cambronne and Roy 2007). In addition, it was shown that the signal sequence required for Dot/Icm-dependent translocation was located near the C-terminus of the Ank proteins, which is consistent with the mapping studies performed on several *L. pneumophila* Dot/Icm substrates (Voth et al. 2009; Nagai et al. 2005). The location of this translocation signal has obvious repercussions for Dot/Icm substrates that have acquired frameshift mutations or deletions that create C-terminal truncations, as these truncated proteins would lack the translocation signal. The AnkB proteins predicted for the *C. burnetii* K isolate and the Nine Mile AnkI protein were both shown to have C-terminal truncations in comparison to the longer proteins encoded by the Dugway isolate, and that the loss of these C-terminal sequences prevented these proteins from being translocated by the Dot/Icm secretion system (Voth et al. 2009).

Based on previous data suggesting that *C. burnetii* effectors may have an important role in inhibiting host cell apoptosis, several of the Ank proteins encoded by the Nine Mile strain were tested for anti-apoptotic properties. The Ank proteins were ectopically produced in eukaryotic cells, and apoptosis was assessed after cells were treated with staurosporine, which is a potent agonist of apoptosis. The highly conserved AnkG was shown to protect eukaryotic cells from apoptosis induced by staurosporin treatment (Luhrmann et al. 2010). Most strikingly, this anti-apoptotic activity was also observed for *L. pneumophila* producing the *C. burnetii* AnkG protein. Whereas mouse dendritic cells normally undergo apoptosis in response to *L. pneumophila* infection, when dendritic cells were infected with *L. pneumophila* expressing AnkG, the cells remained healthy and supported *L. pneumophila* intracellular replication (Luhrmann et al. 2010; Nogueira et al. 2009). This indicates that the quantities of AnkG that are delivered by the Dot/Icm system during infection are sufficient to block a cell autonomous host response that promotes cell death and limits pathogen replication. Thus, studies on AnkG provide proof of principle that *C. burnetii* proteins translocated into host cells by the Dot/Icm system have effector activities that help create a permissive environment for replication.

9.7.2 Screens for Additional Dot/Icm Effector Proteins

The repertoire of effectors in *C. burnetii* is predicted to include more proteins than the Ank family of effectors. Approaches to identify additional effectors in *C. burnetii* have included targeted screens examining proteins with domains that are similar to regions found in *L. pneumophila* effectors, examining proteins that have PmrA regulatory sequences upstream of their coding regions, examining proteins that interact with the DotF protein, proteomic approaches to identify secreted proteins, and random genetic screens to identify proteins having a Dot/Icm dependent translocation signal (Carey et al. 2011; Chen et al. 2010; Samoilis et al. 2010). These studies have revealed a large number of *C. burnetii* proteins that have the potential to function as effectors based on their ability to be translocated into host cells by the Dot/Icm system (Table 9.2). Future studies are needed to determine the biochemical functions of these proteins to better understand how they may be manipulating host cell functions during infection. Determining the activities of these proteins will provide novel insight into the mechanisms by which *C. burnetii* modulates vesicular transport in the cell and evades intrinsic cell processes that mammalian hosts have evolved to restrict the replication of intracellular pathogens.

9.7.3 Effector Plasticity Between Coxiella burnetii Isolates

The discovery of a large repertoire of effector proteins has revealed that there is a substantial degree of variation of individual effectors when the genomes of different *C. burnetii* isolates are compared. Only four of the Ank proteins (AnkC, AnkF, AnkG and AnkK) and 1 of the 18 Dot/Icm substrates identified by the random screen approach (CBU0077) are conserved in the four sequenced genomes (Table 9.2). Interesting, the non-pathogenic Dugway isolate has the largest genome and fewest pseudogenes and IS elements, which has led to the suggestion that this isolate represents an earlier stage of pathoadaptation than the other isolates that have been sequenced (Beare et al. 2009). One possible explanation for this phenomenon is that most strains of *C. burnetii* have a commensalistic relationship with their hosts, and many of the effector proteins are involved in modulating host responses to limit host pathology. This would explain why Dugway and strains of *C. burnetii* associated with chronic infection tend to show less pseudogenization than strains such as the Nine Mile isolate that cause acute disease. The benefit to the loss of effectors may be more rapid replication within a host organism, which would likely necessitate more efficient transmission between infected and uninfected hosts. Consistent with the hypothesis, the growth rate of the Nine Mile isolate in cultured mammalian cells is faster than that observed for Dugway or the isolates from chronic infection (Roman et al. 1991). It is possible modern large-scale ranching practices have provided habitats that promote efficient transmission of *C. burnetii*, and that this is leading to changes in the effector repertoire where proteins that are involved in maintaining a commensalistic relationship are being lost, resulting in strains that replicate faster and are more pathogenic. Thus, examining *C. burnetii* effector plasticity may provide an interesting model system to determine how changes in host dynamics may result in evolutionary adaptations that enhance virulence.

9.8 Concluding Remarks

Many questions remain about the biogenesis and structure of the *C. burnetii* Dot/Icm apparatus and the process by which effector protein translocation occurs. Recent advances have provided a growing list of *C. burnetii* effectors, however a comprehensive approach is required to understand all the functions mediated by the Dot/Icm system and the individual effectors. The development of genetic tools for *C. burnetii* will inevitably aid this process, as we now have the ability to analyze isogenic mutants deficient in a single effector protein or in a functional Dot/Icm system. Additionally, the ability to introduce functional alleles encoding an effector protein from isolates such as Dugway into the Nine Mile strain will provide gain-of-function phenotypes that will aid in determining the role of these individual effector proteins. Thus, additional studies on the Dot/Icm effectors from *C. burnetii* will provide novel information on the infection strategies this organism has evolved to survive and grow inside mammalian host cells.

References

Altman E, Segal G (2008) The response regulator CpxR directly regulates expression of several *Legionella pneumophila icm/dot* components as well as new translocated substrates. J Bacteriol 190:1985–1996

Alvarez-Martinez CE, Christie PJ (2009) Biological diversity of prokaryotic type IV secretion systems. Microbiol Mol Biol Rev 73:775–808

Bachman MA, Swanson MS (2001) RpoS co-operates with other factors to induce *Legionella pneumophila* virulence in the stationary phase. Mol Microbiol 40:1201–1214
Beare PA, Unsworth N, Andoh M et al (2009) Comparative genomics reveal extensive transposon-mediated genomic plasticity and diversity among potential effector proteins within the genus *Coxiella*. Infect Immun 77:642–656
Berger KH, Isberg RR (1993) Two distinct defects in intracellular growth complemented by a single genetic locus in *Legionella pneumophila*. Mol Microbiol 7:7–19
Beron W, Gutierrez MG, Rabinovitch M et al (2002) *Coxiella burnetii* localizes in a Rab7-labeled compartment with autophagic characteristics. Infect Immun 70:5816–5821
Burstein D, Zusman T, Degtyar E et al (2009) Genome-scale identification of *Legionella pneumophila* effectors using a machine learning approach. PLoS Pathog 5:e1000508
Cambronne ED, Roy CR (2006) Recognition and delivery of effector proteins into eukaryotic cells by bacterial secretion systems. Traffic 7:929–939
Cambronne ED, Roy CR (2007) The *Legionella pneumophila* IcmSW complex interacts with multiple Dot/Icm effectors to facilitate type IV translocation. PLoS Pathog 3:e188
Carey et al. (2011) PLoS Pathog. May;7(5):e1002056. Epub 2011 May 26. PMID: 21637816
Carlsson F, Joshi SA, Rangell L et al (2009) Polar localization of virulence-related Esx-1 secretion in *mycobacteria*. PLoS Pathog 5:e1000285
Cazalet C, Rusniok C, Bruggemann H et al (2004) Evidence in the *Legionella pneumophila* genome for exploitation of host cell functions and high genome plasticity. Nat Genet 36:1165–1173
Charpentier X, Oswald E (2004) Identification of the secretion and translocation domain of the enteropathogenic and enterohemorrhagic *Escherichia coli* effector Cif, using TEM-1 beta-lactamase as a new fluorescence-based reporter. J Bacteriol 186:5486–5495
Chen J, de Felipe KS, Clarke M et al (2004) *Legionella* effectors that promote nonlytic release from protozoa. Science 303:1358–1361
Chen C, Banga S, Mertens K et al (2010) Large-scale identification and translocation of type IV secretion substrates by *Coxiella burnetii*. Proc Natl Acad Sci U S A 107:21755–21760
Chien M, Morozova I, Shi S et al (2004) The genomic sequence of the accidental pathogen *Legionella pneumophila*. Science 305:1966–1968
Christie PJ, Vogel JP (2000) Bacterial type IV secretion: conjugation systems adapted to deliver effector molecules to host cells. Trends Microbiol 8:354–360
Cianciotto NP (2005) Type II secretion: a protein secretion system for all seasons. Trends Microbiol 13:581–588
Coers J, Kagan JC, Matthews M et al (2000) Identification of Icm protein complexes that play distinct roles in the biogenesis of an organelle permissive for *Legionella pneumophila* intracellular growth. Mol Microbiol 38:719–736
Coleman SA, Fischer ER, Howe D et al (2004) Temporal analysis of *Coxiella burnetii* morphological differentiation. J Bacteriol 186:7344–7352
De Buck E, Lammertyn E, Anne J (2008) The importance of the twin-arginine translocation pathway for bacterial virulence. Trends Microbiol 16:442–453
de Felipe KS, Pampou S, Jovanovic OS et al (2005) Evidence for acquisition of *Legionella* type IV secretion substrates via interdomain horizontal gene transfer. J Bacteriol 187:7716–7726
de Felipe KS, Glover RT, Charpentier X et al (2008) *Legionella* eukaryotic-like type IV substrates interfere with organelle trafficking. PLoS Pathog 4:e1000117
Dumenil G, Montminy TP, Tang M et al (2004) IcmR-regulated membrane insertion and efflux by the *Legionella pneumophila* IcmQ protein. J Biol Chem 279:4686–4695
Ensminger AW, Isberg RR (2009) *Legionella pneumophila* Dot/Icm translocated substrates: a sum of parts. Curr Opin Microbiol 12:67–73
Feldman M, Segal G (2004) A specific genomic location within the icm/dot pathogenesis region of different *Legionella* species encodes functionally similar but nonhomologous virulence proteins. Infect Immun 72:4503–4511

Feldman M, Zusman T, Hagag S et al (2005) Coevolution between nonhomologous but functionally similar proteins and their conserved partners in the *Legionella* pathogenesis system. Proc Natl Acad Sci U S A 102:12206–12211

Ferhat M, Atlan D, Vianney A et al (2009) The TolC protein of *Legionella pneumophila* plays a major role in multi-drug resistance and the early steps of host invasion. PLoS One 4:e7732

Gal-Mor O, Segal G (2003) Identification of CpxR as a positive regulator of icm and dot virulence genes of *Legionella pneumophila*. J Bacteriol 185:4908–4919

Hager AJ, Bolton DL, Pelletier MR et al (2006) Type IV pili-mediated secretion modulates Francisella virulence. Mol Microbiol 62:227–237

Heinzen RA, Scidmore MA, Rockey DD et al (1996) Differential interaction with endocytic and exocytic pathways distinguish parasitophorous vacuoles of *Coxiella burnetii* and *Chlamydia trachomatis*. Infect Immun 64:796–809

Hobbs M, Mattick JS (1993) Common components in the assembly of type 4 fimbriae, DNA transfer systems, filamentous phage and protein-secretion apparatus: a general system for the formation of surface-associated protein complexes. Mol Microbiol 10:233–243

Horwitz MA (1983) The Legionnaires' disease bacterium (*Legionella pneumophila*) inhibits phagosome-lysosome fusion in human monocytes. J Exp Med 158:2108–2126

Horwitz MA, Silverstein SC (1980) Legionnaires' disease bacterium (*Legionella pneumophila*) multiples intracellularly in human monocytes. J Clin Invest 66:441–450

Howe D, Heinzen RA (2006) *Coxiella burnetii* inhabits a cholesterol-rich vacuole and influences cellular cholesterol metabolism. Cell Microbiol 8:496–507

Howe D, Melnicakova J, Barak I et al (2003a) Fusogenicity of the *Coxiella burnetii* parasitophorous vacuole. Ann N Y Acad Sci 990:556–562

Howe D, Melnicakova J, Barak I et al (2003b) Maturation of the *Coxiella burnetii* parasitophorous vacuole requires bacterial protein synthesis but not replication. Cell Microbiol 5:469–480

Huang L, Boyd D, Amyot WM et al (2011) The E Block motif is associated with *Legionella pneumophila* translocated substrates. Cell Microbiol 13:227–245

Hubber A, Roy CR (2010) Modulation of host cell function by *Legionella pneumophila* type IV effectors. Annu Rev Cell Dev Biol 26:261–283

Jain S, van Ulsen P, Benz I et al (2006) Polar localization of the autotransporter family of large bacterial virulence proteins. J Bacteriol 188:4841–4850

Jaumouille V, Francetic O, Sansonetti PJ et al (2008) Cytoplasmic targeting of IpaC to the bacterial pole directs polar type III secretion in *Shigella*. EMBO J 27:447–457

Kagan JC, Roy CR (2002) Legionella phagosomes intercept vesicular traffic from endoplasmic reticulum exit sites. Nat Cell Biol 4:945–954

Krall L, Wiedemann U, Unsin G et al (2002) Detergent extraction identifies different VirB protein subassemblies of the type IV secretion machinery in the membranes of *Agrobacterium tumefaciens*. Proc Natl Acad Sci U S A 99:11405–11410

Lee PA, Tullman-Ercek D, Georgiou G (2006) The bacterial twin-arginine translocation pathway. Annu Rev Microbiol 60:373–395

Lima T, Auchincloss AH, Coudert E et al (2009) HAMAP: a database of completely sequenced microbial proteome sets and manually curated microbial protein families in UniProtKB/Swiss-Prot. Nucleic Acids Res 37:D471–D478

Liu M, Conover GM, Isberg RR (2008) *Legionella pneumophila* EnhC is required for efficient replication in tumour necrosis factor alpha-stimulated macrophages. Cell Microbiol 10:1906–1923

Luhrmann A, Roy CR (2007) *Coxiella burnetii* inhibits activation of host cell apoptosis through a mechanism that involves preventing cytochrome c release from mitochondria. Infect Immun 75:5282–5289

Luhrmann A, Nogueira CV, Carey KL et al (2010) Inhibition of pathogen-induced apoptosis by a *Coxiella burnetii* type IV effector protein. Proc Natl Acad Sci U S A 107(44):18997–19001

Lynch D, Fieser N, Gloggler K et al (2003) The response regulator LetA regulates the stationary-phase stress response in *Legionella pneumophila* and is required for efficient infection of *Acanthamoeba castellanii*. FEMS Microbiol Lett 219:241–248

Madariaga MG, Rezai K, Trenholme GM et al (2003) Q fever: a biological weapon in your backyard. Lancet Infect Dis 3:709–721

Marra A, Blander SJ, Horwitz MA et al (1992) Identification of a *Legionella pneumophila* locus required for intracellular multiplication in human macrophages. Proc Natl Acad Sci U S A 89:9607–9611

McCaul TF, Williams JC (1981) Developmental cycle of *Coxiella burnetii*: structure and morphogenesis of vegetative and sporogenic differentiations. J Bacteriol 147:1063–1076

Melnicakova J, Lukacova M, Howe D et al (2003) Identification of *Coxiella burnetii* RpoS-dependent promoters. Ann N Y Acad Sci 990:591–595

Morgan JK, Luedtke BE, Shaw EI (2010a) Polar localization of the *Coxiella burnetii* type IVB secretion system. FEMS Microbiol Lett 305:177–183

Morgan JK, Luedtke BE, Thompson HA et al (2010b) *Coxiella burnetii* type IVB secretion system region I genes are expressed early during the infection of host cells. FEMS Microbiol Lett 311(1):61–69

Mori H, Ito K (2001) The sec protein-translocation pathway. Trends Microbiol 9:494–500

Mosavi LK, Cammett TJ, Desrosiers DC et al (2004) The ankyrin repeat as molecular architecture for protein recognition. Protein Sci 13:1435–1448

Nagai H, Cambronne ED, Kagan JC et al (2005) A C-terminal translocation signal required for Dot/Icm-dependent delivery of the *Legionella* RalF protein to host cells. Proc Natl Acad Sci U S A 102:826–831

Nakano N, Kubori T, Kinoshita M et al (2010) Crystal structure of *Legionella* DotD: insights into the relationship between type IVB and type II/III secretion systems. PLoS Pathog 6(10):e1001129

Nash TW, Libby DM, Horwitz MA (1984) Interaction between the legionnaires' disease bacterium (*Legionella pneumophila*) and human alveolar macrophages. Influence of antibody, lymphokines, and hydrocortisone. J Clin Invest 74:771–782

Newton HJ, Ang DK, van Driel IR et al (2010) Molecular pathogenesis of infections caused by *Legionella pneumophila*. Clin Microbiol Rev 23:274–298

Ninio S, Zuckman-Cholon DM, Cambronne ED et al (2005) The *Legionella* IcmS-IcmW protein complex is important for Dot/Icm-mediated protein translocation. Mol Microbiol 55:912–926

Nogueira CV, Lindsten T, Jamieson AM et al (2009) Rapid pathogen-induced apoptosis: a mechanism used by dendritic cells to limit intracellular replication of *Legionella pneumophila*. PLoS Pathog 5:e1000478

Ogata H, Renesto P, Audic S et al (2005) The genome sequence of *Rickettsia felis* identifies the first putative conjugative plasmid in an obligate intracellular parasite. PLoS Biol 3:e248

Pan X, Luhrmann A, Satoh A et al (2008) Ankyrin repeat proteins comprise a diverse family of bacterial type IV effectors. Science 320:1651–1654

Rasis M, Segal G (2009) The LetA-RsmYZ-CsrA regulatory cascade, together with RpoS and PmrA, post-transcriptionally regulates stationary phase activation of *Legionella pneumophila* Icm/Dot effectors. Mol Microbiol 72:995–1010

Raychaudhury S, Farelli JD, Montminy TP et al (2009) Structure and function of interacting IcmR-IcmQ domains from a type IVb secretion system in *Legionella pneumophila*. Structure 17:590–601

Rikihisa Y, Lin M (2010) *Anaplasma phagocytophilum* and *Ehrlichia chaffeensis* type IV secretion and Ank proteins. Curr Opin Microbiol 13:59–66

Roman MJ, Crissman HA, Samsonoff WA et al (1991) Analysis of *Coxiella burnetii* isolates in cell culture and the expression of parasite-specific antigens on the host membrane surface. Acta Virol 35:503–510

Rossier O, Cianciotto NP (2005) The *Legionella pneumophila* tatB gene facilitates secretion of phospholipase C, growth under iron-limiting conditions, and intracellular infection. Infect Immun 73:2020–2032

Rowbotham TJ (1980) Preliminary report on the pathogenicity of *Legionella pneumophila* for freshwater and soil amoebae. J Clin Pathol 33:1179–1183

Samoilis G, Aivaliotis M, Vranakis I et al (2010) Proteomic screening for possible effector molecules secreted by the obligate intracellular pathogen *Coxiella burnetii*. J Proteome Res 9:1619–1626

Sauvonnet N, Vignon G, Pugsley AP et al (2000) Pilus formation and protein secretion by the same machinery in *Escherichia coli*. EMBO J 19:2221–2228

Segal G, Shuman HA (1999) Possible origin of the *Legionella pneumophila* virulence genes and their relation to *Coxiella burnetii*. Mol Microbiol 33:669–670

Segal G, Purcell M, Shuman HA (1998) Host cell killing and bacterial conjugation require overlapping sets of genes within a 22-kb region of the *Legionella pneumophila* genome. Proc Natl Acad Sci U S A 95:1669–1674

Seshadri R, Samuel JE (2001) Characterization of a stress-induced alternate sigma factor, RpoS, of *Coxiella burnetii* and its expression during the development cycle. Infect Immun 69:4874–4883

Seshadri R, Paulsen IT, Eisen JA et al (2003) Complete genome sequence of the Q-fever pathogen *Coxiella burnetii*. Proc Natl Acad Sci U S A 100:5455–5460

Sexton JA, Vogel JP (2002) Type IVB secretion by intracellular pathogens. Traffic 3:178–185

Shi C, Forsbach-Birk V, Marre R et al (2006) The *Legionella pneumophila* global regulatory protein LetA affects DotA and Mip. Int J Med Microbiol 296:15–24

Shin S, Roy CR (2008) Host cell processes that influence the intracellular survival of *Legionella pneumophila*. Cell Microbiol 10:1209–1220

Sory MP, Cornelis GR (1994) Translocation of a hybrid YopE-adenylate cyclase from *Yersinia enterocolitica* into HeLa cells. Mol Microbiol 14:583–594

Swanson MS, Isberg RR (1995) Association of *Legionella pneumophila* with the macrophage endoplasmic reticulum. Infect Immun 63:3609–3620

Thanabalu T, Koronakis E, Hughes C et al (1998) Substrate-induced assembly of a contiguous channel for protein export from *E.coli*: reversible bridging of an inner-membrane translocase to an outer membrane exit pore. EMBO J 17:6487–6496

Thieme F, Koebnik R, Bekel T et al (2005) Insights into genome plasticity and pathogenicity of the plant pathogenic bacterium xanthomonas campestris pv. Vesicatoria revealed by the complete genome sequence. J Bacteriol 187:7254–7266

Vincent CD, Friedman JR, Jeong KC et al (2006) Identification of the core transmembrane complex of the *Legionella* Dot/Icm type IV secretion system. Mol Microbiol 62:1278–1291

Vogel JP, Andrews HL, Wong SK et al (1998) Conjugative transfer by the virulence system of *Legionella pneumophila*. Science 279:873–876

Voth DE, Heinzen RA (2007) Lounging in a lysosome: the intracellular lifestyle of *Coxiella burnetii*. Cell Microbiol 9:829–840

Voth DE, Howe D, Heinzen RA (2007) *Coxiella burnetii* inhibits apoptosis in human THP-1 cells and monkey primary alveolar macrophages. Infect Immun 75:4263–4271

Voth DE, Howe D, Beare PA et al (2009) The *Coxiella burnetii* ankyrin repeat domain-containing protein family is heterogeneous, with C-terminal truncations that influence Dot/Icm-mediated secretion. J Bacteriol 191:4232–4242

Voulhoux R, Ball G, Ize B et al (2001) Involvement of the twin-arginine translocation system in protein secretion via the type II pathway. EMBO J 20:6735–6741

Zamboni DS, McGrath S, Rabinovitch M et al (2003) *Coxiella burnetii* express type IV secretion system proteins that function similarly to components of the *Legionella pneumophila* Dot/Icm system. Mol Microbiol 49:965–976

Zuckman DM, Hung JB, Roy CR (1999) Pore-forming activity is not sufficient for *Legionella pneumophila* phagosome trafficking and intracellular growth. Mol Microbiol 32:990–1001

Zusman T, Yerushalmi G, Segal G (2003) Functional similarities between the icm/dot pathogenesis systems of *Coxiella burnetii* and *Legionella pneumophila*. Infect Immun 71:3714–3723

Zusman T, Aloni G, Halperin E et al (2007) The response regulator PmrA is a major regulator of the icm/dot type IV secretion system in *Legionella pneumophila* and *Coxiella burnetii*. Mol Microbiol 63:1508–1523

Chapter 10
Role of Lipids in *Coxiella burnetii* Infection

Stacey D. Gilk

Contents

Abstract Lipids are essential components of both eukaryotic and prokaryotic cells, serving diverse functions including energy metabolism and membrane structure. Intracellular vacuolar pathogens such as *Coxiella burnetii* require lipids for both normal bacterial functions as well as formation of the acidic, phagolysosomal-like parasitophorous vacuole (PV) surrounding the bacteria. As an intracellular pathogen, *C. burnetii* can acquire lipid through both *de novo* bacterial synthesis and subversion of host cell pools. The *C. burnetii* genome encodes enzymes required for *de novo* synthesis of fatty acids and phospholipids. The high percentage of branched fatty acids suggests *C. burnetii* modifies these molecules to generate a bacterial cell envelope

S.D. Gilk (✉)
Coxiella Pathogenesis Section, Laboratory of Intracellular Parasites,
Rocky Mountain Laboratories, National Institute of Allergy and Infectious Diseases,
National Institutes of Health, 903 S. 4th St., Hamilton, MT 59840, USA
e-mail: gilks@niaid.nih.gov

R. Toman et al. (eds.), *Coxiella burnetii: Recent Advances and New Perspectives in Research of the Q Fever Bacterium*, Advances in Experimental Medicine and Biology 984,
DOI 10.1007/978-94-007-4315-1_10,

that can resist the harsh environment of the PV, such as the acidic pH. In addition to fatty acids and their derivatives, *C. burnetii* requires isoprenoids, particularly sterols as the PV membrane is cholesterol-rich. With the exception of two eukaryote-like sterol reductases, *C. burnetii* does not have the capability to generate cholesterol, suggesting sterols are actively diverted from the host cell. While *C. burnetii* utilizes host cell lipids for membrane biogenesis and possibly energy, bacterial manipulation of host cell lipid signaling pathways may support establishment of the intracellular niche. For example, effectors secreted by the *C. burnetii* Type IV secretion system may either directly or indirectly modify host cell lipids. Further understanding of the lipid biosynthetic capabilities of *C. burnetii*, along with *C. burnetii's* manipulation of host cell lipids, will provide insight into the host-pathogen relationship.

Keywords Cholesterol biosynthesis • Fatty acid synthesis • Phospholipid synthesis • Sterol reductase • Membrane fluidity • Type IV secretion system • Statin • Membrane composition

10.1 Introduction

Lipids are a large group of structurally and functionally diverse molecules, with their hydrophobicity being the major common characteristic. Most lipids are composed of long hydrocarbon chains (*e.g.*, fatty acids) or interconnected rings (*e.g.*, sterols), and play essential biological roles including membrane formation, signaling pathway regulation, and generation and storage of energy. Despite a simplified cellular organization as compared to mammalian cells, bacteria exhibit a larger variety of lipid species. This diversity is at least partially due to the wide range of conditions such as temperature and pH that bacteria encounter in their respective environments. While lipids are essential in all organisms, key differences exist between the lipids of eukaryotic and prokaryotic cells. For example, cholesterol, an essential component of eukaryotic organisms, is largely absent from bacteria. Other lipids such as lipopolysaccharide (LPS) are unique to bacteria and function to protect the bacteria from the environment.

For intracellular pathogens such as *Coxiella burnetii,* lipids are critical for bacterial survival and pathogenesis. First, all organisms require fatty acids and phospholipids for membrane structure. As a gram-negative bacterium, the envelope surrounding *C. burnetii* is comprised of two membranes; both leaflets of the inner membrane are composed of phospholipids, while the outer leaflet of the outer membrane is dominated by lipopolysaccharide (LPS), a hallmark of gram-negative bacteria. In addition, the environmentally stable small cell variant (SCV) of *C. burnetii* has a thicker envelope and internal, multi-lamellar membranes (McCaul and Williams 1981). Second, lipid breakdown of either *de novo* or exogenous fatty acids could provide an important source of carbon or ATP. Third, host cell lipids are essential components of the parasitophorous vacuole (PV) membrane, which not only serves as a permeability barrier to sequester ions and nutrients essential for growth, but also as an interface between the bacteria and the host cell cytoplasm. And finally, many intracellular pathogens manipulate host cell trafficking, apoptosis, and other critical pathways by subverting signaling lipids such as cholesterol and phosphoinositides (Cossart and Roy 2010; Weber et al. 2009).

The role of lipids in bacterial pathogenesis has only recently garnered attention and more in depth investigation, especially in organisms like *C. burnetii* that have been historically difficult to study in the laboratory setting. In this chapter, we will first examine the lipid biosynthetic abilities and composition of *C. burnetii*. Second, we will look at the role of sterols in *C. burnetii*-host cell interactions, and discuss how this pathogen may hijack host cell lipids.

10.2 *C. burnetii* Lipid Metabolism

10.2.1 Fatty Acid Synthesis

Fatty acids consist of a polar carbon chain with a reactive, hydrophilic carboxyl head group (Fig. 10.1). They serve as an energy source, as building blocks for more complex lipids such as membrane phospholipids, and are used for post-translational modification of proteins. The carbon chain may be saturated or unsaturated; unsaturated fatty acids contain one or more double bonds, creating a distinctive "kink" and altering the packing ability of the carbon chain. In addition to saturation state, fatty acids may be classified as straight chain or branched, meaning they have a methyl group attached at either the penultimate carbon (*iso-*), or the third carbon from the end (*anteiso-*). In most bacteria, the majority of fatty acids are straight chain (Kaneda 1991). Branched fatty acids are usually less abundant than straight fatty acids, though some bacteria such as *Bacillus subtilis* and *Mycobacteria spp.* have a high proportion of mono- and multi-methyl branched fatty acids (Clejan et al. 1986; Jackson et al. 2007).

Fatty acid synthesis is highly conserved in bacteria. Indeed, genomic analysis of *C. burnetii* indicates that the bacterium contains most of the enzymes required to synthesize fatty acids through the type II pathway. Acyl carrier protein (ACP), a critical enzyme in this pathway, binds all fatty acyl intermediates including both the primer and elongation units. Chain elongation is carried out two carbons at a time, and is performed by a conserved set of enzymes. In *C. burnetii*, the majority of these enzymes (FabH, FabD, FabG, ACP, and FabF) are clustered in the genome, while FabZ is genomically isolated. Interestingly, *C. burnetii* does not have one of the major classes of bacterial enoyl-acyl carrier protein reductase (FabK, I, or L), the enzyme catalyzing the final step of the elongation cycle. Instead, it appears to encode an enoyl reductase (CBU0270) more closely related to the recently described FabV from *Vibrio cholera* (Massengo-Tiasse and Cronan 2008). Like FabV, CBU0270 is predicted to be much larger (~45 kDa) than other classes of bacterial enoyl reductases (~27 kDa), and contains a predicted active site of Y-(X_8)-K (Massengo-Tiasse and Cronan 2009). The FabV class has a strong preference for NADH as a cofactor

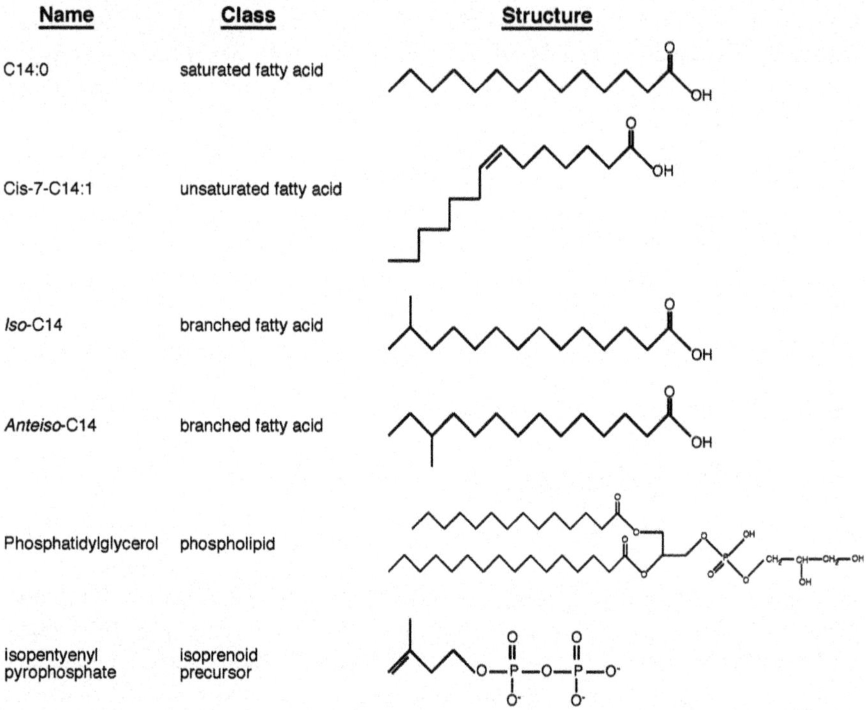

Fig. 10.1 Structures of representative lipids found in *C. burnetii*

(Massengo-Tiasse and Cronan 2008), FabL prefers NADPH (Heath et al. 2000) and FabI has no preference (Bergler et al. 1996). *C. burnetii* lacks the oxidative branch of the pentose phosphate pathway, an important source of NADPH (Beare et al. 2009), suggesting CBU0270 and other enzymes may use NADH as a co-factor. The physiological and pathogenic relevance of the CBU0270 cofactor preference remains to be elucidated.

Several studies have determined that the majority of *C. burnetii* fatty acids are branched (Chan et al. 1976; Amano and Williams 1984; Tzianabos et al. 1981). Moreover, the overall fatty acid profile is most similar to *Legionella* and some gram-positive bacteria (Amano et al. 1984). A large percentage of *anteiso*-branched fatty acids may be an adaptation by *C. burnetii* to persist in the acidic environment of the PV (Amano and Williams 1984; Tzianabos et al. 1981). For example, *Listeria monocytogenes* has higher proportions of *anteiso* fatty acids when grown under acidic conditions, presumably to maintain membrane fluidity (Giotis et al. 2007). Interestingly, despite marked morphological and stability differences between the SCV and large cell variant (LCV) developmental forms, very little variation is detected in their fatty acid composition (Amano et al. 1984). Virulent phase I and avirulent phase II also have minor and most likely insignificant differences in fatty acid profiles (Amano and Williams 1984; Wollenweber et al. 1985).

10.2.2 Phospholipid Synthesis

Fatty acid chains are transferred to membrane-bound glycerol-3-phosphate to form lysophosphatidic acid (LPA), which is further acylated to form phosphatidic acid (PA), the key intermediate in membrane phospholipid synthesis. Due to their amphipathic nature, phospholipids are the major component of cell membranes, and variation occurs in both the head group and the acyl chains. Phospholipids are critical determinants of membrane-protein topology; indeed, mutants lacking specific head groups have defects in solute transport, electron transport, and cell division (Shibuya et al. 1985; Zhang et al. 2005; Bogdanov et al. 2010; Miyazaki et al. 1985). *C. burnetii* contains the predicted enzymes for PA head group modification to form phosphatidylserine (PS), phosphatidylethanolamine (PE), phosphatidylglycerol (PG) and cardiolipin (CL).

Because phospholipids determine membrane-protein topology, the composition of zwitterionic (*e.g.,* PE) and anionic (*e.g.,* PG and cardiolipin) phospholipids in membranes is tightly controlled. Regulation of the bacterial envelope phospholipid composition may play a critical role in the environmental stability of SCV versus LCV forms, or the ability of *C. burnetii* to grow in the acidic environment of the PV. For example, it has been shown that *Bacillus stearothermophilus* increases the PG and cardiolipin content in order to adapt to an acidic environment (Card and Trautman 1990).

Biochemical analysis identified PA, PE, PS, PI (phosphatidylinositol), PC (phosphatidylcholine), and PG in virulent phase I organisms; avirulent phase II organisms, with the exception of a lack of PI, had a similar profile (Domingues et al. 2002). The truncated LPS of phase II organisms is speculated to alter the membrane environment and lead to a change in the phospholipid composition of the bacterial envelope. *C. burnetii* possesses the enzymes required for cardiolipin synthesis; however, the phospholipid was not identified (Domingues et al. 2002). Conversely, despite experimental data demonstrating their presence in *C. burnetii,* the genes predicted to be involved in the synthesis of PI and PC are absent from the organism's genome. Further analysis of phospholipids, particularly from the SCV and LCV developmental forms, as well as under different growth conditions in the newly developed axenic media (Omsland et al. 2009), will provide important insight into the *C. burnetii* membrane plasticity and composition.

10.2.3 Isoprenoid Synthesis

Isoprenoids (or terpenoids) are a large and diverse class of organic chemicals, and include sterols, antimicrobials, peptidoglycan precursors, vitamins, and coenzyme Q. The backbone and key intermediate of the isoprenoid biosynthetic pathway, isopentenyl diphosphate, is synthesized by one of two pathways: (1) through mevalonic acid and HMG-CoA (mevalonate pathway), or (2) the MEP/DOXP non-mevalonate pathway. Unlike the majority of gram-negative bacteria, *C. burnetii* does not utilize the MEP/DOXP pathway. Instead, the bacterium encodes the mevalonate pathway that is more common in gram-positive bacteria and eukaryotes (Beare et al. 2009). Interestingly, *C. burnetii's* close relative *Legionella pneumophila* also appears to have the mevalonate pathway, suggesting both bacteria acquired these genes by horizontal transfer from a primitive eukaryote. This hypothesis is further supported by the clustering of the isoprenoid synthesis genes (CBU607, 608, 609, and 610) in the *C. burnetii* genome.

Once the isoprenoid backbone is generated, it can be modified to form a wide variety of small molecules and steroids. Based on the genome, *C. burnetii's* primary use of isopentenyl diphosphate is generation of peptidoglycans for the bacterial cell envelope. Like the majority of bacteria, *de novo* cholesterol synthesis by *C. burnetii* is unlikely as the majority of the terminal enzymes in the biosynthetic pathway are missing. However, the bacterium does encode two eukaryotic-like sterol reductases that may modify cholesterol precursors (see Sect. 10.3.2).

10.2.4 Fatty Acid Catabolism

Fatty acids can be broken down for energy or carbon through a process call beta-oxidation (or lipolysis). In this process, long carbon chains are split into acetyl-CoA, which can then enter the tricarboxylic acid (TCA) cycle for ATP production. The *C. burnetii* genome contains the necessary enzymes to degrade fatty acids to acetyl-CoA, including long-chain fatty acid-CoA ligase, acyl-CoA dehydrogenase, enoyl-CoA hydrogenase, and acetyl-CoA C-acyltransferase. Alternatively, the acetyl-CoA subunits can be converted to carbohydrate through the glyoxylate shunt of the TCA cycle, a process required for growth of bacteria using acetate or fatty acids as a sole carbon source. *C. burnetii* is missing the two enzymes on the glyoxylate shunt (isocitrate lyase and malate synthase), indicating that the bacterium does not utilize fatty acids as a carbon source.

10.2.5 Modification of Bacterial Membrane Fluidity

When bacteria encounter a change in their environment, such as temperature or pH, one adaptive strategy is to modify membrane fluidity, thus altering membrane functions such as permeability and protein-protein interactions. To alter membrane fluidity, bacteria typically modify the phospholipid acyl chain saturation or the ratio of branch to unbranched amino acids, as opposed to the high energy cost of phospholipid turnover. Saturated straight chain fatty acids pack tightly, creating membranes that are less fluid and less permeable (Zhang and Rock 2008). Unsaturated fatty acids, because of the kink introduced by the double bond, do not pack tightly, thus increasing the membrane fluidity. *C. burnetii* possesses a putative fatty acid desaturase, CBU0920, which may serve to introduce a double bond into an existing fatty acid, thus increasing membrane fluidity.

Anteiso-chains, because the methyl group is further down the acyl chain, increase membrane fluidity as compared to *iso*-chains (Zhang and Rock 2008). Bacteria that produce *anteiso*- and *iso*- fatty acids typically modify the *anteiso:iso* ratio in response to temperature and pH (Kaneda 1991). For example, *Listeria monocytogenes* modifies the *anteiso:iso* ratio in response to temperature and pH changes (Giotis et al. 2007). A second method bacteria use is introduction of a methyl group to form a cyclopropane ring; *C. burnetii* does not have a predicted cyclopropane fatty acid synthase nor have cyclopropanes been detected experimentally (Chan et al. 1976).

10.3 *C. burnetii* Manipulation of Host Cell Lipids

10.3.1 Cholesterol

Cholesterol is a critical component of cellular membranes, providing structural stability, platforms for signaling proteins, and precursors for signaling molecules. Cholesterol is emerging as a critical player in microbial pathogenesis, particularly in the biogenesis of PV membranes where cholesterol levels may affect mechanical strength, ionic permeability, and oxygen level (Needham et al. 1988; Corvera et al. 1992). Both protozoan parasites and bacterial pathogens have been shown to recruit and/or modify host cell cholesterol during their intracellular life cycle.

De novo cholesterol synthesis in mammalian cells occurs through the mevalonate pathway using acetyl-CoA. The first sterol intermediate in the pathway, lanosterol, is further modified by 19 enzymatic reactions of demethylation, hydroxylation, and double bond reduction to generate the final sterol product, cholesterol (Fig. 10.2a). The majority of these steps are catalyzed by enzymes found in the ER, with a key step being the reduction of the C24 double bond catalyzed by the Δ24 sterol reductase. In the absence of this enzyme, cholesterol is replaced by its precursor, desmosterol. The mammalian Δ24 sterol reductase, DHCR24/Seladin, is a bifunctional protein that reduces sterol intermediates and confers resistance to oxidative stress and subsequent apoptotic death through a non-enzymatic process (Wu et al. 2004; Lu et al. 2006, 2008).

The *C. burnetii* PV membrane is structurally resilient and impermeable to compounds as small as 500 Daltons (Heinzen and Hackstadt 1997), characteristics that may reflect the high cholesterol content of the PV membrane (Fig. 10.3; Howe and Heinzen 2006). *C. burnetii*-infected cells have 70% more total cholesterol than uninfected cells, as well as a significant upregulation of genes involved in both host cell cholesterol biosynthesis and exogenous cholesterol uptake. For example, host low density lipoprotein (LDL) receptor and HMG-CoA reductase (3-hydroxy-3-methyl-glutaryl-CoA reductase) gene expression levels peak at 4–5 days post infection when the PV is at its maximum size and bacteria are in mid to late log phase. However, it is not clear if the upregulation is simply a response by the host cell to

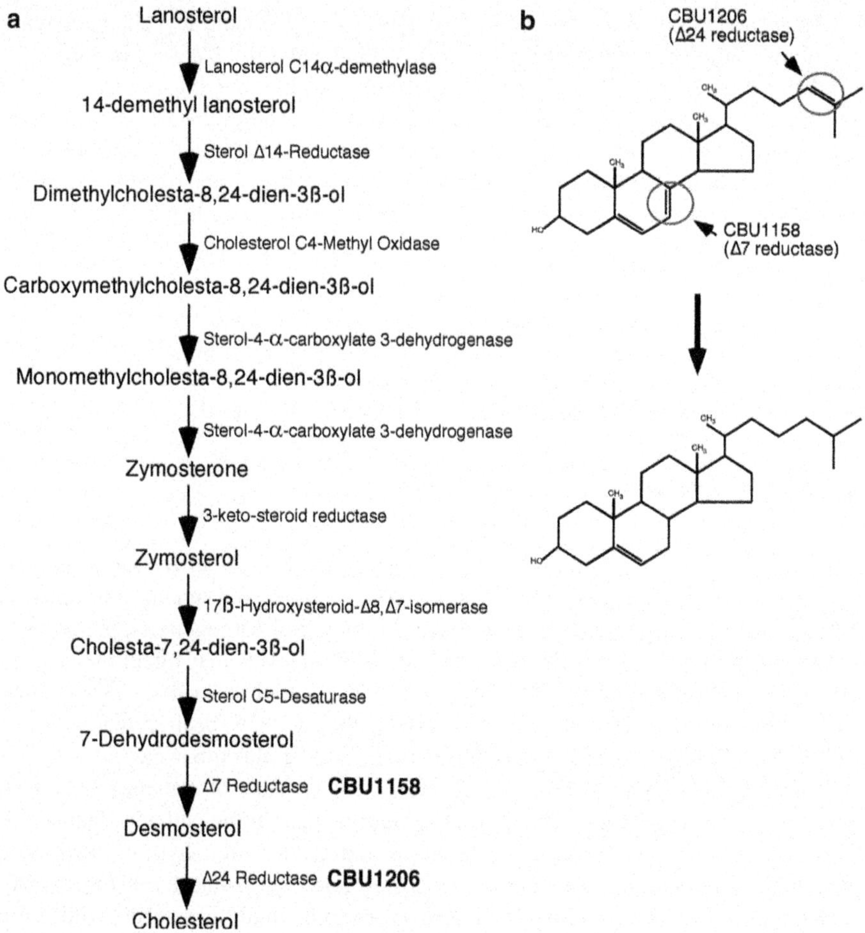

Fig. 10.2 The *C. burnetii* sterol reductases. (**a**) The cholesterol biosynthetic pathway in mammalian cells requires a series of enzymatic modifications of the first recognizable cholesterol intermediate, lanosterol. *C. burnetii* lacks the majority of the enzymes in the pathway, with the exception of the final two reductases. (**b**) The predicted sites of action of CBU1158, a putative Δ7 sterol reductase, and CBU1206, a putative Δ24 sterol reductase

maintain cholesterol homeostasis, or if the bacteria are actively modulating host gene expression. In addition to providing mechanical stability, cholesterol, along with sphingolipids, may form lipid raft microdomains in the PV membrane that partition signaling proteins important in PV biogenesis. The presence of the lipid raft proteins flotillin-1 and flotillin-2 on the PV supports this hypothesis (Howe and Heinzen 2006). While *C. burnetii* does not have the full complement of genes to synthesize cholesterol *de novo*, the genome does encode two sterol reductase enzymes (Fig. 10.2b, see also Sect. 10.3.2), which in mammalian cells function in the very final steps of cholesterol biosynthesis.

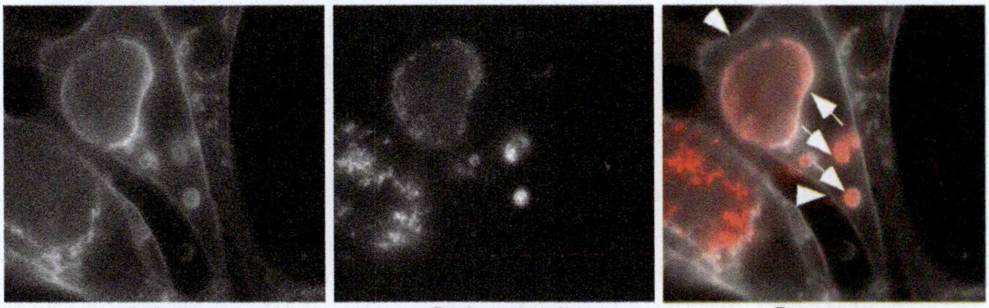

Fig. 10.3 The *C. burnetii* vacuole is sterol-rich. Filipin, a sterol-binding fluorophore, labels the *C. burnetii* parasitophorous vacuole membrane (*arrows*). The labeling intensity is similar to that of the plasma membrane (*arrowhead*), which contains 70–90% of total cellular cholesterol (Lange et al. 1989)

The high cholesterol content of the *C. burnetii* PV membrane suggests that cholesterol is critical for pathogen replication in the host cell. Indeed, pharmacological reagents that block host cell cholesterol biosynthesis or LDL uptake dramatically impact PV development and bacterial replication (Howe and Heinzen 2006). However, pleiotropic effects of these inhibitors on multiple targets of cholesterol trafficking and synthesis make interpretation of these data difficult. For example, U18666A has been shown to inhibit both trafficking of LDL (Liscum and Faust 1989; Lange et al. 1997) and *de novo* cholesterol synthesis (Sexton et al. 1983). In addition, the *de novo* synthesis inhibitors typically target cholesterol synthesis immediately upstream and downstream of lanosterol (Fig. 10.1a), therefore blocking synthesis of all sterols. Thus, these studies point towards the importance of sterols in the *C. burnetii* PV, but not specifically to cholesterol.

To examine the specific requirement for cholesterol without the use of pharmaceutical inhibitors or cholesterol-depleting agents, we developed cholesterol-free tissue culture model systems using J774 murine macrophages and DHCR24$^{-/-}$ mouse embryonic fibroblasts. Both cell lines lack DHCR24, the final enzyme in the cholesterol biosynthetic pathway (Fig. 10.2; Wechsler et al. 2003; Rodriguez-Acebes et al. 2008). When adapted to serum-free media with no exogenous source of cholesterol, desmosterol replaces cholesterol as the major sterol. While desmosterol can substitute for cholesterol in cellular functions such as cell growth (Rodriguez-Acebes et al. 2008, 2009), due to the difference in packing in membranes, desmosterol cannot substitute for cholesterol in lipid rafts, membrane microdomains that serve as signaling platforms (Singh et al. 2009; Vainio et al. 2006; Rog et al. 2008).

Surprisingly, when compared to wild type cells, *C. burnetii* infection, PV formation, and replication in DHCR24$^{-/-}$ macrophages and fibroblast cells lacking cholesterol was similar. (Gilk et al. 2010). Furthermore, it does not appear that *C. burnetii* generates cholesterol from host cell sterols. These observations indicate that in an *in vitro* cell culture system, *C. burnetii* does not require cholesterol. However, this does not rule out the possibility that *C. burnetii* requires cholesterol either under specific cell conditions (*e.g.*, oxidative burst), *in vivo*, or in different host species.

10.3.2 C. burnetii Sterol Reductases

While *C. burnetii* does not have the biosynthetic machinery necessary to generate cholesterol *de novo*, the genome encodes two putative sterol reductases, CBU1158 and CBU1206 (Seshadri et al. 2003). In eukaryotic cells, these reductases function in the final stages of cholesterol biosynthesis by reducing the carbon 7 and 24 double bonds of sterol precursors, respectively (Fig. 10.2b). While the presence of a Δ24 sterol reductase in prokaryotes appears unique to *C. burnetii*, two other bacteria, *Legionella drancortii* and *Protochlamydia amoebophila*, possess a Δ7 sterol reductase. Both organisms are found in amoeba, where it has been proposed these genes were acquired from a eukaryotic host by inter-domain horizontal gene transfer (Moliner et al. 2009). In fact, CBU1206 is most closely related to the predicted Δ24 sterol reductase of the soil amoeba *Naegleria gruberii*. This raises the possibility that in the natural environment the amoeba can serve as a reservoir host for *C. burnetii*. There is one report of *C. burnetii* surviving but not replicating within *Acanthamoeba* (La Scola and Raoult 2001), but the potential relationship between *C. burnetii* and amoeba remains to be fully explored.

The fact that *C. burnetii* has maintained these genes while undergoing genomic reduction suggests they serve a critical function in the bacterium's life cycle. One possibility is *C. burnetii* utilizes these enzymes to generate cholesterol in situations where the sterol is limiting. This infers both CBU1158 and CBU1206 are active enzymes capable of modifying sterols. This hypothesis was tested for CBU1206 by expressing the protein in a mutant of the *Saccharomyces cerevisiae* Δ24 sterol reductase homolog, *erg4*. Yeast produce structurally and functionally similar ergosterol in the place of cholesterol. The *erg4* knockout accumulates ergosta-5,7,22,24(28)-tetraen-3ß-ol in its cell wall in the place of ergosterol, making the cell hypersensitive to divalent cations and drugs such as brefeldin A, cycloheximide, and fluconazol (Zweytick et al. 2000). When expressed in an *erg4* mutant, CBU1206 not only rescued the drug sensitivity but also resulted in the generation of ergosterol (Gilk et al. 2010). The full range of CBU1206 substrates, as well as the enzymatic activity of CBU1158, have not yet been determined. In addition, it is unknown if CBU1206 can rescue $DHCR24^{-/-}$ cells.

The role(s) of *C. burnetii* sterol reductases in pathogenesis is not clear. In mammalian cells, the high sterol content of host cells makes it unlikely that the bacteria would synthesize more cholesterol. This raises the possibility that these enzymes generate novel sterol species that have signaling or structural roles; alternatively, they may serve a non-enzymatic function. The mammalian homolog of CBU1206, DHCR24, interacts with the tumor suppressor p53 and is postulated to directly bind reactive oxygen species (Lu et al. 2006, 2008; Wu et al. 2004). It is also possible CBU1158 and CBU1206 have critical functions in environmental reservoirs such as ticks (Baca and Paretsky 1983).

10.3.3 A Role for Sterols In Vivo?

While *in vitro* models suggest *C. burnetii* does not require cholesterol, the role of cholesterol *in vivo* remains to be elucidated. In other pathogens, *in vivo* data on the relationship between cholesterol and disease is also limited. Cholesterol synthesis inhibitors such as lovastatin and atorvastatin decrease *Salmonella* infection in an *in vivo* mouse model system (Catron et al. 2004). However, in the case of *Anaplasma phagocytophilum*, atorvastatin appears to increase the rate of infection and disease in mice, although the authors speculate that atorvastatin does not reduce the already low levels of LDL found in rodents, and that an increase in LDL receptors on *Anaplasma*-infected cells helps support bacterial infection (Xiong et al. 2009). Pre-treatment with statins improve survival from *Steptococcus pneumonia* challenge in mice, where it has been shown that statins protect against cholesterol-dependent cytotoxins (Rosch et al. 2010). In humans, a relationship between statins and infectious disease has been suggested, with recent clinical studies demonstrating statins protect from bacterial respiratory infections (Chalmers et al. 2008; Thomsen et al. 2008). Statins decrease mortality caused by *Staphylococcus aureus* by 28% (Liappis et al. 2001); a recent study suggests statin treatment increase the secretion of extracellular traps by phagocytic cells, leading to the killing of the bacteria (Chow et al. 2010). The potential for statins to protect or treat chronic Q fever, which is characterized by endocarditis, has not been tested.

10.3.4 Proteins that Modulate Host Cell Lipids

C. burnetii possesses a Dot/Icm type IV secretion system (T4SS) predicted to secrete a diverse set of proteins with effector functions into the host cytosol (Voth and Heinzen 2009). The repertoire of *C. burnetii* effector proteins is likely responsible for modulating multiple host processes; however, the function of only AnkG, which inhibits apoptosis, has been demonstrated (Luhrmann et al. 2010). However, *L. pneumophila* has multiple T4SS substrates that manipulate host cell lipids, specifically phosphoinositides. Several *L. pneumophila* effectors, such as LidA, SidC, and Drra/SidM, bind phosphatidylinositol-4-phosphate, or PI(4)P (Weber et al. 2006; Brombacher et al. 2009; Ragaz et al. 2008). PI(4)P is found in the trans-Golgi network and promotes trafficking along the secretory pathway, and has also been localized to the *L. pneumophila* replicative vacuole (Weber et al. 2006). To date, the identified *C. burnetii* type IV effectors have no strong homology to lipid-binding or modifying proteins (Beare et al. 2009); however, such motifs, especially for sterol-binding proteins, are poorly defined.

In addition to type IV secretion, *C. burnetii* may secrete lipid-binding/modifying proteins through a Sec-dependent pathway. Other bacteria secrete lipid-modifying proteins through non T4SS's, including phospholipases and phosphoinositide kinases (Weber et al. 2009). For example, *L. pneumophila* secretes two phospholipase A enzymes (PlaA and PlaC) via a type II secretion system, while a third phospholipase,

PlaD, is secreted by an unknown mechanism (Banerji et al. 2008). *C. burnetii* encodes two putative phospholipases, CBU0467 (a putative phospholipase A1) and CBU0920 (phospholipase D). Phospholipase D enzymes cleave phosphatidylcholine to generate the signaling molecule phosphatidic acid (PA) and a free choline group, while phospholipase A-mediated cleavage of phospholipids generates a free fatty acid and lysophospholipid, a second messenger molecule that can regulate such cellular pathways as cell growth and morphology (Wymann and Schneiter 2008).

Other potential lipid modifying enzymes in *C. burnetii* include several putative inositol-phosphate phosphatases: CBU0599, CBU0701, CBU1133. These proteins might modulate host cell inositol phosphates, soluble signaling and secondary messengers involved in multiple host cell processes, including calcium release, gene expression, and cytoskeleton assembly (Tsui and York 2010). The possibility that these proteins, along with unidentified secreted *C. burnetii* proteins, might manipulate host cell functions via lipid interactions is an important and exciting area of further investigation.

10.4 Summary

In this chapter, we explored the composition and synthesis of *C. burnetii* lipids, along with ways the bacteria might manipulate host cell lipids to facilitate establishment of the intracellular niche. Despite the importance of lipids in *C. burnetii* pathogenesis, the subject remains largely unexplored. Evidence suggests *C. burnetii* has a unique fatty acid profile, an attribute that may promote *C. burnetii* survival in the extracellular environment and within the acidic and hydrolytic PV. An intriguing question is whether *C. burnetii*-infected cells contain unusual lipids, perhaps through the activity of the unique *C. burnetii* sterol reductases. Also of interest are mechanisms by which *C. burnetii* subverts host cell lipids to enable PV development. A better understanding of the role of lipids will provide important insights into the complex interplay between *C. burnetii* and its host.

Acknowledgements I would like to thank Audrey Chong, Bob Heinzen, and Seth Winfree for critical reading of the manuscript. DHCR24$^{+/-}$ mice were generously provided by Quark Biotech (Israel). This research was supported by the Intramural Research Program of the National Institute of Allergy and Infectious Diseases.

References

Amano K, Williams JC (1984) Chemical and immunological characterization of lipopolysaccharides from phase I and phase II *Coxiella burnetii*. J Bacteriol 160:994–1002

Amano K, Williams JC, Mccaul TF, Peacock MG (1984) Biochemical and immunological properties of *Coxiella burnetii* cell wall and peptidoglycan-protein complex fractions. J Bacteriol 160:982–988

Baca OG, Paretsky D (1983) Q fever and *Coxiella burnetii*: a model for host-parasite interactions. Microbiol Rev 47:127–149

Banerji S, Aurass P, Flieger A (2008) The manifold phospholipases A of *Legionella pneumophila* – identification, export, regulation, and their link to bacterial virulence. Int J Med Microbiol 298:169–181

Beare PA, Unsworth N, Andoh M, Voth DE, Omsland A, Gilk SD, Williams KP, Sobral BW, Kupko JJ 3rd, Porcella SF, Samuel JE, Heinzen RA (2009) Comparative genomics reveal extensive transposon-mediated genomic plasticity and diversity among potential effector proteins within the genus *Coxiella*. Infect Immun 77:642–656

Bergler H, Fuchsbichler S, Hogenauer G, Turnowsky F (1996) The enoyl-[acyl-carrier-protein] reductase (FabI) of *Escherichia coli*, which catalyzes a key regulatory step in fatty acid biosynthesis, accepts NADH and NADPH as cofactors and is inhibited by palmitoyl-CoA. Eur J Biochem 242:689–694

Bogdanov M, Heacock P, Guan Z, Dowhan W (2010) Plasticity of lipid-protein interactions in the function and topogenesis of the membrane protein lactose permease from *Escherichia coli*. Proc Natl Acad Sci U S A 107:15057–15062

Brombacher E, Urwyler S, Ragaz C, Weber SS, Kami K, Overduin M, Hilbi H (2009) Rab1 guanine nucleotide exchange factor SidM is a major phosphatidylinositol 4-phosphate-binding effector protein of *Legionella pneumophila*. J Biol Chem 284:4846–4856

Card GL, Trautman JK (1990) Role of anionic lipid in bacterial membranes. Biochim Biophys Acta 1047:77–82

Catron DM, Lange Y, Borensztajn J, Sylvester MD, Jones BD, Haldar K (2004) *Salmonella enterica* serovar Typhimurium requires nonsterol precursors of the cholesterol biosynthetic pathway for intracellular proliferation. Infect Immun 72:1036–1042

Chalmers JD, Singanayagam A, Murray MP, Hill AT (2008) Prior statin use is associated with improved outcomes in community-acquired pneumonia. Am J Med 121(1002–1007):e1

Chan ML, Mcchesney J, Paretsky D (1976) Further characterization of a lipopolysaccharide from *Coxiella burneti*. Infect Immun 13:1721–1727

Chow OA, Von Kockritz-Blickwede M, Bright AT, Hensler ME, Zinkernagel AS, Cogen AL, Gallo RL, Monestier M, Wang Y, Glass CK, Nizet V (2010) Statins enhance formation of phagocyte extracellular traps. Cell Host Microbe 8:445–454

Clejan S, Krulwich TA, Mondrus KR, Seto-Young D (1986) Membrane lipid composition of obligately and facultatively alkalophilic strains of *Bacillus spp*. J Bacteriol 168:334–340

Corvera E, Mouritsen OG, Singer MA, Zuckermann MJ (1992) The permeability and the effect of acyl-chain length for phospholipid bilayers containing cholesterol: theory and experiment. Biochim Biophys Acta 1107:261–270

Cossart P, Roy CR (2010) Manipulation of host membrane machinery by bacterial pathogens. Curr Opin Cell Biol 22:547–554

Domingues P, Palkovic P, Toman R (2002) Analysis of phospholipids from *Coxiella burnetii* by fast atom bombardment mass spectrometry. A rapid method for differentiation of virulent phase I and low virulent phase II cells. Acta Virol 46:121–124

Gilk SD, Beare PA, Heinzen RA (2010) *Coxiella burnetii* expresses a functional {Delta}24 sterol reductase. J Bacteriol 192:6154–6159

Giotis ES, Mcdowell DA, Blair IS, Wilkinson BJ (2007) Role of branched-chain fatty acids in pH stress tolerance in *Listeria monocytogenes*. Appl Environ Microbiol 73:997–1001

Heath RJ, Su N, Murphy CK, Rock CO (2000) The enoyl-[acyl-carrier-protein] reductases FabI and FabL from *Bacillus subtilis*. J Biol Chem 275:40128–40133

Heinzen RA, Hackstadt T (1997) The *Chlamydia trachomatis* parasitophorous vacuolar membrane is not passively permeable to low-molecular-weight compounds. Infect Immun 65:1088–1094

Howe D, Heinzen RA (2006) *Coxiella burnetii* inhabits a cholesterol-rich vacuole and influences cellular cholesterol metabolism. Cell Microbiol 8:496–507

Jackson M, Stadthagen G, Gicquel B (2007) Long-chain multiple methyl-branched fatty acid-containing lipids of *Mycobacterium tuberculosis*: biosynthesis, transport, regulation and biological activities. Tuberculosis (Edinb) 87:78–86

Kaneda T (1991) *Iso-* and *anteiso*-fatty acids in bacteria: biosynthesis, function, and taxonomic significance. Microbiol Rev 55:288–302

La Scola B, Raoult D (2001) Survival of *Coxiella burnetii* within free-living amoeba *Acanthamoeba castellanii*. Clin Microbiol Infect 7:75–79

Lange Y, Swaisgood MH, Ramos BV, Steck TL (1989) Plasma membranes contain half the phospholipid and 90% of the cholesterol and sphingomyelin in cultured human fibroblasts. J Biol Chem 264:3786–3793

Lange Y, Ye J, Chin J (1997) The fate of cholesterol exiting lysosomes. J Biol Chem 272:17018–17022

Liappis AP, Kan VL, Rochester CG, Simon GL (2001) The effect of statins on mortality in patients with bacteremia. Clin Infect Dis 33:1352–1357

Liscum L, Faust JR (1989) The intracellular transport of low density lipoprotein-derived cholesterol is inhibited in Chinese hamster ovary cells cultured with 3-beta-[2-(diethylamino)ethoxy] androst-5-en-17-one. J Biol Chem 264:11796–11806

Lu X, Kambe F, Cao X, Yoshida T, Ohmori S, Murakami K, Kaji T, Ishii T, Zadworny D, Seo H (2006) DHCR24-knockout embryonic fibroblasts are susceptible to serum withdrawal-induced apoptosis because of dysfunction of caveolae and insulin-Akt-Bad signaling. Endocrinology 147:3123–3132

Lu X, Kambe F, Cao X, Kozaki Y, Kaji T, Ishii T, Seo H (2008) 3beta-Hydroxysteroid-delta24 reductase is a hydrogen peroxide scavenger, protecting cells from oxidative stress-induced apoptosis. Endocrinology 149:3267–3273

Luhrmann A, Nogueira CV, Carey KL, Roy CR (2010) Inhibition of pathogen-induced apoptosis by a *Coxiella burnetii* type IV effector protein. Proc Natl Acad Sci U S A 107:18997–19001

Massengo-Tiasse RP, Cronan JE (2008) *Vibrio cholerae* FabV defines a new class of enoyl-acyl carrier protein reductase. J Biol Chem 283:1308–1316

Massengo-Tiasse RP, Cronan JE (2009) Diversity in enoyl-acyl carrier protein reductases. Cell Mol Life Sci 66:1507–1517

Mccaul TF, Williams JC (1981) Developmental cycle of *Coxiella burnetii*: structure and morphogenesis of vegetative and sporogenic differentiations. J Bacteriol 147:1063–1076

Miyazaki C, Kuroda M, Ohta A, Shibuya I (1985) Genetic manipulation of membrane phospholipid composition in *Escherichia coli*: *pgsA* mutants defective in phosphatidylglycerol synthesis. Proc Natl Acad Sci U S A 82:7530–7534

Moliner C, Raoult D, Fournier PE (2009) Evidence that the intra-amoebal *Legionella drancourtii* acquired a sterol reductase gene from eukaryotes. BMC Res Notes 2:51

Needham D, Mcintosh TJ, Evans E (1988) Thermomechanical and transition properties of dimyristoylphosphatidylcholine/cholesterol bilayers. Biochemistry 27:4668–4673

Omsland A, Cockrell DC, Howe D, Fischer ER, Virtaneva K, Sturdevant DE, Porcella SF, Heinzen RA (2009) Host cell-free growth of the Q fever bacterium *Coxiella burnetii*. Proc Natl Acad Sci U S A 106:4430–4434

Ragaz C, Pietsch H, Urwyler S, Tiaden A, Weber SS, Hilbi H (2008) The *Legionella pneumophila* phosphatidylinositol-4 phosphate-binding type IV substrate SidC recruits endoplasmic reticulum vesicles to a replication-permissive vacuole. Cell Microbiol 10:2416–2433

Rodriguez-Acebes S, De La Cueva P, Ferruelo AJ, Fernandez-Hernando C, Lasuncion MA, Martinez-Botas J, Gomez-Coronado D (2008) Dose-dependent dual effects of cholesterol and desmosterol on J774 macrophage proliferation. Biochem Biophys Res Commun 377:484–488

Rodriguez-Acebes S, De La Cueva P, Fernandez-Hernando C, Ferruelo AJ, Lasuncion MA, Rawson RB, Martinez-Botas J, Gomez-Coronado D (2009) Desmosterol can replace cholesterol in sustaining cell proliferation and regulating the SREBP pathway in a sterol-Delta24-reductase-deficient cell line. Biochem J 420:305–315

Rog T, Vattulainen I, Jansen M, Ikonen E, Karttunen M (2008) Comparison of cholesterol and its direct precursors along the biosynthetic pathway: effects of cholesterol, desmosterol and 7-dehydrocholesterol on saturated and unsaturated lipid bilayers. J Chem Phys 129:154508

Rosch JW, Boyd AR, Hinojosa E, Pestina T, Hu Y, Persons DA, Orihuela CJ, Tuomanen EI (2010) Statins protect against fulminant pneumococcal infection and cytolysin toxicity in a mouse model of sickle cell disease. J Clin Invest 120:627–635

Seshadri R, Paulsen IT, Eisen JA, Read TD, Nelson KE, Nelson WC, Ward NL, Tettelin H, Davidsen TM, Beanan MJ, Deboy RT, Daugherty SC, Brinkac LM, Madupu R, Dodson RJ, Khouri HM, Lee KH, Carty HA, Scanlan D, Heinzen RA, Thompson HA, Samuel JE, Fraser CM, Heidelberg JF (2003) Complete genome sequence of the Q-fever pathogen *Coxiella burnetii*. Proc Natl Acad Sci U S A 100:5455–5460

Sexton RC, Panini SR, Azran F, Rudney H (1983) Effects of 3 beta-[2-(diethylamino)ethoxy] androst-5-en-17-one on the synthesis of cholesterol and ubiquinone in rat intestinal epithelial cell cultures. Biochemistry 22:5687–5692

Shibuya I, Miyazaki C, Ohta A (1985) Alteration of phospholipid composition by combined defects in phosphatidylserine and cardiolipin synthases and physiological consequences in *Escherichia coli*. J Bacteriol 161:1086–1092

Singh P, Saxena R, Paila YD, Jafurulla M, Chattopadhyay A (2009) Differential effects of cholesterol and desmosterol on the ligand binding function of the hippocampal serotonin(1A) receptor: implications in desmosterolosis. Biochim Biophys Acta 1788:2169–2173

Thomsen RW, Riis A, Kornum JB, Christensen S, Johnsen SP, Sorensen HT (2008) Preadmission use of statins and outcomes after hospitalization with pneumonia: population-based cohort study of 29,900 patients. Arch Intern Med 168:2081–2087

Tsui MM, York JD (2010) Roles of inositol phosphates and inositol pyrophosphates in development, cell signaling and nuclear processes. Adv Enzyme Regul 50:324–337

Tzianabos T, Moss CW, Mcdade JE (1981) Fatty acid composition of rickettsiae. J Clin Microbiol 13:603–605

Vainio S, Jansen M, Koivusalo M, Rog T, Karttunen M, Vattulainen I, Ikonen E (2006) Significance of sterol structural specificity. Desmosterol cannot replace cholesterol in lipid rafts. J Biol Chem 281:348–355

Voth DE, Heinzen RA (2009) *Coxiella* type IV secretion and cellular microbiology. Curr Opin Microbiol 12:74–80

Weber SS, Ragaz C, Reus K, Nyfeler Y, Hilbi H (2006) *Legionella pneumophila* exploits PI(4)P to anchor secreted effector proteins to the replicative vacuole. PLoS Pathog 2:e46

Weber SS, Ragaz C, Hilbi H (2009) Pathogen trafficking pathways and host phosphoinositide metabolism. Mol Microbiol 71:1341–1352

Wechsler A, Brafman A, Shafir M, Heverin M, Gottlieb H, Damari G, Gozlan-Kelner S, Spivak I, Moshkin O, Fridman E, Becker Y, Skaliter R, Einat P, Faerman A, Bjorkhem I, Feinstein E (2003) Generation of viable cholesterol-free mice. Science 302:2087

Wollenweber HW, Schramek S, Moll H, Rietschel ET (1985) Nature and linkage type of fatty acids present in lipopolysaccharides of phase I and II *Coxiella burnetii*. Arch Microbiol 142:6–11

Wu C, Miloslavskaya I, Demontis S, Maestro R, Galaktionov K (2004) Regulation of cellular response to oncogenic and oxidative stress by Seladin-1. Nature 432:640–645

Wymann MP, Schneiter R (2008) Lipid signalling in disease. Nat Rev Mol Cell Biol 9:162–176

Xiong Q, Lin M, Rikihisa Y (2009) Cholesterol-dependent *Anaplasma phagocytophilum* exploits the low-density lipoprotein uptake pathway. PLoS Pathog 5:e1000329

Zhang YM, Rock CO (2008) Membrane lipid homeostasis in bacteria. Nat Rev Microbiol 6:222–233

Zhang W, Campbell HA, King SC, Dowhan W (2005) Phospholipids as determinants of membrane protein topology. Phosphatidylethanolamine is required for the proper topological organization of the gamma-aminobutyric acid permease (GabP) of *Escherichia coli*. J Biol Chem 280:26032–26038

Zweytick D, Hrastnik C, Kohlwein SD, Daum G (2000) Biochemical characterization and subcellular localization of the sterol C-24(28) reductase, Erg4p, from the yeast *Saccharomyces cerevisiae*. FEBS Lett 470:83–87

Chapter 11
Axenic Growth of *Coxiella burnetii*

Anders Omsland

Contents

A. Omsland (✉)
Coxiella Pathogenesis Section, Laboratory of Intracellular Parasites,
Rocky Mountain Laboratories, National Institute of Allergy and Infectious Diseases,
National Institutes of Health, 903 South 4th Street, Hamilton, MT 59840, USA
e-mail: omslanda@niaid.nih.gov

R. Toman et al. (eds.), *Coxiella burnetii: Recent Advances and New Perspectives in Research of the Q Fever Bacterium*, Advances in Experimental Medicine and Biology 984,
DOI 10.1007/978-94-007-4315-1_11,

Abstract Early metabolic studies of *C. burnetii* showed minimal metabolic activity of axenic (host cell-free) organisms in buffers adjusted to neutral pH. However, our understanding of the organism's physiology was greatly improved upon the discovery that *C. burnetii* requires an acidic pH for metabolic activation. Indeed, information gained from acid activation studies coupled with contemporary analyses using transcription microarrays, metabolic pathway reconstruction and metabolite typing, led to an axenic culture system that supports robust growth of *C. burnetii*. While axenic culture of *C. burnetii* can present some technical challenges, the technique is currently facilitating new lines of investigation and development of genetic tools. Axenic cultivation of *C. burnetii* should also prove useful in clinical settings.

Keywords Axenic cultivation • Culture system • Extracellular • Metabolic activity • Microaerobic

11.1 Introduction

The discovery of *C. burnetii* as the causative agent of human Q fever in the late 1930s was followed almost immediately by unsuccessful attempts to cultivate the pathogen in axenic (host cell-free) medium (Cox 1938). Microscopic evaluation established an intracellular vacuolar niche for the organism and contributed to *C. burnetii*'s designation as an obligate intracellular bacterium. Subsequent analyses of the organism's metabolic abilities in buffers adjusted to physiological pH showed substantial activity of central metabolic pathways (e.g., glycolysis and RNA synthesis) in bacterial extracts, but not intact cells (Baca and Paretsky 1983). The demonstration that the *C. burnetii* parasitophorous vacuole (PV) exhibits phagolysosomal characteristics (Burton et al. 1971, 1978) led to the discovery that a moderately acidic pH mimicking that of a phagolysosome (~pH 4.5) is required for metabolic activation of *C. burnetii* (Hackstadt and Williams 1981a). Indeed, *C. burnetii* transport, catabolism and incorporation of both glucose and glutamate are dependent on acidic pH (Hackstadt 1983). Moreover, pH dependent transport of carbon and energy sources is required for *de novo* protein synthesis (Zuerner and Thompson 1983), stability of the ATP pool (Hackstadt and Williams 1981b), and membrane energization (Hackstadt 1983). Acid activation of *C. burnetii* metabolism was presented as a "biochemical stratagem" (Hackstadt and Williams 1981a) manifested as metabolic quiescence and stability at the neutral pH found in the extracellular environment, and metabolic activation and bacterial replication at the acidic pH found in the organism's PV (Akporiaye et al. 1983). With improved knowledge of *C. burnetii* metabolic requirements, researchers revisited the possibility of *C. burnetii* axenic cultivation using more complex buffer systems. While *de novo* nucleic acid and protein biosynthesis were observed for several hours, net biomass increases indicated *C. burnetii* was not replicating (Chen et al. 1990).

11.2 Genomic Insight into *C. burnetii* Metabolic Capacity

The sequencing of multiple *C. burnetii* strains allowed a comprehensive look at *C. burnetii*'s metabolic potential (Beare et al. 2009b; Seshadri et al. 2003). Overall, the ~2 Mb genome of *C. burnetii* is roughly twice the size of most genomes of obligate intracellular rickettsiae (Ellison et al. 2008; McLeod et al. 2004) and chlamydiae (Stephens et al. 1998). Moreover, analysis of coding ORFs shows that *C. burnetii* central metabolic pathways, such as glycolysis, the tricarboxylic acid cycle (TCA), and pathways for nucleic acid biosynthesis, are largely intact. As discussed below, *C. burnetii* does have predicted metabolic peculiarities that may reflect adaptation to intracellular parasitism, including some that may impact overall metabolism.

11.2.1 Glycolysis, TCA, and Pentose Phosphate Pathways

Biochemical studies show that *C. burnetii* can catabolize glucose (Hackstadt and Williams 1981a, b). However, the organism does not encode the initial enzyme of glycolysis (glucokinase or hexokinase) needed to convert glucose to glucose-6-phosphate. Because *C. burnetii* lysates can phosphorylate glucose in the presence of carbamoyl phosphate (Paretsky et al. 1962), glucose may be phosphorylated by an alternative carbamoyl phosphate-dependent mechanism that subsequently allows oxidation of glucose via glycolysis. *C. burnetii* encodes a complete TCA cycle and readily metabolizes glutamate, succinate, and pyruvate (Hackstadt and Williams 1981b). However, the organism lacks the glyoxylate bypass enzymes isocitrate lyase and malate synthase, and consequently cannot utilize fatty acids as a sole source of carbon. This is in contrast to the intracellular bacterium *Mycobacterium tuberculosis* that relies on fatty acids as a source of carbon during intracellular growth (Schnappinger et al. 2003). The non-oxidative branch of the pentose phosphate pathway is complete in *C. burnetii*, allowing generation of pentose sugars for nucleic acid biosynthsis. However, the oxidative branch of the pathway is defective due to the lack of glucose-6-phosphate dehydrogenase and 6-phosphogluconate dehydrogenase. Thus, *C. burnetii* cannot rely on this pathway for synthesis of NADPH, a major source of reducing equivalents for biosynthetic reactions and anti-oxidant mechanisms. Overall, biochemical and genomic data suggest oxidation of glycolytic and TCA cycle intermediates, and substrates that funnel directly into these pathways, as critical for *C. burnetii* energy production.

11.2.2 Amino Acid Acquisition

Based on reconstruction of amino acid biosynthetic pathways, *C. burnetii* appears auxotrophic for 11 amino acids (Beare et al. 2009b; Seshadri et al. 2003). Thus, during intracellular growth, *C. burnetii* must acquire these amino acids via transport of free amino acids or peptides from the PV lumen. Autophagosomes, carrying defunct organelles and protein aggregates, and fluid phase endosomes, fuse with the *C. burnetii* PV (Beron et al. 2002; Heinzen et al. 1996). This trafficking likely provides key protein precursors required for *C. burnetii* growth. Indeed, the *C. burnetii* PV lumen is clearly a proteolytic environment as evidenced by the rapid breakdown of bovine serum albumin and the presence of active cathepsins (Howe et al. 2010). This activity is predicted to convert proteins to short peptides and free amino acids that can be transported by *C. burnetii* using several amino acid permeases, di-, tri- and oligo-peptide transporters, and 13 major facilitator superfamily transporters (Beare et al. 2009b). *C. burnetii*'s numerous amino acid auxotrophies suggest the PV has a steady supply of protein precursors, at least during the organism's replicative phase.

11.2.3 Energy Production

The mechanism(s) of *C. burnetii* acid activation is poorly defined. Data from early acid activation studies suggest extracellular acidic pH has multiple beneficial effects on *C. burnetii* physiology including membrane energization (Hackstadt 1983) and activation of metabolite transport (Hackstadt and Williams 1983). With respect to membrane energization, a proton gradient imposed by incubation of host cell-free organisms at acidic pH (i.e., pH 3–5) results in a transient increase of the *C. burnetii* ATP pool (Hackstadt and Williams 1981b). This effect is consistent with proton gradient driven synthesis of ATP via the organism's F_1F_0 ATP synthase. However, maintenance of the ATP pool is dependent on an oxidizable substrate, such as glutamate. Acidic pH clearly assists *C. burnetii* metabolite transport as demonstrated by enhanced transport of glutamate at pH 4.5 (Hackstadt and Williams 1983). Thus, a working model of *C. burnetii* acid activation of metabolism involves stimulation of proton symporters that acquire critical carbon and energy sources.

C. burnetii pH-dependent transport of energy sources and their subsequent oxidation generates reducing equivalents (e.g., NADH) that drives respiratory chain extrusion of protons. This process results in a proton motive force that promotes ATP synthesis. *C. burnetii* encodes a branched respiratory chain with distinct terminal ubiquinol cytochrome *o* (encoded by *cyoABCDE*) or cytochrome *bd* (encoded by *cydAB*) oxidases (Fig. 11.1). Cytochrome *bd*, with its increased affinity for oxygen, is used by bacteria under microaerobic conditions. Thus, *C. burnetii* may adapt to microaerobic levels of oxygen during intracellular growth. In fact, the intracellular oxygen tension of cultured eukaryotic cells and their compartments can be much lower than the extracellular oxygen tension (Hu et al. 1992). Furthermore, the level of oxygen in the spleen, a tissue that can contain *C. burnetii* infected macrophages, has been measured at approximately 3% (Braun et al. 2001).

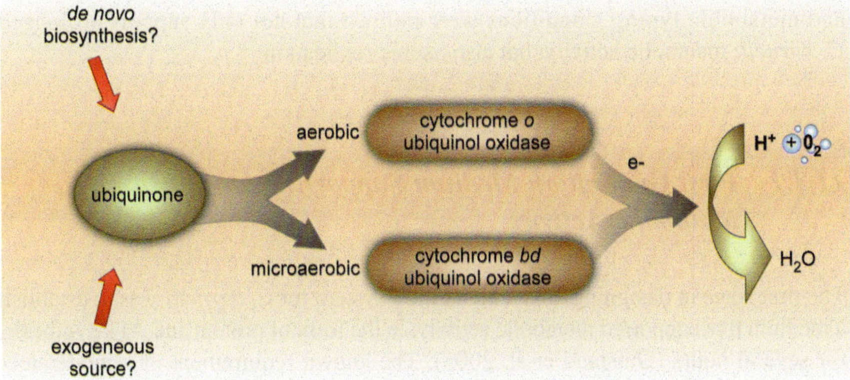

Fig. 11.1 Schematic overview of the *C. burnetii* aerobic respiratory chain. *C. burnetii*'s branched respiratory chain has distinct terminal ubiquinol oxidases incorporating either cytochrome *o* or cytochrome *bd* that are typically used under aerobic and microaerobic conditions, respectively. It is unclear how *C. burnetii* acquires ubiquinone, the predicted major quinone cofactor in electron transport activities, as key biosynthetic enzymes are missing from the pathogen's genome

11.2.4 pH Homeostasis

C. burnetii cytoplasmic pH drops when the organism is incubated in an acidic buffer without an oxidizable energy source. Associated with this pH drop is depletion of the ATP pool and loss of viability (Hackstadt and Williams 1981b). Thus, like other more extreme acidophiles, maintenance of cytoplasmic neutrality by *C. burnetii* is key for survival and the organism likely deploys several mechanisms to maintain pH homeostasis. As described above, proton extrusion via the electron transport chain could contribute to the regulation of cytoplasmic pH. Moreover, energy requiring expulsion of protons could occur by the F_0F_1-ATPase reverse reaction, and by K^+/H^+ and Na^+/H^+ antiporters. Proton consumption via decarboxylation of amino acids can also be utilized as a mechanism to prevent cytoplasmic acidification (Kashiwagi et al. 1992). Ammonia generation during conversion of glutamine to glutamate (Harth et al. 1994) is another example of how *C. burnetii* might couple amino acid metabolism to cytoplasmic pH homeostasis.

11.3 Establishment of a *C. burnetii* Axenic Culture System

Early work establishing conditions that support transient acid activation of *C. burnetii* metabolism provided a foundation on which to further define nutritional and physiochemical conditions that support prolonged axenic metabolic activity. Using contemporary tools, such as biochemical pathway reconstruction, transcriptomics, and metabolite typing, conditions were defined that not only support prolonged *C. burnetii* metabolic activity, but also axenic replication.

11.3.1 First Generation Medium Supporting Sustained Metabolic Activity

The first stage in designing an axenic culture system for *C. burnetii* was to establish a medium that supported metabolic activity, in the form of protein and ATP synthesis, for several hours (Omsland et al. 2008). The known requirement of a moderately acidic environment (~pH 4.5) for *C. burnetii* metabolic activation (Hackstadt and Williams 1981a) prompted screening of several buffers to identify one with both good buffering capacity at acidic pH and the ability to support *C. burnetii* metabolic activity. Buffers based on acetate and succinate were deleterious to *C. burnetii de novo* protein synthesis. However, a citrate buffer supported high metabolic activity and provided excellent buffering capacity.

Sodium (Na^+), potassium (K^+) and chloride (Cl^-) make up the principal ion gradients used by microbes to transport nutrients and also help establish physiological osmotic pressure in living systems. The eukaryotic cell cytoplasm contains high concentrations of K^+ (~140 mM) relative to Na^+ (~15 mM) and Cl^- (~15 mM). Conversely, these levels are roughly reversed (K^+, ~5 mM; Na^+, ~145 mM; and Cl^-, ~110 mM) in interstitial fluid. Surprisingly, although the *C. burnetii* PV is located in the host cell cytoplasm, pathogen metabolic activity was enhanced in a citrate salts buffer (CSB) containing ion levels resembling interstitial fluid (Omsland et al. 2008). This metabolic effect may reflect PV interactions with fluid phase endosomes that traffic material from the extracellular space to the PV lumen (Heinzen et al. 1996).

CSB was then supplemented with three complex nutrient sources to compensate for *C. burnetii*'s auxotrophies and to exploit the organism's otherwise sophisticated metabolic machinery: (1) neopeptone, a combination of peptides, nucleotides and vitamins, (2) fetal bovine serum (FBS), and (3) RPMI cell culture medium. Supplementation of CSB with individual nutrient sources only marginally improved *C. burnetii* metabolic activity. However, a considerable increase in metabolic activity was observed when the nutrient sources were combined, creating a first generation medium called Complex *Coxiella* Medium (CCM) (Omsland et al. 2008). *C. burnetii* metabolic activity was optimal within a narrow range of nutrient concentrations. In fact, addition of extra nutrients, such as succinate or pyruvate, actually caused a decline in *C. burnetii* ATP pools and *de novo* protein synthesis. Incomplete oxidation of excess substrate resulting in generation of inhibitory intermediates may explain *C. burnetii* sensitivity to certain carbon sources (Teixeira de Mattos and Neijssel 1997). The negative effects of succinate and pyruvate on

C. burnetii metabolism may also reflect suboptimal biochemical routing of the oxidized substrate (Tesh et al. 1983).

11.3.2 Second Generation Medium Supporting Axenic Replication

Although CCM supported prolonged *C. burnetii* metabolic activity, increases in genome equivalents (GE) indicating cell division were not observed. Therefore, transcription microarray technology was used to assess the overall metabolic state of *C. burnetii* in CCM with the goal of defining potential medium nutritional deficiencies (Omsland et al. 2009). When compared to the transcriptome of *C. burnetii* replicating in Vero host cells, ribosomal gene expression of *C. burnetii* in CCM was dramatically downregulated, suggesting suboptimal protein synthesis could be blocking replication. Thus, an additional source of amino acids (casamino acids) was added to CCM in an attempt to boost pathogen *de novo* protein synthesis (Omsland et al. 2009). A high concentration of L-cysteine was also added as this supplement is necessary for axenic growth of *Legionella pneumophila* (Ewann and Hoffman 2006), a close phylogenetic relative of *C. burnetii*. The new medium called Acidified Citrate Cysteine Medium (ACCM), supported a considerable increase in *C. burnetii de novo* protein synthesis. However, cell division was still not observed.

As discussed, the presence of a branched respiratory chain in *C. burnetii* suggests the organism might adjust to alterations in oxygen concentration during intracellular growth. Previous testing of media had been conducted under ambient (~20%) oxygen conditions. Therefore, the effect of oxygen availability on *C. burnetii* metabolic activity was probed using Phenotype Microarrays incubated in 20, 5 or 2.5% oxygen environments. The number of oxidized substrates increased as oxygen level decreased, suggesting *C. burnetii* is a microaerophile. Subsequent incubation of *C. burnetii* in ACCM in a 2.5% oxygen environment resulted in robust growth with approximately a 3 $\log_{10}$ increase in *C. burnetii* GE observed over a 6 day incubation (Omsland et al. 2009). Importantly, the infectivity of axenically propagated bacteria for Vero cells was similar to host cell-propagated bacteria. *C. burnetii* colony growth was also achieved using semi-solid ACCM-agarose plates.

11.4 Practical Aspects of *C. burnetii* Axenic Culture

Axenic cultivation of *C. burnetii* is enabling new lines of investigation that are difficult or impossible to conduct with bacteria propagated in host cells. However, axenic growth can be challenging and attention to detail is important for success. In this section, technical issues that are important to successful axenic culture of *C. burnetii* are discussed.

11.4.1 Preparation of Acidified Citrate Cysteine Medium (ACCM)

ACCM supports substantial growth of *C. burnetii*. However, the organism is highly sensitive to medium variations so care must be taken during medium preparation. For example, *C. burnetii* tolerates a moderate range of nutrient concentrations with elevated concentrations of certain nutrients, including some carbon sources, readily inhibiting *C. burnetii* metabolic activity and growth (Omsland et al. 2008). Moreover, ion conditions, especially the concentration of chloride, have considerable effects on metabolic activity (Omsland et al. 2008). The ability of *C. burnetii* to grow in ACCM is even affected by the water used to prepare the medium. Reproducible culture can be achieved using regular deionized water. The sensitivity of *C. burnetii* to ion and nutrient concentrations also means that a humidified incubator should be used for cultivation of *C. burnetii* in ACCM to prevent medium evaporation. Finally, *C. burnetii* acidophilic metabolism is highly sensitive to extracellular pH (Hackstadt and Williams 1981a). Therefore, care should be taken to adjust ACCM to pH 4.75 as moderate (e.g., 0.5 pH units) deviations from this value may compromise growth. The acidic pH of ACCM may contribute to medium instability as culture is inconsistent if the medium is more than 1 week old.

11.4.2 Modifications of ACCM

ACCM is a complex nutrient medium containing chemically undefined components, namely FBS, neopeptone, and casamino acids. We recently found that 1% FBS can be replaced with 1 mg/ml of synthetic methyl-β-cyclodextrin (Mβ-CD) (Omsland et al. 2011). Mβ-CD is used in other bacteriologic media where it may sequester toxic metabolites (Marchini et al. 1995; Imaizumi et al. 1983). The modified medium termed ACCM-2 supports improved growth of *C. burnetii* in liquid medium and as colonies in solid medium.

11.4.3 Growth of *C. burnetii* in ACCM

C. burnetii cultures can be established in ACCM with or without shaking. Gentle shaking (75 RPM) improves *C. burnetii* growth in ACCM while vigorous shaking can cause protein precipitation. Robust growth occurs in shaken 125 ml Erlenmeyer flasks containing 20 ml of medium (Fig. 11.2a) or 250 ml Erlenmeyer flasks containing 40 ml of medium. *C. burnetii* appears highly sensitive to several detergents (A. Omsland and R. Heinzen, unpublished observations). Thus, disposable polyethylene terepthalate glycol (PETG) Erlenmeyer flasks with 0.2 μm filtered screw top caps are preferred. Without shaking, culture vessels should provide a large air-liquid interface to promote gas exchange. T-25 and T-75 tissue culture flasks containing 7 and 20 ml of medium, respectively, work well in this manner. Although liquid cultures

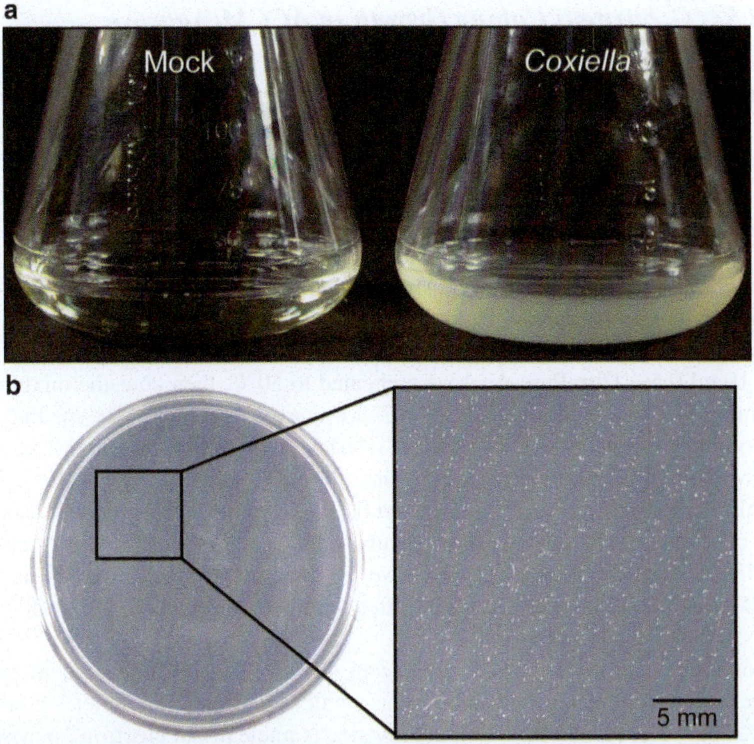

Fig. 11.2 Axenic culture of *C. burnetii*. (**a**) Mock and *C. burnetii* Nine Mile (phase II)-inoculated ACCM-2 following 6 days of incubation with shaking in a 2.5% oxygen environment. (**b**) Representative colony formation by *C. burnetii* Nine Mile (phase II) in ACCM-agarose following 6 days of incubation in a 2.5% oxygen environment

can be started using fewer than 10 organisms, an inoculum of 1.0×10^6 organisms/ml will develop to stationary phase (~10^9 organisms/ml) in approximately 6 days (Omsland et al. 2009), similar to the organism's growth kinetics in cultured cells (Coleman et al. 2004). A critical parameter that affects *C. burnetii* axenic growth is oxygen concentration. *C. burnetii* replicates in ACCM when oxygen levels are between 1% and 5%. Declines in growth are observed in 7.5% oxygen while no growth is observed in 10% oxygen. An incubator with automatic monitoring and gas adjustments ensures a reliable oxygen environment.

11.4.4 Storage of ACCM Cultured C. burnetii

ACCM-cultivated *C. burnetii* are sensitive to repeated freeze-thaw cycles, but excellent viability is observed when stocks are stored in ACCM containing a cryoprotectant, such as 10% dimethyl sulfoxide. Nonetheless, for optimal viability, *C. burnetii* stocks should be stored as single-use aliquots.

11.4.5 *C. burnetii* Colony Growth in ACCM-Agarose

An important advance enabled by axenic growth is the ability to obtain *C. burnetii* clones by colony formation in ACCM-agarose. Minute colonies form when organisms are embedded in semi-solid medium using a soft agarose overlay method (Fig. 11.2b). *C. burnetii* will not grow directly on the surface of solid ACCM. This behavior may reflect a requirement by *C. burnetii* to be submersed in an acid bath for proper membrane energization and transporter function. The use of a thin overlay medium promotes outgrowth of *C. burnetii* colonies in one focal plane, thus facilitating identification and picking of isolated colonies.

To prepare ACCM-agarose plates, combine 10 ml 2× ACCM (pH 4.75, 37°C) with 10 ml 0.5% UltraPure Agarose preheated to 50°C, then pour the mixture into 100 mm Petri dish creating a 0.25% ACCM-agarose base. After the medium base has solidified, prepare the 0.1% agarose overlay medium by combining 1.25 ml 2× ACCM (37°C), 0.75 ml water (37°C), and 0.5 ml 0.5% UltraPure Agarose (50°C). Add the *C. burnetii* inoculum (maximum 0.1 ml) to the overlay medium, mix, then pour over the ACCM-agarose base. Incubate plates at 4°C for 30 min to promote solidification of the low percentage overlay medium. Remove excess liquid in ACCM plates by placing them in a biosafety cabinet, lids cracked, for 20–30 min or until condensation is gone.

Colonies develop in ACCM-agarose in 4–7 days when incubated in a 2.5% oxygen environment. Colonies may be as small as 0.05 mm in diameter (Omsland et al. 2009). The reason for small colony size is unclear, but electron microscopic analysis reveals that 6 day old colonies are primarily comprised of *C. burnetii* small cell variants (SCV) which are non-replicative, stationary phase cells (Coleman et al. 2004; Omsland et al. 2011). Nutrient depletion in the colony microenvironment may drive *C. burnetii* into stationary phase and halt further colony expansion.

11.4.6 Selection and Recovery of *C. burnetii* Genetic Transformants Using ACCM-Agarose

Axenic culture in both ACCM and ACCM-agarose greatly facilitates isolation of *C. burnetii* genetic transformants. Compared to cell culture-based methods, where clonal isolation of *C. burnetii* transformants can take 8–12 weeks (Beare et al. 2009a), transformants can be isolated in as little as 16 days using axenic techniques (Beare et al. 2010; Omsland et al. 2011). Picking of *C. burnetii* colonies is facilitated by first locating isolated colonies using a dissecting microscope. Colonies can then be extracted from solid ACCM-agarose with a micro-pipette tip with an enlarged opening (approximately 0.5 mm). The clonal populations of *C. burnetii* can then be expanded by culture in ACCM following release of embedded organisms by repeated pipetting.

11.4.7 Determination of C. burnetii Viability Using ACCM-Agarose

Enumeration of *C. burnetii* in stock preparations is commonly conducted using an infectious focus-forming unit (FFU) assay or by determining the number of GE by quantitative PCR (Q-PCR) (Coleman et al. 2004). The FFU assay relies on immunofluorescent detection of PVs in cultured host cells. This method is complicated by the fact that phase I and phase II *C. burnetii* have dramatically different infectivity for cultured host cells. For example, the infection efficiency of phase II *C. burnetii* for Vero cells is approximately 5 GE per FFU while several hundred phase I *C. burnetii* are required to produce a single FFU (Cockrell et al. 2008; R. A. Heinzen, unpublished data). While the GE assay is sensitive and accurate, it does not distinguish between live and dead bacteria. The plating efficiency of the *C. burnetii* Nine Mile phase II strain in ACCM-agarose is nearly 1 GE per colony (Beare et al. 2010). Thus, quantification of *C. burnetii* by a conventional colony forming unit assay using ACCM-agarose plates is an alternative method to accurately enumerate the number of live organisms in a given preparation.

11.5 Summary

Axenic culture has removed the severe experimental obstacles associated with *C. burnetii*'s obligate reliance on a eukaryotic host cell for growth. Perhaps most importantly, ACCM culture techniques are aiding development of new genetic tools for *C. burnetii* (Chen et al. 2010; Voth et al. 2011; Beare et al. 2010). Moreover, biochemical and physiological studies can now be conducted without the metabolic background of the host cell.

C. burnetii axenic growth requirements have *in vivo* correlates. Oxygen gradients that occur within biological systems, including across biological membranes, can be associated with oxygen concentrations that are considerably lower than atmospheric oxygen levels (i.e., ~20%) (Hu et al. 1992; Khan et al. 2003). *C. burnetii* microaerophilic metabolism was hypothesized based on the organism's branched respiratory chain containing terminal oxidases typically associated with bacterial responses to normoxic and microaerobic levels of oxygen. The physiological basis of enhanced *C. burnetii* metabolism in a microaerobic environment is unclear, but may include sensitivity to oxidative stress. Technical difficulties in measuring ion levels in acidic vacuoles have prevented direct measurement of PV Na^+, K^+, and Cl^- concentrations. However, *C. burnetii*'s requirement for specific levels of these ions in ACCM likely correlates with their *in vivo* roles in physiological processes such as metabolite transport and pH homeostasis. Finally, a pH of ~4.75 is firmly established for the *C. burnetii* PV (Akporiaye et al. 1983; Maurin et al. 1992), a value that correlates with optimal pathogen growth in ACCM.

Several advances in *C. burnetii* genetic transformation have been made using ACCM (Chen et al. 2010; Voth et al. 2011; Beare et al. 2010). For example, secretion into the host cell cytoplasm of several potential *C. burnetii* effector molecules was recently demonstrated with a shuttle vector developed using axenic procedures (Chen et al. 2010; Voth et al. 2011). Continued development of genetic methods should soon allow fulfillment of molecular Koch's postulate for putative *C. burnetii* virulence genes. ACCM culture will also aid investigations of mechanisms used by *C. burnetii* to persist in the harsh conditions of its PV, and the biological bases of developmental form transitions. For instance, *C. burnetii* is highly sensitive to certain short chain fatty acids, such as propionate, during growth in ACCM (unpublished data). In *L. pneumophila*, nutritional cues including amino acid availability and exposure to short chain fatty acids can trigger a stringent response that drives transition of replicative to transmissive forms that are biologically reminiscent of *C. burnetii* large cell variant (LCV) and SCV developmental forms (Sauer et al. 2005; Edwards et al. 2009; Dalebroux et al. 2009). Thus, these molecules may also trigger stringent response genes that regulate *C. burnetii* morphological transitions, a hypothesis that is testable with ACCM culture techniques.

C. burnetii can be grown to high numbers in embryonated eggs (Williams et al. 1981). However, bacteria purified from eggs or mammalian host cells (Cockrell et al. 2008) contain eucaryotic host cell contamination, making these preparations suboptimal for biochemical analyses or vaccine production. Additionally, the homogenization and centrifugation steps involved in purification can generate infectious aerosols, making the procedure potentially hazardous. Purification of *C. burnetii* from ACCM is simple, safe and results in organisms free of host cell contamination.

ACCM cultures of the avirulent *C. burnetii* Nine Mile phase II isolate can be started from tissue samples of infected SCID mice that contain as few as 10 organisms (Omsland et al. 2011). Virulent *C. burnetii* isolates also grow well in ACCM; thus, it should be possible to isolate these organisms from infected immunocompetent animals and human patients using ACCM culture techniques. Recent Q fever outbreaks (Gilsdorf et al. 2008; Schimmer et al. 2009) highlight the need for improved clinical and epidemiological methods. Axenic isolation of *C. burnetii* should prove valuable in these settings.

Acknowledgements I would like to thank Robert A. Heinzen for critical review of this manuscript, and Anita Mora and Austin Athman for graphic illustrations. This work was funded by the Intramural Research Program of the National Institutes of Health, National Institute of Allergy and Infectious Diseases

References

Akporiaye ET, Rowatt JD, Aragon AA, Baca OG (1983) Lysosomal response of a murine macrophage-like cell line persistently infected with *Coxiella burnetii*. Infect Immun 40:1155–1162

Baca OG, Paretsky D (1983) Q fever and *Coxiella burnetii*: a model for host-parasite interactions. Microbiol Rev 47:127–149

Beare PA, Howe D, Cockrell DC, Omsland A, Hansen B, Heinzen RA (2009a) Characterization of a *Coxiella burnetii ftsZ* mutant generated by *Himar1* transposon mutagenesis. J Bacteriol 191:1369–1381

Beare PA, Unsworth N, Andoh M, Voth DE, Omsland A, Gilk SD, Williams KP, Sobral BW, Kupko JJ 3rd, Porcella SF, Samuel JE, Heinzen RA (2009b) Comparative genomics reveal extensive transposon-mediated genomic plasticity and diversity among potential effector proteins within the genus *Coxiella*. Infect Immun 77:642–656

Beare PA, Omsland A, Cockrell DC, Howe D, Heinzen RA (2010) Expansion of the *Coxiella burnetii* genetics tool box using axenic culture. *Abstracts of the 24th Meeting of the American Society for Rickettsiolgy, abstracts 94,* Stevenson, WA

Beron W, Gutierrez MG, Rabinovitch M, Colombo MI (2002) *Coxiella burnetii* localizes in a Rab7-labeled compartment with autophagic characteristics. Infect Immun 70:5816–5821

Braun RD, Lanzen JL, Snyder SA, Dewhirst MW (2001) Comparison of tumor and normal tissue oxygen tension measurements using OxyLite or microelectrodes in rodents. Am J Physiol Heart Circ Physiol 280:H2533–H2544

Burton PR, Kordova N, Paretsky D (1971) Electron microscopic studies of the rickettsia *Coxiella burnetii*: entry, lysosomal response, and fate of rickettsial DNA in L-cells. Can J Microbiol 17:143–150

Burton PR, Stueckemann J, Welsh RM, Paretsky D (1978) Some ultrastructural effects of persistent infections by the rickettsia *Coxiella burnetii* in mouse L cells and green monkey kidney (Vero) cells. Infect Immun 21:556–566

Chen SY, Vodkin M, Thompson HA, Williams JC (1990) Isolated *Coxiella burnetii* synthesizes DNA during acid activation in the absence of host cells. J Gen Microbiol 136:89–96

Chen C, Banga S, Mertens K, Weber MM, Gorbaslieva I, Tan Y, Luo ZQ, Samuel JE (2010) Large-scale identification and translocation of type IV secretion substrates by *Coxiella burnetii*. Proc Natl Acad Sci U S A 107:21755–21760

Cockrell DC, Beare PA, Fischer ER, Howe D, Heinzen RA (2008) A method for purifying obligate intracellular *Coxiella burnetii* that employs digitonin lysis of host cells. J Microbiol Methods 72:321–325

Coleman SA, Fischer ER, Howe D, Mead DJ, Heinzen RA (2004) Temporal analysis of *Coxiella burnetii* morphological differentiation. J Bacteriol 186:7344–7352

Cox HR (1938) A filter-passing infectious agent isolated from ticks. III. Description of organism and cultivation experiments. Public Health Report 53:2270–2276

Dalebroux ZD, Edwards RL, Swanson MS (2009) SpoT governs *Legionella pneumophila* differentiation in host macrophages. Mol Microbiol 71:640–658

Edwards RL, Dalebroux ZD, Swanson MS (2009) *Legionella pneumophila* couples fatty acid flux to microbial differentiation and virulence. Mol Microbiol 71:1190–1204

Ellison DW, Clark TR, Sturdevant DE, Virtaneva K, Porcella SF, Hackstadt T (2008) Genomic comparison of virulent *Rickettsia rickettsii* Sheila Smith and avirulent *Rickettsia rickettsii* Iowa. Infect Immun 76:542–550

Ewann F, Hoffman PS (2006) Cysteine metabolism in *Legionella pneumophila*: characterization of an L-cystine-utilizing mutant. Appl Environ Microbiol 72:3993–4000

Gilsdorf A, Kroh C, Grimm S, Jensen E, Wagner-Wiening C, Alpers K (2008) Large Q fever outbreak due to sheep farming near residential areas, Germany, 2005. Epidemiol Infect 136:1084–1087

Hackstadt T (1983) Estimation of the cytoplasmic pH of *Coxiella burnetii* and effect of substrate oxidation on proton motive force. J Bacteriol 154:591–597

Hackstadt T, Williams JC (1981a) Biochemical stratagem for obligate parasitism of eukaryotic cells by *Coxiella burnetii*. Proc Natl Acad Sci U S A 78:3240–3244

Hackstadt T, Williams JC (1981b) Stability of the adenosine 5′-triphosphate pool in *Coxiella burnetii*: influence of pH and substrate. J Bacteriol 148:419–425

Hackstadt T, Williams JC (1983) pH dependence of the *Coxiella burnetii* glutamate transport system. J Bacteriol 154:598–603

Harth G, Clemens DL, Horwitz MA (1994) Glutamine synthetase of *Mycobacterium tuberculosis*: extracellular release and characterization of its enzymatic activity. Proc Natl Acad Sci U S A 91:9342–9346

Heinzen RA, Scidmore MA, Rockey DD, Hackstadt T (1996) Differential interaction with endocytic and exocytic pathways distinguish parasitophorous vacuoles of *Coxiella burnetii* and *Chlamydia trachomatis*. Infect Immun 64:796–809

Howe D, Shannon JG, Winfree S, Dorward DW, Heinzen RA (2010) *Coxiella burnetii* phase I and II variants replicate with similar kinetics in degradative phagolysosome-like compartments of human macrophages. Infect Immun 78:3465–3474

Hu H, Sosnovsky G, Swartz HM (1992) Simultaneous measurements of the intra- and extra-cellular oxygen concentration in viable cells. Biochim Biophys Acta 1112:161–166

Imaizumi A, Suzuki Y, Ono S, Sato H, Sato Y (1983) Heptakis(2,6-O-dimethyl)beta-cyclodextrin: a novel growth stimulant for *Bordetella pertussis* phase I. J Clin Microbiol 17:781–786

Kashiwagi K, Miyamoto S, Suzuki F, Kobayashi H, Igarashi K (1992) Excretion of putrescine by the putrescine-ornithine antiporter encoded by the *potE* gene of *Escherichia coli*. Proc Natl Acad Sci U S A 89:4529–4533

Khan N, Shen J, Chang TY, Chang CC, Fung PC, Grinberg O, Demidenko E, Swartz H (2003) Plasma membrane cholesterol: a possible barrier to intracellular oxygen in normal and mutant CHO cells defective in cholesterol metabolism. Biochemistry 42:23–29

Marchini A, D'apolito M, Massari P, Atzeni M, Copass M, Olivieri R (1995) Cyclodextrins for growth of *Helicobacter pylori* and production of vacuolating cytotoxin. Arch Microbiol 164:290–293

Maurin M, Benoliel AM, Bongrand P, Raoult D (1992) Phagolysosomes of *Coxiella burnetii*-infected cell lines maintain an acidic pH during persistent infection. Infect Immun 60:5013–5016

Mcleod MP, Qin X, Karpathy SE, Gioia J, Highlander SK, Fox GE, Mcneill TZ, Jiang H, Muzny D, Jacob LS, Hawes AC, Sodergren E, Gill R, Hume J, Morgan M, Fan G, Amin AG, Gibbs RA, Hong C, Yu XJ, Walker DH, Weinstock GM (2004) Complete genome sequence of *Rickettsia typhi* and comparison with sequences of other rickettsiae. J Bacteriol 186:5842–5855

Omsland A, Cockrell DC, Fischer ER, Heinzen RA (2008) Sustained axenic metabolic activity by the obligate intracellular bacterium *Coxiella burnetii*. J Bacteriol 190:3203–3212

Omsland A, Cockrell DC, Howe D, Fischer ER, Virtaneva K, Sturdevant DE, Porcella SF, Heinzen RA (2009) Host cell-free growth of the Q fever bacterium *Coxiella burnetii*. Proc Natl Acad Sci U S A 106:4430–4434

Omsland A, Beare PA, Hill J, Cockrell DC, Howe D, Hansen B, Samuel JE, Heinzen RA (2011) Isolation from animal tissue and genetic transformation of *Coxiella burnetii* are facilitated by an improved axenic growth medium. Appl Environ Microbiol 77:3720–3725

Paretsky D, Consigli RA, Downs CM (1962) Studies on the physiology of rickettsiae. III. Glucose phosphorylation and hexokinase activity in *Coxiella burnetii*. J Bacteriol 83:538–543

Sauer JD, Bachman MA, Swanson MS (2005) The phagosomal transporter A couples threonine acquisition to differentiation and replication of *Legionella pneumophila* in macrophages. Proc Natl Acad Sci U S A 102:9924–9929

Schimmer B, Dijkstra F, Vellema P, Schneeberger PM, Hackert V, Ter Schegget R, Wijkmans C, Van Duynhoven Y, Van Der Hoek W (2009) Sustained intensive transmission of Q fever in the south of the Netherlands, 2009. Euro Surveill 14:19210

Schnappinger D, Ehrt S, Voskuil MI, Liu Y, Mangan JA, Monahan IM, Dolganov G, Efron B, Butcher PD, Nathan C, Schoolnik GK (2003) Transcriptional adaptation of *Mycobacterium tuberculosis* within macrophages: insights into the phagosomal environment. J Exp Med 198:693–704

Seshadri R, Paulsen IT, Eisen JA, Read TD, Nelson KE, Nelson WC, Ward NL, Tettelin H, Davidsen TM, Beanan MJ, Deboy RT, Daugherty SC, Brinkac LM, Madupu R, Dodson RJ, Khouri HM, Lee KH, Carty HA, Scanlan D, Heinzen RA, Thompson HA, Samuel JE, Fraser CM, Heidelberg JF (2003) Complete genome sequence of the Q-fever pathogen *Coxiella burnetii*. Proc Natl Acad Sci U S A 100:5455–5460

Stephens RS, Kalman S, Lammel C, Fan J, Marathe R, Aravind L, Mitchell W, Olinger L, Tatusov RL, Zhao Q, Koonin EV, Davis RW (1998) Genome sequence of an obligate intracellular pathogen of humans: *Chlamydia trachomatis*. Science 282:754–759

Teixeira De Mattos MJ, Neijssel OM (1997) Bioenergetic consequences of microbial adaptation to low-nutrient environments. J Biotechnol 59:117–126

Tesh MJ, Morse SA, Miller RD (1983) Intermediary metabolism in *Legionella pneumophila*: utilization of amino acids and other compounds as energy sources. J Bacteriol 154:1104–1109

Voth DE, Beare PA, Howe D, Sharma UM, Samoilis G, Cockrell DC, Omsland A, Heinzen RA (2011) The *Coxiella burnetii* cryptic plasmid is enriched in genes encoding type IV secretion system substrates. J Bacteriol 193:1493–1503

Williams JC, Peacock MG, Mccaul TF (1981) Immunological and biological characterization of *Coxiella burnetii*, phases I and II, separated from host components. Infect Immun 32:840–851

Zuerner RL, Thompson HA (1983) Protein synthesis by intact *Coxiella burnetii* cells. J Bacteriol 156:186–191

Chapter 12
Developmental Biology of *Coxiella burnetii*

Michael F. Minnick and Rahul Raghavan

Contents

Abstract The biphasic developmental cycle of *Coxiella burnetii* is central to the pathogen's natural history and survival. A small, dormant cell morphotype (the small-cell variant or SCV) allows this obligate intracellular bacterium to persist for extended periods outside of host cells, resist environmental conditions that would be lethal to most prokaryotes, and is the major infectious stage encountered by eukaryotic hosts. In contrast, a large, metabolically-active morphotype (the large-cell variant or LCV) provides for replication of the agent within acidified parasitophorous vacuoles (PVs) of a host cell. The marked physiological changes,

M.F. Minnick (✉)
Division of Biological Sciences, The University of Montana, Missoula, MT 59812, USA
e-mail: mike.minnick@mso.umt.edu

R. Raghavan
Department of Ecology and Evolutionary Biology, Yale University, New Haven, CT 06520, USA
e-mail: r.raghavan@yale.edu

R. Toman et al. (eds.), *Coxiella burnetii: Recent Advances and New Perspectives in Research of the Q Fever Bacterium*, Advances in Experimental Medicine and Biology 984, DOI 10.1007/978-94-007-4315-1_12,

differential gene expression, and the regulatory and structural components involved in *Coxiella's* morphogenesis from LCV to SCV and back to the LCV are fascinating attributes of the pathogen and are reviewed in this chapter.

Keywords Developmental cycle • *Coxiella* • Large-cell variant (LCV) • Small-cell variant (SCV) • Morphogenesis

12.1 *Coxiella* Cell Morphotypes and the Developmental Cycle

The life cycle of *C. burnetii* intertwines an alternation of SCV and LCV morphotypes with a markedly slow growth rate [~10–12 h generation time; (Coleman et al. 2004)]. The pleomorphic nature of *C. burnetii* cells was actually noted in the original paper describing the pathogen, in 1938, and included descriptions of coccoid, granular and bacillary forms of the bacterium (Davis and Cox 1938). Almost five decades later, remarkable electron microscopy work by McCaul and Williams (1981) was used as a foundation to describe *Coxiella* SCV and LCV morphotypes, and to propose a biphasic developmental cycle that includes both vegetative and sporogenic differentiation. LCVs were observed to be typical gram-negative cells, larger in size (in excess of 1 μm in length vs. 0.2–0.5 μm SCVs), more pleomorphic, and that the nucleoid was much less electron-dense and dispersed in comparison to SCVs. SCVs were also found to contain what appeared to be an extensive intracellular membrane system that was absent in LCVs. Ultrastructural work also led the authors to propose that both LCVs and SCVs replicate by transverse binary fission and that LCVs generate a polar, endospore-like structure (ELS) of ~130–170 nm in diameter, surrounded by a coat of at least four layers. However, biochemical work in this study showed that *C. burnetii* lacks dipicolinic acid; a hallmark component of endospores (McCaul and Williams 1981), and subsequent analysis of the *C. burnetii* genome showed that genes which are typically involved in sporulation by other bacteria are absent (Seshadri et al. 2003). The ELS's role in *Coxiella's* biology and whether the particle is actually infectious remains to be determined but is certainly intriguing. Interestingly, a pressure-resistant sub-population of SCVs, termed small dense cells (SDCs) can be obtained by treating preparations of the bacterium with 20,000–50,000 Lb/in^2 (McCaul and Williams 1981; McCaul et al. 1991).

The development of a protocol for generating synchronous *Coxiella* cultures (Coleman et al. 2004), wherein purified SCVs are used as the inoculum to establish an infection in cultured host cells, has provided a superior venue to elucidate stage-specific events and increase our understanding of the pathogen's life cycle and concomitant metamorphosis. This approach was used to show that *C. burnetii* undergoes a biphasic development which alternates between SCV and LCV morphotypes and that LCVs are the sole replicative cell type (Coleman et al. 2004) (Fig. 12.1). In addition, synchronous *C. burnetti* infections exhibit all the features of a bacterial growth cycle examined in a batch culture (closed) system. Specifically, SCVs undergo metamorphosis to LCVs during an approximately 2-day lag phase within a

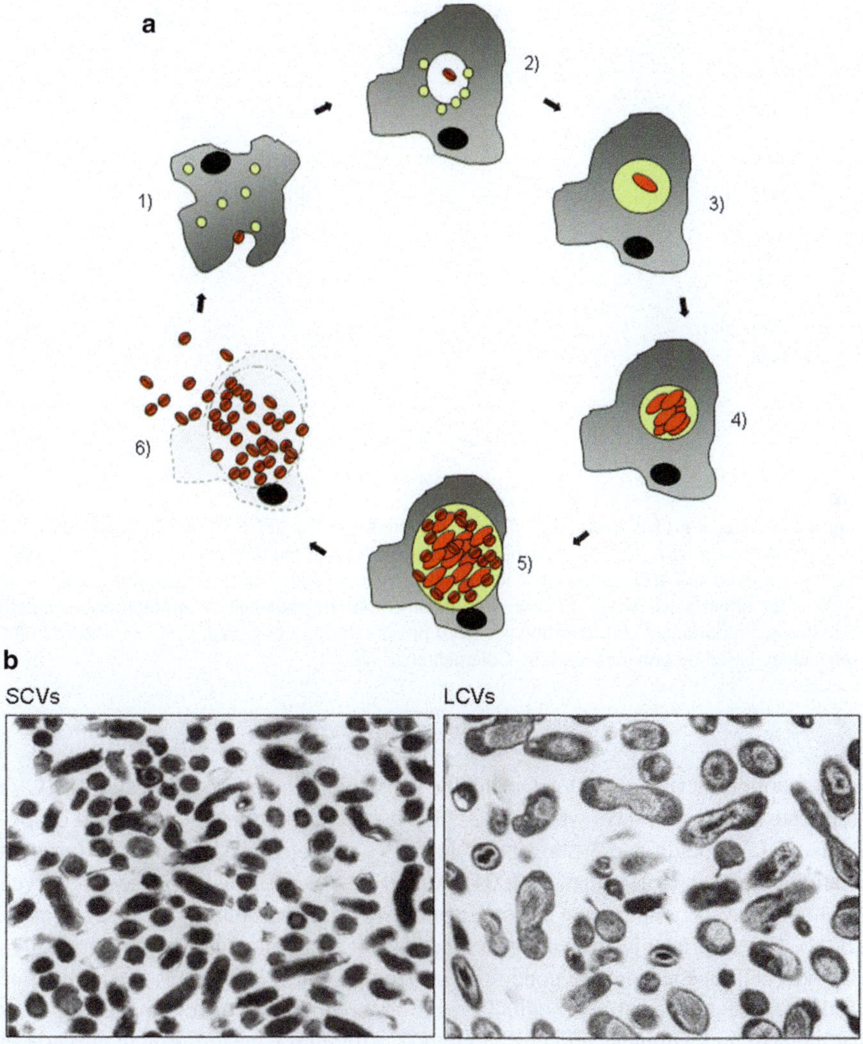

Fig. 12.1 (**a**) Developmental cycle of C. *burnetii*. (*1*) Passive internalization of an SCV into a typical host cell such as an alveolar macrophage, (*2*) SCV within the confines of a phagosome, with lysosomes (*small yellow granules*) approaching the periphery of the vacuole. (*3*) PV or phagolysosome following acidification (*yellow*) and subsequent morphogenesis of SCV to LCV. (*4*) LCV replication within the PV. (*5*) Metamorphosis of LCV to SCV at late log/early stationary phase. (*6*). Lysis of host cell with release of SCVs to repeat cycle. (**b**) Purified SCVs and LCVs of *C. burnetii* (courtesy of R. Heinzen)

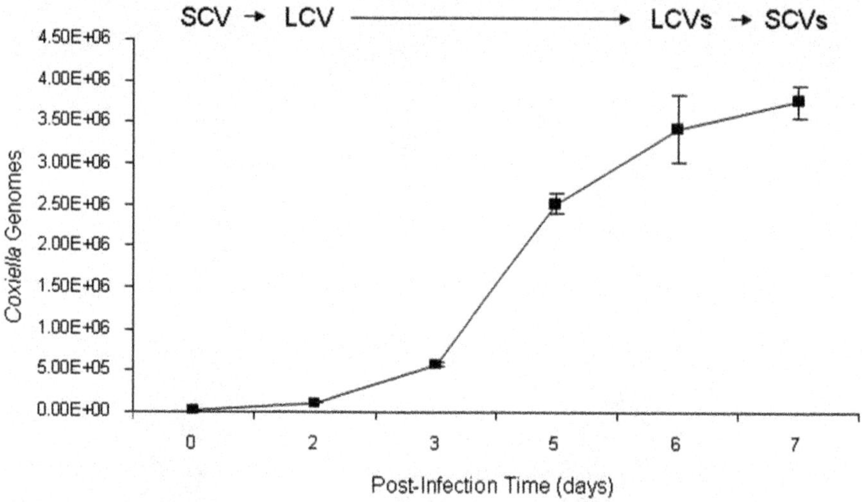

Fig. 12.2 Typical growth cycle of C. *burnetii* as determined by qPCR. A synchronous culture of *C. burnetii* (strain RSA 439) was established using SCVs and cultured Vero cells (*green monkey kidney epithelial cells*). Genomic DNA was assayed over the course of infection with qPCR, using *rpoS* as the target gene. Values represent the means of six independent determinations ± standard deviations. Predominant cell morphotypes (*top*) present during the growth cycle are shown at the top and are based on previous work by Coleman et al. (2004)

parasitophorous vacuole (PV). During the subsequent log phase, the resulting LCVs replicate for about 4 days with a 10.2–11.7-h generation time. Near the end of log-phase, LCVs begin a morphogenesis to SCVs, effectively signaling the onset of stationary phase (Coleman et al. 2004). A final senescence (death) phase is observed at 12–14 days post-infection in long-term, synchronized cultures (Raghavan et al. 2008). Not surprisingly, growth cycle kinetics and bacterial yields vary depending on the host cell type and culture conditions utilized [e.g., *Coxiella* grows at a higher rate and to a greater density in epithelial cells as compared to macrophage cell lines (Raghavan and Minnick, unpublished data)]. Although *Coxiella* growth curves can be generated by a fluorescent focus-forming unit assay (Coleman et al. 2004), quantitative PCR (qPCR) has proven to be a more rapid and sensitive technique for quantifying the number of bacterial genomes in a given culture. Resulting data can also serve as a backdrop for transcriptional profiles of various genes over the course of development. The qPCR protocol typically involves a "housekeeping" gene such as *dotA* or *rpoS* as the target for genome quantification (Coleman et al. 2004; Raghavan et al. 2008). Typical results of the protocol are shown in Fig. 12.2, with the phases of the growth cycle (to stationary phase) clearly seen and the respective, predominant cell morphotype that would be present, based on previous work (Coleman et al. 2004).

The developmental cycle of *Coxiella* undoubtedly evolved to counter two distinct and markedly different environments encountered by the bacterium; i.e., the intracellular and extracellular environments. As a result, LCVs are metabolic / replication machines that allow *C. burnetii* to live within the confines of a PV; a highly specialized intracellular niche that would not be suitable for most prokaryotes, owing to the acidic pH (pH ~4.5), various reactive oxygen species and the degradative enzymes that are present. This extraordinary specialization strategy and intracellular niche obviously decreased competition with other bacterial pathogens but required a considerable list of bacterial adaptations. Not surprisingly, *Coxiella* is an acidophile whose SCV dormancy is broken and metabolism induced primarily in response to acidic pH (Hackstadt and Williams 1981), although other intracellular cues may be involved. In sharp contrast, SCVs are metabolically dormant, long-lived and extremely durable morphotypes that enhance survival and persistence outside of host cells in both the host (extracellular fluids, serum, etc.) and the environment (barnyard dust, amniotic fluids, etc.). Thus, in many respects SCVs resemble endospores in other bacteria, in that they provide a durable and dormant cell type for nutritional overwintering and persistence during environmentally harsh conditions. Clearly, SCVs confer a tremendous advantage to an obligate intracellular bacterium, as most obligates cannot survive for extended periods of time outside of host cells. A notable exception is seen in the convergent evolution of *Chlamydia* with its superficially analogous elementary bodies (EBs) and reticulate bodies (RBs). EBs are small, dormant and can persist outside of host cells, while RBs are larger and metabolically active. Interestingly, while EBs are the only infectious cell morphotype in *Chlamydia*, both SCVs and LCVs of *Coxiella* are infectious (Heinzen 1997; Wiebe et al. 1972). Nevertheless, it is thought that SCVs are the primary infectious stage of *C. burnetii* as a direct result of the biphasic life cycle (Fig. 12.1a) and SCV durability in the environment.

12.2 Unique Attributes of the Cell Morphotypes

In addition to differences in cell size, SCVs and LCVs are markedly different in structure. Whereas SCVs are strong enough to withstand extremes of pressure, heat, osmotic stress, dessication, UV light, disinfectants and sonication, LCVs are surprisingly fragile (Babudieri 1959; McCaul et al. 1981). Early work done to purify cell morphotypes often involved mixed *Coxiella* cell populations, sonication (or some other lysis protocol) and differential centrifugation, but often resulted in damaged LCVs. Most work today capitalizes on the growth cycle (Fig. 12.2) coupled with a synchronously-infected culture model. Using this approach, SCVs can be obtained by infecting host cells monolayers and incubating for 4 weeks without replenishing the medium. The first week of growth is done at 37°C and 5% CO_2, followed by 3 weeks at room temperature with the culture flask lids tightened down (Coleman et al. 2004). The resulting population is entirely SCV in composition. In contrast, LCVs can be obtained from synchronously-infected cultures at approximately mid-log phase (4 days; Fig. 12.2) by using digitonin, a gentle detergent that dissolves the cholesterol-rich PV and host cell membranes, thereby releasing the intracellular bacteria (Cockrell et al. 2008).

The highly condensed nucleoid of SCVs versus the extended, fibrillar nucleoid of LCVs is one of the most obvious structural differences seen in *C. burnetii* (McCaul and Williams 1981). To be sure, the smaller cell size and metabolic dormancy of SCVs correlate with a compact genome, whereas the LCV's larger size and increased transcriptional activity would provide more volume and require a more dispersed chromosome, respectively. Two basic proteins, Hq1 and ScvA, have been described and may be involved in altering *Coxiella's* DNA topology during LCV-to-SCV morphogenesis and are discussed below.

Structural components and attributes that provide the extraordinary sturdiness and stability to SCVs, are poorly defined. Early work showed that SCVs contain comparatively more surface-exposed LPS than LCVs, when analyzed by immunoelectron microscopy (McCaul et al. 1991). However, a potential role for LPS in conferring physical strength to SCVs seems unlikely. Using endospores of *Bacillus* and *Clostridium* as a reference point, it is possible that the sturdiness of SCVs results from a cell wall that is analogous, at least in function, to an endospore "coat". Endospore coats contain over 70 proteins, of which several are cysteine-rich and cross-linked (Henriques and Moran 2007). In addition, mineralization and a dehydrated state confer additional resistance to environmental insults upon the endospore. However, the possible involvement of cell wall protein cross-linkage, mineralization or cytosolic dehydration in SCVs has not been fully examined, to date.

Perhaps the most striking and disparate attribute of *Coxiella* cell morphotypes is their metabolic state. Classical work by Hackstadt and Williams (1981) demonstrated an acid-dependent uptake of ^{14}C-labeled carbon sources, with a preference for glutamate over glucose. Intracellular growth of *Coxiella* was inhibited if the pH of the lysosome was increased by adding lysosomotropic amines to the culture medium (Hackstadt and Williams 1981). A subsequent report showed that the relative metabolic state of LCVs was much greater than SCVs in an axenic medium (Heinzen 1997). More recent work by Coleman et al. (2004) indicates that SCVs are essentially dormant, and that they require about a 2-day lag phase before reaching log-phase growth in synchronously-infected cultures (see Fig. 12.2). While acid pH (pH 4.7–4.8), and possibly other cues such as host cell nutrients or enzymes trigger SCV-to-LCV morphogenesis and activate metabolism, identification of environmental signals that initiate LCV-to-SCV morphogenesis and dormancy are less clear. One possibility that has been proposed is depletion of one or more important cellular nutrients, as a result of a heavily-infected host cell (Heinzen et al. 1996a), in a fashion similar to the nutrient-depletion trigger for endospore formation in *Bacillus* and *Clostridium* species. The physical constraints imparted by PVs to the host cell are also noteworthy, and they may be involved in helping to induce morphogenesis and dormancy. Namely, *Coxiella* densities can reach thousands of bacteria per cell. As a result, the PV is enlarged to a point where host cell organelles are literally pushed to the cell membrane, and may therefore be functionally impaired. In support of this notion is that host cells containing a high density of *C. burnetii* may have a reduced ability to traffic nutrient-laden vesicles to the PV (Heinzen et al. 1996b). Thus, organelle impairment may indirectly trigger *Coxiella's* morphogenesis through host cell stress and/or nutrient deprivation.

12.3 Transcriptional and Protein Profiles as a Function of Developmental Stage

There are a number of reports describing proteins of *C. burnetii* that are present in either SCVs or LCVs. However, since assigning a protein to a specific *Coxiella* cell morphotype is predicated on the purity of the cell preparation and the sensitivity of a given detection protocol, we will refer to "stage-specific" proteins as $LCV^{Hi/Lo}$ or $SCV^{Hi/Lo}$, below.

12.3.1 SCV^{Hi}/ LCV^{Lo} Proteins

Early work by Heinzen and Hackstadt showed that *Coxiella* synthesizes a ~13.2-kDa histone-like protein, termed Hq1, which is homologous to eukaryotic H1 proteins and unique to SCVs (Heinzen and Hackstadt 1996) and encoded by the *hcbA* gene. Like most histones, Hq1 is extremely basic (pI 13.1), owing to its high lysine content (29%). The stage-specific bias of Hq1 expression suggests it may serve as an effector of chromatin structure and DNA condensation in SCVs. Interestingly, the Hc1 orthologue of *Chlamydia* is also developmentally regulated with maximal expression in EB's (Hackstadt et al. 1991) and has been shown to condense the nucleoid when heterologously expressed in *Escherichia coli* (Barry et al. 1992). An additional role for Hq1 in global gene regulation during LCV-to-SCV morphogenesis is a distinct possibility, especially considering the change in DNA topology that occurs in the presence of these molecules (Barry et al. 1993). Surprisingly, analysis of *hcbA* transcripts over the course of a synchronous infection showed that maximal expression occurs at 3 days post-infection (Coleman et al. 2004), suggesting that Hq1 translation may be delayed until LCV-to-SCV morphogenesis is triggered.

Another potential effector of chromatin structure and/or function is ScvA, a small (~3.6 kDa), highly basic (pI ~11) protein that binds double-stranded DNA and is unique to the SCV when analyzed by immunoblots and immunoelectron microscopy (Heinzen et al. 1996a). Interestingly, qRT-PCR analysis over the course of a synchronously-infected culture showed that *scvA* mRNA is at high levels (~10^3 copies per genome) and stable throughout the 0–8 days post-infection period (Coleman et al. 2004). Further analysis revealed that the ScvA protein is degraded during SCV-to-LCV morphogenesis and during LCV replication, suggesting that regulation of this gene product occurs post-transcriptionally (Coleman et al. 2004). An intriguing possibility is that ScvA and Hq1 may provide amino acid and nutrient sources during the SCV-to-LCV transition, since the proteins degrade during this phase of infection (Heinzen et al. 1999).

Another SCV^{Hi} / LCV^{Lo} protein is P1; an OMP with a predicted beta-barrel structure that has been shown to possess porin-like activity (McCaul et al. 1991; Varghees et al. 2002). Although originally thought to be a single protein, P1 is actually a

Table 12.1 SCV^{Hi}/ LCV^{Lo} proteins of *Coxiella* (in alphabetical order)

Protein (Description)	Gene(s)[a]	Citation(s)
Era (GTP-binding Era-like protein)	CBU1502	Coleman et al. (2007)
Hq1 (Histone H1-like protein, HcbA)	CBU0456	Coleman et al. (2004), Heinzen and Hackstadt (1996)
Hypothetical protein	CBU2079	Coleman et al. (2007)
MetC (cystathionine gamma-lyase)	CBU2025	Coleman et al. (2007)
Omp34 (34-kDa omp)		Heinzen et al. (1999)
P1 (P1A Omp)	CBU0311	Coleman et al. (2004)
P1 paralogues (P1B-D, OmpA1-4)	CBU0307, 1260,1600,1814, 0307, 1412, 1413, 1414	Battisti and Minnick (unpublished data)
ScvA (Small cell variant protein A)	CBU1267a	Coleman et al. (2004), Heinzen et al. (1996a)
TolB (TolB protein)	CBU0090	Coleman et al. (2007)

[a]Designations from GenBank accession number AE016828 (Strain RSA 493)

constituent of a paralagous family of eight OMPs, including P1A-P1D [CBU0311, CBU1414, CBU1413 and CBU1412, respectively] and OmpA1-OmpA4 [CBU0307, CBU1260, CBU1600 and CBU1814, respectively]. Genes for each paralogue are upregulated during log-phase growth when synchronous-infection cultures are analyzed by qRT-PCR (Coleman et al. 2004; Battisti and Minnick, unpublished data). Although P1's porin substrate specificity is unknown, an intriguing possibility is that each P1 family member possesses differential specificity. It is important to note that while transcript levels for the *p1/ompA* gene family uniformly increase during exponential-phase growth, both OmpA1 and P1A proteins appear more prominent in SCVs than LCVs (Battisti and Minnick, unpublished data).

In addition to a relatively greater quantity of surface-exposed LPS (26), SCVs also possess a greater quantity of a 34-kDa outer membrane protein (omp), termed OMP34 (Heinzen et al. 1999). Characterization and function of OMP34 has not been determined.

More recent proteomic work on purified LCVs and SCVs has revealed four additional SCV^{Hi}/ LCV^{Lo} proteins, including TolB (Involved in TonB-independent uptake of proteins), cystathionine gamma-lyase (L-methionine biosynthesis), a hypothetical protein (CBU2079) and a GTP-binding Era-like protein (GTPase, cell growth) homolog (Coleman et al. 2007).

Taken together, the results in Table 12.1 suggest that SCV^{Hi}/ LCV^{Lo} proteins are induced to: (a) facilitate cell cycling and differentiation (Era homolog), (b) promote uptake of an unknown substrate(s) (TolB, P1/OmpA), (c) facilitate nucleoid consensation during LCV-to-SCV morphogenesis (Hq1, ScvA), and (d) provide for specialized biosynthetic processes (MetC).

12.3.2 LCV^{Hi} / SCV^{Lo} Proteins

Many of the LCV^{Hi}/SCV^{Lo} proteins reflect the enhanced metabolic state of the LCV relative to the SCV. For example, elongation factors Tu and Ts (EF-Tu and EF-Ts, respectively) are essential cofactors involved in ribosome function and are markedly upregulated in LCVs (Seshadri et al. 1999). EF genes are undoubtedly induced to meet the demand of increased mRNA synthesis and translation during log-phase growth.

The *C. burnetii* genome encodes three sigma factors, including RpoD (σ^{70}), RpoH (σ^{32}) and RpoS (σ^{38}). Orthologues of these sigma factors serve as global regulators of transcription for genes expressed during normal growth or housekeeping activities (*rpoD*), heat / stress shock (RpoH), and starvation or stationary phase (RpoS). However, work by Seshadri and Samuel showed that RpoS was actually highest in LCVs and barely detectable in SCVs, following immunoblotting with anti-RpoS antibodies (Seshadri and Samuel 2001). These results are antithetical to the more typical role for RpoS in regulating genes as the cell approaches stationary phase. In keeping with protein work, the *rpoS* transcript was shown to increase early in log-phase growth (1–3 days post-infection) and wane towards stationary phase (6–8 days post-infection) (Coleman et al. 2004). The actual role of RpoS in development is unclear but intriguing, especially considering that if *Coxiella rpoS* is heterologously expressed in *E. coli*, its expression is highest upon entry into stationary phase (Seshadri and Samuel 2001). Further analysis by qRT-PCR has revealed that all three *Coxiella* sigma factors are induced by 1 day post-infection in synchronized cultures, with approximately 834 and 998 *rpoH* transcripts per genome on days 1 and 2, respectively; ~60 *rpoD* mRNA copies per genome on day 1 with a steady decline to <10 by day 4; and roughly 30 copies of *rpoS* mRNA on day 1 with a slight decrease in transcripts by day 4 (Coleman and Minnick, unpublished data). These results suggest that RpoH serves as a sigma factor during SCV-to-LCV morphogenesis. Interestingly, heat shock does not significantly increase the amount of *rpoH* transcript in synchronized infections, suggesting that some other cue(s) is responsible for *rpoH* induction (Coleman and Minnick, unpublished data).

Recent work has shown that *C. burnetii* maximally expresses a DNA-binding peroxiredoxin (a bacterioferritin comigratory protein; BCP) during log-phase growth (2–3 days post-infection). *Coxiella's* BCP demonstrates peroxidase activity in vitro, is dependent upon thioredoxin-thioredoxin reductase, and is likely involved in detoxifying endogenous hydroperoxide byproducts of metabolism during intracellular replication (Hicks et al. 2010). The DNA-binding activity of the *C. burnetii* BCP has not been previously reported for other BCP's, and may serve to protect the pathogen's supercoiled DNA from oxidative damage during growth. BCP was previously observed to be an LCV^{Hi} / SCV^{Lo} protein during a comparative proteomic analysis of LCVs and SCVs (Coleman et al. 2007).

Although LPS is the only known bona fide virulence factor of *Coxiella* (Moos and Hackstadt 1987), the bacterium's type IV secretion system (T4SS) and its translocated effectors are currently under close scrutiny as candidate determinants of virulence. The *Coxiella* T4SS is similar to the Dot/Icm system of *Legionella pneumophila* (Seshadri et al. 2003), and in many instances *C. burnetii* genes can complement the corresponding mutants of *L. pneumophila* when ectopically expressed (Zamboni et al. 2003; Zusman et al. 2003). Work examining expression of four T4SS genes during a synchronously-infected culture showed that *icmQ*, *icmS*, *icmW* and *dotB* were expressed during exponential-phase growth (24–96 h post-infection). This time period was also shown to coincide with enlargement of the PV (Zamboni et al. 2003). Because of this temporal correlation, *Coxiella's* T4SS and its substrates are implicated as factors that contribute to host cell subversion and biogenesis of the intracellular niche during a host cell infection. To what extent these genes are down-regulated as the LCV transitions to a SCV was not examined, but likely to occur since SCVs form late in the timeline of the PV's maturation. Later work examining transcription of the T4SS gene, *dotA*, over the course of the growth cycle showed maximal expression during log-phase growth (Coleman et al. 2004), in keeping with the other T4SS genes analyzed previously (Zamboni et al. 2003).

Comparative proteomic work involving purified LCVs and SCVs identified 15 LCV^{Hi} / SCV^{Lo} proteins (Coleman et al. 2007), including two proteins discussed above (BCP, EF-Tu), two chaperonins (GroEL and HtpG), two ribosomal proteins (S1 and L9), FtsZ, ParB, the universal stress protein A (UspA), N utilization substance protein A (NusA), stringent starvation protein (SspA), segregation and condensation protein (ScpB), DMRL synthase (RibH), and two hypothetical proteins (CBU0658, CBU1754) (Table 12.2). Curiously, the most abundant LCV^{Hi} / SCV^{Lo} protein was a hypothetical protein, CBU0658, whose function has not yet been determined. Taken as a whole, the results in Table 12.2 suggest that LCV^{Hi} / SCV^{Lo} proteins are induced to: (a) facilitate the marked increase in translation (EF-Tu, NusA, ribosomal proteins S1 and L9), (b) counter the stressful environment imposed by the PV or by toxic byproducts of metabolism (BCP, GroEL, HtpG, UspA), (c) help to regulate transcription of other genes (RpoH, RpoS, RpoD, SspA), (d) provide for biosynthetic processes (RibH), and (e) assist in DNA replication and cell division (FtsZ, ParB, ScpB).

Finally, differential gene / protein expression may exist between strains of *C. burnetii* that possess smooth or rough LPS. Strains possessing rough, phase II (e.g., RSA 439) and intermediate (e.g., Nine Mile Crazy) LPS phenotypes have been employed to investigate stage-specific gene and protein expression in several studies (Briggs et al. 2008; Coleman et al. 2004; Hicks et al. 2010; Samoilis et al. 2007). The underlying, unstated hypothesis is that gene / protein expression results derived from such strains can be extrapolated to smooth LPS strains (e.g., Nine Mile Phase I) of *C. burnetii*. However, since these strains grow at different rates in cell cultures, this hypothesis clearly requires additional research for verification.

12.4 Gene Regulation

Complex, intertwined regulatory networks modulate *Coxiella's* growth within phagolysosomal vacuoles. The genome and gene expression profiles of this obligate intracellular bacterium have the hallmarks of a parasite that has adapted to life in a

Table 12.2 LCV^{Hi} / SCV^{Lo} proteins of *Coxiella* (in alphabetical order)

Protein (Description)	Gene[a]	Citation(s)
BCP (Bacterioferritin comigratory protein, peroxiredoxin)	CBU0963	Coleman et al. (2007), Hicks et al. (2010)
DotA (T4SS component)	CBU1648	Coleman et al. (2004)
DotB (T4SS component)	CBU1645	Zamboni et al. (2003)
EF-Ts (Translation elongation factor- Ts)	CBU1385	Seshadri et al. (1999)
EF-Tu (Translation elongation factor- Tu)	CBU0236	Coleman et al. (2007), Seshadri et al. (1999)
FtsZ (Cell division protein)	CBU0141	Coleman et al. (2007)
GroEL (Heat shock protein, chaperone)	CBU1718	Coleman et al. (2007)
HtpG (Heat shock protein, chaperone)	CBU0309	Coleman et al. (2007)
Hypothetical	CBU0658	Coleman et al. (2007)
Hypothetical	CBU1754	Coleman et al. (2007)
IcmQ (T4SS component)	CBU1634	Zamboni et al. (2003)
IcmS (T4SS component)	CBU1642	Zamboni et al. (2003)
IcmW (T4SS component)	CBU1650	Zamboni et al. (2003)
L9 (Ribosomal protein)	CBU0867	Coleman et al. (2007)
NusA (N utilization substance protein A)	CBU1433	Coleman et al. (2007)
ParB (Chromosomal partitioning protein)	CBU1927	Coleman et al. (2007)
RibH (6,7, dimethyl-8-ribityllumazine synthase)	CBU0648	Coleman et al. (2007)
RpoD (σ^{70})	CBU1596	Coleman and Minnick (unpublished data)
RpoH (σ^{32})	CBU1909	Coleman and Minnick (unpublished data)
RpoS (σ^{38})	CBU1669	Coleman et al. (2004), Seshadri and Samuel (2001)
S1 (Ribosomal protein)	CBU0528	Coleman et al. (2007)
ScpB (Segregation/condensation protein)	CBU1060	Coleman et al. (2007)
SspA (Stringent starvation protein A)	CBU1747	Coleman et al. (2007)
UspA (Universal stress protein)	CBU1916	Coleman et al. (2007)

[a]Designations from GenBank accession number AE016828 (Strain RSA 493)

high-stress environment. However, unlike free-living bacteria that live in ever-fluctuating environments, *Coxiella's* specialized niche provides a higher degree of predictability that renders a large number of regulatory elements essentially superfluous. Accordingly, only 28 (1.5% of 1,818 protein-coding genes) recognizable transcriptional regulators are present in *Coxiella* as opposed to 207 (5% of protein-coding genes) in *E. coli* (Kanehisa et al. 2004). Another adaptation that distinguishes *Coxiella* from its free-living γ-proteobacteria relatives is its slow generation time. While its exact cause is not clear, slow growth is a universal phenomenon observed in bacteria that live in stable but nutrient-poor environments (Merhej et al. 2009; Vieira-Silva and Rocha 2010). As in most other slow-growing bacteria, *Coxiella* has only one rRNA operon, which is also thought to be an adaptation to its long generation time because multiplicity of rRNAs has been shown to be a burden during slow growth (Stevenson and Schmidt 1998). However, since the sole rRNA operon in

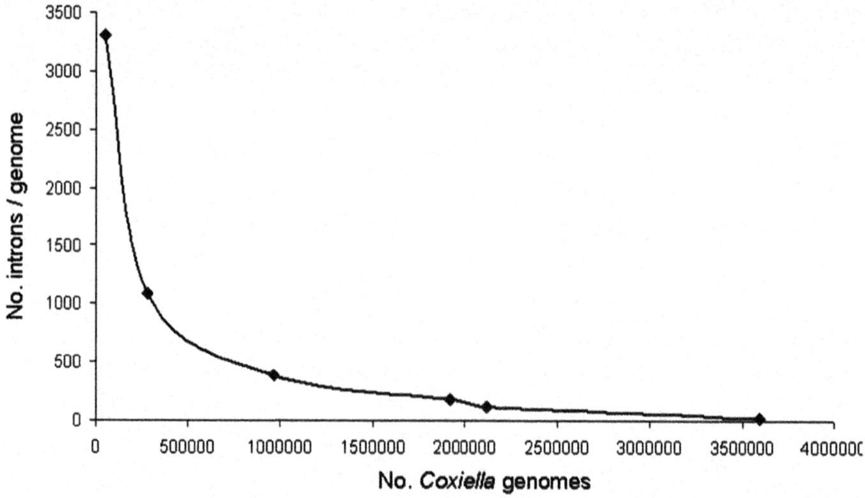

Fig. 12.3 The number of *Coxiella* intron RNA's (L1917 and L1951) per genome as a function of genome number, in a synchronous culture of *C. burnetii* (strain RSA 439) in Vero cells. Genome and intron values were determined on days 3, 4, 5, 6, 8 and 12 using qPCR and qRT-PCR, respectively (n=3)

Coxiella is positioned close to the origin of replication (within the first 16% of the genome), increased gene-dosage during replication would likely provide multiple copies of the rRNA genes during exponential growth (Vieira-Silva and Rocha 2010).

Intriguingly, the 23S rRNA gene of *Coxiella* encodes two group I introns. These self-splicing introns are able to re-associate with ribosomes and retard growth-rates (Raghavan et al. 2007, 2008) with a clear negative correlation between the amount of intron RNA and bacterial growth (Fig. 12.3). Additionally, the 23S rRNA also contains an intervening sequence (IVS), which is likely removed by the action of RNase III, leaving the large subunit rRNA in two pieces. This fragmentation is not thought to affect the function of the molecule but could aid in the efficient degradation of ribosomes when *Coxiella* transforms from LCV to SCV, as suggested for *Salmonella* during log to stationary phase shift (Hsu et al. 1994). There is yet another parasitic element in *Coxiella* that might influence its growth and replication; an intein within the *dnaB* helicase gene (Raghavan et al. 2008). It is possible that, since intein-splicing is a slow process, it reduces the pool of functional DnaB proteins in the cell, thereby retarding bacterial replication rate.

Although conditions inside the endolysomal vacuole are predictable, it is nevertheless a very harsh environment not usually conducive to bacterial growth. The low-nutrient and high-stress status of this niche is highlighted by the expression of the stationary phase sigma factor RpoS (σ^{38}) during exponential growth in *Coxiella*. This global regulator of gene expression is known to accumulate during cellular starvation and modulates the expression of a large number of genes including those responsible for stress response. It has been shown previously that in *L. pneumophila*, a close relative of *Coxiella*, RpoS regulates the expression of genes required for survival in protozoan hosts. Hence, it is possible that, unlike in *E. coli*, where RpoS regulates the expression of stationary-phase genes, this global regulator controls the expression of genes that aid in overcoming the stressful conditions encountered by *Coxiella* within the PV (Seshadri and Samuel 2001). The role of starvation in elevated RpoS levels is clear from elevated levels of stringent starvation protein A (SspA) during exponential phase (Coleman et al. 2004). This protein is known to be induced by starvation for glucose, nitrogen, phosphate and amino acids and in turn increases the synthesis of RpoS. In most bacteria, RpoS function is influenced and controlled by at least four small RNAs (sRNAs), including *arcZ*, *dsrA*, *rprA* and *rydB*. The roles, if any, of these sRNAs in *Coxiella's* unique RpoS transcriptional picture is not clear as the sRNA profile in *Coxiella* remains to be elucidated. However, Seshadri and Samuel (2001) have reported that the *C. burnetti's rpoS* ORF contains a region complementary to the first stem-loop of *E. coli*'s *dsrA* sRNA.

A high-stress, low-nutrient environment also triggers the global stringent response in which the (p)ppGpp alarmone produced by SpoT and RelA, with the assistance of DksA, helps the bacteria to tide over the stressful conditions (Traxler et al. 2008). The RpoS regulon and stringent response pathway are known to interact, with ppGpp inducing RpoS expression, which in turn controls genes that participate in ppGpp metabolism (Karp et al. 2002). In addition, *Coxiella* has a number of potential transcriptional regulators (Table 12.3) that might respond to levels of various nutrients like arginine, tryptophan and phosphonate. The bacterium also possesses TPP and FMN riboswitches; known to regulate gene expression by sensing the amount of their cognate molecules (Griffiths-Jones et al. 2003). Even though iron is not thought to play a very important role in virulence, a Fur-based regulatory system that detects and responds to iron concentrations is also present (Briggs et al. 2008). It has also been shown that *Coxiella* modulates the autophagic pathway in host cells, probably to compensate for a lack of nutrients within the endolysosome (Romano et al. 2007).

In addition to the lack of nutrients, *C. burnetii* also has to deal with acid-induced stress while growing in the phagolysosome. Apparently to compensate for its acidic surroundings, over 60% of *Coxiella's* proteome is basic with more than 45% of proteins having pI values ≥ 9. A good case in point for this alkaline shift in *Coxiella* proteins is the RpoS protein itself, which has a pI of 9.6 in contrast to a pI of 4.6 for the *E. coli* homologue. *Coxiella's* basic proteins may act as a proton 'sink' to buffer protons that enter the cytoplasm (Seshadri and Samuel 2001). *Coxiella* also has four predicted sodium ion/proton exchangers (CBU1590, CBU1259, CBU0459, CBU1582-1588) that may help maintain pH homeostasis (Seshadri et al. 2003). SspA, in addition to its role during starvation, is also important for survival during acid-induced stress in *E. coli* (Karp et al. 2002), underlining the interconnectivity of regulatory networks.

Another source of stress within phagolysosomes is reactive oxygen species (ROS). *Coxiella* has a number of genes including *ahpC*, *bcp*, *sodB* and *sodC*, which are known to protect bacteria from both internal and external ROS. In addition,

Table 12.3 Potential transcriptional regulators of *C. burnetii*. As annotated in GenBank accession no. AE016828 (Strain RSA 493)

Length	PID	Locus tag	COG	Locus	Product
159	29653823	CBU0480	COG1438K	*argR*	Arginine repressor
153	29654658	CBU1361	COG1959K		FeS assembly SUF system regulator
311	29654767	CBU1476	COG0583K	*oxyR*	Hydrogen peroxide inducible genes activator
111	215919326	CBU2041	COG3829KT		PAS domain containing protein
113	29654105	CBU0775	COG1725K		GntR family transcriptional regulator
707	29653654	CBU0303	COG0317TK	*spoT*	GTP pyrophosphokinase
714	29654672	CBU1375	COG0317TK	*relA*	RelA/SpoT family protein
232	215919141	CBU1227	COG0745TK	*qseB*	Transcriptional regulatory protein
214	29654353	CBU1043	COG2197TK	*gacA.4*	LuxR family DNA binding response regulator
217	29654272	CBU0955	COG2197TK	*gacA.3*	LuxR family DNA binding response regulator
216	29654109	CBU0780	COG2197TK	*gacA.2*	LuxR family DNA binding response regulator
211	29654049	CBU0712	COG2197TK	*gacA.1*	GacA family DNA binding response regulator
73	29655255	CBU1972	COG2197TK		LuxR family transcriptional regulator
127	29655092	CBU1805	COG2197TK		LuxR family transcriptional regulator
84	29655091	CBU1804	COG2771K		LuxR family transcriptional regulator
193	29654146	CBU0818	COG1309K		TetR family transcriptional regulator
82	29654104	CBU0774	COG1983KT	*pspC*	PspC domain containing protein
79	29653920	CBU0582	COG5007K	*bolA*	BolA family protein
97	29653653	CBU0302	COG1758K	*rpoZ*	DNA directed RNA polymerase omega subunit
352	29654959	CBU1669	COG0568K	*rpoS*	RNA polymerase sigma factor
288	29655194	CBU1909	COG0568K	*rpoH*	RNA polymerase factor sigma 32
698	29654887	CBU1596	COG0568K	*rpoD*	RNA polymerase sigma factor
1414	29653584	CBU0232	COG0086K	*rpoC*	DNA directed RNA polymerase subunit beta
1377	215918919	CBU0231	COG0085K	*rpoB*	DNA directed RNA polymerase subunit beta
327	29653615	CBU0263	COG0202K	*rpoA*	DNA directed RNA polymerase subunit alpha
114	29653851	CBU0509	COG2973K	*trpR*	Trp operon repressor
248	29653715	CBU0367	COG0745TK	*phoB*	Phosphate regulon transcriptional regulator

Coxiella also contains two universal stress proteins (Usp) (CBU1916 and CBU1983) that are known to be upregulated in response to environmental insults. Samoilis et al. (2007) have shown that at least one of them (CBU1983, UspA) is expressed in *Coxiella* during intracellular growth. Importantly, it has been shown that UspA protects *E. coli* from superoxide-generating agents (Nachin et al. 2005) and is induced by ppGpp (Kvint et al. 2003), the stringent response alarmone; again highlighting the interconnectivity of regulatory networks. Another intriguing stress response protein in *Coxiella* is the 50S ribosomal protein L25/general stress protein CTC (CBU1840). Two types of proteins make up this family: those that contain the N-terminal ribosomal protein L25 domain only (e.g., in *E. coli*) and those that contain an additional C-terminal domain involved in stress response (e.g., in *Bacillus subtilis*). Interestingly, the homologue in *C. burnetii* is the second type, suggesting a role for this protein in controlling stress.

Two-component regulatory systems help bacteria to detect and respond to various environmental cues. In *Coxiella*, only a few two-component systems are recognizable, probably due to the relative stability of its intracellular niche (Beare et al. 2009). Similar to *E. coli*, the PhoB-PhoR system in *Coxiella* may respond to inorganic phosphate in the environment and modulate expression of phosphate regulon genes. Only a RtsB-like protein of the RstB-RstA system is present in *Coxiella*. Mg^{2+} stimulates this signal transduction system in a PhoP-dependent manner in *E. coli*. Since both a PhoP-PhoQ system and RstA are absent in *Coxiella*, it is not clear how the bacterium responds to magnesium levels in the environment. The GacA-GacS system is known to regulate virulence factors in bacteria that are pathogenic for plants (Heeb and Haas 2001). Interestingly, GacA-GacS also has a role in inducing slow-growth variants under stress in *Pseudomonas aeruginosa* (Davies et al. 2007). In *Legionella*, this two-component system interacts with the stringent response pathway and RpoS to initiate the transmission phase (Molofsky and Swanson 2004). Since SCVs are similar to *Legionella*'s transmission phase, it is possible that the GacA-GacS system has a role in controlling the biphasic development of *Coxiella* (Beare et al. 2009). The quorum sensing regulators QseB-QseC, which respond to autoinducers and LuxR-family regulators that sense cell density, may control LCV growth within phagolysosomes. The QseB-QseC regulator has also been classified as a PmrA-PmrB two-component system in *Coxiella* and has been shown to regulate the Dot/Icm type IV secretion system (Zusman et al. 2007). In addition, there are two histidine kinases (CBU0789 and CBU1761) of unknown function that may be the sensory components of two yet-to-be identified two-component systems.

Collectively, much needs to be learned about *C. burnetii*'s regulatory machinery. However, the available evidence suggests the presence of a minimal regulatory apparatus that helps this obligatory intracellular pathogen to thrive in the very harsh but relatively-stable phagolysosome.

References

Babudieri B (1959) Q fever: a zoonosis. Adv Vet Sci 5:81–182

Barry CE III, Hayes SF, Hackstadt T (1992) Nucleoid condensation in *Escherichia coli* that express a chlamydial histone homolog. Science 256:377–379

Barry CE III, Brickman TJ, Hackstadt T (1993) Hc1-mediated effects on DNA structure: a potential regulator of chlamydial development. Mol Microbiol 9:273–283

Beare PA, Unsworth N, Andoh M, Voth DE, Omsland A, Gilk SD, Williams KP, Sobral BW, Kupko JJ 3rd, Porcella SF, Samuel JE, Heinzen RA (2009) Comparative genomics reveal extensive transposon-mediated genomic plasticity and diversity among potential effector proteins within the genus *Coxiella*. Infect Immun 77:642–656

Briggs HL, Pul N, Seshadri R, Wilson MJ, Tersteeg C, Russell-Lodrigue KE, Andoh M, Bäumler AJ, Samuel JE (2008) Limited role for iron regulation in *Coxiella burnetii* pathogenesis. Infect Immun 76:2189–2201

Cockrell DC, Beare PA, Fischer ER, Howe D, Heinzen RA (2008) A method for purifying obligate intracellular *Coxiella burnetii* that employs digitonin lysis of host cells. J Microbiol Methods 72:321–325

Coleman SA, Fischer ER, Howe D, Mead DJ, Heinzen RA (2004) Temporal analysis of *Coxiella burnetii* morphological differentiation. J Bacteriol 186:7344–7352

Coleman SA, Fischer ER, Cockrell DC, Voth DE, Howe D, Mead DJ, Samuel JE, Heinzen RA (2007) Proteome and antigen profiling of *Coxiella burnetii* developmental forms. Infect Immun 75:290–298

Davies JA, Harrison JJ, Marques LL, Foglia GR, Stremick CA, Storey DG, Turner RJ, Olson ME, Ceri H (2007) The GacS sensor kinase controls phenotypic reversion of small colony variants isolated from biofilms of *Pseudomonas aeruginosa* PA14. FEMS Microbiol Ecol 59:32–46

Davis GE, Cox HR (1938) A filter-passing infectious agent isolated from ticks. I. Isolation from *Dermacentor andersoni*, reactions in animals, and filtration experiments. Public Health Rep 53:2259–2276

Griffiths-Jones S, Bateman A, Marshall M, Khanna A, Eddy SR (2003) Rfam: an RNA family database. Nucleic Acids Res 31:439–441

Hackstadt T, Williams JC (1981) Biochemical stratagem for obligate parasitism of eukaryotic cells by *Coxiella burnetii*. Proc Natl Acad Sci U S A 78:3240–3244

Hackstadt T, Baehr W, Ying Y (1991) *Chlamydia trachomatis* developmentally regulated protein is homologous to eukaryotic histone H1. Proc Natl Acad Sci U S A 88:3937–3941

Heeb S, Haas D (2001) Regulatory roles of the GacS/GacA two-component system in plant-associated and other gram-negative bacteria. Mol Plant Microbe Interact 14:1351–1363

Heinzen RA (1997) Intracellular development of *Coxiella burnetii*. In: Anderson B, Friedman H, Bendinelli M (eds) Rickettsial infection and immunity. Plenum Press, New York, pp 99–129

Heinzen RA, Hackstadt T (1996) A developmental stage-specific histone H1 homolog of *Coxiella burnetii*. J Bacteriol 178:5049–5052

Heinzen RA, Howe D, Mallavia LP, Rockey DD, Hackstadt T (1996a) Developmentally regulated synthesis of an unusually small, basic peptide by *Coxiella burnetii*. Mol Microbiol 22:9–19

Heinzen RA, Scidmore MA, Rockey DD, Hackstadt T (1996b) Differential interaction with endocytic and exocytic pathways distinguish parasitophorous vacuoles of *Coxiella burnetii* and *Chlamydia trachomatis*. Infect Immun 64:796–809

Heinzen RA, Hackstadt T, Samuel JE (1999) Developmental biology of *Coxiella burnetii*. Trends Microbiol 7:149–154

Henriques AO, Moran CP Jr (2007) Structure, assembly, and function of the spore surface layers. Annu Rev Microbiol 61:555–588

Hicks LD, Raghavan R, Battisti JM, Minnick MF (2010) A DNA-binding peroxiredoxin of *Coxiella burnetii* is involved in countering oxidative stress during exponential-phase growth. J Bacteriol 192:2077–2084

Hsu D, Shih LM, Zee YC (1994) Degradation of rRNA in *Salmonella* strains: a novel mechanism to regulate the concentrations of rRNA and ribosomes. J Bacteriol 176:4761–4765

Kanehisa M, Goto S, Kawashima S, Okuno Y, Hattori M (2004) The KEGG resource for deciphering the genome. Nucleic Acids Res 32(Database issue):D277–D280

Karp PD, Riley M, Saier M, Paulsen IT, Collado-Vides J, Paley SM, Pellegrini-Toole A, Bonavides C, Gama-Castro S (2002) The EcoCyc database. Nucleic Acids Res 30:56–58

Kvint K, Nachin L, Diez A, Nyström T (2003) The bacterial universal stress protein: function and regulation. Curr Opin Microbiol 6:140–145

McCaul TF, Williams JC (1981) Developmental cycle of *Coxiella burnetii*: structure and morphogenesis of vegetative and sporogenic differentiations. J Bacteriol 147:1063–1076

McCaul TF, Hackstadt T, Williams JC (1981) Ultrastructural and biological aspects of *Coxiella burnetii* under physical disruptions. In: Burgdorfer W, Anacker RL (eds) Rickettsiae and rickettsial diseases. Academic Press, New York, pp 267–280

McCaul TF, Banerjee-Bhatnagar N, Williams JC (1991) Antigenic differences between *Coxiella burnetii* cells revealed by postembedding immunoelectron microscopy and immunoblotting. Infect Immun 59:3243–3253

Merhej V, Royer-Carenzi M, Pontarotti P, Raoult D (2009) Massive comparative genomic analysis reveals convergent evolution of specialized bacteria. Biol Direct 4:13

Molofsky AB, Swanson MS (2004) Differentiate to thrive: lessons from the *Legionella pneumophila* life cycle. Mol Microbiol 53:29–40

Moos A, Hackstadt T (1987) Comparative virulence of intra- and interstrain lipopolysaccharide variants of *Coxiella burnetii* in the guinea pig model. Infect Immun 55:1144–1150

Nachin L, Nannmark U, Nyström T (2005) Differential roles of the universal stress proteins of *Escherichia coli* in oxidative stress resistance, adhesion, and motility. J Bacteriol 187:6265–6272

Raghavan R, Miller SR, Hicks LD, Minnick MF (2007) The unusual 23S rRNA gene of *Coxiella burnetii*: two self-splicing group I introns flank a 34-base-pair exon, and one element lacks the canonical omegaG. J Bacteriol 189:6572–6579

Raghavan R, Hicks LD, Minnick MF (2008) Toxic introns and parasitic intein in *Coxiella burnetii*: legacies of a promiscuous past. J Bacteriol 190:5934–5943

Romano PS, Gutierrez MG, Berón W, Rabinovitch M, Colombo MI (2007) The autophagic pathway is actively modulated by phase II *Coxiella burnetii* to efficiently replicate in the host cell. Cell Microbiol 9:891–909

Samoilis G, Psaroulaki A, Vougas K, Tselentis Y, Tsiotis G (2007) Analysis of whole cell lysate from the intercellular bacterium *Coxiella burnetii* using two gel-based protein separation techniques. J Proteome Res 6:3032–3041

Seshadri R, Samuel JE (2001) Characterization of a stress-induced alternate sigma factor, RpoS, of *Coxiella burnetii* and its expression during the developmental cycle. Infect Immun 69:4874–4883

Seshadri R, Hendrix LR, Samuel JE (1999) Differential expression of translational elements by life cycle variants of *Coxiella burnetii*. Infect Immun 67:6026–6033

Seshadri R, Paulsen IT, Eisen JA, Read TD, Nelson KE, Nelson WC, Ward NL, Tettelin H, Davidsen TM, Beanan MJ, Deboy RT, Daugherty SC, Brinkac LM, Madupu R, Dodson RJ, Khouri HM, Lee KH, Carty HA, Scanlan D, Heinzen RA, Thompson HA, Samuel JE, Fraser CM, Heidelberg JF (2003) Complete genome sequence of the Q-fever pathogen *Coxiella burnetii*. Proc Natl Acad Sci U S A 100:5455–5460

Stevenson BS, Schmidt TM (1998) Growth rate-dependent accumulation of RNA from plasmid-borne rRNA operons in *Escherichia coli*. J Bacteriol 180:1970–1972

Traxler MF, Summers SM, Nguyen HT, Zacharia VM, Hightower GA, Smith JT, Conway T (2008) The global, ppGpp-mediated stringent response to amino acid starvation in *Escherichia coli*. Mol Microbiol 68:1128–1148

Varghees S, Kiss K, Frans G, Braha O, Samuel JE (2002) Cloning and porin activity of the major outer membrane protein P1 from *Coxiella burnetii*. Infect Immun 70:6741–6750

Vieira-Silva S, Rocha EP (2010) The systemic imprint of growth and its uses in ecological (meta) genomics. PLoS Genet 6:e1000808

Wiebe ME, Burton PR, Shankel DM (1972) Isolation and characterization of two cell types of *Coxiella burnetii* phase I. J Bacteriol 110:368–377

Zamboni DS, McGrath S, Rabinovitch M, Roy CR (2003) *Coxiella burnetii* express type IV secretion system proteins that function similarly to components of the *Legionella pneumophila* Dot/Icm system. Mol Microbiol 49:965–976

Zusman T, Yerushalmi G, Segal G (2003) Functional similarities between the icm/dot pathogenesis systems of *Coxiella burnetii* and *Legionella pneumophila*. Infect Immun 71:3714–3723

Zusman T, Aloni G, Halperin E, Kotzer H, Degtyar E, Feldman M, Segal G (2007) The response regulator PmrA is a major regulator of the *icm/dot* type IV secretion system in *Legionella pneumophila* and *Coxiella burnetii*. Mol Microbiol 63:1508–1523

Chapter 13
Genetic Manipulation of *Coxiella burnetii*

Paul A. Beare

Contents

Abstract Until very recently, *Coxiella burnetii* was viewed and studied as an obligate intracellular bacterium that relied exclusively on a eucaryotic host cell for growth. Other medically relevant obligate intracellular bacteria reside in the genera *Anaplasma*, *Chlamydia*, *Ehrlichia*, *Orientia*, and *Rickettsia*. An obligate intracellular

P.A. Beare (✉)
Coxiella Pathogenesis Section, Laboratory of Intracellular Parasites, Rocky Mountain Laboratories, National Institute of Allergy and Infectious Diseases, National Institutes of Health, 903 South 4th Street, Hamilton, MT 59840, USA
e-mail: pbeare@niaid.nih.gov

R. Toman et al. (eds.), *Coxiella burnetii: Recent Advances and New Perspectives in Research of the Q Fever Bacterium*, Advances in Experimental Medicine and Biology 984, DOI 10.1007/978-94-007-4315-1_13, © Springer Science+Business Media Dordrecht 2012

lifestyle presents a significant obstacle to genetic transformation. Procedures that are straightforward with free-living bacteria, such as antibiotic selection and cloning, can be very difficult when growth of transformants is restricted to a host cell. Long-term passage in host cells to expand small transformant populations can further complicate the procedure. Despite these and other obstacles, at least rudimentary systems are currently available for genetic transformation of most obligate intracellular bacterial pathogens. Dramatically aiding the development of new genetic methods for *C. burnetii* is the recent discovery of a medium that supports host cell-free growth of the organism in liquid, and importantly, on solid media as clonal colonies. The expanded *C. burnetii* genetics toolbox now includes transposon systems for random mutagenesis and single-copy, site-specific chromosomal gene knock-ins, as well as a shuttle vector for heterologous gene expression and *in trans* complementation. A reliable method of targeted gene inactivation remains a challenge. Advances in *C. burnetii* genetic manipulation will allow identification of genes essential for intracellular parasitism and disease pathogenesis, and undoubtedly fuel new interest in this minimally studied bacterial pathogen.

Keywords Transposon mutagenesis • Electroporation • Antibiotic selection • Allelic exchange • Genetic transformation • Virulence factors • Shuttle vector • Complementation • Targeted gene inactivation

13.1 Introduction

Replication of the human Q fever bacterium *Coxiella burnetii* in an acidic (~pH 4.8) and highly degradative phagolysosome-like compartment is a fascinating example of host cell parasitism (Howe et al. 2010). During the infectious cycle, developmental transitions occur in this hostile parasitophorous vacuole (PV) that generate replicative large cell variant (LCV) and non-replicative small cell variant (SCV) forms, with the SCV appearing specifically adapted for long term extracellular survival (Heinzen et al. 1999). *C. burnetii*'s pronounced extracellular stability, low aerosol infectious dose, and ability to cause a debilitating influenza-like illness (Maurin and Raoult 1999), have warranted its classification as a U.S. Centers for Disease Control and Prevention (CDC) category B select agent with potential for illegitimate use (Atlas 2003).

The historic genetic intractability of *C. burnetii* has severely limited molecular dissection of virulence factors associated with intracellular parasitism and pathogenic mechanisms that promote both acute and chronic Q fever. Many genes encoding potential *C. burnetii* virulence determinants were revealed by genome sequencing (Seshadri et al. 2003; Beare et al. 2009b). However, molecular Koch's postulates (Falkow 2004) were impossible to fulfill for these genes because methods for gene inactivation/complementation were missing. By necessity, *C. burnetii* gene function and regulation have been primarily examined by using heterologous expression in surrogate hosts (Voth et al. 2009; Hicks et al. 2010; Briggs et al. 2008).

The first report of *C. burnetii* genetic transformation was published in 1996 by Suhan et al. (1996). Knowledge gained from this success, and achievements in transforming other obligate intracellular bacteria, helped Beare et al. (2009a) transform *C. burnetii* again in 2009 using the *mariner*-based *Himar1* transposon (Tn). This chapter provides a general overview of technologies used to transform obligate intracellular bacteria. Obstacles encountered during development of the *C. burnetii* *Himar1* transformation system specific to host cell-dependent growth are discussed, as are recent advances in *C. burnetii* genetic manipulation enabled by axenic (host cell-free) growth.

13.2 First Genetic Transformation of an Obligate Intracellular Bacterium (*Chlamydia trachomatis*)

Although transient in nature, Tam et al. in 1994 (1994) were the first to genetically transform an obligate intracellular bacterium. *Chlamydia trachomatis* was transformed to chloramphenicol resistance by electroporation with pPBW100. This 10.6 kb plasmid is a chimeric construct comprised of the majority of an endogenous plasmid carried by *C. trachomatis* serovar E, and an *Escherichia coli* cloning vector with a ColE1 origin of replication. Inserted immediately downstream from a native chlamydial promoter was a chloramphenicol acetyltransferase (CAT) gene. Following electroporation of the infectious elementary body (EB) developmental form of chlamydiae and infection of McCoy cells, inclusions containing chloramphenicol-resistant chlamydia were detected. However, resistant organisms were lost following several tissue culture passes, possibly due to incompatibility of pPBW100 with the endogenous chlamydial plasmid. Nonetheless, the study provided proof-of-principle that an obligate intracellular bacterium could be genetically transformed using electroporation.

13.3 Genetic Transformation of *C. burnetii*

In what proved to be the first stable genetic transformation of an obligate intracellular bacterium, Suhan et al. (1996) in 1996 successfully transformed *C. burnetii* to ampicillin resistance. This was accomplished using a potential *C. burnetii-E. coli* shuttle vector termed pSKO(+)1000 that was generated by cloning a 5.8 kb *C. burnetii* autonomous replication sequence, *ars,* into the *E. coli* cloning vector pBluescript which contains a β-lactamase gene (*blaM*) conferring ampicillin resistance. The *ars* was previously identified as a *C. burnetii* chromosomal DNA fragment that, when circularized with a kanamycin cassette, stably replicated in *E. coli* as evidenced by kanamycin resistant bacteria (Suhan et al. 1994). Interestingly, *ars* does not appear to function as the *C. burnetii* chromosomal origin of replication (Suhan et al. 1994).

BHK-21 fibroblasts were infected with *C. burnetii* electroporated with pSKO(+)1000 and selection for ampicillin resistant organisms conducted for 2–3 months in cell culture. Southern blotting of transformants for *blaM* revealed both chromosomal integration and autonomous replication of pSKO(+)1000, with integration occurring via homologous recombination between the plasmid *ars* and the corresponding region in the *C. burnetii* chromosome (Suhan et al. 1996). However, Southern blot signal strength indicated about a seven-fold greater level of the pSKO(+)1000 integrant, suggesting the *ars* replicon in *C. burnetii* is unstable (Thompson and Suhan 1996). The pBluscript ColE1 replicon of pSKO(+)1000 was also likely non-functional as ampicillin-resistant organisms were not obtained following transformation with pBluescript alone. Cloning from the transformant pool by limiting dilution revealed a population of ampicillin-resistant organisms without *blaM*, indicating spontaneous mutation to ampicillin resistance had occurred during the lengthy selection procedure. Nonetheless, similar to initial results with chlamydia (Tam et al. 1994), this study represented a breakthrough in genetic transformation of *C. burnetii* by showing (1) electroporation as method to introduce DNA into *C. burnetii*, (2) ampicillin as a potential selectable marker, (3) the occurrence of homologous recombination, and (4) the possible utility of pSKO(+)1000 as a shuttle vector.

13.4 Success in Genetic Transformation of Other Obligate Intracellular Bacteria

Transformation of *C. trachomatis* and *C. burnetii* was followed by similar success in transformation of the obligate intracellular bacterial pathogens *Rickettsia prowazekii* (Driskell et al. 2009; Liu et al. 2007; Qin et al. 2004; Rachek et al. 1998, 2000), *R. montanensis* (Baldridge et al. 2010), *R. monacensis* (Baldridge et al. 2005, 2007), *R. typhi* (Troyer et al. 1999), *R. conorii* (Renesto et al. 2002), *R. rickettsii* (Kleba et al. 2010; Clark et al. 2011), *Anaplasma marginale* (Felsheim et al. 2010), *Anaplasma phagocytophilum* (Felsheim et al. 2006), and *Chlamydia psittaci* (Binet and Maurelli 2009).

Leading the way in transformation of *Rickettsia* spp. is the laboratory of David Wood at the University of South Alabama. Initial success was achieved in transforming *R. prowazekii* to rifampin or erythromycin resistance, thereby establishing two methods of positive selection of transformants (Rachek et al. 1998, 2000). A portion of the *rpoB* gene from an *R. prowazekii* strain containing a mutation conferring resistance to rifampin was cloned into a ColE1-based suicide plasmid. Following electroporation, a single crossover event occurred between the plasmid and chromosome at the *rpoB* locus to result in allelic exchange and genetic transformation to rifampin resistance (Rachek et al. 1998). A similar approach was employed to generate erythromycin resistant transformants using the *ereB* gene, but in this case, homologous recombination was mediated by a plasmid copy of the entire rickettsial *gltA* gene, encoding citrate synthase, to result in two chromosomal copies of *gltA* flanking heterologous *ereB* (Rachek et al. 2000).

The Wood lab then developed two systems for random Tn mutagenesis of *R. prowazekii* (Liu et al. 2007; Qin et al. 2004). First, an Epicentre EZ::Tn5 was constructed carrying *arr-2* that encodes an enzyme that inactivates rifampin by ADP-ribosylation. Here, the codon content of *arr-2* was optimized for rickettsial codon usage and the gene cloned downstream from a strong rickettsial ribosomal subunit promoter to achieve optimal expression in *R. prowazekii.* On average, electroporation experiments using the transposome complex (i.e., Tn and purified transposase) yielded 1–2 unique insertions per experiment with no transformants generated in some experiments (Qin et al. 2004). Variability in generating active transposome complexes was speculated as a reason for low transformation efficiency (Qin et al. 2004). A *mariner*-based *Himar1* Tn system (Lampe et al. 1999) was then tested for efficacy in *R. prowazekii.* Two important reasons made this system potentially more favorable for transforming rickettsia. First, only expression of the *Himar1* transposase is required for transposition (Lampe et al. 1999). Second, the *R. prowazekii* chromosome with 71% A+T content has abundant *Himar1* TA dinucleotide insertion sites (Andersson et al. 1998; Lampe et al. 1999). In addition, *Himar1* had previously proven effective in transforming the obligate intracellular bacterium, *A. phagocytophilum* (Felsheim et al. 2006). The *R. prowazekii* system employed a single suicide plasmid (pMW1650) that encoded *Himar1* containing the *arr-2* cassette described above, a green fluorescent protein (GFP) gene under control of the rickettsial *ompA* promoter (Baldridge et al. 2005), and a ColE1 origin of replication for rescue cloning. Synthesis of pMW1650 transposase was directed by the *Borrelia burdorferi* flagellin promoter (Stewart et al. 2004). Electroporation with pMW1650 generated brightly fluorescent rickettsia and numerous transposition events (Liu et al. 2007).

Transformation studies by laboratory of Ted Hackstadt using pMW1650 have recently provided novel insight into virulence mechanisms of *R. rickettsii* (Kleba et al. 2010; Clark et al. 2011). In one study, a small plaque-forming mutant of *R. rickettsii* was isolated with a Tn insertion in *sca2,* which encodes an autotransporter protein. Characterization of the mutant revealed that Sca2 is involved in actin-based motility and is required for full virulence in guinea pigs (Kleba et al. 2010). A second study used *Himar1* to complement *in trans* a defective *relA/spoT*-like gene carried by a lytic *R. rickettsii* strain that produces clear plaques (Clark et al. 2011). Complementation resulted in conversion to a "turbid" plaque phenotype, suggesting involvement of the *relA/spoT*-like protein in host cell lysis. The isogenic strains were equally virulent for guinea pigs, thereby resolving a running debate on whether plaque clarity is associated with virulence (Clark et al. 2011; Hackstadt 1996b).

The Wood lab also achieved the first and currently only success in targeted inactivation of a putative virulence gene encoded by an obligate intracellular bacterium. Electroporation with linear DNA containing the *R. prowazekii* phospholipase D gene (*pld*) disrupted with the *arr-2* cassette resulted in inactivation of the wild type gene by a double crossover event. The mutant showed attenuated virulence for guinea pigs (Driskell et al. 2009).

The EZ::Tn5 system has also proven effective in transforming two non-pathogenic rickettsia, i.e., *R. monacensis* and *R. montanensis*. In these cases, expression of Tn-encoded CAT and GFP genes was driven by a rickettsial outer membrane protein promoter. Only a single transformant clone was obtained for each organism following extensive tissue culture passage, indicating transformation frequencies were quite low.

Both *A. phagocytophilum* (Felsheim et al. 2006) and *A. marginale* (Felsheim et al. 2010) have been transformed using *Himar1*. These studies utilized separate suicide plasmids encoding the *Himar1* transposase on one vector and the Tn on the other. The *A. marginale tr* promoter (Barbet et al. 2005) drove expression of transposase, and a Tn-encoded operon of GFP and spectinomycin resistance genes. Co-electroporation with both plasmids resulted in several Tn integration events in *A. phagocytophilum*. However, only a single event was identified in *A. marginale* that resulted from integration of plasmid DNA by single crossover at the *tr* promoter region, and not transposition (Felsheim et al. 2006, 2010).

Finally, and 15 years after the first transformation of *Chlamydia* spp., Binet and Maurelli (Binet and Maurelli 2009) transformed *C. psittaci* to kasugamycin and spectinomycin resistance. Electroporation DNA consisted of both linear and circular molecules containing an allele of the organism's16s rRNA with engineered mutations conferring antibiotic resistance. Electroporated EBs were subjected directly to a plaque assay using L929 mouse fibroblasts, and antibiotic resistant plaques picked and molecularly typed. Transformation occurred via allelic exchange mediated by a minimum of two crossover events, with circular DNA substrates giving higher transformation frequencies than linear substrates. Importantly, and as now demonstrated for *Coxiella, Rickettsia* and *Anaplasma* spp., the study showed that modified DNA could be incorporated into the chlamydial chromosome by homologous recombination.

13.5 Optimization of *C. burnetii* Genetic Transformation Using the *Himar1* Tn

Transformation of multiple obligate intracellular bacteria, including *C. burnetii*, proves that the substantial hurdles to genetic manipulation of these organisms can be overcome. Electroporation effectively introduces DNA into these bacteria and positive selection of transformants growing in cell culture is feasible using resistance to antibiotics. Moreover, homologous recombination occurs in all tested genera.

Previous success in transformation of obligates assisted Beare et al. (2009a) in development of a second genetic transformation system for *C. burnetii*. The system employed *Himar1* and the *C. burnetii* Nine Mile RSA439 (phase II, clone 4) strain. This strain is considered avirulent due to a large chromosomal deletion that eliminates multiple genes involved in lipopolysaccharide biosynthesis and is suitable for work at biosafety level-2 (Hackstadt 1996a). Other *C. burnetii* strains are biosafety level-3 organisms that are regulated by the CDC, Division of Select Agents and Toxins (DSAT) (Atlas 2003). Although *C. burnetii* RSA439 is avirulent in animal models of Q fever, interactions between this strain and human macrophages are indistinguishable from those of the virulent Nine Mile RSA493 (phase I) strain, which was used in the original *C. burnetii* transformation study (Suhan et al. 1996; Howe et al. 2010). Thus, RSA439 biologically recapitulates host cell interactions of virulent *C. burnetii* while at the same time providing an easier system for developing genetic tools.

As described above for *Anaplasma* spp., *C. burnetii* transformation with *Himar1* utilized two different suicide plasmids, encoding either transposon or transposase, as a delivery system. These plasmids contain a ColE1 *ori* because Suhan et al. (1996) demonstrated this replicon is unstable in *C. burnetii*. Synthesis of CAT and mCherry red fluorescent protein genes as an operon, and transposase alone, were all driven by the small heat shock protein Hsp20 (CBU1169) promoter (*1169*^{P}). *C. burnetii* co-electroporated with the *Himar1* suicide plasmids was used to infect Vero cells. After passage of organisms five times in fresh Vero cells every 1–2 weeks, PV harboring chloramphenicol-resistant bacteria were clearly visible by phase contrast light microscopy. Thirty-five unique Tn insertion sites were identified by rescue cloning of the Tn-encoded ColE1 *ori*. Micromanipulation was used to clone a transformant designated B2c with a Tn insertion in the cell division gene *ftsZ* (CBU0141) (Beare et al. 2007, 2009a). The FtsZ mutant had a slower overall growth rate than wild type *C. burnetii* that correlated with formation of filamentous forms having incomplete division septae.

13.5.1 Antibiotic Selection

Himar1 mutagenesis yielded the first *C. burnetii* clone with a defined gene mutation (*ftsZ*) generated by genetic transformation. This accomplishment required optimization of several steps of the transformation procedure. First, and critically important, was establishment of an effective means of positive selection of transformants using resistance to antibiotics. Ampicillin resistance was used in the first genetic transformation of *C. burnetii* (Suhan et al. 1996). However, complicating the system was emergence of significant numbers of background ampicillin resistant, non-transformed bacteria that spontaneously developed following long-term selection in cell culture. Another problem with β-lactam antibiotics is that treated, non-transformed *C. burnetii* can still synthesize protein that direct biosynthesis of the pathogen's characteristic large and spacious PV (Howe et al. 2003). This makes screening of transformed bacteria by PV formation ambiguous. Another issue unique to *C. burnetii* is that many antibiotics show reduced activities in the acidic confines of the PV (Maurin et al. 1992). Further reducing the set of antibiotics for potential use in transformation of *C. burnetii* are restrictions based on therapeutic efficacy in treatment of human infections. In the US, the CDC, DSAT, ultimately approves the use of antibiotic resistance genes in *C. burnetii* transformation studies due to its select agent status. Based on these criteria, our lab has received approval to use genes conferring resistance to ampicillin, kanamycin, and chloramphenicol for *C. burnetii* genetic transformation studies.

Resistance to a relatively low concentration (5 μg/ml in tissue culture media) of chloramphenicol is an effective means of selecting *C. burnetii* genetic transformants (Beare et al. 2009a). Spontaneous development of non-transformed, resistant mutants was not detected, nor was host cell toxicity, which can be a problem with higher concentrations of chloramphenicol that can inhibit mitochondrial function (Li et al. 2010). As an alternative selection scheme, resistance to kanamycin was evaluated. A minimal inhibitory concentration (MIC) of 250 μg/ml of kanamycin was determined for *C. burnetii* in a Vero cell infection model and genetic transformants have been generated using a Kan resistance marker. However, unlike the situation with chloramphenicol, kanamycin resistant non-transformed *C. burnetii* spontaneously arise at low frequency (P. A. Beare and R. A. Heinzen, unpublished data). However, cloning procedures (discussed below) can derive pure transformants from mixed populations containing resistant, but non-transformed bacteria. Also, because chloramphenicol is bacteriostatic, a significant number of inhibited but viable non-transformed, or "carryover", bacteria are present in cultures following electroporation. These are eventually diluted over passage, but again, depending on the experiment, cloning may be desirable to ensure a clonal transformant. The establishment of CAT and Kan as alternative selectable markers is an important development that allows two transformation procedures in a single strain.

13.5.2 Construct Optimization

Constructs used in transformation generally have antibiotic resistance genes and other screenable markers, such as genes encoding fluorescent proteins, under control of a characterized pathogen promoter. The promoter *1169*^{P} was selected to drive synthesis of CAT, mCherry red fluorescent protein and transposase genes in the *C. burnetii Himar1* system because transcription microarray data indicated CBU1169 is constitutively expressed at a moderate level throughout the pathogen's infectious cycle (P. A. Beare and R. A. Heinzen, unpublished data). *C. burnetii* undergoes a biphasic developmental cycle inside host cells (Coleman et al. 2004), so promoters of developmentally regulated genes should be avoided in transformation constructs. Chemically synthesized genes that more closely match the codon usage of *Rickettsia* spp. have been used to optimize expression of heterologous genes in this pathogen (Qin et al. 2004). However, unlike rickettsia, with their unusual genomic 29% G+C content, the *C. burnetii* genome is 43% G+C with no obvious codon biases that may negatively impact expression of foreign genes.

13.5.3 Electroporation Parameters

The presence of multiple eucaryotic-like genes in the *C. burnetii* genome suggests ancestral acquisition of DNA by horizontal transfer (Beare et al. 2009b; Seshadri et al. 2003). Naturally competent bacteria acquire DNA through direct uptake of exogenous DNA and integration into the bacteria's genetic repertoire. This mechanism of genetic exchange occurs in many bacteria, including *C. burnetii's* close relative *Legionella pneumophila* (Charpentier et al. 2011), and is exploited in laboratory transformation procedures (Silva Ode and Blokesch 2010). However, *C. burnetii* ORFs involved in a natural competence, i.e., *comA* (CBU0855), *comF* (CBU0464) and *comE* (CBU0532), contain frameshift mutations, suggesting the organism is no longer naturally competent (Beare et al. 2009b; Seshadri et al. 2003). Fortunately, electroporation is effective in introducing DNA into *C. burnetii* (Beare et al. 2009a; Suhan et al. 1996).

Prior to electroporation, obligate intracellular bacteria must be purified from host cells, a process that typically involves several centrifugation steps (Williams et al. 1981; Shannon and Heinzen 2007). While tedious and time consuming, this step is critically important in providing purified bacteria, free of host cell contaminants, that can be suspended at high concentration (approx. 10^{10} bacteria per ml) in a suitable low ionic strength buffer. In the initial transformation of *C. burnetii,* a buffer consisting of 272 mM sucrose and 10% glycerol as osmoprotectants was employed. Electrical field strengths of 12.5 or 16 kV/cm for 6, 12, or 24 ms were tested along with 0.5 μg of pSKO(+)1000. Relative to *E. coli,* higher field strengths are required for electroporation of *C. burnetii* due to its smaller size (≤1 μm or less). Transformation was most efficient when organisms were electroporated at 12.5 and 16 kV/cm for 12 ms (Suhan et al. 1996).

Using 272 mM sucrose/10% glycerol electroporation buffer, conditions for efficient electroporation of *C. burnetii* were further explored before *Himar1* transformation using a DNase I protection assay. *C. burnetii* was electroporated with 1 μg of the cloning vector pCR2.1 using field strengths of 5, 16 or 20 kV/cm, a resistance of 500 Ω, and a capacitance of 25 μF. Organisms were subsequently treated with DNase I to remove extracellular plasmid, thereby allowing DNA protected by incorporation into *C. burnetii* to be detected by Q-PCR. Mock and 5 kV/cm electroporation samples had similar levels of DNaseI resistant plasmid while ten-fold higher levels were observed with 16 kV/cm and 20 kV/cm samples (P. A. Beare and R. A. Heinzen, unpublished data). A similar strategy was used to confirm electroporation-induced uptake of DNA by *C. trachomatis* (Tam et al. 1994), *C. psittaci* (Binet and Maurelli 2009) and *R. prowazekii* (Rachek et al. 1998) in initial transformation experiments.

Another possibility taken into consideration was the potential that SCV and LCV developmental forms might be differentially permissive to transformation. Lacking information on this point, *C. burnetii* used in electroporation was harvested from host cells at 6 days post-infection, when vacuoles contain roughly equal number of SCV and LCV (Coleman et al. 2004).

13.5.4 Expansion of Transformants

Immediately following electroporation of *C. burnetii* with *Himar1*, bacterial suspensions were diluted with RPMI tissue culture medium and used to infect Vero cells. To allow expression of resistance markers by slow growing *C. burnetii*, antibiotic was not added to culture media until 22 h post-infection, and media plus antibiotic changed weekly throughout the selection process. Because individual infected host cells can contain *Coxiella*-laden PV for weeks without pathogen release, expansion of transformants was facilitated by disrupting host cell monolayers every 1–2 weeks by sonication, then re-infecting fresh Vero cells with released *C. burnetii*. The expansion of *C. burnetii* transformants early in the selection process was monitored by Q-PCR of the *Himar1*-containing CAT sequence. The CAT signal increased over time relative to that of a chromosomal housekeeping gene, indicating transformants were expanding and non-transformed bacteria were being eliminated. After 5 weeks of selection, and a few tissue culture passages, abundant PV harboring transformed *C. burnetii* were obvious in monolayers by phase-contrast light microscopy and the presence of chromosomally-integrated *Himar1* sequences confirmed by Southern blotting. Moreover, fluorescence microscopy revealed red fluorescent *C. burnetii* expressing the mCherry transgene (Beare et al. 2009a).

13.5.5 Cloning of Transformants

Success in *C. burnetii Himar1* transformation highlighted the importance of an efficient method to clone transformants for phenotypic analyses. Cloning of *C. burnetii* by plaque formation is exceedingly difficult owing to the pathogen's nonlytic infectious process. In fact, only two *C. burnetii* clones derived by this procedure are described in the literature (Williams et al. 1981). Clones of *C. burnetii* have also been obtained by the tedious and time-consuming method of end-point limiting dilution (Suhan et al. 1996).

Fortunately, a new method of cloning *C. burnetii* from infected cells using micromanipulation has been developed (Beare et al. 2007). *C. burnetii* infections conducted at a low multiplicity of infection (e. g., 0.1) generate PV easily visible by phase contrast light microscopy (Beare et al. 2007). These PV contain clonal populations and are therefore the biological equivalent of a very small bacterial colony. The micromanipulation method employs a micromanipulator to place a sterile, disposable glass micropipette with a 20 μm opening directly into an isolated PV. PV contents (i.e., *C. burnetii*) are then extracted using a hydraulically-driven microinjector that is operated in reverse to generate suction. PV contents are then placed in a sterile microfuge tube by operating the microinjector in the forward direction. Cloned *C. burnetii* are then diluted in tissue culture media and stored at −80°C or expanded immediately by infecting fresh cell culture monolayers. Prior to expansion, an aliquot of harvested *C. burnetii* can be genotyped by PCR and DNA sequencing to identify interesting clones (Beare et al. 2006). Because PV typically contain a just few hundred organisms, a 4 week expansion in tissue culture is required to generate enough bacteria for purification. While the method is somewhat technical, a skilled worker can pick one vacuole roughly every 3 min, thereby generating dozens of clones in a few hours (Beare et al. 2007). Indeed, as described above, the procedure yielded the first clone of *C. burnetii* mutated by genetic transformation (Beare et al. 2009a).

13.6 Development of New *C. burnetii* Genetic Tools Aided by Axenic Culture

Adaptation of the *Himar1* system to random mutagenesis of *C. burnetii* was a significant technological advance that occurred despite the experimental obstacles intrinsic to a host cell-based transformation protocol. Using host cells, 8–12 weeks are required for clonal isolation and expansion of transformants, severely slowing the rate at which new technologies can be tested. Moreover, obligatism eliminates transformants that cannot infect and subsequently grow in host cells, thereby precluding identification of many genes required for virulence. Finally, cloning, while doable, is technically challenging.

These and other hurdles have now been overcome by the recent description of a method for axenic (host cell-free) propagation of *C. burnetii* (Omsland et al. 2009, 2011). A growth medium called Acidified Citrate Cysteine Medium (ACCM) was developed that supports roughly 3 logs ($\log_{10}$) of growth of *C. burnetii* over 6 d in a 2.5% oxygen environment. *C. burnetii* also plates with high efficiency in semi-solid ACCM agarose to form minute colonies, an important advance that dramatically facilitates cloning (Omsland et al. 2011). Significant advantages of using ACCM versus Vero host cells in *C. burnetii* genetic transformation studies include expanded clonal populations of transformants in 16 days instead of 2–3 months, recovery of transformants incapable of intracellular growth, and establishment of transformation frequencies.

13.6.1 An Improved Himar1 System

ACCM culture techniques were used to develop an optimized *Himar1*-based Tn mutagenesis system for *C. burnetii*. In the original *Himar1* system, *1169*^{P} was used to drive expression of CAT and mCherry red fluorescent protein genes as a single transcriptional unit (Beare et al. 2011). Expression was sufficient to confer resistance to 5 μg/ml of chloramphenicol, but mCherry fluorescence was low. Therefore, a new *Himar1* transposon plasmid (pITR-CAT:: *311*^{P}-MC) was constructed that employed the CBU0311 (outer membrane porin P1) promoter (*311*^{P}) to drive expression of the mCherry red fluorescent protein gene independent of *1169*^{P}-driven CAT (Beare et al. 2011). CBU0311 is constitutively expressed at an approximately six-fold higher level than CBU1169 (P. A. Beare and R. A. Heinzen, unpublished data) (Fig. 13.1a).

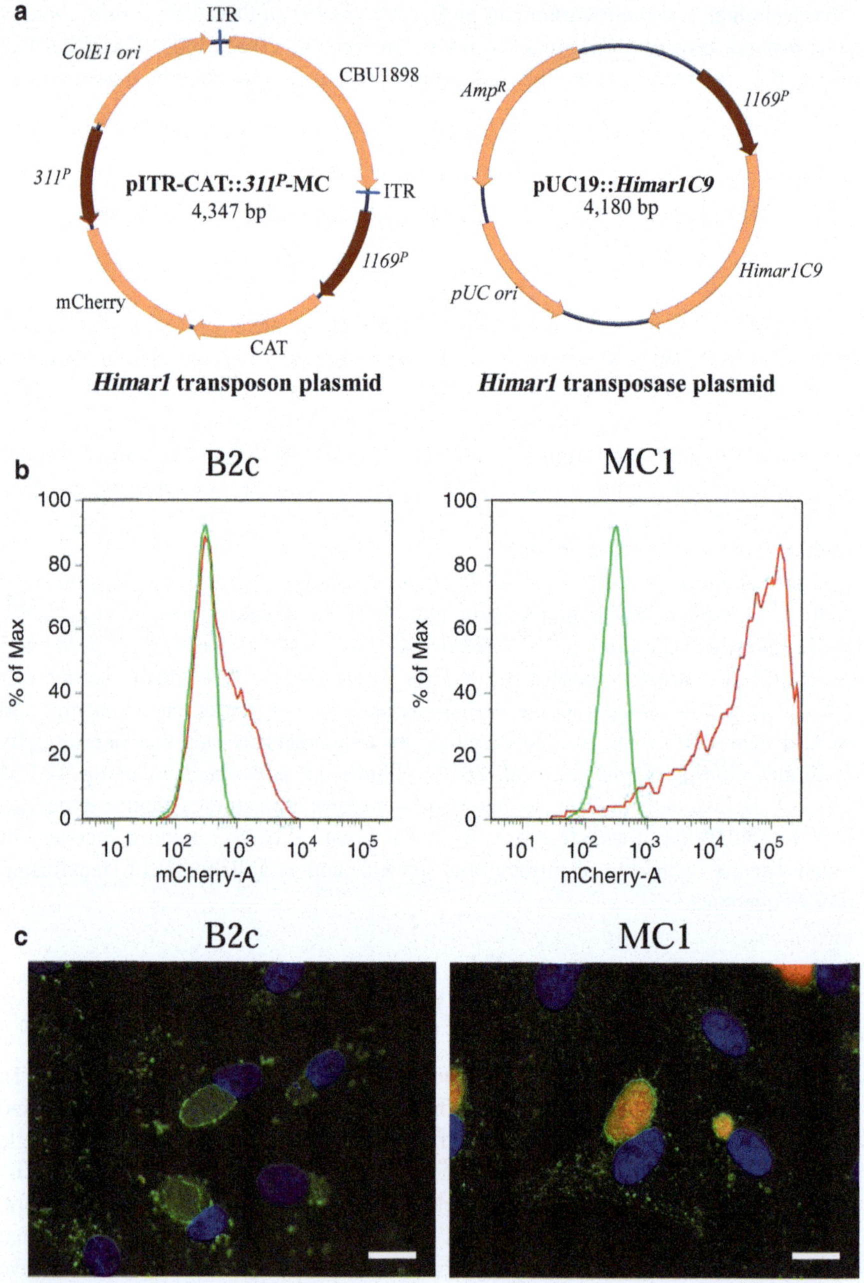

Fig. 13.1 A two plasmid *C. burnetii Himar1* transposon system. (**a**) Plasmid maps of transposon suicide plasmid pITR-CAT::*311*^{P}-MC and the transposase suicide plasmid pUC19::*Himar1*C9. The plasmid pITR-CAT::*311*^{P}-MC contains the *Himar1* transposon with integral mCherry red fluorescent protein and CAT genes under control of the CBU0311 (outer membrane porin P1) (*311*^{P}) and CBU1169 (*hsp20*) (*1169*^{P}) promoters, respectively. Also within the *Himar1* inverted

Testing of new *Himar1* constructs was conducted entirely in the absence of *C. burnetii* host cell propagation. Bacteria for electroporation were cultured for 6 days in ACCM, pelleted and washed once in 10% glycerol, then suspended in 10% glycerol at a cell density of approximately 4×10^9 organisms/ml. Ten micrograms of each *Himar1* Tn and transposase plasmid was added to 50 μl of *C. burnetii* suspension (2×10^8 organisms) and the mixture placed in a 0.1 cm electroporation cuvette. Cells were electroporated at 18 kV, 500 Ω, 25 μF for 7 –13 ms which results in approximately 50% killing (P. A. Beare and R. A. Heinzen, unpublished data). Following electroporation, *C. burnetii* was diluted in RPMI cell culture medium and added to ACCM. Cultures were incubated 24 h prior to addition of antibiotic. The MIC of kanamycin in ACCM is higher than in cell culture media (375 μg/ml versus 250 μg/ml) while the MIC of chloramphenicol is lower (3 μg/ml versus 5 μg/ml). ACCM cultures were incubated an additional 3 days to allow antibiotic selection of transformants, then organisms plated in semi-solid ACCM agarose containing antibiotic. Plates were incubated 6 days to allow colony formation, then agarose plugs containing individual colonies picked and added to 1.0 ml of ACCM in a sterile 1.5 ml microfuge tube. Organisms were released from the agarose by repeated pipetting, then added to ACCM which was incubated 6 days to allow expansion of the clone. A transformant termed MC1 containing an intergenic *Himar1* insertion between CBU0316 and CBU0317 was further characterized. MC1 has no growth defect in Vero cells relative to wild-type *C. burnetii.* Furthermore, by both fluorescence microscopy and flow cytometry, MC1 exhibits significantly higher mCherry red protein fluorescence than the previously described B2c clone, where the mCherry gene is the second gene in an operon with *cat* and expression is driven by 1169^P (Beare et al. 2009a, 2011) (Fig. 13.1b, c). *C. burnetii* MC1 is a valuable strain to study pathogen-host cell interactions using fluorescence microscopy and other fluorescence-based techniques.

Himar1 can also be employed to complement, *in cis*, naturally occurring mutants of *C. burnetii*. This application was recently employed by Mertens et al. (2010) who complemented a defective catalase gene (*katE*) of the Nine Mile isolate with the

terminal repeats (ITR) is a ColE1 origin of replication that allows rescue cloning of the Tn in *E. coli*. In pUC19::*Himar1*C9, 1169^P drives expression of the C9 variant of the *Himar1* transposase. (**b**) Flow cytometry of live Vero cells infected for 5 days with the *Himar1* B2c Nine Mile (phase II) transformant, generated as previously described (Beare et al. 2009a), or the *Himar1* MC1 Nine Mile (phase II) transformant, generated using the plasmids in panel A. The red trace in flow cytometry histograms shows mCherry red fluorescent protein fluorescence of Vero cells infected with *C. burnetii* transformants for 5 days. The green trace shows autofluorescence of Vero cells infected with wild type *C. burnetii* for 5 days. (**c**) Confocal fluorescence microscopy of Vero cells infected for 5 days with the *Himar1* B2c or the MC1 transformant (red). Cells were fixed with 4% paraformaldehyde, then immunostained for the lysosomal protein CD63 (*green*). Cell nuclei (*blue*) are stained with DAPI. Images of B2c and MC1-infected cells were captured for 500 and 100 ms, respectively (bar, 5 μm). Both flow cytometry and immunofluorescence microscopy reveal considerably more mCherry red protein fluorescence from *C. burnetii* MC1, where expression of mCherry is driven independently by 311^P, then from *C. burnetii* B2c, where expression of mCherry is driven by 1169^P as the second gene in an operon with CAT (Beare et al. 2009a) wild type gene from the Dugway 5J108-111 isolate. With both CAT and Kan available as selectable markers in *C. burnetii*, a Tn mutant generated with a *Himar1* construct expressing one maker could conceivably be complemented *in cis* with a *Himar1* construct expressing the other marker. This strategy was recently employed by Ristow et al. (2007) to define the virulence function of an OmpA-like protein of *Leptospira interrogans*. Libraries of *C. burnetii Himar1* mutants should be attainable to allow wholesale testing of individual mutant clones for defects in macrophage invasion and/or replication. Saturation-level Tn mutagenesis would further permit identification of the global gene repertoire required for animal infection by in vivo expression (Rediers et al. 2005), signature-tagged mutagenesis (Hensel et al. 1995), or deep sequencing technologies (Gawronski et al. 2009). Irrespective of application, that ability to obtain *C. burnetii Himar1* Tn clones in about 2 weeks clearly illustrates the power of axenic culture in development of new genetic tools. Indeed, as discussed below, subsequent development of a *C. burnetii* shuttle vector and a Tn7 transformation system was accelerated by using ACCM.

13.6.2 RSF1010 ori Based Shuttle Vector

An important recent discovery aided by ACCM culture is that vectors derived from the IncQ plasmid RSF1010 autonomously replicate in *C. burnetii* (Chen et al. 2010; Voth et al. 2011). Shuttle vectors containing the RSF1010 *ori* appear compatible with the endogenous *C. burnetii* plasmid (Voth et al. 2011; Chen et al. 2010) and have a copy number of roughly 3–6 (P. A. Beare and R. A. Heinzen, unpublished data). This new genetic tool was used in construction of reporter plasmids that allow identification of proteins secreted by *C. burnetii* into the host cell cytosol, presumably via the activity of the organism's specialized Dot/Icm type IV secretion system (T4SS). The secretion assay detects cytosolic delivery of β-lactamase (BlaM) C-terminally fused in-frame to a *C. burnetii* protein containing a translocation signal. The detection scheme is based on BlaM cleavage of the β-lactam ring of a cell-loaded fluorescent compound, CCF4/AM. Cleavage results in a change in cytosolic CCF4/AM fluorescence from green to blue (Campbell 2004; de Felipe et al. 2008). Independently, Chen et al. (2010) and Voth et al. (2011) constructed the RSF1010 *ori* BlaM plasmids pCBTEM and pJB-CAT-BlaM (Fig. 13.2a), respectively. In pCBTEM, *blaM* expression is under control of IPTG-inducible tac^{P} with the vector also containing mCherry red fluorescent protein and CAT genes under control of *C. burnetii groES* (CBU1719) and *com1* (CBU1910) promoters, respectively. In pJB-CAT-BlaM, *blaM* and CAT are driven independently by constitutive 1169^{P}. Use of each vector verified cytoplasmic translocation of a unique set of 6 suspected *C. burnetii* secreted proteins (Chen et al. 2010; Voth et al. 2011). In particular, Voth et al. (2011) demonstrated that the cryptic QpH1 plasmid of *C. burnetii* encodes six secreted proteins, including CBUA0015 (Fig. 13.2b).

The backbone of pJB-CAT-BlaM (i.e., pJB-CAT) has been further modified to contain a cloning site that allows 1169^{P}-directed synthesis of proteins with N- or C-terminal epitope tags of 2xHA or 3xFLAG (Fig. 13.2c). A version encoding

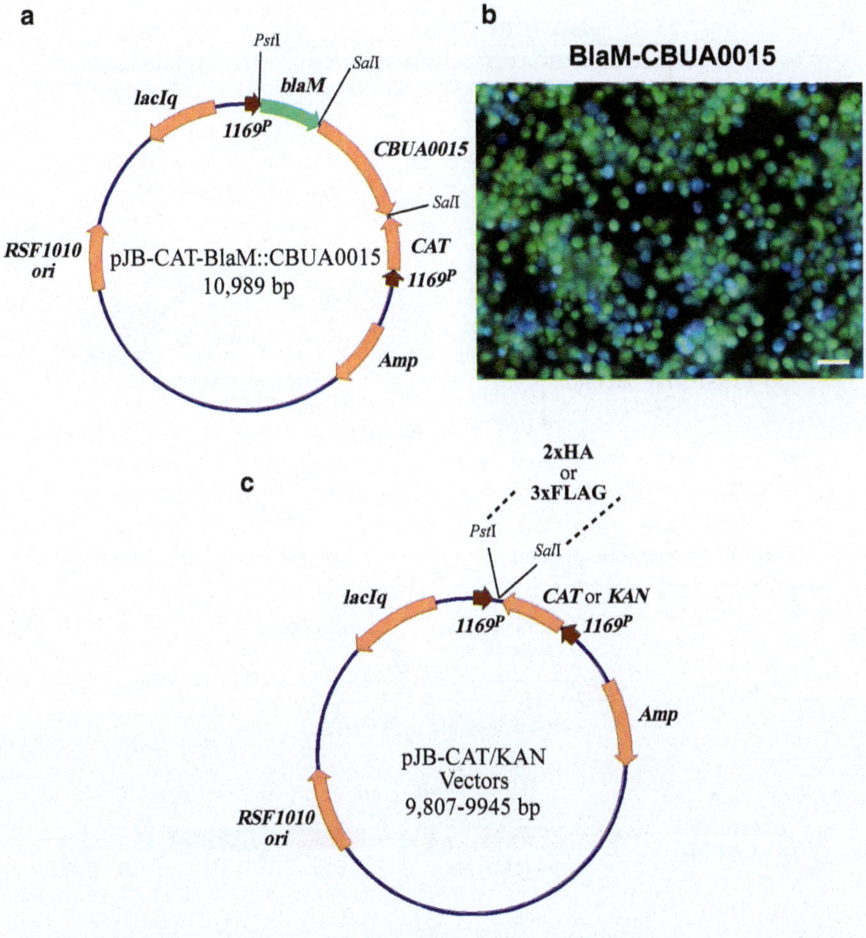

Fig. 13.2 RSF1010 *ori*-based fusion protein vectors. (**a**) Plasmid map of pJB-CAT-BlaM::CBUA0015. Genes encoding suspected type IV secretion substrates, such as CBUA0015, are cloned downstream and in-frame with *blaM* using the unique SalI restriction site. Fusion protein expression is under of control of the constitutive promoter, *1169*^{P}. (**b**) BlaM translocation assay showing cytosolic delivery of BlaM-CBUA0015. THP-1 cells were infected with *C. burnetii* Nine Mile (phase II) transformed with pJB-CAT-BlaM::CBUA0015. At 48 h post-infection, cells were incubated for 1 h with the fluorescent substrate CCF4/AM. Cytosolic delivery of BlaM::CBUA0015 results in BlaM cleavage of CCF4/AM and the generation of blue fluorescent cells. Green fluorescent cell are uninfected (bar, 30 μm). (**c**) Composite plasmid map of 4 different pJB-CAT and pJB-KAN vectors. Individual plasmids allow N- or C-terminal in-frame addition of the epitope tags 2xHA or 3xFLAG by cloning into unique SalI or PstI restriction sites

kanamycin instead of chloramphenicol resistance (pJB-KAN) has also been constructed (Fig. 13.2c). Applications of these vectors include expression of dominant/negative proteins for functional studies, expression of epitope tagged proteins to assess protein binding partners and intracellular trafficking, and *in trans* complementation of mutants generated by Tn or other mutagenesis techniques to fulfill

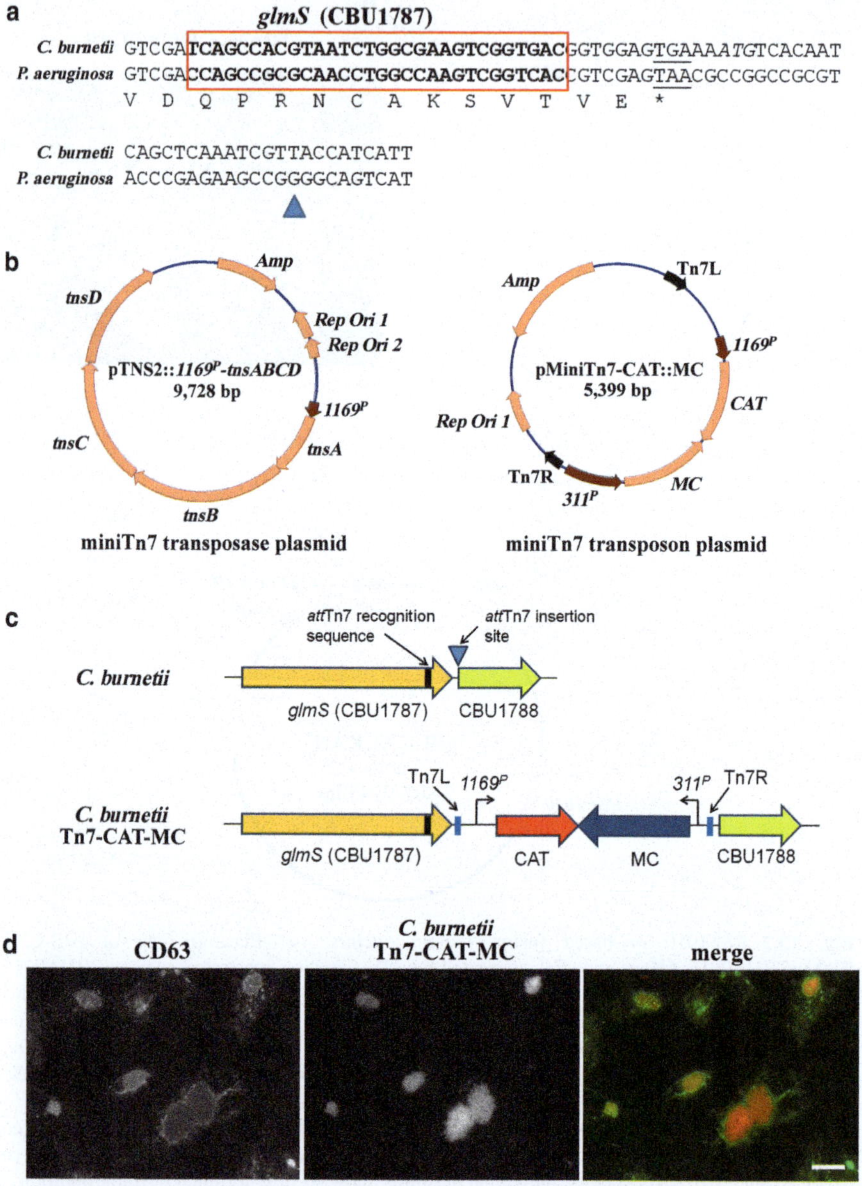

Fig. 13.3 A system for site-specific, single-copy gene integration in *C. burnetii* using Tn7. (**a**) Alignment of the *glmS* regions *Pseudomonas aeruginosa* and *C. burnetii*. The Tn7 transposase recognizes a specific 30 bp site termed *att*Tn7 that, in *P. aeruginosa* and other gram-negative bacteria, is located in the extreme 3′ end of *glmS* encoding glucosamine-6-phophosphate synthetase. The *C. burnetii glmS* (CBU1787) region shares 24 nucleotides with the *P. aeruginosa att*Tn7 site (*red box*). Tn7 transposition is orientation specific and occurs 36 bp downstream of the *att*Tn7 site in an intergenic region (*blue arrowhead*). (**b**) Maps of the two plasmids comprising the *C. burnetii* miniTn7 transposon system. The suicide plasmid pTNS2::*1169*^{P}-*tnsABCD* encodes the *tnsABCD*

molecular Koch's postulates (Falkow 2004). *C. burnetii* antibiotic resistant colony formation on ACCM-agarose following transformation with *E. coli* propagated pJB-CAT recently established a transformation frequency (transformants/protocol) of approximately 5×10^{-5} for *C. burnetii* (Omsland et al. 2011). If a significantly higher frequency is observed with *C. burnetii*-propagated plasmid, a pathogen restriction/modification system may be active that might be circumvented by propagating transformation constructs in a methylation defective *E. coli* strain (Binet and Maurelli 2009). However, review of the *C. burnetii* genome sequence reveals no identifiable restriction/modification systems.

13.6.3 Tn7

The establishment of pJB-CAT/KAN and other RSF1010 *ori*-based based shuttle vectors is a significant advance in *C. burnetii* genetic manipulation. However, a potential downside of their use is unwanted gene dosage effects due to multiple plasmid copies. Therefore, a system for site-specific single-copy *C. burnetii* gene knock-ins using Tn7 was developed with the aid of axenic culture (Beare et al. 2011). As discussed above, *Himar1* can also be used to chromosomally integrate transgenes (Beare et al. 2010; Clark et al. 2011; Mertens et al. 2010); however, its highly random nature of integration necessitates mapping and phenotyping of clones to ensure a Tn insertion has not generated an unintended phenotype.

The Tn7 transposase, encoded by *tnsABCD,* recognizes a specific 30 bp site termed *att*Tn7 that, in *Pseudomonas aeruginosa* and other gram-negative bacteria, is located in the extreme 3′ end of *glmS* encoding glucosamine-6-phophosphate synthetase (Choi et al. 2005). Transposition is orientation specific and occurs 36 bp downstream of *att*Tn7 in an intergenic region (Choi et al. 2005). Analysis of the 3′ end *C. burnetii glmS* (CBU1787) revealed a putative *att*Tn7 site (Fig. 13.3a). To test whether Tn7 can utilize this site, a two-suicide plasmid Tn7 system was constructed (Choi et al. 2005). The plasmid pTNS2::*1169^P-tnsABCD* carries the transposase genes *tnsABCD* under control of *1169^P*. The plasmid pMiniTn7T-CAT::MC carries Tn7 with CAT and mCherry red fluorescent protein genes driven independently by *1169^P* and *311^P*, respectively (Fig. 13.3b). Co-electroporation with both plasmids and selection of chloramphenicol resistant *C. burnetii* in ACCM yielded transformants with Tn7 inserted 36 bp downstream of the predicted *att*Tn7 site in an intergenic region between *glmS* (CBU1787) and CBU1788, which encodes a hypothetical

transposase operon driven by *1169^P*. The suicide plasmid pMiniTn7T-CAT::MC encodes CAT and mCherry red fluorescent protein genes under control of *311^P* and *1169^P*, respectively. The CAT/mCherry (MC) cassette is flanked by Tn7L and Tn7R elements. (**c**) Schematic of the *glmS* regions in *C. burnetii* Nine Mile (phase II) and *C. burnetii* Nine Mile (phase II)/Tn7-CAT-MC. Tn insertion occurred 36 bp downstream from the predicted *att*Tn7 site. (**d**) Confocal fluorescence microscopy of Vero cells infected for 5 days with *C. burnetii* Nine Mile (phase II)/Tn7-CAT-MC (*red*). Cells were fixed with 4% paraformaldehyde, then immunostained for the lysosomal protein CD63 (*green*). Bar, 5 μm

protein (Fig. 13.3c). The transformant exhibits bright mCherry red protein fluorescence in infected Vero cells at 5 days post infection (Fig. 13.3d) with no obvious growth defect relative to wild-type bacteria.

13.6.4 Targeted Gene Inactivation

A remaining challenge in *C. burnetii* genetic manipulation is development of reliable methods of targeted gene disruption. Allelic exchange generally relies on homologous recombination between a mutant and wild type allele. An important finding of Suhan et al. (1996) is that homologous recombination can occur between the 5.8 kb *ars* fragment in pSKO(+)1000 and the *C. burnetii* chromosome. Interestingly, *C. burnetii* genome sequencing indicates smaller DNA segments are sufficient substrate for homologous recombination. This is evidenced by the numerous *C. burnetii* chromosomal rearrangements mediated by recombination between insertion sequence (IS) elements of approximately 1.4 kb in size. In fact, the Tn plasmid used in initial *Himar1* mutagenesis of *C. burnetii* contained an *IS1111A* element (CBU1898; 1,092 bp) outside the Tn to mediate chromosomal integration of the suicide plasmid by a single crossover in the event of failed *Himar1* transposition (Beare et al. 2009a). This event was detected at a low frequency, proving roughly 1,000 bp of homologous sequence is sufficient for recombination in *C. burnetii* (P. A. Beare and R. A. Heinzen, unpublished data).

Two relatively simple strategies for targeted gene inactivation that would exploit *C. burnetii*'s recombination ability involve (1) a linear DNA fragment containing a target gene disrupted by an antibiotic resistance marker, and (2) a suicide plasmid containing a truncated internal gene fragment. Linear fragments require a low-frequency double crossover event for gene inactivation, making this approach difficult. More promising might be insertional duplication to result in target gene copies with 5′ and 3′ deletions. This can occur via single crossover recombination with a suicide plasmid carrying an internal target gene fragment (Vagner et al. 1998; Tao et al. 1992). A drawback of this system is that the integrated plasmid can spontaneously delete, resulting in reversion back to wild type. Continuous antibiotic selection dramatically lowers the probability of this event. Also, small genes may be difficult to inactivate due to suboptimal recombination substrate. A more sophisticated method of gene disruption exploits both single crossover plasmid integration and a counter-selectable marker. The first step involves chromosomal integration of a suicide plasmid carrying upstream and downstream regions of a target gene flanking a counter-selectable marker. In the next step, the "co-integrant" is resolved by a second recombination event between the plasmid-encoded flanking region and the reciprocal region in the chromosome, resulting in removal the wild type gene. A counter-selectable marker, such as *sacB* conferring sucrose sensitivity, allows selection of the second crossover event (Reyrat et al. 1998). In *C. burnetii*, *sacB*-mediated sucrose sensitivity may work as method of counter-selection as the organism is sensitive to 5% sucrose in ACCM (P. A. Beare and R. A. Heinzen, unpublished data).

Methods of targeted gene inactivation employing heterologous recombination functions should be tested for efficacy in *C. burnetii*. Success in prokaryotes has been achieved using the heterologous site-specific recombinases Flp and Cre from *Saccharomyces cerevisiae* and P1 phage, respectively. Flp promotes recombination between two 34 bp Flp recombinase target (*FRT*) sites (Schweizer 2003) whereas Cre promotes recombination between two 34 bp *loxP* sites (Song et al. 2009). Traditionally, these systems have been used to create conditional gene deletions in mice but are now commonly used in bacteria to generate unmarked mutations and large chromosomal deletions (Schweizer 2003; Song et al. 2009; Bestor et al. 2010; Kato and Hashimoto 2008). *FRT* or *LoxP* sites flanking a *C. burnetii* chromosomal region (gene) of interest would first be generated by homologous recombination, mediated by a single crossover event, with two suicide plasmids carrying these sites, and CAT or Kan resistance markers. Subsequent transformation with a suicide plasmid expressing the appropriate recombinase should then result in target excision. A recombineering approach using bacteriophage λ Red recombination functions should also be tested (Thomason et al. 2007). λ Red functions promote high efficiency in vivo homologous recombination of sequences with homologies as short as 40 bp. Site-directed gene disruption might be accomplished by co-transforming with a linear targeting sequence containing an integral antibiotic resistance gene and a suicide plasmid expressing the λ Red recombination enzymes. Finally, mobile group II introns (targetrons) should be evaluated for effectiveness in *C. burnetii*. This technique works with other fastidious bacteria (Rodriguez et al. 2008) and relies on retargeting of the intron to a target gene by a process referred to as "retro-homing" (Karberg et al. 2001).

13.7 Summary

Several milestones have been reached in developing genetic methods for obligate intracellular bacterial pathogens. However, owing largely to the recent discovery of an axenic growth system, the repertoire of *C. burnetii* genetic tools has now surpassed that of other obligates to include *Himar1* Tn mutagenesis, RSF1010 *ori*-based shuttle vectors, and a Tn7 system for single-copy site-specific gene delivery. A robust method of targeted gene inactivation in *C. burnetii* is a remaining hurdle. Once this impediment is overcome, traditional methods of site-specific gene inactivation and complementation to define gene function will be available. A system of inducible gene expression is also desirable for establishing the temporal requirement of a particular *C. burnetii* virulence determinant. Tightly regulated gene expression in intracellular *L. pneumophila* (Roy et al. 1998) and *Listeria monocytogenes* (Dancz et al. 2002) is attainable using IPTG induction of the *lac* repressor/operator system. IPTG gene induction works with ACCM grown *C. burnetii* and might similarly be effective with host cell grown organisms (Chen et al. 2010).

The past lack of methods to genetically manipulate *C. burnetii* has significantly impeded progress in understanding the genetic basis of the pathogen's unique lifestyle and virulence. Nonetheless, the increasing genetic tractability of *C. burnetii* will inevitably lead to novel insight into intracellular parasitism and disease pathogenesis, in addition to enabling development of new pathogen countermeasures, such as rationally designed attenuated or subunit vaccines. Ultimately, a paradigm shift should occur with respect to how *C. burnetii* is viewed and studied, *i.e.*, a tractable *facultative* intracellular parasite amenable to genetic manipulation.

Acknowledgements I would like to thank Robert Heinzen for critical reading of the manuscript. This research was supported by the Intramural Research Program of the National Institute of Allergy and Infectious Diseases.

References

Andersson SG, Zomorodipour A, Andersson JO, Sicheritz-Ponten T, Alsmark UC, Podowski RM, Naslund AK, Eriksson AS, Winkler HH, Kurland CG (1998) The genome sequence of *Rickettsia prowazekii* and the origin of mitochondria. Nature 396:133–140

Atlas RM (2003) Bioterrorism and biodefence research: changing the focus of microbiology. Nat Rev Microbiol 1:70–74

Baldridge GD, Burkhardt N, Herron MJ, Kurtti TJ, Munderloh UG (2005) Analysis of fluorescent protein expression in transformants of *Rickettsia monacensis*, an obligate intracellular tick symbiont. Appl Environ Microbiol 71:2095–2105

Baldridge GD, Burkhardt NY, Felsheim RF, Kurtti TJ, Munderloh UG (2007) Transposon insertion reveals pRM, a plasmid of *Rickettsia monacensis*. Appl Environ Microbiol 73: 4984–4995

Baldridge GD, Burkhardt NY, Oliva AS, Kurtti TJ, Munderloh UG (2010) Rickettsial *ompB* promoter regulated expression of GFPuv in transformed *Rickettsia montanensis*. PLoS One 5:e8965

Barbet AF, Agnes JT, Moreland AL, Lundgren AM, Alleman AR, Noh SM, Brayton KA, Munderloh UG, Palmer GH (2005) Identification of functional promoters in the *msp2* expression loci of *Anaplasma marginale* and *Anaplasma phagocytophilum*. Gene 353:89–97

Beare PA, Samuel JE, Howe D, Virtaneva K, Porcella SF, Heinzen RA (2006) Genetic diversity of the Q fever agent, *Coxiella burnetii*, assessed by microarray-based whole-genome comparisons. J Bacteriol 188:2309–2324

Beare PA, Howe D, Cockrell DC, Heinzen RA (2007) Efficient method of cloning the obligate intracellular bacterium *Coxiella burnetii*. Appl Environ Microbiol 73:4048–4054

Beare PA, Howe D, Cockrell DC, Omsland A, Hansen B, Heinzen RA (2009a) Characterization of a *Coxiella burnetii ftsZ* mutant generated by *Himar1* transposon mutagenesis. J Bacteriol 191:1369–1381

Beare PA, Unsworth N, Andoh M, Voth DE, Omsland A, Gilk SD, Williams KP, Sobral BW, Kupko JJ 3rd, Porcella SF, Samuel JE, Heinzen RA (2009b) Comparative genomics reveal extensive transposon-mediated genomic plasticity and diversity among potential effector proteins within the genus *Coxiella*. Infect Immun 77:642–656

Beare PA, Omsland A, Cockrell DC, Howe D, Heinzen RA (2010) Expansion of the *Coxiella burnetii* genetics tool box using axenic culture. Abstracts of the 24th Meeting of American Society of Rickettsiology, abstract 94 Stevenson, WA

Beare PA, Sandoz KM, Omsland A, Rockey DD, Heinzen RA (2011) Advances in genetic manipulation of obligate intracellular bacterial pathogens. Front Microbiol 2:97

Bestor A, Stewart PE, Jewett MW, Sarkar A, Tilly K, Rosa PA (2010) Use of the Cre-*lox* recombination system to investigate the lp54 gene requirement in the infectious cycle of *Borrelia burgdorferi*. Infect Immun 78:2397–2407

Binet R, Maurelli AT (2009) Transformation and isolation of allelic exchange mutants of *Chlamydia psittaci* using recombinant DNA introduced by electroporation. Proc Natl Acad Sci USA 106:292–297

Briggs HL, Pul N, Seshadri R, Wilson MJ, Tersteeg C, Russell-Lodrigue KE, Andoh M, Baumler AJ, Samuel JE (2008) Limited role for iron regulation in *Coxiella burnetii* pathogenesis. Infect Immun 76:2189–2201

Campbell RE (2004) Realization of beta-lactamase as a versatile fluorogenic reporter. Trends Biotechnol 22:208–211

Charpentier X, Kay E, Schneider D, Shuman HA (2011) Antibiotics and UV radiation induce competence for natural transformation in *Legionella pneumophila*. J Bacteriol 193:1114–1121

Chen C, Banga S, Mertens K, Weber MM, Gorbaslieva I, Tan Y, Luo ZQ, Samuel JE (2010) Large-scale identification and translocation of type IV secretion substrates by *Coxiella burnetii*. Proc Natl Acad Sci U S A 107:21755–21760

Choi KH, Gaynor JB, White KG, Lopez C, Bosio CM, Karkhoff-Schweizer RR, Schweizer HP (2005) A Tn7-based broad-range bacterial cloning and expression system. Nat Methods 2:443–448

Clark TR, Ellison DW, Kleba B, Hackstadt T (2011) Complementation of *Rickettsia rickettsii* RelA/SpoT restores a non-lytic plaque phenotype. Infect Immun 79:1631–1637

Coleman SA, Fischer ER, Howe D, Mead DJ, Heinzen RA (2004) Temporal analysis of *Coxiella burnetii* morphological differentiation. J Bacteriol 186:7344–7352

Dancz CE, Haraga A, Portnoy DA, Higgins DE (2002) Inducible control of virulence gene expression in *Listeria monocytogenes*: temporal requirement of listeriolysin O during intracellular infection. J Bacteriol 184:5935–5945

De Felipe KS, Glover RT, Charpentier X, Anderson OR, Reyes M, Pericone CD, Shuman HA (2008) *Legionella* eukaryotic-like type IV substrates interfere with organelle trafficking. PLoS Pathog 4:e1000117

Driskell LO, Yu XJ, Zhang L, Liu Y, Popov VL, Walker DH, Tucker AM, Wood DO (2009) Directed mutagenesis of the *Rickettsia prowazekii pld* gene encoding phospholipase D. Infect Immun 77:3244–3248

Falkow S (2004) Molecular Koch's postulates applied to bacterial pathogenicity–a personal recollection 15 years later. Nat Rev Microbiol 2:67–72

Felsheim RF, Herron MJ, Nelson CM, Burkhardt NY, Barbet AF, Kurtti TJ, Munderloh UG (2006) Transformation of *Anaplasma phagocytophilum*. BMC Biotechnol 6:42

Felsheim RF, Chavez AS, Palmer GH, Crosby L, Barbet AF, Kurtti TJ, Munderloh UG (2010) Transformation of *Anaplasma marginale*. Vet Parasitol 167(167–1):74

Gawronski JD, Wong SM, Giannoukos G, Ward DV, Akerley BJ (2009) Tracking insertion mutants within libraries by deep sequencing and a genome-wide screen for *Haemophilus* genes required in the lung. Proc Natl Acad Sci U S A 106:16422–16427

Hackstadt T (1996a) Biosafety concerns and *Coxiella burnetii*. Trends Microbiol 4:341–342

Hackstadt T (1996b) The biology of rickettsiae. Infect Agents Dis 5:127–143

Heinzen RA, Hackstadt T, Samuel JE (1999) Developmental biology of *Coxiella burnetii*. Trends Microbiol 7:149–154

Hensel M, Shea JE, Gleeson C, Jones MD, Dalton E, Holden DW (1995) Simultaneous identification of bacterial virulence genes by negative selection. Science 269:400–403

Hicks LD, Raghavan R, Battisti JM, Minnick MF (2010) A DNA-binding peroxiredoxin of *Coxiella burnetii* is involved in countering oxidative stress during exponential-phase growth. J Bacteriol 192:2077–2084

Howe D, Melnicakova J, Barak I, Heinzen RA (2003) Maturation of the *Coxiella burnetii* parasitophorous vacuole requires bacterial protein synthesis but not replication. Cell Microbiol 5:469–480

Howe D, Shannon JG, Winfree S, Dorward DW, Heinzen RA (2010) *Coxiella burnetii* phase I and II variants replicate with similar kinetics in degradative phagolysosome-like compartments of human macrophages. Infect Immun 78:3465–3474

Karberg M, Guo H, Zhong J, Coon R, Perutka J, Lambowitz AM (2001) Group II introns as controllable gene targeting vectors for genetic manipulation of bacteria. Nat Biotechnol 19:1162–1167

Kato J, Hashimoto M (2008) Construction of long chromosomal deletion mutants of *Escherichia coli* and minimization of the genome. Methods Mol Biol 416:279–293

Kleba B, Clark TR, Lutter EI, Ellison DW, Hackstadt T (2010) Disruption of the *Rickettsia rickettsii* Sca2 autotransporter inhibits actin-based motility. Infect Immun 78:2240–2247

Lampe DJ, Akerley BJ, Rubin EJ, Mekalanos JJ, Robertson HM (1999) Hyperactive transposase mutants of the *Himar1* mariner transposon. Proc Natl Acad Sci U S A 96:11428–11433

Li CH, Cheng YW, Liao PL, Yang YT, Kang JJ (2010) Chloramphenicol causes mitochondrial stress, decreases ATP biosynthesis, induces matrix metalloproteinase-13 expression, and solid-tumor cell invasion. Toxicol Sci 116:140–150

Liu ZM, Tucker AM, Driskell LO, Wood DO (2007) *Mariner*-based transposon mutagenesis of *Rickettsia prowazekii.* Appl Environ Microbiol 73:6644–6649

Maurin M, Raoult D (1999) Q fever. Clin Microbiol Rev 12:518–553

Maurin M, Benoliel AM, Bongrand P, Raoult D (1992) Phagolysosomal alkalinization and the bactericidal effect of antibiotics: the *Coxiella burnetii* paradigm. J Infect Dis 166:1097–1102

Mertens K, Berg J, Brennan BE, Chen C, Weber M, Hill J, Samuel JE (2010) Evolutionary loss of *katE* as balance mechanism for antioxidant gene expression in *Coxiella burnetii.* Abstracts of the 24th Meeting of American Society of Rickettsiology, abstract 49

Omsland A, Cockrell DC, Howe D, Fischer ER, Virtaneva K, Sturdevant DE, Porcella SF, Heinzen RA (2009) Host cell-free growth of the Q fever bacterium *Coxiella burnetii.* Proc Natl Acad Sci U S A 106:4430–4434

Omsland A, Beare PA, Hill J, Cockrell DC, Howe D, Hansen B, Samuel JE, Heinzen RA (2011) Isolation from animal tissue and genetic transformation of *Coxiella burnetii* facilitated by an improved axenic growth medium. Appl Environ Microbiol 77:3720–3725

Qin A, Tucker AM, Hines A, Wood DO (2004) Transposon mutagenesis of the obligate intracellular pathogen *Rickettsia prowazekii.* Appl Environ Microbiol 70:2816–2822

Rachek LI, Tucker AM, Winkler HH, Wood DO (1998) Transformation of *Rickettsia prowazekii* to rifampin resistance. J Bacteriol 180:2118–2124

Rachek LI, Hines A, Tucker AM, Winkler HH, Wood DO (2000) Transformation of *Rickettsia prowazekii* to erythromycin resistance encoded by the *Escherichia coli ereB* gene. J Bacteriol 182:3289–3291

Rediers H, Rainey PB, Vanderleyden J, De Mot R (2005) Unraveling the secret lives of bacteria: use of in vivo expression technology and differential fluorescence induction promoter traps as tools for exploring niche-specific gene expression. Microbiol Mol Biol Rev 69:217–261

Renesto P, Gouin E, Raoult D (2002) Expression of green fluorescent protein in *Rickettsia conorii.* Microb Pathog 33:17–21

Reyrat JM, Pelicic V, Gicquel B, Rappuoli R (1998) Counterselectable markers: untapped tools for bacterial genetics and pathogenesis. Infect Immun 66:4011–4017

Ristow P, Bourhy P, Da Cruz Mcbride FW, Figueira CP, Huerre M, Ave P, Girons IS, Ko AI, Picardeau M (2007) The OmpA-like protein Loa22 is essential for leptospiral virulence. PLoS Pathog 3:e97

Rodriguez SA, Yu JJ, Davis G, Arulanandam BP, Klose KE (2008) Targeted inactivation of *Francisella tularensis* genes by group II introns. Appl Environ Microbiol 74:2619–2626

Roy CR, Berger KH, Isberg RR (1998) *Legionella pneumophila* DotA protein is required for early phagosome trafficking decisions that occur within minutes of bacterial uptake. Mol Microbiol 28:663–674

Schweizer HP (2003) Applications of the *Saccharomyces cerevisiae* Flp-*FRT* system in bacterial genetics. J Mol Microbiol Biotechnol 5:67–77

Seshadri R, Paulsen IT, Eisen JA, Read TD, Nelson KE, Nelson WC, Ward NL, Tettelin H, Davidsen TM, Beanan MJ, Deboy RT, Daugherty SC, Brinkac LM, Madupu R, Dodson RJ, Khouri HM, Lee KH, Carty HA, Scanlan D, Heinzen RA, Thompson HA, Samuel JE, Fraser CM, Heidelberg JF (2003) Complete genome sequence of the Q-fever pathogen *Coxiella burnetii.* Proc Natl Acad Sci U S A 100:5455–5460

Shannon JG, Heinzen RA (2007) Infection of human monocyte-derived macrophages with *Coxiella burnetii.* Methods Mol Biol 431:189–200

Silva Ode S, Blokesch M (2010) Genetic manipulation of *Vibrio cholerae* by combining natural transformation with FLP recombination. Plasmid 64:186–195

Song H, Wolschendorf F, Niederweis M (2009) Construction of unmarked deletion mutants in mycobacteria. Methods Mol Biol 465:279–295

Stewart PE, Hoff J, Fischer E, Krum JG, Rosa PA (2004) Genome-wide transposon mutagenesis of *Borrelia burgdorferi* for identification of phenotypic mutants. Appl Environ Microbiol 70:5973–5979

Suhan M, Chen SY, Thompson HA, Hoover TA, Hill A, Williams JC (1994) Cloning and characterization of an autonomous replication sequence from *Coxiella burnetii*. J Bacteriol 176:5233–5243

Suhan ML, Chen SY, Thompson HA (1996) Transformation of *Coxiella burnetii* to ampicillin resistance. J Bacteriol 178:2701–2708

Tam JE, Davis CH, Wyrick PB (1994) Expression of recombinant DNA introduced into *Chlamydia trachomatis* by electroporation. Can J Microbiol 40:583–591

Tao L, Leblanc DJ, Ferretti JJ (1992) Novel streptococcal-integration shuttle vectors for gene cloning and inactivation. Gene 120:105–110

Thomason LC, Costantino N, Shaw DV, Court DL (2007) Multicopy plasmid modification with phage lambda Red recombineering. Plasmid 58:148–158

Thompson HA, Suhan ML (1996) Genetics of *Coxiella burnetii*. FEMS Microbiol Lett 145:139–146

Troyer JM, Radulovic S, Azad AF (1999) Green fluorescent protein as a marker in *Rickettsia typhi* transformation. Infect Immun 67:3308–3311

Vagner V, Dervyn E, Ehrlich SD (1998) A vector for systematic gene inactivation in *Bacillus subtilis*. Microbiology 144:3097–3104

Voth DE, Howe D, Beare PA, Vogel JP, Unsworth N, Samuel JE, Heinzen RA (2009) The *Coxiella burnetii* ankyrin repeat domain-containing protein family is heterogeneous, with C-terminal truncations that influence Dot/Icm-mediated secretion. J Bacteriol 191:4232–4242

Voth DE, Beare PA, Howe D, Sharma UM, Samoilis G, Cockrell DC, Omsland A, Heinzen RA (2011) The *Coxiella burnetii* cryptic plasmid is enriched in genes encoding type IV secretion system substrates. J Bacteriol 193:1493–1503

Williams JC, Peacock MG, Mccaul TF (1981) Immunological and biological characterization of *Coxiella burnetii*, phases I and II, separated from host components. Infect Immun 32:840–851

Chapter 14
Role of Innate and Adaptive Immunity in the Control of Q Fever

Christian Capo and Jean-Louis Mege

Contents

Abstract Acute Q fever is commonly resolved without an antibiotic regimen, but a primary infection may develop into a chronic infection in a minority of cases. *Coxiella burnetii*, the causative agent of Q fever, is known to infect macrophages both *in vitro* and *in vivo*. It has been observed that the intracellular survival of *C. burnetii* requires the subversion of the microbicidal properties of macrophages. Adaptive immunity is also essential to cure *C. burnetii* infection, as demonstrated by clinical studies and animal models. Indeed, the control of infection in patients with primary Q fever involves a systemic cell-mediated immune response and

C. Capo • J.-L. Mege (✉)
Unité de Recherche sur les Maladies Infectieuses Transmissibles et Emergentes,
CNRS-IRD UMR 6236, Institut Fédératif de Recherche 48, Faculté de Médecine,
Université de la Méditerranée, Marseille, France
e-mail: Jean-Louis.Mege@univmed.fr

R. Toman et al. (eds.), *Coxiella burnetii: Recent Advances and New Perspectives in Research of the Q Fever Bacterium*, Advances in Experimental Medicine and Biology 984,
DOI 10.1007/978-94-007-4315-1_14,

granuloma formation with an essential role for interferon-γ in the protection against *C. burnetii*. In contrast, chronic Q fever is characterized by defective cell-mediated immunity with the defective formation of granulomas and over-production of interleukin-10, an immunoregulatory cytokine. Finally, epidemiological data demonstrate that age and gender are risk factors for Q fever. The analysis of gene expression programs in mice reveals the importance of sex-related genes in *C. burnetii* infection because only 14% of the modulated genes are sex-independent, while the remaining 86% are differentially expressed in males and females. These results open a new field to understand how host metabolism controls *C. burnetii* infection in humans.

Keywords Age • Antibodies • *Coxiella burnetii* • Cytokines • Gender • Granulomas • Macrophages • Q fever • T lymphocytes

14.1 The Natural History of Q Fever

Q fever is caused by *Coxiella burnetii*, an obligate intracellular bacterium phylogenetically related to *Legionellae* species and *Francisella tularensis* (Weisburg et al. 1989). Q fever has a wide spectrum of clinical manifestations, including fever, pneumonia, hepatitis and, less frequently, neurological manifestations (Raoult et al. 2005). Naïve patients usually contract Q fever via aerosols and develop a primary infection (Fig. 14.1). More than half of patients will not exhibit any symptoms beyond a seroconversion, and only 2% will develop a severe disease that leads to their hospitalization. Acute Q fever almost always resolves without a specific antibiotic treatment. However, *C. burnetii* persists in patients who are apparently cured (Harris et al. 2000). In certain hosts, the primary infection may develop into a chronic infection (Raoult 1990). The latency between acute and chronic infection may last from months to years. Patients with valvular damage, patients with cancer (such as lymphomas) and pregnant women are at a high risk of a developing chronic infection (Fenollar et al. 2001). The main clinical manifestation of chronic Q fever is endocarditis (Brouqui et al. 1993), which represents 48% of the blood culture-negative cases of infective endocarditis (Houpikian and Raoult 2005). Q fever endocarditis, that may lead to death (Maurin and Raoult 1999), is characterized by fibrosis, calcification, slight inflammation and vascularization, small or absent vegetations, and a high level of antibodies specific for *C. burnetii* (Lepidi et al. 2003). Combination of doxycycline and chloroquine has changed the prognosis of the disease; now, less than 5% of patients experience a relapse after 18 months of treatment (Houpikian et al. 2002). There is a correlation between circulating concentrations of doxycycline and decreased titers of anti-*C. burnetii* antibodies (Rolain et al. 2003). Besides the two major clinical presentations of Q fever, two additional forms have recently been reported. The hyperinflammatory syndrome is associated with hepatitis and autoantibodies in middle-aged male patients; corticosteroids improve the patient outcome (Maurin and Raoult 1999). Persistent asthenia lasting for months or

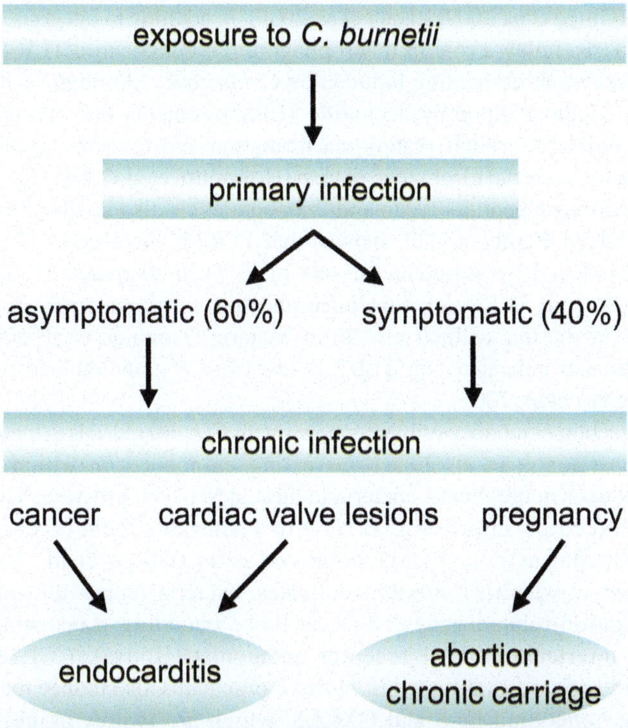

Fig. 14.1 The natural history of Q fever. The primary infection by *C. burnetii* is unapparent or symptomatic (acute Q fever). Months or years later, some individuals may develop a chronic infection that principally manifests as endocarditis

years has been reported in some patients with acute Q fever (Ayres et al. 1998). Finally, Q fever is deemed a category B biological terrorist agent (Madariaga et al. 2003). In this review, we will focus on the recent knowledge of innate and adaptive immune responses to *C. burnetii* infection.

14.2 Innate Immunity

14.2.1 Monocytes and Macrophages

Monocytes and macrophages are the major targets of *C. burnetii*. The intracellular survival of *C. burnetii* organisms requires the subversion of the microbicidal properties of these cells through mechanisms that are described in the chapters on the bacterial persistence and intracellular trafficking of *C. burnetii*. Briefly, *C. burnetii* interacts with monocytes and macrophages via αvβ3 integrin, induces strong remodeling of the actin cytoskeleton, and activates protein tyrosine kinase pathways and cytokine production (Ghigo et al. 2009; Raoult et al. 2005). In addition to integrins, the Toll-like receptors (TLRs) play a role in macrophage infection because TLR4 is involved in the cytoskeleton remodeling induced by *C. burnetii*. Although TLR4 (which is known to recognize lipopolysaccharide (LPS)) controls the immune response against *C. burnetii* through granuloma formation and cytokine production, it is dispensable for bacterial clearancze *in vivo* (Honstettre et al. 2004). TLR2 (which is known to recognize peptidoglycan and lipopeptides) is also involved in *C. burnetii* infection. Indeed, Zamboni et al. showed that TLR2 is involved in TNF and IFN-γ production induced by avirulent variants of *C. burnetii* (phase II organisms) in macrophages from TLR2-deficient mice and that these macrophages are highly permissive for the intracellular growth of bacteria (Zamboni et al. 2004). *In vivo* experiments also indicated that TLR2 is involved in granuloma formation, as is TLR4 (Meghari et al. 2005).

The interaction of monocytes/macrophages with *C. burnetii* leads to their functional polarization (Benoit et al. 2008). In resting monocytes in which *C. burnetii* survives but does not replicate, bacteria induce an M1-type program. This program, usually induced by interferon (IFN)-γ or bacterial products, contributes to the microbicidal activity of M1-polarized cells (Ghigo et al. 2004). In the macrophages in which *C. burnetii* replicates, bacteria induce the expression of M2 polarization-related genes, such as those encoding transforming growth factor β1, interleukin (IL)-1 receptor antagonist (IL-1ra), CCL18, mannose receptor and arginase 1, although *C. burnetii* organisms also induce the expression of the genes encoding IL-6 and CXCL8, which are two cytokines associated with an M1 program (Benoit et al. 2008). These findings indicate that the polarization of macrophages into M1- or M2-like phenotypes controls the intracellular life of *C. burnetii*.

14.2.2 Dendritic Cells

Dendritic cells (DCs) are immune sentinels that are necessary to orchestrate the immune response to infection. It has been reported that phase I *C. burnetii* (virulent bacteria) infects and replicates within human DCs without inducing maturation or inflammatory cytokine production. In contrast, phase II bacteria induce a dramatic maturation of DCs (Shannon et al. 2005). However, these findings raise several points. First, it is difficult to understand how Th1 polarization of T cells in acute Q fever is associated with a major defect of DC maturation. Second, we have found that DCs stimulated with phase I *C. burnetii* were able to induce T cell proliferation in an allogenic reaction. Third, the transcriptomic program of *C. burnetii*-stimulated DC clusters (as well as that of DCs stimulated with LPS from enterobacteria) is known to induce DC maturation (unpublished data). We believe that, compared to canonical activators, the maturation of DCs is only partial in response to phase I *C. burnetii,* and DCs retain the potential to activate T cells.

14.2.3 Others

The role of natural killer (NK) cells in *C. burnetii* infection remains poorly understood. Although histopathological lesions are more severe in SCIDβγ mice (which are devoid of T, B and NK cells) compared to SCID mice, the bacterial burden is similar in both types of mice (Andoh et al. 2007). It is likely that γδ T cells are not required for controlling primary pulmonary *C. burnetii* infection because mice depleted of both αβ $CD4^+$ and $CD8^+$ T cells are similarly susceptible to infection as SCID mice (Read et al. 2010).

14.3 Adaptive Immunity

It is well established that adaptive immunity is required to cure the infectious diseases that are caused by intracellular bacteria (Collins and Kaufmann 2002). Several lines of evidence (arising from both animal models and clinical studies) support the idea that adaptive immunity is essential to cure *C. burnetii* infection (Fig. 14.2). Hence,

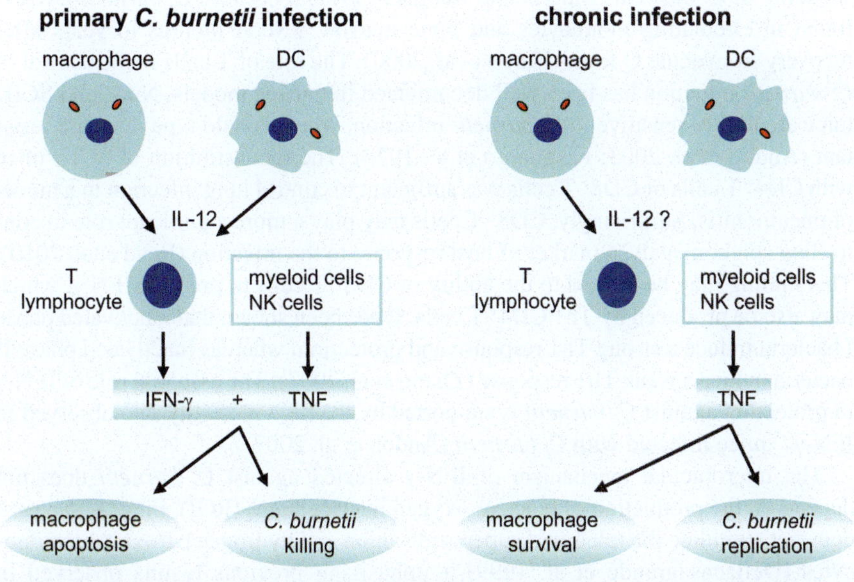

Fig. 14.2 The immune response to *C. burnetii* infection. In primary *C. burnetii* infection, the uptake of organisms by macrophages and DCs leads to the presentation of bacterial antigens to T lymphocytes. The production of IFN-γ by T lymphocytes (and NK cells) and that of TNF by immune cells including macrophages and T lymphocytes induce the apoptosis of infected macrophages and *C. burnetii* killing. In patients who are unable to mount an IFN-γ response (after defective presentation of antigen or other mechanisms), T lymphocytes are unable to produce IFN-γ. As a consequence, macrophages survive and their microbicidal activity is impaired

SCID mice that are devoid of T and B cells succumb to infection, demonstrating the major role of the adaptive immune response for the control of *C. burnetii* infection (Andoh et al. 2003). Patients that are experiencing immunosuppression (due to immunosuppressive treatment, pregnancy, and lymphomas) are at risk to develop chronic Q fever and to mount inefficient anti-*C. burnetii* immunity (Maurin and Raoult 1999).

14.3.1 T Lymphocytes and Cytokines

The control of infection in patients with primary Q fever involves a systemic cell-mediated immune response and granuloma formation. A cell-mediated immune response, which manifests as a marked proliferative response to *C. burnetii* antigen, is observed in patients who have convalesced from acute Q fever and in patients with acute Q fever hepatitis (Koster et al. 1985a). Individuals vaccinated with formalin-inactivated *C. burnetii* exhibit specific lymphoproliferation and IFN-γ production in response to *C. burnetii* challenge (Izzo and Marmion 1993; Izzo et al. 1988). The combination of IFN-γ production and granuloma formation suggests a Th1-type polarization of the immune response. Nevertheless, immune control of Q fever does not lead to *C. burnetii* eradication because animals exhibit persistent shedding of *C. burnetii* (Maurin and Raoult 1999) and because *C. burnetii* DNA is found in circulating monocytes and bone marrow several months to years after recovery from acute Q fever (Harris et al. 2000). The role of T cells in the control of *C. burnetii* infection has been well documented in murine models. Nude and SCID mice are highly sensitive to *C. burnetii* infection, whereas wild type mice are resistant (Andoh et al. 2003; Kishimoto et al. 1978). The reconstitution of SCID mice with $CD4^+$ T cells or $CD8^+$ T cells was sufficient to control lung infection in a model of pneumonitis. Surprisingly, $CD8^+$ T cells may play a more significant role in controlling splenomegaly, a marker of host response to the infection (Read et al. 2010). This finding may be related to the ability of $CD8^+$ T cells to produce IFN-γ, which may also be produced by Th1 $CD4^+$ T cells. It has been shown that inactivated phase I bacteria induce a strong Th1 response and protection, whereas inactivated phase II bacteria induce a weak Th1 response (Zhang et al. 2007). The essential role of IFN-γ in protection against *C. burnetii* is supported by the high mortality rate observed in IFN-$\gamma^{-/-}$ mice infected with *C. burnetii* (Andoh et al. 2007).

The microbicidal mechanism of IFN-γ directed against *C. burnetii* does not depend on the production of reactive oxygen intermediates (ROI). First, *C. burnetii* does not stimulate the release of superoxide anion and hydrogen peroxide by monocytes (Dellacasagrande et al. 1999), emphasizing previous results observed in neutrophils (Akporiaye et al. 1990). Second, the survival of *C. burnetii* is similar between control monocytes and monocytes from patients with chronic granulomatous disease (Akporiaye et al. 1990), in which the NADPH oxidase complex is not functional. The data on the role of reactive nitrogen intermediates are less clear. *C. burnetii* induces the production of nitric oxide (NO) by murine alveolar macrophages, but inhibitors of NO production do not modify the infection rate of macrophages

(Yoshiie et al. 1999). IFN-γ controlled *C. burnetii* infection despite the absence of NO synthase in one report (Zamboni and Rabinovitch 2003) and less efficiently in another report (Brennan et al. 2004). In THP-1 macrophages, *C. burnetii* does not induce NO production, even in the presence of IFN-γ, and L-arginine inhibitors have no effect on the survival of *C. burnetii* (Dellacasagrande et al. 1999).

Experiments have indicated that IFN-γ stimulates the microbicidal program directed against *C. burnetii* (Fig. 14.2) through an oxygen-independent mechanism. Indeed, *C. burnetii* can survive in a phagosome that shares the properties of late endosomes but not those of lysosomes. IFN-γ restores phagosome-lysosome fusion and affects phagosomal pH (Ghigo et al. 2002). It is likely that this mechanism accounts for the restoration of phagosome-lysosome fusion in monocytes in patients with acute Q fever (Ghigo et al. 2004). In addition, changing phagosomal pH may have some therapeutic consequences, as demonstrated by the efficiency of the combination of doxycycline and chloroquine in the treatment of patients with chronic Q fever. This therapeutic regimen promotes vacuolar alkalization and restores the *in vitro* bactericidal activity of antibiotics against *C. burnetii* (Maurin and Raoult 1999).

While the survival of obligate intracellular organisms usually requires the prevention of host cell death (Clifton et al. 1998), IFN-γ is able to promote the apoptosis of *C. burnetii*-infected macrophages (Fig. 14.2). The apoptotic effect of IFN-γ on *C. burnetii*-infected cells is dependent on tumor necrosis factor (TNF). Indeed, IFN-γ up-regulates TNF production and induces the expression of membrane TNF. Neutralizing TNF with specific antibodies prevents macrophage apoptosis and the eradication of *C. burnetii* (Dellacasagrande et al. 1999). Concomitantly, IFN-γ induces homotypic adherence of *C. burnetii*-infected macrophages, which depends on β2 integrins and CD54. When adherence is disrupted by mechanical dissociation or by blocking the integrin receptors, cell apoptosis and the bacterial killing induced by IFN-γ are inhibited (Dellacasagrande et al. 2002). These findings may help to explain the mechanisms of granuloma formation in acute Q fever. In contrast, decreased IFN-γ production may impair the aggregation and the microbicidal activity of monocytes, as observed during chronic Q fever.

Finally, IFN-γ may control *C. burnetii* infection by different mechanisms such as cytokine production and/or regulation of the nutrient supply. *C. burnetii*-infected monocytes stimulated with IFN-γ release large amounts of TNF (Dellacasagrande et al. 1999). Besides its role in apoptosis, TNF may contribute to the microbicidal activity of macrophages. TNF affects phagocytosis but not the later steps of the microbicidal process. Indeed, neutralizing anti-TNF antibodies decreased *C. burnetii* internalization by monocytes from patients with Q fever endocarditis, but they have no effect on the long term survival of bacteria (Dellacasagrande et al. 2000a). In addition, IFN-γ controls iron metabolism in macrophages through the down-modulation of transferrin receptors, resulting in a decreased assimilation of iron. *C. burnetii* up-regulates the expression of transferrin receptors in murine macrophage cell lines, which results in an increased cell iron content and bacterial burden. An intracellular iron chelator, desferoxamine, suppresses the replication of *C. burnetii* (Howe and Mallavia 1999). It remains to be determined whether IFN-γ-mediated killing of *C. burnetii* involves a decrease in cellular iron content.

The major role of cell-mediated immunity in the protection against *C. burnetii* is emphasized by defective cell-mediated immunity during chronic Q fever (Fig. 14.2). Lymphocytes from patients with Q fever endocarditis do not proliferate in response to the *C. burnetii* antigen, in contrast to lymphocytes from patients with acute Q fever (Koster et al. 1985a). The mechanisms of this specific unresponsiveness may include alterations in T cell subsets, but $CD4^+$ T cell lymphopenia was observed in patients with Q fever endocarditis and in cured patients who exhibited a normal immune response (Sabatier et al. 1997). This suppression is most likely mediated by immunoregulatory mediators such as the prostaglandin E2 (Koster et al. 1985b) or cytokines. IL-10, an immunoregulatory cytokine that is overproduced in chronic Q fever (Capo et al. 1996a; Honstettre et al. 2003), may be involved in Q fever-associated immunosuppression, perhaps via the induction of regulatory T cells. Finally, a defective imbalance of cytokines and chemokines may result in an impaired migration of immunocompetent cells to target organs (Meghari et al. 2006). Indeed, *C. burnetii*-infected monocytes exhibited defective transmigration through endothelium activated by TNF (Dellacasagrande et al. 2000b).

Immunosuppression in chronic Q fever is associated with an exacerbated inflammatory response. Severe inflammation is found in almost every patient with Q fever endocarditis and consists of up-regulation of the circulating levels of TNF, IL-6, two inflammatory cytokines, type II TNF receptors and IL-1ra (Capo et al. 1999). While IL-1ra levels are significantly higher in acute Q fever than in chronic Q fever, the levels of soluble CD23, a leukocyte activation marker also known as the low affinity receptor for immunoglobulin E, are specifically increased during chronic Q fever (Capo et al. 1999). In addition, TNF and IL-1β production is increased in monocytes from patients with Q fever endocarditis, whereas it remains low in patients with uncomplicated acute Q fever (Capo et al. 1996b; Dellacasagrande et al. 2000a). TNF production is related to the disease activity. First, it is higher in patients with Q fever recently diagnosed than both in those who have been monitored for more than 12 months and in cured patients. Second, there is a correlation between high levels of TNF production and titers of immunoglobulin G directed against *C. burnetii* (Capo et al. 1996b). Production of the chemokines CCL2 and CL5 is increased in monocytes from patients with Q fever endocarditis (Meghari et al. 2006). IL-6 production is high in both acute and chronic Q fever (Capo et al. 1999), which supports the overproduction of IL-6 that was reported in patients with post-Q fever fatigue syndrome (Penttila et al. 1998). Clearly, chronic Q fever is associated with immunosuppression and exacerbated inflammation.

14.3.2 Granulomas

The control of infection in patients with primary Q fever involves systemic cell-mediated immune responses and granuloma formation. The granulomatous lesions have a central open space and a fibrin ring, and they are referred to as doughnut granulomas. They consist of macrophages with epithelioid morphology and multinucleated giant cells, and they are paucibacillary in Q fever (Pellegrin et al. 1980; Srigley et al. 1985; Voigt et al. 1983). During chronic Q fever, granulomas are scarce and are replaced by lymphocyte infiltration and necrosis foci in the liver (Raoult et al. 1990). We recently adapted the *in vitro* method described by the Altare group to analyze mycobacterial granulomas (Puissegur et al. 2004) to that of *C. burnetii* granulomas (Delaby et al. 2010). The formation of *C. burnetii* granulomas is studied by incubating peripheral blood mononuclear cells with *C. burnetii*-coated sepharose beads. In the first step, monocytes induce granuloma formation, and in the second step, T lymphocytes are recruited by the nascent granuloma. The application of this method to mononuclear cells from patients with Q fever reveals that a large number of patients are unable to form granulomas due to a defective initial interaction of monocytes with bacterial antigens (Delaby et al. 2012).

14.3.3 B Lymphocytes and Antibodies

Antibodies are considered dispensable in Q fever, according to the paradigm of protection that is mediated by T cells in chronic infection. It is likely that the reality is less simple. First, large amounts of antibodies are produced in humans and animals infected with *C. burnetii*. Antibodies develop within 3–4 weeks of the onset of symptoms of primary infection: the majority of the antibodies are directed against phase II antigens and a minority are directed against phase I antigens. A similar model accounts for the response to the Q fever vaccine. Nevertheless, past studies have reported that not all vaccinated or convalescent patients have detectable levels of antibodies (Marmion et al. 1990). Increased levels of antibodies directed against phase I antigens are related to the chronic development of Q fever, with a diagnostic value for titers higher than 800 (Fournier et al. 1998). We have shown that mice that overexpress the human IL-10 transgene in macrophages produce high amounts of antibodies directed against phase I and II bacteria, similar to chronic human infection (Meghari et al. 2008). Second, initial descriptions of *C. burnetii* infection have indicated that anti-*C. burnetii* serum confers passive protection in mice and guinea pigs (Shannon and Heinzen 2009). Vaccination with phase I cellular vaccine induces both an antibody response and protection (Waag et al. 2008). Third, wild type mice and mice deficient in the receptors for the Fc fragment of IgG (FcγR) are equally protected by passive immunization, suggesting that antibodies and their receptors are not essential in *C. burnetii* clearance (Shannon and Heinzen 2009). This conclusion is strengthened by the finding that athymic mice and SCID mice are not protected by passive immunization (Humphres and Hinrichs 1981; Zhang et al. 2007). However, B cell-deficient mice are more severely affected than are wild type mice after *C. burnetii* infection, indicating that antibodies play a regulatory role in the infection even if bacterial clearance is not affected (Andoh et al. 2007). One can hypothesize that immunoregulatory cytokines are produced by macrophages only when FcγRs are engaged in the presence of an infectious trigger, as documented in leishmaniosis (Humphres and Hinrichs 1981; Zhang et al. 2007). We showed that macrophages incubated with *C. burnetii* that has been opsonized with specific IgG release higher amounts of IL-10 than those incubated with unopsonized bacteria (Desnues et al. 2009). The fate of microorganisms in this environment is a source of debate. In our experimental conditions, *C. burnetii* opsonization favors the formation of multibacillary vacuoles and increased bacterial replication (Desnues et al. 2009). The Heinzen group confirmed the opsonizing activity of anti-*C. burnetii* antibodies but did not find any effect of these antibodies on the growth rate of bacteria (Shannon and Heinzen 2009). Taken together, these results indicate that anti-*C. burnetii* antibodies may have a dual role in Q fever: they confer protection in synergy with T cells in acute Q fever and exacerbate immune depression in chronic Q fever through the production of IL-10.

14.4 Host Metabolism and Control of *C. burnetii* Infection

It is well known that age is a risk factor for Q fever. Indeed, symptomatic Q fever occurs more frequently in people who are over 15 years of age than in people who are under 15 years. The prevalence of clinical cases in children significantly increases with age (Maltezou and Raoult 2002). We have recently found that the bacterial burden and the number of granuloma are increased in the tissues of 14-month-old mice compared to 1-month-old mice (Leone et al. 2007).

Gender also affects *C. burnetii* infection. Men are more frequently symptomatic than are women (with a man:woman ratio of 2.5), despite comparable exposure and seroprevalence (Maltezou and Raoult 2002; Tissot-Dupont et al. 1992). The predisposition for infection in men may be explained by differences in sex hormones such as 17β-oestradiol. Indeed, female C56BL/6 mice exhibit fewer granulomas and a lower bacterial burden than do males, and ovariectomized mice exhibit disease rates that are comparable to those of males. The administration of 17β-oestradiol prevents the effect of ovariectomy on host response and tissue burden (Leone et al. 2004). The study of gene expression programs in mice reveals the importance of sex-related genes during *C. burnetii* infection. Multiclass analysis has identified 2,777 probes for which expression is specifically modulated by *C. burnetii* infection. Only 14% of the modulated genes are sex-independent, and the remaining 86% are differentially expressed in males and females. Castration of males and females has indicated that sex hormones are responsible for more than 60% of the observed gene modulation, and this reduction is most pronounced in males. Using functional annotation of modulated genes, we have identified four clusters that are enriched in males and are related to cell-cell adhesion, signal transduction, defensins and cytokine/Jak-Stat pathways. Up-regulation of the IL-10 and Stat-3 genes may account for the high susceptibility of men with Q fever to *C. burnetii* infection and to autoantibody production. Two clusters have been identified in females, including the circadian rhythm pathway, which consists of a feedback loop with positive (Clock, Arntl) and negative (Per) limbs. We have found that Clock and Arntl are down-regulated, whereas Per is up-regulated. These changes may be associated with efficient bacterial elimination in females but not in males, in which an exacerbated host response would be prominent (Textoris et al. 2010). Preliminary results in humans suffering from Q fever indicate that these gene pathways are differentially modulated in men and women.

14.5 Conclusions: New Basis for Vaccine Development

The immune response to *C. burnetii* is organized within granulomas. A better understanding of this response is becoming possible through the development of new approaches, including live imaging. However, this response does not protect the patients from the risk of relapse. The innate immune response is sufficient to control the infection, and the adaptive immune response allows for bacterial clearance. Both responses are also involved in features of the disease such as the uncontrolled inflammatory response in some patients with acute Q fever and in patients with chronic Q fever. The fact that a minority of exposed patients develop chronic Q fever suggests that certain genetic factors related to host immune response are critical. The relationship between the clinical expression of Q fever and gender is an excellent illustration of this interplay. The next step in understanding the immune response to *C. burnetii* will consist of the evaluation of individual host responses by systemic approaches to relate each clinical expression of Q fever with a specific genetic signature. These relations will be essential to analyze treatment failures.

References

Akporiaye ET, Stefanovich D, Tsosie V et al (1990) *Coxiella burnetii* fails to stimulate human neutrophil superoxide anion production. Acta Virol 34:64–70

Andoh M, Naganawa T, Hotta A et al (2003) SCID mouse model for lethal Q fever. Infect Immun 71:4717–4723

Andoh M, Zhang G, Russell-Lodrigue KE et al (2007) T cells are essential for bacterial clearance, and γ interferon, tumor necrosis factor α, and B cells are crucial for disease development in *Coxiella burnetii* infection in mice. Infect Immun 75:3245–3255

Ayres JG, Flint N, Smith EG et al (1998) Post-infection fatigue syndrome following Q fever. QJM 91:105–123

Benoit M, Barbarat B, Bernard A et al (2008) *Coxiella burnetii*, the agent of Q fever, stimulates an atypical M2 activation program in human macrophages. Eur J Immunol 38:1065–1070

Brennan RE, Russell K, Zhang G et al (2004) Both inducible nitric oxide synthase and NADPH oxidase contribute to the control of virulent phase I *Coxiella burnetii* infections. Infect Immun 72:6666–6675

Brouqui P, Tissot-Dupont H, Drancourt M et al (1993) Chronic Q fever. Ninety-two cases from France, including 27 cases without endocarditis. Arch Intern Med 153:642–648

Capo C, Zaffran Y, Zugun F et al (1996a) Production of interleukin-10 and transforming growth factor beta by peripheral blood mononuclear cells in Q fever endocarditis. Infect Immun 64:4143–4147

Capo C, Zugun F, Stein A et al (1996b) Upregulation of tumor necrosis factor-α and interleukin-1β in Q fever endocarditis. Infect Immun 64:1638–1642

Capo C, Amirayan N, Ghigo E et al (1999) Circulating cytokine balance and activation markers of leucocytes in Q fever. Clin Exp Immunol 115:120–123

Clifton DR, Goss RA, Sahni SK et al (1998) NF-κB-dependent inhibition of apoptosis is essential for host cell survival during *Rickettsia rickettsii* infection. Proc Natl Acad Sci U S A 95:4646–4651

Collins H, Kaufmann SHE (2002) Acquired immunity against bacteria. In: Kaufmann SHE, Sher A, Ahmed R (eds) Immunology of infectious diseases. ASM Press, Washington, pp 207–221

Delaby A, Espinosa L, Lepolard C et al (2010) 3D reconstruction of granulomas from transmitted light images implemented for long-time microscope applications. J Immunol Methods 360:10–19

Delaby A, Gorvel L, Espinosa L et al (2012) Defective monocyte dynamics in Q fever granuloma deficiency. J Infect Dis 205:1086–1094

Dellacasagrande J, Capo C, Raoult D et al (1999) IFN-γ-mediated control of *Coxiella burnetii* survival in monocytes: the role of cell apoptosis and TNF. J Immunol 162:2259–2265

Dellacasagrande J, Ghigo E, Capo C et al (2000a) *Coxiella burnetii* survives in monocytes from patients with Q fever endocarditis: involvement of tumor necrosis factor. Infect Immun 68:160–164

Dellacasagrande J, Moulin PA, Guilianelli C et al (2000b) Reduced transendothelial migration of monocytes infected by *Coxiella burnetii*. Infect Immun 68:3784–3786

Dellacasagrande J, Ghigo E, Raoult D et al (2002) IFN-γ-induced apoptosis and microbicidal activity in monocytes harboring the intracellular bacterium *Coxiella burnetii* require membrane TNF and homotypic cell adherence. J Immunol 169:6309–6315

Desnues B, Imbert G, Raoult D et al (2009) Role of specific antibodies in *Coxiella burnetii* infection of macrophages. Clin Microbiol Infect 15(Suppl 2):161–162

Fenollar F, Fournier PE, Carrieri P et al (2001) Risks factors and prevention of Q fever endocarditis. Clin Infect Dis 33:312–316

Fournier PE, Marrie TJ, Raoult D (1998) Diagnosis of Q fever. J Clin Microbiol 36:1823–1834

Ghigo E, Capo C, Tung CH et al (2002) *Coxiella burnetii* survival in THP-1 monocytes involves the impairment of phagosome maturation: IFN-γ mediates its restoration and bacterial killing. J Immunol 169:4488–4495

Ghigo E, Honstettre A, Capo C et al (2004) Link between impaired maturation of phagosomes and defective *Coxiella burnetii* killing in patients with chronic Q fever. J Infect Dis 190:1767–1772

Ghigo E, Pretat L, Desnues B et al (2009) Intracellular life of *Coxiella burnetii* in macrophages. Ann N Y Acad Sci 1166:55–66

Harris RJ, Storm PA, Lloyd A et al (2000) Long-term persistence of *Coxiella burnetii* in the host after primary Q fever. Epidemiol Infect 124:543–549

Honstettre A, Imbert G, Ghigo E et al (2003) Dysregulation of cytokines in acute Q fever: role of interleukin-10 and tumor necrosis factor in chronic evolution of Q fever. J Infect Dis 187:956–962

Honstettre A, Ghigo E, Moynault A et al (2004) Lipopolysaccharide from *Coxiella burnetii* is involved in bacterial phagocytosis, filamentous actin reorganization, and inflammatory responses through Toll-like receptor 4. J Immunol 172:3695–3703

Houpikian P, Raoult D (2005) Blood culture-negative endocarditis in a reference center: etiologic diagnosis of 348 cases. Medicine (Baltimore) 84:162–173

Houpikian P, Habib G, Mesana T et al (2002) Changing clinical presentation of Q fever endocarditis. Clin Infect Dis 34:E28–E31

Howe D, Mallavia LP (1999) *Coxiella burnetii* infection increases transferrin receptors on J774A. 1 cells. Infect Immun 67:3236–3241

Humphres RC, Hinrichs DJ (1981) Role of antibody in *Coxiella burnetii* infection. Infect Immun 31:641–645

Izzo AA, Marmion BP (1993) Variation in interferon-γ responses to *Coxiella burnetii* antigens with lymphocytes from vaccinated or naturally infected subjects. Clin Exp Immunol 94:507–515

Izzo AA, Marmion BP, Worswick DA (1988) Markers of cell-mediated immunity after vaccination with an inactivated, whole-cell Q fever vaccine. J Infect Dis 157:781–789
Kishimoto RA, Rozmiarek H, Larson EW (1978) Experimental Q fever infection in congenitally athymic nude mice. Infect Immun 22:69–71
Koster FT, Williams JC, Goodwin JS (1985a) Cellular immunity in Q fever: specific lymphocyte unresponsiveness in Q fever endocarditis. J Infect Dis 152:1283–1289
Koster FT, Williams JC, Goodwin JS (1985b) Cellular immunity in Q fever: modulation of responsiveness by a suppressor T cell-monocyte circuit. J Immunol 135:1067–1072
Leone M, Honstettre A, Lepidi H et al (2004) Effect of sex on *Coxiella burnetii* infection: protective role of 17β-estradiol. J Infect Dis 189:339–345
Leone M, Bechah Y, Meghari S et al (2007) *Coxiella burnetii* infection in C57BL/6 mice aged 1 or 14 months. FEMS Immunol Med Microbiol 50:396–400
Lepidi H, Houpikian P, Liang Z et al (2003) Cardiac valves in patients with Q fever endocarditis: microbiological, molecular, and histologic studies. J Infect Dis 187:1097–1106
Madariaga MG, Rezai K, Trenholme GM et al (2003) Q fever: a biological weapon in your backyard. Lancet Infect Dis 3:709–721
Maltezou HC, Raoult D (2002) Q fever in children. Lancet Infect Dis 2:686–691
Marmion BP, Ormsbee RA, Kyrkou M et al (1990) Vaccine prophylaxis of abattoir-associated Q fever: eight years' experience in Australian abattoirs. Epidemiol Infect 104:275–287
Maurin M, Raoult D (1999) Q fever. Clin Microbiol Rev 12:518–553
Meghari S, Honstettre A, Lepidi H et al (2005) TLR2 is necessary to inflammatory response in *Coxiella burnetii* infection. Ann N Y Acad Sci 1063:161–166
Meghari S, Desnues B, Capo C et al (2006) *Coxiella burnetii* stimulates production of RANTES and MCP-1 by mononuclear cells: modulation by adhesion to endothelial cells and its implication in Q fever. Eur Cytokine Netw 17:253–259
Meghari S, Bechah Y, Capo C et al (2008) Persistent *Coxiella burnetii* infection in mice overexpressing IL-10: an efficient model for chronic Q fever pathogenesis. PLoS Pathog 4:e23
Pellegrin M, Delsol G, Auvergnat JC et al (1980) Granulomatous hepatitis in Q fever. Hum Pathol 11:51–57
Penttila IA, Harris RJ, Storm P et al (1998) Cytokine dysregulation in the post-Q-fever fatigue syndrome. QJM 91:549–560
Puissegur MP, Botanch C, Duteyrat JL et al (2004) An in vitro dual model of mycobacterial granulomas to investigate the molecular interactions between mycobacteria and human host cells. Cell Microbiol 6:423–433
Raoult D (1990) Host factors in the severity of Q fever. Ann N Y Acad Sci 590:33–38
Raoult D, Raza A, Marrie TJ (1990) Q fever endocarditis and other forms of chronic Q fever. In: Marrie TJ (ed) Q fever. The disease. CRC Press, Boca Raton, pp 3784–3786
Raoult D, Marrie TJ, Mege JL (2005) Natural history and pathophysiology of Q fever. Lancet Infect Dis 5:219–226
Read AJ, Erickson S, Harmsen AG (2010) Role of $CD4^+$ and $CD8^+$ T cells in clearance of primary pulmonary infection with *Coxiella burnetii*. Infect Immun 78:3019–3026
Rolain JM, Mallet MN, Raoult D (2003) Correlation between serum doxycycline concentrations and serologic evolution in patients with *Coxiella burnetii* endocarditis. J Infect Dis 188:1322–1325
Sabatier F, Dignat-George F, Mege JL et al (1997) $CD4^+$ T-cell lymphopenia in Q fever endocarditis. Clin Diagn Lab Immunol 4:89–92
Shannon JG, Heinzen RA (2009) Adaptive immunity to the obligate intracellular pathogen *Coxiella burnetii*. Immunol Res 43:138–148
Shannon JG, Howe D, Heinzen RA (2005) Virulent *Coxiella burnetii* does not activate human dendritic cells: role of lipopolysaccharide as a shielding molecule. Proc Natl Acad Sci U S A 102:8722–8727
Srigley JR, Vellend H, Palmer N et al (1985) Q-fever. The liver and bone marrow pathology. Am J Surg Pathol 9:752–758

Textoris J, Ban LH, Capo C et al (2010) Sex-related differences in gene expression following *Coxiella burnetii* infection in mice: potential role of circadian rhythm. PLoS One 5:e12190
Tissot-Dupont H, Raoult D, Brouqui P et al (1992) Epidemiologic features and clinical presentation of acute Q fever in hospitalized patients: 323 French cases. Am J Med 93:427–434
Voigt JJ, Delsol G, Fabre J (1983) Liver and bone marrow granulomas in Q fever. Gastroenterology 84:887–888
Waag DM, England MJ, Bolt CR et al (2008) Low-dose priming before vaccination with the phase I chloroform-methanol residue vaccine against Q fever enhances humoral and cellular immune responses to *Coxiella burnetii*. Clin Vaccine Immunol 15:1505–1512
Weisburg WG, Dobson ME, Samuel JE et al (1989) Phylogenetic diversity of the *Rickettsiae*. J Bacteriol 171:4202–4206
Yoshiie K, Matayoshi S, Fujimura T et al (1999) Induced production of nitric oxide and sensitivity of alveolar macrophages derived from mice with different sensitivity to *Coxiella burnetii*. Acta Virol 43:273–278
Zamboni DS, Rabinovitch M (2003) Nitric oxide partially controls *Coxiella burnetii* phase II infection in mouse primary macrophages. Infect Immun 71:1225–1233
Zamboni DS, Campos MA, Torrecilhas AC et al (2004) Stimulation of toll-like receptor 2 by *Coxiella burnetii* is required for macrophage production of pro-inflammatory cytokines and resistance to infection. J Biol Chem 279:54405–54415
Zhang G, Russell-Lodrigue KE, Andoh M et al (2007) Mechanisms of vaccine-induced protective immunity against *Coxiella burnetii* infection in BALB/c mice. J Immunol 179:8372–8380

Chapter 15
Immune Response and *Coxiella burnetii* Invasion

Amira Ben Amara, Yassina Bechah, and Jean-Louis Mege

Contents

Abstract *Coxiella burnetii*, the causative agent of Q fever, has evolved a wealth of mechanisms in order to persist within hosts. Two tissues, namely adipose tissue and placenta, are candidates to house *C. burnetii*, but the mechanisms governing *C. burnetii* survival in these tissues are still unknown. In contrast, monocytes and macrophages are well-known targets of *C. burnetii*. First, *C. burnetii* has developed a specific strategy of phagocytosis subversion that consists of the inhibition of integrin interplay. Second, *C. burnetii* persistence is associated with macrophage activation profiles. Indeed, monocytes (in which *C. burnetii* survives without replication) exhibit a proinflammatory M1-type response, whereas macrophages

A. Ben Amara • Y. Bechah • J.-L. Mege (✉)
Unité de Recherche sur les Maladies Infectieuses Transmissibles et Emergentes, CNRS-IRD UMR 6236, Institut Fédératif de Recherche 48, Faculté de Médecine, Université de la Méditerranée, Marseille, France
e-mail: Jean-Louis.Mege@univmed.fr

R. Toman et al. (eds.), *Coxiella burnetii: Recent Advances and New Perspectives in Research of the Q Fever Bacterium*, Advances in Experimental Medicine and Biology 984, DOI 10.1007/978-94-007-4315-1_15,

(in which *C. burnetii* slowly replicates) are polarized towards an M2-type. Third, interleukin-10 produced by monocytes is a main factor of the chronic development of Q fever, and murine models confirm the key role of interleukin-10 in *C. burnetii* persistence. Fourth, apoptotic cells may play a key role in chronic Q fever. The uptake of apoptotic cells by circulating monocytes increases *C. burnetii* replication by redirecting monocytes toward a non-protective M2 profile. In the presence of interferon-γ, apoptotic cell engulfment is inhibited and monocytes polarized toward an M1 program are able to kill *C. burnetii*; this is the situation observed in patients with uncomplicated acute Q fever. Finally, we cannot exclude that regulatory T cells may play a role in *C. burnetii* persistence because their number is increased in patients with chronic Q fever.

Keywords Adipose tissue • Apoptosis • *Coxiella burnetii* • Interleukin-10 • Macrophage polarization • Monocytes • Placenta • Q fever • Regulatory T cells

15.1 Introduction

Persistent infections may be clinical and related to chronic infection, as observed in chronic Q fever, or subclinical and corresponding to latent infection, as found in mycobacterial and viral infections (Munz-Elias and Mc Kinney 2002). Persistent pathogens, such as *Coxiella burnetii*, have evolved a wealth of mechanisms to antagonize the bactericidal potential of the effectors of immune responses. This includes interference with the microbicidal activity of myeloid cells, dampening of inflammatory signals and inactivation of immune cells (Roy and Mocarski 2007; Sansonetti and Di Santo 2007). In this chapter, we will describe the mechanisms used by *C. burnetii* to control the immune response. This control is directly due to bacterial products that inactivate or avoid effector functions of immunocompetent cells but also indirectly due to regulatory cytokines and regulatory T cells (Belkaid and Rouse 2005). In addition, one strategy used by *C. burnetii* and already described for several viruses could rely on the choice of targeted tissues and cell types. Finally, *C. burnetii* persistence may be related to the route of infection, as some have reported for the initiation of systemic or mucosal responses (Martinoli et al. 2007).

15.2 The Choice of Niche for *C. burnetii* Persistence: Cells and Tissues

C. burnetii must initially choose convenient cells and tissues to avoid efficient immune responses.

15.2.1 Cells Targeted by C. burnetii

Monocytes and macrophages are the major targets of *C. burnetii*. As these cells have strong microbicidal activity, microorganisms have to subvert these mechanisms to create favorable conditions for survival. We have described strategies used by *C. burnetii* to make monocytes and macrophages permissive for bacterial survival. For that purpose, we have used *C. burnetii* organisms found in natural conditions (virulent organisms also called phase I organisms) and variants obtained upon serial passages in culture (avirulent organisms also called phase II organisms). We clearly showed that *C. burnetii* has developed a strategy of phagocytic subversion (Capo et al. 1999). The phagocytosis of phase I *C. burnetii* requires the engagement of $\alpha v\beta 3$ integrin and leads to a low phagocytosis rate and intracellular survival. The phagocytosis of phase II *C. burnetii* is mediated by $\alpha v\beta 3$ integrin and CR3 ($\alpha M\beta 2$ integrin, CD11b/CD18), resulting in high phagocytosis rate and bacterial elimination. The low efficiency of virulent bacteria uptake is probably critical for the persistence of *C. burnetii* in monocytes and macrophages. Inefficient uptake results from the uncoupling of the $\alpha v\beta 3$ integrin from CR3, which is secondary to the inappropriate activation of macrophages and to actin cytoskeleton reorganization (Meconi et al. 1998, 2001).

Besides low bacterial uptake, a relative deficiency in cytokine production and subsequent macrophage activation is necessary for *C. burnetii* persistence. Our initial reports have indicated that virulent *C. burnetii* organisms are less efficient than avirulent organisms at inducing the synthesis and release of tumor necrosis factor (TNF). This limited efficiency is not the consequence of poor activity of bacterial lipopolysaccharide (LPS) because purified LPS from virulent organisms is more potent than LPS from avirulent organisms at stimulating TNF release (Dellacasagrande et al. 2000b). In fact, virulent organisms, which bind poorly to monocytes, likely present fewer LPS molecules to target cells than do avirulent bacteria, which efficiently bind to monocytes (Dellacasagrande et al. 2000b). The effect of activation on uptake efficiency is strengthened by the finding that neutralizing anti-TNF antibodies decrease *C. burnetii* internalization by monocytes from patients with Q fever endocarditis but do not affect the long-term survival of bacteria (Dellacasagrande et al. 2000a).

The concept of macrophage polarization illustrates their functional heterogeneity. Hence, polarized macrophages have been classified as classically activated (M1) macrophages that support microbicidal activity and alternatively activated (M2) macrophages that are not competent to eliminate pathogens (Martinez et al. 2009; Mege et al. 2011). The identification of M1 and M2 macrophages only results from a combination of markers including membrane receptors, cytokines, chemokines and effector mediators (Benoit et al. 2008b). The activation state of monocytes and macrophages after encountering *C. burnetii* are apparently different (Fig. 15.1). While monocytes exhibit a pro-inflammatory M1-type response, macrophages are polarized toward an M2-type. *C. burnetii*-infected macrophages release the M2-associated molecules interleukin (IL)-10, transforming growth factor (TGF)-$\beta 1$ and CCL18, and they express mannose receptor (MR) and the active form of

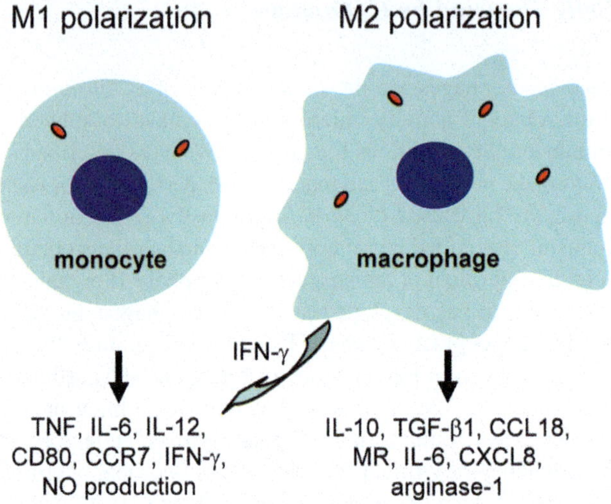

Fig. 15.1 *C. burnetii* and polarization of myeloid cells. In circulating monocytes, *C. burnetii* stimulates an M1 profile that control bacterial replication. In macrophages, *C. burnetii* induces an atypical M2 profile that is unable to control *C. burnetii* replication, as observed during acute Q fever. The treatment of macrophages with IFN-γ reorients them toward a M1 profile

arginase-1. They also secrete high levels of IL-6 and CXCL8, two molecules associated with M1 polarization. However, *C. burnetii*-infected macrophages do not express the M1 molecules (TNF, IL-12, CD80 and CCR7) and fail to produce nitric oxide (NO) (Benoit et al. 2008a). These differences in macrophage activation status are associated with variations in bacterial survival because *C. burnetii* survives in monocytes without replication and slowly replicates in macrophages. We cannot exclude that the *C. burnetii* replication within macrophages is the consequence of increased uptake associated with M2 polarization (Benoit et al. 2008b).

The macrophage activation state may also affect the intracellular trafficking of *C. burnetii*. We have revisited this topic in our reports, but a lack of consensus exists among the researchers who study this aspect of the intracellular life of *C. burnetii*. This point is discussed in a dedicated chapter. The inability of naïve macrophages and monocytes from patients with chronic Q fever (in which *C. burnetii* persists) to induce phagosome-lysosome fusion makes the intracellular trafficking of *C. burnetii* a critical parameter of bacterial persistence. Intracellular trafficking is under the control of IL-10, a cytokine that is responsible for bacterial persistence and the chronic development of Q fever (Ghigo et al. 2004).

15.2.2 Tissues Targeted by C. burnetii

The persistence of *C. burnetii* has been clearly documented in the bone marrow of patients who have been apparently cured of Q fever (Harris et al. 2000). It is likely that *C. burnetii* survives in bone marrow macrophages. Once differentiated, macrophages could carry the microorganisms to peripheral sites where immune control is impaired, allowing reactivation. We cannot exclude that stromal cells may be targeted by *C. burnetii* and become bacterial reservoirs. In addition to bone marrow cells, two tissues, namely adipose tissue and placenta, are candidates to house *C. burnetii*. Beyond its metabolic role, adipose tissue has recently emerged as a target for intracellular microorganisms (Desruisseaux et al. 2007). Indeed, adipose tissue is rich in cells that are highly reactive to infection, such as adipocytes and macrophages that express several features of non-microbicidal macrophages (Lumeng et al. 2007). Adipocytes have been recognized as reservoirs for *Trypanosoma cruzi* (Combs et al. 2005) and *Mycobacterium tuberculosis* (Neyrolles et al. 2006), the agents of Chagas disease and tuberculosis, respectively. We found that adipose tissue acts as a reservoir for latent *Rickettsia prowazekii*, the agent of epidemic typhus, and may account for the reactivation of *R. prowazekii* infection in a murine model of Brill-Zinsser disease (Bechah et al. 2010). We also found that *C. burnetii* is present in adipose tissue for several months following infection in mice, whereas it is absent from the blood, liver and spleen (unpublished data). The nature of the cell types that are targeted by *C. burnetii* remains to determine. Adipose tissue macrophages are the expected targets of *C. burnetii* but adipocytes may also harbor *C. burnetii*. We recently found that murine fibroblastic cells in vitro differentiated into adipocytes are permissive to *C. burnetii* infection (unpublished data). The mechanisms involved in the tropism of *C. burnetii* for adipose tissue cells are unknown but it is important to note that adipose-secreted cytokines (adipokines) have both inflammatory and immunoregulatory activities (Ouchi et al. 2011), suggesting that *C. burnetii* survival is controlled by the microenvironment of adipose tissue. In addition, it has been recently demonstrated that the adipokine adiponectin modulates the polarization of murine peritoneal macrophages (Ohashi et al. 2010). As the polarization status of macrophages controls the intracellular fate of *C. burnetii* (see above), we can hypothesize that the polarization status of adipose tissue macrophages governs their microbicidal activity against *C. burnetii*. Hence, the imbalance in the microenvironment of adipose tissue may control or favor *C. burnetii* infection.

Placenta is also a temporary reservoir for numerous pathogens including *C. burnetii*. Pregnancy consists in the tolerance of a semi-allogeneic graft that needs the regulation of the maternal immune response. This regulation occurs through the interaction between the fetal trophoblasts and the maternal immune system essentially consisting of uterine natural killer cells, dendritic cells and macrophages. In the presence of pathogens such as *C. burnetii*, the tolerogenic phenotype of maternal immune cells is changed into a phenotype that may lead to fetal expulsion. In addition, pregnancy is considered to be an important factor for the chronic development of Q fever and its bacterial persistence (Maurin and Raoult 1999). This phenomenon has been reproduced in animal models. The infection of C57/BL6 mice followed by repeated pregnancies results in abortion, perinatal death and endocarditis in some animals (Stein et al. 2000). The placentas of the infected animals contained up to 10^9 organisms/gram of tissue, and it is likely that the heavily infected placentas contaminated the environment at the time of parturition, leading to the persistence of viable organisms in the soil for several months (Langley et al. 1988). Aerosols from the secretions and excretions of ruminants are the major source of contamination for humans (Tissot-Dupont et al. 2004). The site of *C. burnetii* replication within the placenta remains unknown. We have recently observed that *C. burnetii* infects and persists within a trophoblastic cell line. The bacteria induce an original transcriptional program that modulates genes involved in inflammatory responses and in the maintenance of pregnancy (Ben Amara et al. 2010). It is likely that *C. burnetii*-infected trophoblasts interact with immune cells present in the placenta and interfere with fetal tolerance. In addition, the placenta is a tissue rich in macrophages and we recently found that placental macrophages harbor *C. burnetii* (unpublished data). It is possible that the tolerogenic status of pregnancy favors the polarization of placental macrophages toward a M2 phenotype unable to control *C. burnetii* infection.

15.3 Role of Immunoregulatory Cytokines in *C. burnetii* Persistence

The persistence of *C. burnetii* in monocytes/macrophages and within mice and humans is related to the cytokine IL-10 (Table 15.1). IL-10 shares with IL-4, IL-13 and TGF-β the ability to down-modulate the microbicidal activity of macrophages and redirect their polarization status. Moreover, IL-10 has been associated with an increased susceptibility of the host to intracellular organisms and with chronic bacterial persistence (Mege et al. 2006). During the past few years, we have provided converging evidence that IL-10 is the main factor in the chronic development of Q fever. First, IL-10 is produced by mononuclear cells from patients with Q fever endocarditis and Q fever with valvulopathy who were at risk for developing chronic Q fever (Capo et al. 1996; Honstettre et al. 2003). Second, IL-10 induces *C. burnetii* replication in monocytes. IL-10 acts through the down-modulation of TNF production

Table 15.1 Role of IL-10 in *C. burnetii* infection

Evidence	Reference
IL-10 specifically increases *C. burnetii* replication in naïve monocytes	Ghigo et al. (2001)
IL-10 produced by patients with chronic Q fever	Capo et al. (1996)
IL-10 produced by patients with valvulopathy and at risk for developing chronic Q fever	Honstettre et al. (2003)
C. burnetii infection is controlled in patients with low IL-10 production but not in patients with high IL-10 production. Adding IL-10 to monocytes impairs *C. burnetii* killing that is restored by IL-10 neutralization	Ghigo et al. (2004)
IL-10 neutralization corrects the defective transendothelial migration of leukocytes observed in chronic Q fever	Meghari et al. (2006)
C. burnetii infection is persistent in mice that overexpress IL-10 in the macrophage compartment	Meghari et al. (2008)

because TNF restores the microbicidal activity against *C. burnetii* in IL-10-treated monocytes (Ghigo et al. 2004). IL-10 also upregulates the release of type II receptors of TNF (TNF-RII). During chronic Q fever, the expression and release of monocyte TNF-RII are increased (Ghigo et al. 2000). It is likely that soluble TNF-RII interferes with TNF-stimulated microbicidal activity of monocytes, thus sustaining bacterial replication. The effect of IL-10 on *C. burnetii* replication is specific because IL-4 and TGF-β1 (two other immunoregulatory cytokines that act on monocytes) fail to induce *C. burnetii* replication (Ghigo et al. 2001). Third, IL-10 is involved in the microbicidal defect of monocytes observed in chronic Q fever. Indeed, *C. burnetii* is eliminated in monocytes from patients with acute Q fever and low IL-10 production, whereas it replicates in monocytes from patients with chronic Q fever and high IL-10 production. The microbicidal activity of monocytes from patients with chronic Q fever is restored by neutralizing IL-10 (Ghigo et al. 2004). As the fusion of *C. burnetii*-containing phagosomes with lysosomes is a key component of the microbicidal activity of monocytes (Ghigo et al. 2002), we have studied the co-localization of *C. burnetii* with cathepsin D (a marker of phagosome-lysosome fusion) in monocytes from patients with chronic Q fever. Phagosome-lysosome fusion is impaired in these patients relative to patients with acute Q fever, and the neutralization of endogenous IL-10 reestablishes phagosome-lysosome fusion (Ghigo et al. 2002). Fourth, IL-10 affects the migration of immune cells to peripheral tissues. Indeed, the transendothelial migration of mononuclear cells is defective in patients with Q fever endocarditis, and this defect is corrected by neutralizing IL-10 (Meghari et al. 2006). Defective transendothelial migration of mononuclear cells partly accounts for the defective granuloma formation in patients with chronic Q fever (Maurin and Raoult 1999).

Murine models of *C. burnetii* infection also confirm the key role of IL-10 in bacterial persistence. Several murine models, including immunodefective mice, have permitted the reproduction of acute *C. burnetii* infection but not chronic infection or *C. burnetii* persistence (Maurin and Raoult 1999). An efficient model for chronic Q fever pathogenesis has been developed using transgenic mice that constitutively over-express IL-10 in the macrophage compartment (Meghari et al. 2008). Transgenic mice infected via an intraperitoneal injection exhibited sustained tissue infection and a strong antibody response in contrast to wild-type mice. Thus, bacterial persistence is IL-10-dependent, as in chronic Q fever. In addition, the number of granulomas is low in the spleen and liver of infected transgenic mice, as in patients with chronic Q fever. Interestingly, macrophages from transgenic mice are unable to kill *C. burnetii* and are characterized by a non-microbicidal M2-type transcriptional program consisting of increased expression of arginase-1, MR and Ym1/2, in contrast to wild-type macrophages in which inducible NO synthase (iNOS) and inflammatory cytokines are overproduced. In the spleen and liver of transgenic mice infected with *C. burnetii*, the expression of arginase-1 is increased while the microbicidal pathway of IL-12p40, IL-23p19 and iNOS is depressed. Finally, the infection of transgenic mice through the intratracheal route, which is close to the natural route of infection, results in pulmonary lesions that consist of mixed interstitial and mild alveolar mononuclear cell pneumonia with a persistent infection of the lungs.

Hence, the IL-10-mediated polarization of macrophages is necessary to increase *C. burnetii* susceptibility and persistence in host cells and tissues. There are likely multiple mechanisms leading to IL-10 repolarization of macrophages. We reported a candidate mechanism involving redox partners using vanin-1 deficient mice (Meghari et al. 2007). Vanin-1 is a membrane-anchored pantetheinase that controls tissue inflammation via local cysteamine release and control of the levels of glutathione. The lack of vanin-1 decreases the formation of granulomas in the spleen and liver and significantly affects the activation pattern of bone marrow-derived macrophages stimulated by *C. burnetii*. These macrophages exhibit decreased iNOS and CCL2 gene expression with increased IL-10 and arginase-1 expression, as found in the livers of infected mice. These results indicate that vanin-1 plays a role in granuloma formation in response to *C. burnetii* by skewing macrophage activation toward an M2 program.

The control of *C. burnetii* infection may also be related to the interaction of macrophages with T cells. In mice deficient for CD28 (a co-stimulatory molecule that exerts a regulatory effect on T cell responses), we observed a decreased *C. burnetii* burden in infected tissues and an impaired production of IL-10 by peritoneal macrophages (Honstettre et al. 2006). Taken together, these results indicate that a precise regulation of macrophage activation through IL-10 controls macrophage polarization and *C. burnetii* persistence.

15.4 Role of Apoptosis in *C. burnetii* Persistence

Apoptosis plays a role in interferon (IFN)-γ-mediated immune protection against *C. burnetii* (Dellacasagrande et al. 1999), and it has been established that preventing apoptosis is a survival strategy of *C. burnetii*. Two reports have clearly documented this point. Epithelial cell lines infected with *C. burnetii* are resistant to chemical inducers of apoptosis due to interference with the intrinsic pathway of cell death (Luhrmann and Roy 2007). *C. burnetii* infection of macrophages inhibits cell apoptosis via the upregulation of c/IAP2 and A1/bfl1, which prevents the activation of caspase 3 and the BH3-only proteins (Voth et al. 2007). This mechanism of apoptosis prevention was recently analyzed: a *C. burnetii*-type IV effector system enabled proteins belonging to the ankyrin repeat family to be delivered into host cells, thus inhibiting cell apoptosis (Luhrmann et al. 2010).

The role of cellular apoptosis in *C. burnetii* persistence has recently been reevaluated (Fig. 15.2). The engulfment of apoptotic lymphocytes by monocytes and macrophages increases both *C. burnetii* replication and the maturation of bacterial phagosomes into phago-lysosomes (Benoit et al. 2008c). *C. burnetii* replication is associated with an M2 program characterized by the up-regulation of CD14, IL-10 and IL-1 receptor antagonist (IL1-ra) in monocytes and IL-10, TGF-β1, IL1-ra and MR in macrophages. The neutralization of IL-10 and TGF-β1 prevents the replication of *C. burnetii* due to the engulfment of apoptotic lymphocytes, suggesting that IL-10 and TGF-β1 are critically involved in bacterial replication. Finally, IFN-γ

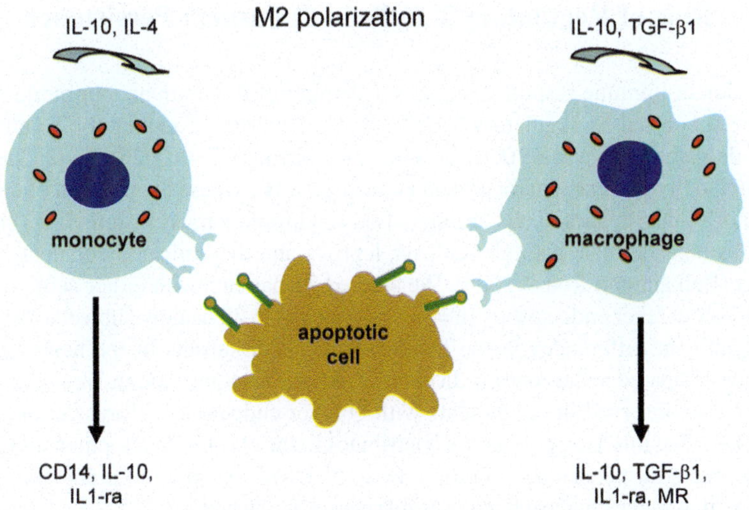

Fig. 15.2 Modulation of macrophage polarization and *C. burnetii* replication. M2 cytokines and the ingestion of apoptotic cells reorient the polarization of monocytes and macrophages toward a M2 phenotype that allows intense *C. burnetii* replication. This mechanism may occur during the chronic evolution of Q fever. The M2 polarization induced by the uptake of apoptotic cells is reverted by IFN-γ and/or inflammatory context

reverses the immune deactivation that is induced by apoptotic cells; it prevents the replication of *C. burnetii* by re-polarizing monocytes and macrophages toward an M1 program (Benoit et al. 2008c). We hypothesize that a similar mechanism may account for *C. burnetii* persistence in endocarditis, the major manifestation of chronic Q fever that affects altered valves. Cardiac valve lesions are associated with a pathological fluid shear stress, which increases the apoptosis of neutrophils, platelets and monocytes. This association suggests that leukocyte apoptosis may be related to cardiac valvulopathy. We monitored the circulating levels of apoptotic leukocytes in patients with valvulopathy and/or Q fever. Patients with valvulopathy exhibited increased levels of circulating apoptotic leukocytes, independently of Q fever (Benoit et al. 2008c). We have hypothesized that the uptake of apoptotic cells by phagocytes may be responsible for the immune impairment observed in Q fever endocarditis, and we propose a model in which apoptotic cells play a key role in the establishment of Q fever endocarditis. The engulfment of apoptotic cells by monocytes and macrophages increases *C. burnetii* replication by redirecting their activation states toward M2 profiles that are non-protective against most pathogens. In an inflammatory context (such as that found in the presence of IFN-γ), the effect of apoptotic cell engulfment is inhibited. Monocytes and macrophages polarized toward a M1 program are able to kill *C. burnetii*, as is observed in patients with acute Q fever without valvulopathy.

15.5 Role of Regulatory T Cells in *C. burnetii* Persistence

An efficient immune response requires a shaping of the immune amplitude, which can be supported by regulatory T cells. Several subsets of regulatory T cells have been described (Shevach 2006). Naturally occurring $CD4^+CD25^+$ T cells develop within the thymus early in the life and constitutively express the α-chain of the IL-2 receptor (CD25). These cells are increased in chronic persistent infections such as schistosomiasis, filariasis, tuberculosis, leprosy and disseminated cutaneous leishmania (Belkaid and Rouse 2005). We studied 71 patients with either acute Q fever (n=17) or Q fever endocarditis (n=54), in addition to 27 healthy subjects. We found no significant differences in the levels of all T cell subsets in the acute Q fever patients compared to the controls. In contrast, a significant increase in Tregs expressing Foxp3 was observed in the patients with Q fever endocarditis (Layez et al. 2012). It is likely that this Treg population contributes to the chronic development of Q fever and to bacterial persistence via the release of IL-10 or via an immunosuppressive process that makes macrophages permissive for *C. burnetii*.

15.6 Conclusions

C. burnetii establishes a survival niche within macrophages by preventing cell death and inducing an immunoregulatory loop based on IL-10 and regulatory T cells. Understanding the persistence of *C. burnetii* may provide clues for the development of chronic Q fever treatments and for reducing the risk of relapse while undergoing adapted treatment. Another problem is to understand how *C. burnetii* establishes a chronic infection that manifests as endocarditis. The immunosuppression associated with chronic Q fever is systemic and cannot account for the valvular localization of *C. burnetii*, even in the presence of valve lesions. We have recently investigated the transcriptional response in valves from infectious endocarditis patients, and we found that genes related to cell death are upregulated (Benoit et al. 2010). We suggest that valve dysfunction (mediated by the release of apoptotic cells) creates a local immunosuppressive environment that favors bacterial persistence. However, the presence of *C. burnetii* in adipose tissue may be essential for accounting for Q fever relapses, even if the presence of *C. burnetii* in human adipose tissue has yet to be reported.

References

Bechah Y, Paddock CD, Capo C et al (2010) Adipose tissue serves as a reservoir for recrudescent *Rickettsia prowazekii* infection in a mouse model. PLoS One 5:e8547

Belkaid Y, Rouse BT (2005) Natural regulatory T cells in infectious disease. Nat Immunol 6:353–360

Ben Amara A, Ghigo E, Le Priol Y et al (2010) *Coxiella burnetii*, the agent of Q fever, replicates within trophoblasts and induces a unique transcriptional response. PLoS One 5:e15315

Benoit M, Barbarat B, Bernard A et al (2008a) *Coxiella burnetii*, the agent of Q fever, stimulates an atypical M2 activation program in human macrophages. Eur J Immunol 38:1065–1070
Benoit M, Desnues B, Mege JL (2008b) Macrophage polarization in bacterial infections. J Immunol 181:3733–3739
Benoit M, Ghigo E, Capo C et al (2008c) The uptake of apoptotic cells drives *Coxiella burnetii* replication and macrophage polarization: a model for Q fever endocarditis. PLoS Pathog 4:e1000066
Benoit M, Thuny F, Le Priol Y et al (2010) The transcriptional programme of human heart valves reveals the natural history of infective endocarditis. PLoS One 5:e8939
Capo C, Zaffran Y, Zugun F et al (1996) Production of interleukin-10 and transforming growth factor beta by peripheral blood mononuclear cells in Q fever endocarditis. Infect Immun 64:4143–4147
Capo C, Lindberg FP, Meconi S et al (1999) Subversion of monocyte functions by *Coxiella burnetii*: impairment of the cross-talk between αvβ3 integrin and CR3. J Immunol 163:6078–6085
Combs TP, Nagajyothi MS et al (2005) The adipocyte as an important target cell for *Trypanosoma cruzi* infection. J Biol Chem 280:24085–24094
Dellacasagrande J, Capo C, Raoult D et al (1999) IFN-γ-mediated control of *Coxiella burnetii* survival in monocytes: the role of cell apoptosis and TNF. J Immunol 162:2259–2265
Dellacasagrande J, Ghigo E, Capo C et al (2000a) *Coxiella burnetii* survives in monocytes from patients with Q fever endocarditis: involvement of tumor necrosis factor. Infect Immun 68:160–164
Dellacasagrande J, Ghigo E, Hammami SM et al (2000b) αvβ3 integrin and bacterial lipopolysaccharide are involved in *Coxiella burnetii*-stimulated production of tumor necrosis factor by human monocytes. Infect Immun 68:5673–5678
Desruisseaux MS, Nagajyothi TME et al (2007) Adipocyte, adipose tissue, and infectious disease. Infect Immun 75:1066–1078
Ghigo E, Capo C, Amirayan N et al (2000) The 75-kD tumour necrosis factor (TNF) receptor is specifically up-regulated in monocytes during Q fever endocarditis. Clin Exp Immunol 121:295–301
Ghigo E, Capo C, Raoult D et al (2001) Interleukin-10 stimulates *Coxiella burnetii* replication in human monocytes through tumor necrosis factor down-modulation: role in microbicidal defect of Q fever. Infect Immun 69:2345–2352
Ghigo E, Capo C, Tung CH et al (2002) *Coxiella burnetii* survival in THP-1 monocytes involves the impairment of phagosome maturation: IFN-γ mediates its restoration and bacterial killing. J Immunol 169:4488–4495
Ghigo E, Honstettre A, Capo C et al (2004) Link between impaired maturation of phagosomes and defective *Coxiella burnetii* killing in patients with chronic Q fever. J Infect Dis 190:1767–1772
Harris RJ, Storm PA, Lloyd A et al (2000) Long-term persistence of *Coxiella burnetii* in the host after primary Q fever. Epidemiol Infect 124:543–549
Honstettre A, Imbert G, Ghigo E et al (2003) Dysregulation of cytokines in acute Q fever: role of interleukin-10 and tumor necrosis factor in chronic evolution of Q fever. J Infect Dis 187:956–962
Honstettre A, Meghari S, Nunes JA et al (2006) Role for the CD28 molecule in the control of *Coxiella burnetii* infection. Infect Immun 74:1800–1808
Langley JM, Marrie TJ, Covert A et al (1988) Poker players' pneumonia. An urban outbreak of Q fever following exposure to a parturient cat. N Engl J Med 319:354–356
Layez C, Brunet C, Lépolard C et al (2012) Foxp3+CD4+CD25+ regulatory T cells are increased in patients with *Coxiella burnetii* endocarditis. FEMS Immunol Med Microbiol 64:137–139
Luhrmann A, Roy CR (2007) *Coxiella burnetii* inhibits activation of host cell apoptosis through a mechanism that involves preventing cytochrome c release from mitochondria. Infect Immun 75:5282–5289
Luhrmann A, Nogueira CV, Carey KL et al (2010) Inhibition of pathogen-induced apoptosis by a *Coxiella burnetii* type IV effector protein. Proc Natl Acad Sci U S A 107:18997–19001

Lumeng CN, Bodzin JL, Saltiel AR (2007) Obesity induces a phenotypic switch in adipose tissue macrophage polarization. J Clin Invest 117:175–184
Martinez FO, Helming L, Gordon S (2009) Alternative activation of macrophages: an immunologic functional perspective. Annu Rev Immunol 27:451–483
Martinoli C, Chiavelli A, Rescigno M (2007) Entry route of *Salmonella typhimurium* directs the type of induced immune response. Immunity 27:975–984
Maurin M, Raoult D (1999) Q fever. Clin Microbiol Rev 12:518–553
Meconi S, Jacomo V, Boquet P et al (1998) *Coxiella burnetii* induces reorganization of the actin cytoskeleton in human monocytes. Infect Immun 66:5527–5533
Meconi S, Capo C, Remacle-Bonnet M et al (2001) Activation of protein tyrosine kinases by *Coxiella burnetii*: role in actin cytoskeleton reorganization and bacterial phagocytosis. Infect Immun 69:2520–2526
Mege JL, Meghari S, Honstettre A et al (2006) The two faces of interleukin 10 in human infectious diseases. Lancet Infect Dis 6:557–569
Mege JL, Mehraj V, Capo C (2012) Macrophage polarization and bacterial infections. Curr Opin Infect Dis 24:230–234
Meghari S, Capo C, Raoult D et al (2006) Deficient transendothelial migration of leukocytes in Q fever: the role played by interleukin-10. J Infect Dis 194:365–369
Meghari S, Berruyer C, Lepidi H et al (2007) Vanin-1 controls granuloma formation and macrophage polarization in *Coxiella burnetii* infection. Eur J Immunol 37:24–32
Meghari S, Bechah Y, Capo C et al (2008) Persistent *Coxiella burnetii* infection in mice overexpressing IL-10: an efficient model for chronic Q fever pathogenesis. PLoS Pathog 4:e23
Munz-Elias EJ, Mc Kinney JD (2002) Bacterial persistence: strategies for survival. In: Kaufmann SHE, Sher A, Ahmed R (eds) Immunology of infectious diseases. ASM Press, Washington, DC, pp 331–355
Neyrolles O, Hernandez-Pando R, Pietri-Rouxel F et al (2006) Is adipose tissue a place for *Mycobacterium tuberculosis* persistence? PLoS One 1:e43
Ohashi K, Parker JL, Ouchi N et al (2010) Adiponectin promotes macrophage polarization toward an anti-inflammatory phenotype. J Biol Chem 285:6153–6160
Ouchi N, Parker JL, Lugus JJ et al (2011) Adipokines in inflammation and metabolic disease. Nat Rev Immunol 11:85–97
Roy CR, Mocarski ES (2007) Pathogen subversion of cell-intrinsic innate immunity. Nat Immunol 8:1179–1187
Sansonetti PJ, Di Santo JP (2007) Debugging how bacteria manipulate the immune response. Immunity 26:149–161
Shevach EM (2006) From vanilla to 28 flavors: multiple varieties of T regulatory cells. Immunity 25:195–201
Stein A, Lepidi H, Mege JL et al (2000) Repeated pregnancies in BALB/c mice infected with *Coxiella burnetii* cause disseminated infection, resulting in stillbirth and endocarditis. J Infect Dis 181:188–194
Tissot-Dupont H, Amadei MA, Nezri M et al (2004) Wind in November, Q fever in December. Emerg Infect Dis 10:1264–1269
Voth DE, Howe D, Heinzen RA (2007) *Coxiella burnetii* inhibits apoptosis in human THP-1 cells and monkey primary alveolar macrophages. Infect Immun 75:4263–4271

Chapter 16
Antigenic Analysis for Vaccines and Diagnostics

Laura R. Hendrix and Chen Chen

Contents

L.R. Hendrix (✉) • C. Chen
Department of Microbial and Molecular Pathogenesis, College of Medicine,
Texas A&M Health Science Center, 3107 Medical Research and Education Building,
Bryan, TX 77807-3260, USA
e-mail: lhendrix@tamhsc.edu

R. Toman et al. (eds.), *Coxiella burnetii: Recent Advances and New Perspectives in Research of the Q Fever Bacterium*, Advances in Experimental Medicine and Biology 984,
DOI 10.1007/978-94-007-4315-1_16,

Abstract *Coxiella burnetii* infection is frequently unrecognized or misdiagnosed, as symptoms generally mimic an influenza-like illness. However, the disease (Q fever) may result in chronic infection, usually manifesting as potentially fatal endocarditis. The development of a chronic fatigue-like sequela may also occur. Infected ruminants are the major reservoir for infection in humans, primarily through exposure to birth products or aerosols that transmit the bacterium over wide regions. A vaccine against *C. burnetii* infection has been in use in Australia for abattoir and agricultural workers for many years. The possibility of adverse reactions in those with previous exposure to the agent has prevented its use elsewhere. Subunit vaccines, utilizing chemical extraction of components thought to cause adverse reactions, are in development, but none are yet licensed. Others have sought to combine immunogenic peptides with or without selected lipopolysaccharide components to produce a vaccine without the possibility of adverse reactions. Selected immunogenic proteins have been shown to induce both humoral and cellular immune responses. Although current diagnosis of infection relies on serological testing, the presentation of specific antibody occurs 7–15 days following the onset of symptoms, delaying treatment that may result in prolonged morbidity. PCR detection of DNA to specific *C. burnetii* antigens in the blood is possible early in infection, but PCR may become negative when PII IgG antibodies appear. PCR is useful for early diagnosis when Q fever is suspected, as in large epidemics, and shortens the delay in the identification of Q fever endocarditis. Others have combined PCR with ELISA or other methods to increase the ability to detect infection at any stage. The search for new diagnostic reagents and vaccines has utilized new methods for discovery of immunoreactive proteins. DNA analysis of the heterogeneity of *C. burnetii* isolates has led to a greater understanding of the diversity of isolates and a means to determine whether there is a correlation between strain and disease severity. 2-D SDS PAGE of immunogenic proteins reactive with human or animal infection sera and mass spectrometric analysis of specific secreted or outer membrane proteins have identified candidate antigens. Microarrays have allowed the analysis of peptide libraries of open reading frames to evaluate the immunogenicity of complete genomes.

Keywords *Coxiella burnetii* • Antigens • Diagnostics • Immunogenic proteins • Vaccines

16.1 *Coxiella burnetii* Infection

Coxiella burnetii causes a worldwide zoonotic disease known as Q fever, an occupational hazard for abattoir and agricultural workers with exposure to cattle, sheep and goats. While disease in animals is generally not apparent, infected animals represent a major reservoir for the dissemination of *C. burnetii* to humans, mainly via exposure to birth products for farm workers or through aerosols that may transmit the bacterium over wide regions. The bacterium exists as two infectious forms, small spore-like cells and larger vegetative cells, with the small cell variants able to survive for long periods in the environment. In addition to domesticated animals, *C. burnetii* is known to infect wild vertebrates and arthropods. In humans, Q fever is generally a self-limiting, flu-like illness with characteristic high fever and headache, but may also present as pneumonia, hepatitis, meningoencephalitis, peripheral thrombosis, myopericarditis, or bone marrow aplasia (Ormsbee and Marmion 1990). The disease may persist after overt clinical symptoms have subsided and reactivate with appropriate stress (Baca and Paretsky 1983). Commonly, the disease is unrecognized or undiagnosed, and in many areas, a high percentage of the population has serological evidence of previous or latent infection (Luoto et al. 1965; Meiklejohn et al. 1981; Richardus et al. 1987; Wisniewski et al. 1969). Acute disease is readily treated with antibiotics, although the response to antibiotics may be slow. However, the disease may manifest as chronic infection in the form of endocarditis or hepatitis. These cases may have a poor prognosis and may be associated with suppression of cell-mediated immunity. Chronic infections have not responded well to antibiotic regimens, but the best outcome is seen following a combination of chloroquine and doxycycline administered over 1–2 years (Raoult et al. 1999; Million et al. 2010). A post-Q fever debility syndrome, with symptoms similar to chronic fatigue syndrome, is also recognized and may follow in up to 10% of acute disease cases (Marmion et al. 2005).

A vaccine to *C. burnetii* infection is licensed in Australia where large numbers of infections occurred annually before the implementation of a national vaccination program. However, the vaccine is not routinely used elsewhere due to adverse reactions that may occur in those with previous immunity, often from undiagnosed contact with the agent. While a variety of guidelines have been defined for specific sero-diagnostic strategies, diagnosis of disease is often hampered by serological evidence of past or latent infection. New methods for detection of acute cases of *C. burnetii* in blood using PCR (polymerase chain reaction) are currently restricted to the limited period before the appearance of IgG antibodies.

16.2 Early Vaccines

Vaccine development against Q fever, primarily prompted by the need to prevent disease in abattoir and agricultural workers, veterinarians and laboratory researchers, has been an active area of research for several decades. Early vaccines prepared from formalin-inactivated laboratory stocks passed many times in hen's eggs were found to be 100–300 times less effective than vaccines made from fresh isolates (Ormsbee et al. 1964). These findings were corroborated by observations that the bacterium exists as two antigenic phase variants, known as virulent phase I (PI) and avirulent phase II (PII), that differ in their ability to protect against infection in a guinea pig model (Stoker and Fiset 1956; Moos and Hackstadt 1987). Phase variation in the Nine Mile (NM) strain was found to relate to a large chromosomal deletion (Vodkin and Williams 1986; Hoover et al. 2002; Denison et al. 2007a) leading to an irreversible smooth-to-rough mutation in lipopolysaccharide (LPS) O-side chain expression (Schramek and Mayer 1982) that occurs following passage of the organism in non-immunologically competent hosts, such as hen's eggs used for laboratory propagation (Lukacova et al. 2008).

Early research on Q fever was characterized by outbreaks of disease in laboratory workers. The ability to grow large numbers of organisms in chick embryo yolk sacs allowed the production of a formalin-inactivated whole cell vaccine to protect laboratory workers. A formalin-inactivated, ether-extracted 10% suspension of yolk sac (Smadel vaccine) from the Henzerling strain was highly immunogenic in man and animals (Smadel et al. 1948). This PI whole cell vaccine (WCV-PI) has been tested extensively for use in animals and humans, but its use is restricted in most countries due to the prevalence of local and systemic adverse reactions that can occur primarily in previously sensitized individuals. Nevertheless, studies have shown WCV-PI to be able to induce humoral and cellular immune responses to *C. burnetii* antigens and to prevent Q fever in humans and animals (Benenson 1959). These vaccines have also prevented shedding of *C. burnetii* into the environment by infected animals and offered protection against infertility, abortion and low birth weight in livestock (Bilberstein et al. 1977; Schmeer et al. 1987; Arricau-Bouvery et al. 2005). Vaccines utilizing killed PII organisms have not been shown to be effective to prevent natural infection in animals (Fishbein and Raoult 1992), but WCV-PI use in animals is not widespread since there is little economic incentive to eradicate what is generally a subclinical infection (Sawyer et al. 1987). Increased outbreaks in France, Germany, and the Netherlands have emphasized the importance of Q fever detection in livestock (Fishbein and Raoult 1992; Hellenbrand et al. 2001; Carrieri et al. 2002; Tissot-Dupont et al. 2007; Enserink 2010). Measures taken to curb infection of livestock in the Netherlands have included massive vaccinations of non-pregnant goats with an inactivated NM PI vaccine shown to reduce shedding and abortion in non-infected animals that unfortunately requires prior culling of pregnant ewes (CEVA, Libourne, France, Frankel et al. 2011).

An attenuated *C. burnetii* strain designated Grita M-44, thought to be in PII, was tested as a vaccine in the former Soviet Union (Genig 1968). It was later abandoned when studies showed this strain was able to persist in guinea pigs with mild lesions occurring in heart, spleen and liver, suggesting there could be a reactivation of infection in vaccinated individuals due to residual PI organisms (Johnson et al. 1976, 1977).

A new generation vaccine for Q fever is required to confer protection against infection without the necessity to screen for prior immunity before vaccination, as skin testing of potential vaccinees may result in adverse reactions and is a barrier to approval for use in the US and elsewhere. Chemical extraction methods were developed to reduce the adverse effects of whole killed cell vaccines while retaining good immunogenicity. Initial efforts focused on developing a vaccine from whole killed cells with the removal of components responsible for sensitization, thought to be due to PI-LPS. Extraction methods utilized included phenol-water mixtures, dimethylsulfoxide (DMSO), formamide, trichloroacetic acid (TCA) and chloroform-methanol (CM). Kazar et al. tested a TCA-extracted NM PI vaccine in workers at high risk for exposure to *C. burnetii* infection in the former Czechoslovakia.

While a significant antibody response occurred in vaccinated individuals, severe local and systemic reactions occurred in individuals who had been previously infected with *C. burnetii* (Kazar et al. 1982).

A CM-extraction residue vaccine (CMR) from PI Henzerling strain *C. burnetii* was developed through a joint effort between Rocky Mountain Laboratories and the US Army Research Institute for Infectious Disease (Williams and Cantrell 1982; Williams et al. 1986). This vaccine could be used at higher doses in mice than the WCV-PI without causing severe adverse reactions. Efficacy of the CMR vaccine has been shown so far in rodents, sheep and non-human primates (Brooks et al. 1986; Waag et al. 1997, 2002). Phase one human trials showed it could be safely administered to human volunteers unscreened for prior immunity at doses of 30 or 60 μg. However, the CMR vaccine was found to still contain some components responsible for adverse reactions, as higher doses in human volunteers caused reactions qualitatively similar to those seen with 30 μg doses of WCV-PI (Fries et al. 1993). A more recent study showed that a small priming dose given before vaccination with CMR vaccine is able to prime the immune system to mount an anamnestic response to *C. burnetii* (Waag et al. 2008). The production of specific antibody and the activation of peripheral blood cells by recall antigen in vitro indicated the CMR vaccine could offer protection to infection in humans.

16.3 Licensed *Coxiella* Vaccine

Since there are no *C. burnetii* vaccines that completely eliminate adverse reactions, none are currently licensed for use in the US. Australia, however, with large numbers of Q fever cases annually, has inoculated at-risk abattoir workers with formalin-inactivated WCV-PI made from PI Henzerling strain (Q-Vax, Commonwealth Serum Laboratories (CSL)) following testing in the 1980s. In 1994, Ackland et al. showed that the protective efficacy of Q-Vax is nearly 100%, with the duration of protection greater than 5 years (Ackland et al. 1994). However, at the end of 2005, production of the vaccine halted as new regulations for biocontainment facilities were put into effect. With government support, CSL was able to restart production and extend the program to include at-risk rural communities. Results indicate that broadening the population offered the vaccine has resulted in a significant decline in case numbers (Marmion 2007).

Marmion (2007) described the components of Q-Vax that may allow it to be an effective vaccine, suggesting these components would likely be required for new non-cellular vaccines to be effective. First, the highly immunogenic PI LPS or a component of it may be required for the resulting protective qualities of the vaccine. Second, *C. burnetii* proteins that interact with monocytes-macrophages stimulate the cellular immune system to produce interferon-γ and other cytokines, resulting in restriction of *C. burnetii* intracellular growth. Marmion postulated that antibody to PI LPS blocks the occurrence of other interactions of the bacterium with monocytes-macrophages that might down-regulate the cellular immune response, including down-regulation of interferon-γ formation. The third component required for vaccine efficacy may be the small-cell variants of *C. burnetii.* Due to their slow biodegradability, they may display both PI LPS and protein antigens as a continuing antigenic stimulation.

16.4 Diagnostics

The natural antibody response in early stages of *C. burnetii* infection is primarily to protein components. In late stage acute infection or in chronic disease, the antibody response is primarily to the LPS O-antigen. This is detected clinically through the differential response to fixed whole-cell PI and PII antigen at different stages of infection. Patient serum from early acute disease reacts initially with PII IgM antibody followed by PII IgG (Geddes 1983). Previous infections are characterized by the presence of PII and low PI IgG antibody levels. High IgG titers to PI antigen, originally set at>800 but now revised to>1,600, signal possible chronic infection (Dupont et al. 1994; Fournier et al. 1996; Frankel et al. 2011). While current platforms for serological testing include indirect immunofluorescence, complement fixation, enzyme-linked immunosorbent assay (ELISA), and microagglutination (La Scola 2002), commercially available products all currently require the use of whole-cell antigen.

LPS antigen was shown to have interstrain heterogeneity in silver-stained SDS-PAGE (sodium dodecyl sulfate polyacrylamide gel electrophoresis) at the 10–17 kDa size range among *C. burnetii* isolates from acute and chronic infection. LPS antigenic types were separated into three distinct groups – two groups from chronic disease isolates and one from acute disease isolates (Hackstadt 1986). These results suggest that heterogeneity of LPS may affect PI titers, depending on the infecting strain, as organisms used for testing have generally been from one strain (Nine Mile).

Likewise, protein heterogeneity between strains may affect PII titers, as strains have been shown to vary by DNA banding patterns, by PAGE patterns of stained or immunoblotted proteins and other methods, with protein heterogeneity between strains confirmed by genomic sequencing data (Beare et al. 2009). These results suggest testing patient sera using NM PI and PII cells may not always give an accurate picture of the possibility and stage of disease in patients having symptoms compatible to *C. burnetii* infection. In practice, this is not considered to be an issue. For example, tests of the response to past infection or vaccination with Q-Vax show equivalent protection to Priscilla, Henzerling and NM strains as measured by proliferation of blood mononuclear lymphocytes on antigen challenge, with T lymphocytes being the major responders (Izzo et al. 1991). However, eight PI *C. burnetii* isolates from four genomic groups were shown in mouse and guinea pig models of acute Q fever to cause genogroup-specific disease manifestations while vaccine challenge between isolate groups were cross-protective (Russell-Lodrigue et al. 2009). Also, a study of the antigenic variation of polypeptides from 18 strains of *C. burnetii* showed a 28 kDa protein was immunodominant in acute disease isolates but not in strains from human cases of chronic Q fever, suggesting this polypeptide could be a marker for acute disease (To et al. 1998; Zhang et al. 2005). Nevertheless, discovery of new diagnostic reagents have focused on proteins unique to *C. burnetii* that are conserved among all isolates.

16.4.1 Commercially Available Diagnostics

The lack of characteristic symptoms for Q fever infection often leads to a misdiagnosis of an influenza-like illness. Identification of the infecting organism from clinical specimens is not advised as *C. burnetii* is highly infectious and requires a biosafety level three laboratory for cultivation. Since a firm diagnosis of the disease relies on evidence of a four-fold rise in serum antibody titer approximately 14 days after the first serum sample was taken (Derrick 1937; Wisniewski et al. 1969; Brown 1973), serologic testing is a necessary supplement to clinical history. In 1937, Burnet and Freeman demonstrated macroscopic agglutination reactions between patient sera and spleen cell emulsions from infected mice (Burnet and Freeman 1937). The microagglutination test (MAA) developed from this early observation that whole bacterial cells clump or agglutinate in the presence of positive sera. Since PII cells spontaneously auto-agglutinate in normal serum, the creation of particulate PII antigens from TCA treatment of PI cells is required.

Commercially available assays to detect Q fever include the complement fixation test (CFT) and immunofluorescence assay (IFA). The CFT was first developed in 1941 (Bengtson 1941) and entails the addition of *C. burnetii* antigen to patient sera from which any naturally occurring complement has been depleted. A standard amount of complement is added back, followed by the addition of sheep erythrocytes pre-bound to anti-erythrocyte antibodies. If antibodies to *C. burnetii* are present in the patient sera, they will bind to *C. burnetii* antigen and complement will be depleted from the assay due to its binding to the antigen-antibody complexes. The assay is made quantitative by diluting patient sera until antibody levels are low enough for complement to bind to and lyse the erythrocytes. The CFT has been shown to be less sensitive than other assays and has largely been replaced by newer assays, including IFA.

The most common serological test currently used is the indirect IFA (Fenollar et al. 2004). This test is a primary binding assay, similar to the ELISA, and utilizes phase variation of *C. burnetii* LPS to distinguish the acute convalescent response from chronic infection responses. Dilutions of sera are reacted with PI and PII organisms in each slide well and labeled with an immunoglobulin class-specific fluorescent conjugate. Results are viewed using fluorescence microscopy and compared to positive and negative control wells. Positive sera are rerun to determine the end-point titer (the reciprocal of the antibody dilution) to compare to earlier or later samples from the same patient. In acute cases, titers to PI are accompanied by equal or higher titers to PII. This is reversed in chronic disease. PII titers over 1:256 are evidence of active or recent infection. PI IgG is often below detectable levels until convalescence or in the presence of chronic disease. Chronic hepatitis may have high and equal PI and PII titers, while endocarditis frequently has higher PI titers. The IFA is the current reference method for Q fever serodiagnosis (La Scola 2002).

The enzyme-linked immunosorbent assay (ELISA) is a sensitive and reproducible colorimetric assay that can easily process large numbers of serum samples determined by optical density (OD) units on a microplate reader (spectrophotometer). PI or PII whole killed cells, LPS, or individual purified proteins may serve as antigens absorbed to wells of a microtiter plate. Dilutions of patient sera are added to wells and anti-human class specific antisera conjugated to an enzyme, usually peroxidase or alkaline phosphatase, is used to detect sera containing *C. burnetii* antibodies following the addition of the enzyme substrate. Titers are determined by comparison of OD differences between antigen wells and non-antigen control wells. The ELISA has been shown to be more sensitive than CFT, IFA or MAA with fewer false positives (Peter et al. 1987; Waag et al. 1991). However, all current serological tests may require the determination of cutoff values for selected locations or test samples. A commercial ELISA detecting *C. burnetii*-specific IgG was evaluated in 2002 for pre-vaccination screening and diagnosis in Australia, but results suggested negative reactions by ELISA be confirmed by IFA or CFT (Field et al. 2002). A comparison of four commercially available assays to detect IgM PII antibodies to *C. burnetii* included three IFA tests and one ELISA. Sera from 23 patients with clinical symptoms of acute Q fever were tested along with 88 control sera from blood donors. The specificity of results varied from 97.7% to 100% for the IFAs and 98.9% for the ELISA. Sensitivity was 95.65% for the ELISA and 100% for all IFAs (Frangoulidis et al. 2006).

The ELISA may be modified to a fluorescence assay (ELIFA), as was done by Schmeer et al., resulting in a 50-fold higher sensitivity for the analysis of humoral immune responses in vaccinated and naturally infected cattle to several *C. burnetii* proteins (Schmeer et al. 1988). Interestingly, the results showed that the early humoral response in cattle immunized with either chloroform-methanol extracted cell residues or a 27 kDa protein (Com1) was directed against the 27 kDa protein. They were also able to distinguish vaccinated from naturally infected cattle by measuring IgG1 and IgG2 subclass responses with ELIFA.

16.4.2 Diagnostics in Development

16.4.2.1 Sero-Diagnostics

Due to its sensitivity and ability to screen large numbers of samples, investigators have modified the Q fever ELISA for specific situations. Guatteo et al. applied the test to compare the detection of antibodies to an ovine *C. burnetii* strain in milk or serum from six infected dairy cattle herds. Of the 264 positive serum samples, 249 were also positive in milk. Of the 184 serum negative samples, 176 were negative in milk.

These results showed that a more convenient milk-ELISA could be comparable to serum-ELISA in lactating dairy cattle (Guatteo et al. 2007).

Many investigators have looked for specific antigens to replace whole killed cells for use in the ELISA platform. Sekeyova et al. identified candidate *C. burnetii* protein antigens through the creation of three monoclonal antibodies to NM strain and one to Priscilla strain (Sekeyova et al. 2010). They tested the monoclonal antibodies using one- and two-dimensional SDS-PAGE, immunoblot analysis and mass spectrometry and found they identified three specific *C. burnetii* proteins, CBU_0937 (unknown function), CBU_1910 (Com1), and CBU_0236 (elongation factor Tu). They purified Com1 and CBU_0937 proteins through the expression of PCR amplified clones and used them in an ELISA to screen sera from Q fever endocarditis and acute disease patients. Their results showed CBU_0937 was a better diagnostic tool for Q fever ELISA than Com1 protein.

A monoclonal antibody to virenose, an unusual sugar in *C. burnetii* PI LPS O antigen, was generated using a virenose-rich O-polysaccharide LPS fraction conjugated to tetanus toxoid. Testing of the monoclonal antibody at a dilution of 1:1,000 in an ELISA and by immunoblot analysis showed it to be a sensitive and specific method for the detection of virulent *C. burnetii* (Palkovicova et al. 2009).

An IgG ELISA microplate assay using *C. burnetii* heat shock protein, HspB (GroEL), was developed and tested in goats. Sera from naturally and experimentally infected goats were tested. The purified rHspB was recognized by 18 days post-infection (dpi) and was positive at 39–60 dpi in 80–90% of the animals tested. Also, animals presenting with reactivation of infection had statistically higher titers than those in latent infection (Fernandes et al. 2009).

16.4.2.2 DNA-Based Diagnostics

PCR assays detect genes encoding bacterial proteins or non-transcribed regions independent of viability, although non-viable forms are likely to be more readily degraded over time. A real-time PCR assay for the detection of *C. burnetii* DNA in serum samples targeting IS*1111* transposase sequences (also known as htpAB-associated repetitive element (Hoover et al. 1992)) was used in the Netherlands to diagnose acute Q fever early in infection to avoid delay in treatment that can lead to increased hospital admission rates and prolonged morbidity. Researchers found the latest time point after onset of disease in which *C. burnetii* could be detected by PCR of serum samples was day 17. They found 98% of seronegative sera, 90% of sera with IgM-PII antibodies, 23% with IgM-PII/IgG-PII antibodies, 5% with IgM-PII/IgG-PII/IgM-PI antibodies, and 0% with IgG-PI antibodies in combination with any other antibody types were positive by PCR (Schneeberger et al. 2010).

Previously, Fournier and Raoult used a similar IS*1111* probe to compare LightCycler nested-PCR (LCN-PCR) to serology by immunofluorescence for early diagnosis of acute Q fever. While the time line for detection of *C. burnetii* by PCR was similar to Schneeberger et al., the sensitivity in their assay was 26% for seronegative samples, and 5% for seropositive patients (Fournier and Raoult 2003).

The sensitivity of the assays of Schneeberger et al. was greater, possibly due to (1) an increase in numbers of IS*1111* in those isolates, (2) the use of a smaller sized amplification product or (3) a larger initial serum sample. These results indicate that PCR detection of *C. burnetii* can be a useful tool for early diagnosis when Q fever is suspected, such as in large epidemics that have occurred since 2007 in the Netherlands. PCR detection overcomes the lag in antibody response of 7–15 days following onset of symptoms that hampers the standard serological diagnosis (Schneeberger et al. 2010).

Fenollar et al. have also used LCN-PCR with an IS*1111* probe to shorten the diagnostic delay in the identification of Q fever endocarditis and vascular infections. They retrospectively and prospectively tested samples from 48 Q fever endocarditis or vascular infection patients and compared results to 100 controls with endocarditis caused by other agents. They prospectively used the same technique to test 30 patients treated for Q fever endocarditis and 13 patients convalescent for acute Q fever. Results showed that sera stored more than 1 month exhibited 16.2% positive results, while 64% of fresh samples analyzed were positive. High IgG PI titers decreased the sensitivity of the PCR assay. None of 15 serum samples with a PI IgG titer equal to or greater than 1:25,600 were PCR positive, compared to 39.4% positive when the PI IgG titer was less than 1:25,600. If only the fresh sera were counted in samples that had titers less than 1:25,600, 100% were positive. All serum samples from patients treated for Q fever endocarditis or who were diagnosed with acute Q fever and whose IgG PI titers were 1:800 or greater, were negative by PCR. All control patients with endocarditis from other organisms were negative. For 12 patients the first serum sample was positive, but after 1 month of treatment, all subsequent samples were PCR negative. Positivity of all PCR products was confirmed by sequencing (Fenollar et al. 2004).

A high throughput real-time PCR assay with an internal control system and automated DNA preparation was developed as a diagnostic tool to be used during biological weapons attacks or local epidemics (Panning et al. 2008). Use of the IS*1111* transposase results in a probe with a high specificity for *Coxiella*, although individual strains vary as to the number of IS*1111* elements in their genomes, and strains lacking the transposase have been described (Denison et al. 2007b).

A recent study evaluated real-time PCR assays and DNA extraction methods for Q fever diagnosis in use in seven diagnostic or reference laboratories in the Netherlands. All laboratories in the study used real-time PCR assays targeting the IS*1111a* insertion element using different DNA extraction methods. While a low degree of variation in the sensitivities of most of the developed assays was observed, PCR assays that amplified short DNA fragments (70–87 bp) yielded better results than those using larger fragments (202 bp). It was postulated that a shorter sequence would present a better target, due to the likelihood of bacterial DNA degradation in serum. Sequence polymorphisms at target sequences in isolates tested may also affect PCR performance. DNA extraction methods using certain commercial kits yielded better results than others, despite similar chemistries involved (Tilburg et al. 2010).

The detection of *C. burnetii* in tissues has been demonstrated using commercial human *C. burnetii* positive control serum as the primary antibody source (PanBio, Inc, Columbia, MD, USA) and stained using a peroxidase immunohistochemistry method (Vectastain Elite ABC Kit, Vector Laboratories, Inc., Burlingame, CA, USA). Results were compared with those obtained from an experimental fluorescent *in situ* hybridization test targeting the 16 S ribosomal RNA to detect *C. burnetii* in placentas from ruminant abortions. Both methods performed equally well, giving equivalent results (Jensen et al. 2007). Immunohistochemistry alone had been used earlier to detect *C. burnetii* in bovine placentas (Dilbeck and McElwain 1994; Bildfell et al. 2000).

16.5 Searches for Immunogenic and Protective Antigens

The likelihood that Q-Vax would not be approved for licensing in other countries has led to a search for specific immunogenic and protective proteins for the development of a protein subunit vaccine and improved diagnostics. Several early studies suggested that subunit protein vaccines could provide protection. Williams *et al.* showed the 29 kDa *C. burnetii* porin protein P1 conferred protection from lethal challenge in mice (Williams et al. 1990). A protein of similar size was extracted from outer membrane preparations with the detergent Empigen BB and found to be immunogenic and protective in mice against *C. burnetii* challenge. The Empigen-extract was shown to also contain *C. burnetii* LPS (Lukacova et al. 1994). A partially purified 67 kDa GroEL protein conferred full protection in guinea pigs and mice (Zhang et al. 1994). However, because these proteins were not homogeneous, the vaccines may have contained other components not fully identified.

Further studies focused on the identification of immunogenic proteins defined by strong reactivity to infection-derived sera. Muller et al. identified an immunogenic 27 kDa *C. burnetii* antigen on SDS-PAGE gels purified from infected cell cultures by guanidinium hydrochloride treatment and chloroform/methanol extraction (Muller et al. 1987). Techniques including immunoprecipitation of surface-iodinated *C. burnetii*, Western blot analysis of bacterial outer membrane fractions separated on SDS-PAGE, or bacterial products released into acidic media were used to identify surface-exposed candidate antigens (Hendrix et al. 1993; Redd and Thompson 1995). With these techniques, protein isolation and purification was required in order to test immunogenicity, which was time-consuming and, in most cases, not easily done before the advent of PCR and genome sequencing. Methods using immune sera to select immunoreactive antigens from *E. coli* gene banks of surface-exposed *C. burnetii* proteins or N-terminal and internal sequencing of proteins from detergent fractions of infected cells on SDS-PAGE gels were possible when DNA sequencing techniques were available. These methods identified several predominant antigens that were similar in size, including Com1 and P1 (Hendrix et al. 1993; Varghees et al. 2002). Cloned antigens allowed DNA sequencing to be done to positively identify selected antigens and to express and purify them from *E. coli.* In vitro transcription and translation protocols also made possible the conversion of cloned DNA into protein.

More recently, four candidate antigens (Com1, P1, P28 and Cb-Mip) have been cloned and characterized through testing in a sublethal challenge model in BALB/c mice using protection from the development of severe splenomegaly as an indicator of vaccinogenic activity. None of the identified antigens individually conferred significant protection in this model, suggesting multiple antigens may be required to generate protection or that another delivery system may be more appropriate (Zhang and Samuel 2003). A comprehensive search that made use of antisera obtained from an experimental infection at various times post infection resulted in the identification of these proteins and an increased number of immunoreactive antigens by SDS-PAGE, providing a time course of proteins involved in immunity (Zhang et al. 2004).

In a study that was able to generate protection, a purified recombinant protein containing a fusion of *C. burnetii* antigens P1 and HspB (GroEL) was used to immunize mice. Detection of antibodies, T-cell proliferation and cytokine secretion indicated animals immunized with the fusion protein exhibited stronger humoral and cellular immune responses than those immunized with either protein alone. Spleens removed from animals challenged with *C. burnetii* showed those mice that received the fusion protein had smaller spleens that contained significantly lower bacterial loads (Li et al. 2005). These results suggested a promising method for generating a subunit vaccine may reside in the protein-folding properties of the chaperonin protein.

Projects to produce a vaccine through the combined cloning of portions of the O antigen of the LPS molecule heterologously with immunogenic proteins are currently in development. A prerequisite to the development of such a strategy is the determination of all genes and their enzymatic activities required for the assembly of PI *C. burnetii* O antigen. This is made more difficult since the *C. burnetii* O antigen contains unique sugar components and the pathways to their synthesis are not fully characterized (Narasaki et al. 2011).

16.6 New Methods to Identify Immunogenic Proteins

16.6.1 Strain Diversity

In order to successfully detect infections of *C. burnetii*, it is important to have some knowledge of the strain diversity of specific antigens to be used in diagnostics and vaccines. Several typing schemes have been used to classify *C. burnetii* isolates to determine their antigenic diversity over time and geographic locations. Such studies may pinpoint areas having increased diversity of isolates that may require polyvalent vaccines or diagnostics specific to the area or indicate antigens specific only to certain isolates are not appropriate for vaccines or diagnostics for common use. Early studies utilized restriction enzyme-digested DNA and agarose or polyacrylamide gel electrophoresis to detect restriction fragment length polymorphisms (RFLP) to compare isolates, with some studies utilizing pulsed field gels (Vodkin et al. 1986; Hendrix et al. 1991; Jager et al. 1998). These early studies separated the available isolates into distinct genomic groups based on DNA restriction fingerprints. Plasmid types correlated with the genomic groups and in some studies, isolates causing chronic infection were grouped together. Isolates that varied only by passage history, including those that differed by LPS phase, showed no apparent differences in restriction enzyme banding patterns when EcoR1 or BamH1 enzymes were used (Hendrix and Williams 1991). NM LPS phase variants were shown to differ if DNA was digested with Hae III (Vodkin et al. 1986). Differentiation of *Coxiella* isolates was also carried out through a repetitive element PCR genotyping method to detect the presence of insertion sequences of the IS*1111* transposase. Differences between isolates in the number of elements per genome and in sequence alterations within and near IS element coding regions allowed them to be divided into five genomic groups that corresponded to groups determined by restriction enzyme banding patterns (Denison et al. 2007b).

Later genotyping schemes were developed to have greater discriminatory ability to distinguish between non-clonal isolates, such as may be required in epidemiological investigations in order to follow outbreaks. Unlike other genera, 16S ribosomal RNA genes and the generally hypervariable 16S–23S ITS (internal transcribed spacer) region are conserved throughout *C. burnetii* isolates (Stein et al. 1997; Thiele et al. 1994). This has led to the development of more powerful assays utilizing regions that are not transcribed and thus likely to be more variable. Multispacer sequence typing, based on intergenic region sequencing, was used to classify 173 *C. burnetii* isolates. Using 10 variable spacers, the authors were able to establish 30 different genotypes characterized into 3 monophyletic groups by phylogenetic analysis inferred from the sequences (Glazunova et al. 2005). Svraka et al. used Tandem Repeats Finder Software (Benson 1999) to search the *C. burnetii* RSA493 genome sequence for tandem repeats and for the development of primer sets for multiple locus variable number of tandem repeat (VNTR) analysis (MLVA) typing. Twenty-one isolates, including five passage variants, were characterized using seven tandem repeat loci. Results indicated that the number of repeats per VNTR locus varied between 2 and 18 and the number of variant alleles per locus varied between 3 and 6. Nine unique marker allele size combinations or MLVA types were observed among the 21 *C. burnetii* samples. Cluster analysis indicated the MLVA types fell into five major clusters of closely related isolates. Analysis of passage history variants was used to determine the stability of the chosen markers. All five NM passage variants, 2S and two Priscilla variants were stable over time and geographic location (Svraka et al. 2006).

Arricau-Bouvery et al. employed infrequent restriction site-PCR (IRS-PCR) and MLVA typing to characterize 42 *C. burnetii* isolates from livestock and ticks. The IRS-PCR typing of 14 isolates by 4 different restriction assays resulted in 6 genotype groups. MLVA typing of the 14 isolates and 28 additional isolates resulted in 22 additional genotypes using 10 minisatellites from NM RSA493 having repeat units greater than 9 base pairs. The authors proposed two separate panels of markers, one using minisatellite markers which can be typed in any laboratory on agarose gels and a second panel of microsatellites which can be typed using capillary electrophoresis. The Microorganisms Tandem Repeats Database at http://minisatellites.u-psud.fr was used to establish the panel of markers. With the availability of additional *Coxiella* genome sequences, additional tandem repeats will be included in typing schemes that may not be present in RSA493 (Arricau-Bouvery et al. 2006).

MLVA typing has also been applied to isolates from five patients and six sheep in the 2007 outbreak in the Netherlands using six microsatellites from NM RSA493. Results indicated at least four *C. burnetii* genotypes were involved in the outbreak (Klaassen et al. 2009).

Beare et al. used comparative genome hybridization (CGH) on DNA microarrays containing all open reading frames of the NM PI reference isolate to assess intrastrain/species whole genome sequence variation to analyze the diversity of 24 *C. burnetii* isolates. The analysis confirmed earlier RFLP groupings (Hendrix et al. 1991) and showed that two previously ungrouped isolates represent distinct genomic groups. Results showed that both small and large genetic changes were responsible for attenuated virulence due to truncated LPS expression and truncation or loss of other genes may similarly affect virulence. CGH assessment of genome content can be useful to identify cross-protective subunit vaccine candidates and aid in the development of new diagnostic reagents (Beare et al. 2006).

16.6.2 2-D SDS PAGE Western Analysis

Early analysis of antigens utilized one-dimensional SDS-PAGE and immunoblotting with infection-derived sera to detect reactive antigens. With the sequencing of the *C. burnetii* genome and advances in mass spectrometry that allowed protein identification from smaller sample sizes, the ability to select and identify antigens on two-dimensional SDS-PAGE was made easier, although 2-D gels remain a labor intensive and technically demanding method. Studies comparing antigens of large and small cell replication variants of *C. burnetii* determined that two highly antigenic proteins of 12.5 and 15 kDa are strong candidates for diagnostics. They were seen in vaccinated and infected animals throughout infection and were associated with LPS, predominately in SCVs (small cell variant), late in infection (Coleman et al. 2007). Deringer et al. identified nine novel immunogenic proteins by mass spectrometry on Western blots of 2-D SDS-PAGE using guinea pig immune sera and 11 additional proteins which were also previously shown to react to human sera in *C. burnetii* whole cell lysates (Deringer et al. 2011). Sekeyova et al. utilized acute stage versus chronic stage infection sera to select antigens on 2-D SDS-PAGE of *C. burnetii* membrane extracts. Results of these studies showed that 29 *C. burnetii* antigens reacted with acute or chronic sera, with two proteins recognized only by chronic patient sera. These proteins were considered possible candidates for serodiagnostic assays to detect Q fever endocarditis (Sekeyova et al. 2009).

16.6.3 Mass Spectrometric Analysis

As *C. burnetii* do not grow on bacteriologic media in common use and require increased containment for growth, rapid methods to identify it as the biological agent responsible during outbreaks are needed. Several different strategies for the rapid identification of intact bacteria using mass spectrometric approaches have been applied to *C. burnetii* isolates. In 2002, Domingues et al. analyzed phospholipids from *C. burnetii* using fast atom bombardment mass spectrometry (FAB-MS). They acquired constant neutral loss scanning mass spectra (CNL) from NM PI and PII cells to identify various phospholipids within phospholipid classes. Results showed that phospholipids from PII *C. burnetii* were less complex than PI phospholipids and lacked the phosphoinositol class of phospholipids. They postulated that this rapid method could be used to differentiate virulent *C. burnetii* from low virulent strains (Domingues et al. 2002).

Shaw et al. used whole PI *C. burnetii* organisms grown and purified at different times and places to identify unique and reproducible biomarkers by matrix-assisted laser desorption/ionization time-of-flight (MALDI-TOF) mass spectrometry. Small cell variant protein A (ScvA) was one of three consistently identified proteins that would be expected to be present in environmental or intentionally released samples (Shaw et al. 2004). Pierce et al. expanded the use of mass spectrometry to identify *C. burnetii* strains and phase. This approach involved the use of MALDI-TOF mass spectrometry and supervised pattern recognition using Partial Least Squares-Discriminant Analysis (PLS-DA) on seven *C. burnetii* isolates grouped into six strain types. The method was validated by the prediction of unknown *C. burnetii* samples with 100% sensitivity and specificity for five out of the six strain classes (Pierce et al. 2007).

Hernychova et al. employed MALDI-TOF MS analyses on acetonitrile and TCA extractions of inactivated *C. burnetii* to obtain specific fingerprints of *C. burnetii* surface-exposed molecules to be used for detection and identification of *C. burnetii*. They also used electrospray tandem mass spectrometry coupled with nanoscale ultra performance liquid chromatography (LC-MS/MS) to identify 20 acetonitrile-extracted *C. burnetii* proteins (Hernychova et al. 2008).

These techniques have also been used for the discovery of vaccine candidates through identification of specific secreted or outer membrane products. Samoilis et al. used mass spectrometry techniques to identify proteins of *C. burnetii* secreted into the cytosol of infected Vero cells (Samoilis et al. 2010). Similarly, Flores-Ramirez et al. used mass spectrometry and bioinformatics techniques to identify 21 predicted outer membrane proteins and nine lipoproteins in NM PI lysates as possible vaccine candidates (Flores-Ramirez et al. 2009).

16.6.4 Microarray Analysis

Knowledge of the genome sequence has made possible the construction of libraries containing all open reading frames to evaluate the vaccinogenic activity of the complete genome using microarray technology. Various studies have used either animal or patient sera to select and analyze immunoreactive proteins.

Beare et al. generated 1,988 *C. burnetii* open reading frames as transcriptionally active PCR products from which full-length proteins were synthesized for 75% of the ORFs using an *E. coli* based in vitro transcription-translation (IVTT) system. Crude IVTT lysates were used to spot nitrocellulose microarrays that were probed with acute Q fever patient sera and sera from Q-Vax vaccinees. Approximately 50 *C. burnetii* proteins reacted strongly with immune sera. Immunoreactivity to IVTT lysates was confirmed by ELISA using recombinant proteins corresponding to selected array-reactive antigens. Positive proteins included previously reactive immunogens, an ankyrin-repeat containing protein and many hypothetical proteins (Beare et al. 2008).

Microarray analysis of 1,901 ORFs from *C. burnetii* using a plasmid-based expression system was used to evaluate humoral immune responses to Q fever patient sera from acute disease cases versus naïve controls for the discovery of specific seroreactive antigens. Positive responses indicated a bias for signal peptide-containing proteins and were validated using Western blot-style immunostrips. The investigators found 13 antigens that were significantly more reactive in Q fever cases among 21 seroreactive antigens. Sixty-four additional antigens were variably reactive in Q fever patients (Vigil et al. 2010).

In a separate protein microarray study, Vigil et al. compared humoral immune responses to acute and chronic disease sera accompanied by greater than 6 months of supportive sero-surveillance and clinical data. Results of the evaluation of IgM and IgG antibody responses at early and late time points of infection showed a limited number of proteins undergo increasing or decreasing seroreactivity. Serological markers able to differentiate between acute and chronic disease states were identified, along with numerous IgG biomarkers of acute infection and novel IgG biomarkers for acute and chronic infection. This study profiled for the first time the IgM antibody repertoire for both acute and chronic Q fever. These results provided a unique perspective on the development of the antibody response, insight into isotype switch, and identified novel seroreactive antigens for the development of recombinant protein-based diagnostics and potential subunit vaccine candidates (Vigil et al. 2011).

16.6.5 Dominant B and T Cell Antigens by Protein ELISA and ELISPOT Analysis

Only one study has examined both the humoral and cellular immune response to Q fever. This study used IFA positive sera from 55 convalescent Q fever patients, five chronic Q fever patients, and 32 IFA negative controls. Humoral immune responses were tested by IgG indirect-ELISA using six previously identified and five array-selected *C. burnetii* proteins expressed as His-tag fusion proteins in *E. coli*. Cellular immune responses were determined using T cells from infected mice, either with or without a humanized MHC locus (Chen et al. 2009), since previous studies showed murine and human MHCs recognize different epitopes (Engelhard 1994). Enzyme-linked immunosorbent spot (ELISPOT) assays were used to test antigen-specific interferon-γ recall of CD4+T cells from vaccinated C57BL/6 or HLA-DR4 transgenic mice. Differences between the mice in antigen recognition confirmed the use of HLA transgenic mice is a more relevant model for screening human T cell antigens. This study showed that most strong antibody responses also induced interferon-γ recall responses in CD4+ T cells of vaccinated mice (Chen et al. 2009).

16.7 Future Developments

The development of safe and effective new vaccines can be guided by understanding the protective immune responses provided by existing vaccines. Zhang et al. showed that adoptive transfer of T cells from mice immunized with inactivated whole cell PI *C. burnetii* vaccine was protective against challenge for recipient mice (Zhang et al. 2007), confirming earlier studies that suggested T cell dependent immune responses are critical for protective immunity (Kazar et al. 1977, 1986). Additionally, low dose *C. burnetii* infection caused death in SCID and T cell deficient mice, but not B cell deficient mice, implying the essential requirement of T cells for host resistance to *C. burnetii* infection (Andoh et al. 2007). Microarray studies show a major portion of the vaccine-derived humoral responses consist of IgG antibodies to *C. burnetii* proteins (Vigil et al. 2010), indicating the relevance of CD4+ helper T cell responses to immunity. Antigen-specific CD4+ T cells direct secretion of cytokines, including IFN-γ and TNF-α. Both are critical for intracellular clearance of *C. burnetii* (Andoh et al. 2007). INF-γ stimulates production of nitric oxide and reactive oxygen molecules in macrophages, contributing to the control of infection, and may promote apoptosis in infected monocytes (Dellacasagrande et al. 2002; Howe et al. 2002; Brennan et al. 2004). IFN-γ was used successfully to treat previously intractable chronic Q fever and Q fever chronic multifocal osteomyelitis (Morisawa et al. 2001; Neth et al. 2011).

In experiments to determine the molecular targets of CD4+ T cell responses to *C. burnetii* PI-WCV for use in the development of a new subunit vaccine, Chen et al. focused on seven antigens previously shown to be targets of immunodominant B cell responses. Candidate peptides were selected from these antigens based on bioinformatic predictions of binding to murine MHC class II H-2 I-A^b. Peptides were screened for recognition by IFN-γ producing CD4+ T cells in PI-WCV vaccinated mice. Eight distinct H-2 I-A^b-restricted CD4+ T cell epitopes from four different proteins were identified, with the frequency of $CBU1910_{45-59}$ (Com1)-specific CD4+ T cells much higher than for other epitopes (Chen et al. 2011).

The search for *C. burnetii* protein antigens appropriate for use as diagnostic reagents and vaccines has been hampered by an inability to make mutants to identify specific virulence determinants in this obligate intracellular pathogen. New developments in the ability to grow *C. burnetii* in axenic media have now allowed for the creation of mutants, especially since the refinement of growth on plated media (Omsland et al. 2008, 2009). This should allow for the discovery of antigens unique to *C. burnetii* that are essential for virulence and are accessible to the immune system. Table 16.1 lists currently identified *C. burnetii* immunodominant protein antigens.

Table 16.1 Identified *Coxiella burnetii* immunodominant protein antigens

ID CBU No.	Gene	Size (AA)	Identification method	Sera reactive species	Annotation	References
CBU0008		62	Protein microarray analysis	Human	Tested for Q fever diagnostic	Beare et al. (2008)
CBU0092	ygbF	285	2D/mass analysis	Human		Sekeyova et al. (2009)
			Protein microarray analysis	Human		Vigil et al. (2010)
			2D/mass analysis	Guinea pig		Deringer et al. (2011)
CBU0098	nadC	274	Protein microarray analysis	Human		Beare et al. (2008)
CBU0103		480	2D/mass analysis	Guinea pig		Deringer et al. (2011)
CBU0109		267	Protein microarray analysis	Human		Vigil et al. (2010)
CBU0115	mraZ	152	2D/mass analysis	Human		Sekeyova et al. (2009)
CBU0140	ftsA	410	2D/mass analysis	Guinea pig		Deringer et al. (2011)
CBU0223	tuf-1	299	2D/mass analysis	Human		Sekeyova et al. (2009)
CBU0229	rplL	126	2D/mass analysis	Human		Coleman et al. (2007)
			Protein microarray analysis	Human		Vigil et al. (2010)
CBU0235	fusA	699	2D/mass analysis	Human		Sekeyova et al. (2009)
CBU0236	tuf-2	397	2D/mass analysis	Human		Coleman et al. (2007)
			2D/mass analysis	Human	Tested for Q fever diagnostic	Sekeyova et al. (2010)
			2D/mass analysis	Guinea pig		Deringer et al. (2011)
CBU0263	rpoA	327	2D/mass analysis	Human		Sekeyova et al. (2009)
CBU0271	ssb	158	2D/mass analysis	Human		Sekeyova et al. (2009)
CBU0299	rph	237	2D/mass analysis	Guinea pig		Deringer et al. (2011)
CBU0307		231	Protein microarray analysis	Human		Beare et al. (2008)
			2D/mass analysis	Human		Sekeyova et al. (2009)

(continued)

Table 16.1 (continued)

ID CBU No.	Gene	Size (AA)	Identification method	Sera reactive species	Annotation	References
CBU0309	htpG	633	2D/mass analysis	Human		Sekeyova et al. (2009)
CBU0311	P1	223	Western blot and N-terminal seq	Mouse		Varghees et al. (2002)
					C57Bl/6 Mouse T cell antigen	Chen et al. (2009)
					Tested for Q fever diagnostic	
CBU0366	phoR	252	Protein microarray analysis	Human		Beare et al. (2008)
CBU0383	tag	204	Protein microarray analysis	Human	Tested for Q diagnostic	Beare et al. (2008)
					C57Bl/6 Mouse T cell antigen	Chen et al. (2009)
					Tested for Q fever diagnostic	
CBU0391	ribF	320	Protein microarray analysis	Human		Beare et al. (2008)
CBU0436		258	Protein microarray analysis	Human		Vigil et al. (2010)
CBU0479	kdsB	249	2D/mass analysis	Human		Sekeyova et al. (2009)
CBU0480	argR	159	2D/mass analysis	Human	Recognized by chronic Q fever sera	Sekeyova et al. (2009)
CBU0497	fabF	414	2D/mass analysis	Human		Coleman et al. (2007)
CBU0528	rpsA	551	2D/mass analysis	Human		Sekeyova et al. (2009)
			2D/mass analysis	Guinea pig		Deringer et al. (2011)
CBU0545	lemA	192	Protein microarray analysis	Human		Beare et al. (2008)
			Protein microarray analysis	Human		Vigil et al. (2010)
CBU0572		458	2D/mass analysis	Human		Sekeyova et al. (2009)
CBU0609		340	Protein microarray analysis	Human		Beare et al. (2008)

(continued)

Table 16.1 (continued)

ID CBU No.	Gene	Size (AA)	Identification method	Sera reactive species	Annotation	References
CBU0612	ompH	165	Protein microarray analysis	Human	Tested for Q fever diagnostic	Beare et al. (2008)
					Tested for Q fever diagnostic	Chen et al. (2009)
			2D/mass analysis	Human	Recognized by chronic Q fever sera	Sekeyova et al. (2009)
			Protein microarray analysis	Human		Vigil et al. (2010)
CBU0615	lpxA	259	Protein microarray analysis	Human		Vigil et al. (2010)
CBU0628	ppa	175	2D/mass analysis	Human		Sekeyova et al. (2009)
CBU0630	mip	230	Protein microarray analysis	Human		Vigil et al. (2010)
CBU0632		99	2D/mass analysis	Human		Sekeyova et al. (2009)
CBU0648	ribH	151	2D/mass analysis	Human		Sekeyova et al. (2009)
CBU0653		198	Protein microarray analysis	Human		Vigil et al. (2010)
CBU0664		381	Protein microarray analysis	Human		Beare et al. (2008)
CBU0718		94	Protein microarray	Human		Beare et al. (2008)
					C57Bl/6 Mouse T cell antigen	Chen et al. (2009)
					Tested for Q fever diagnostic	
CBU0723		55	Protein microarray analysis	Human		Beare et al. (2008)
CBU0737	tig	442	2D/mass analysis	Human		Coleman et al. (2007)
CBU0750		324	2D/mass analysis	Guinea pig		Deringer et al. (2011)
CBU0754		348	Protein microarray analysis	Human		Beare et al. (2008)
CBU0760		124	Protein microarray analysis	Human		Beare et al. (2008)
CBU0774	pspC	82	Protein microarray analysis	Human		Beare et al. (2008)
CBU0776		283	Protein microarray analysis	Human		Beare et al. (2008)

(continued)

Table 16.1 (continued)

ID CBU No.	Gene	Size (AA)	Identification method	Sera reactive species	Annotation	References
CBU0781	ankG	338	Protein microarray analysis	Human	Tested for Q fever diagnostic	Beare et al. (2008)
					C57Bl/6 Mouse T cell antigen	Chen et al. (2009)
					Tested for Q fever diagnostic	
CBU0800		137	Protein microarray analysis	Human		Beare et al. (2008)
CBU0858	nadE	542	2D/mass analysis	Guinea pig		Deringer et al. (2011)
CBU0891		312	Protein microarray analysis	Human		Beare et al. (2008)
					C57Bl/6 Mouse T cell antigen	Chen et al. (2009)
					Tested for Q fever diagnostic	
			Protein microarray analysis	Human		Vigil et al. (2010)
CBU0895		203	Protein microarray analysis	Human		Vigil et al. (2010)
CBU0898		110	Protein microarray analysis	Human		Beare et al. (2008)
CBU0932	gpIK	501	2D/mass analysis	Guinea pig		Deringer et al. (2011)
CBU0935		129	Protein microarray analysis	Human		Beare et al. (2008)
CBU0937		465	2D/mass analysis	Human	Tested for Q fever diagnostic	Sekeyova et al. (2010)
			2D/mass analysis	Guinea pig		Deringer et al. (2011)
CBU0952	AdaA	227	Phage library	Mouse, Guinea pig	Tested for acute disease diagnostic	Zhang et al. (2005)
			2D/mass analysis	Human		Coleman et al. (2007)
					Tested for Q fever diagnostic	Chen et al. (2009)
			2D/mass analysis	Human		Sekeyova et al. (2009)

(continued)

Table 16.1 (continued)

ID CBU No.	Gene	Size (AA)	Identification method	Sera reactive species	Annotation	References
CBU0963	bcp	151	2D/mass analysis	Human		Coleman et al. (2007)
CBU0968		176	Protein microarray analysis	Human		Beare et al. (2008)
CBU1002	birA	323	Protein microarray analysis	Human		Beare et al. (2008)
CBU1065		184	Protein microarray analysis	Human		Beare et al. (2008)
CBU1094		380	Protein microarray analysis	Human		Vigil et al. (2010)
CBU1098		274	Protein microarray analysis	Human		Beare et al. (2008)
CBU1115		105	Protein microarray analysis	Human	Tested for Q fever diagnostic	Beare et al. (2008)
CBU1121		60	Protein microarray analysis	Human		Beare et al. (2008)
CBU1143	yajC	116	Protein microarray analysis	Human	Tested for Q fever diagnostic	Beare et al. (2008)
			Protein microarray analysis	Human		Vigil et al. (2010)
CBU1157		233	Protein microarray analysis	Human	Tested for Q fever diagnostic	Beare et al. (2008)
CBU1184		242	Protein microarray analysis	Human		Beare et al. (2008)
CBU1200	icd	427	2D/mass analysis	Human		Coleman et al. (2007)
CBU1241	mdh	328	2D/mass analysis	Guinea pig		Deringer et al. (2011)
CBU1249		203	Protein microarray analysis	Human	Tested for Q fever diagnostic	Beare et al. (2008)
CBU1260		248	2D/mass analysis	Human		Sekeyova et al. (2009)
CBU1290	dnaK	656	2D/mass analysis	Human		Sekeyova et al. (2009)
			2D/mass analysis	Guinea pig		Deringer et al. (2011)
CBU1385	tsf	296	2D/mass analysis	Guinea pig		Deringer et al. (2011)
CBU1386	rpsB	313	Protein microarray analysis	Human		Vigil et al. (2010)
CBU1396	sucD	294	2D/mass analysis	Guinea pig		Deringer et al. (2011)

(continued)

Table 16.1 (continued)

ID CBU No.	Gene	Size (AA)	Identification method	Sera reactive species	Annotation	References
CBU1398	sucB	405	Protein microarray analysis	Human		Beare et al. (2008)
			Protein microarray analysis	Human		Vigil et al. (2010)
			2D/mass analysis	Guinea pig		Deringer et al. (2011)
CBU1416		216	2D/mass analysis	Human		Coleman et al. (2007)
CBU1433	nusA	503	2D/mass analysis	Human		Sekeyova et al. (2009)
CBU1513		258	Protein microarray	Human		Vigil et al. (2010)
CBU1627	lcmE	1034	Protein microarray analysis	Human		Vigil et al. (2010)
CBU1697	nth	218	Protein microarray analysis	Human		Beare et al. (2008)
CBU1706		200	2D/mass analysis	Human		Sekeyova et al. (2009)
			2D/mass analysis	Guinea pig		Deringer et al. (2011)
CBU1718	groEL	552	2D/mass analysis	Human		Coleman et al. (2007)
			2D/mass analysis	Human		Sekeyova et al. (2009)
			2D/mass analysis	Guinea pig		Deringer et al. (2011)
CBU1719	groES	96	2D/mass analysis	Human		Sekeyova et al. (2009)
			Protein microarray analysis	Human		Vigil et al. (2010)
CBU1768		153	Protein microarray analysis	Human		Vigil et al. (2010)
CBU1770		432	2D/mass analysis	Human		Sekeyova et al. (2009)
CBU1783	gap	334	2D/mass analysis	Human		Sekeyova et al. (2009)
CBU1789		288	2D/mass analysis	Human		Sekeyova et al. (2009)
CBU1835		459	Protein microarray analysis	Human		Beare et al. (2008)
CBU1853		149	Protein microarray analysis	Human	Tested for Q fever diagnostic	Beare et al. (2008)
CBU1863		603	Protein microarray analysis	Human		Vigil et al. (2010)

(continued)

Table 16.1 (continued)

ID CBU No.	Gene	Size (AA)	Identification method	Sera reactive species	Annotation	References
CBU1865		175	Protein microarray analysis	Human		Beare et al. (2008)
CBU1869		217	Protein microarray analysis	Human	Tested for Q fever diagnostic	Beare et al. (2008)
CBU1910	com1	252	Phage library screen, 2D analysis, protein array	Mouse, guinea pig, human		Hendrix et al. (1991)
					C57Bl/6 Mouse T cell antigen Tested for Q fever diagnostic	Chen et al. (2009)
			2D/mass analysis	Human	Tested for Q fever diagnostic	Sekeyova et al. (2010)
			Protein microarray analysis	Human		Vigil et al. (2010)
			2D/mass analysis	Guinea pig		Deringer et al. (2011)
CBU1940	atpE	100	Protein microarray analysis	Human		Beare et al. (2008)
CBU1943	atpA	515	2D/mass analysis	Human		Coleman et al. (2007)
			2D/mass analysis	Guinea pig		Deringer et al. (2011)
CBU1945	atpD	461	2D/mass analysis	Human		Sekeyova et al. (2009)
CBU1958		32	Protein microarray analysis	Human		Beare et al. (2008)
CBU1966	hemA	413	Protein microarray analysis	Human		Beare et al. (2008)
CBU1967		409	Protein microarray analysis	Human		Beare et al. (2008)
CBU1969	dksA	147	Protein microarray analysis	Human		Beare et al. (2008)
CBU2020		476	Protein microarray analysis	Human		Beare et al. (2008)
CBU2029		210	2D/mass analysis	Human		Sekeyova et al. (2009)
CBU2065		212	Protein microarray analysis	Human		Beare et al. (2008)

References

Ackland JR, Worswick DA, Marmion BP (1994) Vaccine prophylaxis of Q fever. A follow-up study of the efficacy of Q-Vax (CSL) 1985–1990. Med J Aust 160:704–708

Andoh M, Zhang G, Russell-Lodrigue KE, Shive HR, Weeks BR, Samuel JE (2007) T cells are essential for bacterial clearance, and gamma interferon, tumor necrosis factor alpha, and B cells are crucial for disease development in *Coxiella burnetii* infection in mice. Infect Immun 75:3245–3255

Arricau-Bouvery N, Souriau A, Bodier C, Dufour P, Rousset E, Rodolakis A (2005) Effect of vaccination with phase I and phase II *Coxiella burnetii* vaccines in pregnant goats. Vaccine 23:4392–4402

Arricau-Bouvery N, Hauck Y, Bejaoui A, Frangoulidis D, Bodier CC, Souriau A, Meyer H, Neubauer H, Rodolakis A, Vergnaud G (2006) Molecular characterization of *Coxiella burnetii* isolates by infrequent restriction site-PCR and MLVA typing. BMC Microbiol 6:38–51

Baca OG, Paretsky D (1983) Q fever and *Coxiella burnetii*: a model for host-parasite interactions. Microbiol Rev 47:127–149

Beare PA, Samuel JE, Howe D, Virtaneva K, Porcella SF, Heinzen RA (2006) Genetic diversity of the Q fever agent. *Coxiella burnetii*, assessed by microarray-based whole-genome comparisons. J Bacteriol 188:2309–2324

Beare PA, Chen C, Bouman T, Pablo J, Unal B, Cockrell DC, Brown WC, Barbian KD, Porcella SF, Samuel JE, Felgner PL, Heinzen RA (2008) Candidate antigens for Q fever serodiagnosis revealed by immunoscreening of a *Coxiella burnetii* protein microarray. Clin Vaccine Immunol 15:1771–1779

Beare PA, Unsworth N, Andoh M, Voth DE, Omsland A, Gilk SD, Williams KP, Sobral BW, Kupko JJ III, Porcella SF, Samuel JE, Heinzen RA (2009) Comparative genomics reveal extensive transposon-mediated genomic plasticity and diversity among potential effector proteins within the genus *Coxiella*. Infect Immun 77:642–656

Benenson AS (1959) Q fever vaccine: efficacy and present status. In: Smadel JE (ed) Walter Reed Army Institute of Medical Science, Publ No 6th edn. US Government Printing Office, Washington, DC, p 47

Bengtson IA (1941) Complement fixation in "Q" fever. Proc Soc Exp Biol Med 46:665

Benson G (1999) Tandem repeats finder: a program to analyze DNA sequences. Nucleic Acids Res 27:573–580

Bilberstein EL, Riemann HP, Franti CE, Behymer DE, Ruppanner R, Bushnell R, Crenshaw G (1977) Vaccination of dairy cattle against Q fever (*Coxiella burnetii*): results of field trials. Am J Vet Res 38:189–193

Bildfell RJ, Thomson GW, Haines DM, McEwen BJ, Smart N (2000) *Coxiella burnetii* infection is associated with placentitis in cases of bovine abortion. J Vet Diagn Invest 12:419–425

Brennan RE, Russell K, Zhang G, Samuel JE (2004) Both inducible nitric oxide synthase and NADPH oxidase contribute to the control of virulent phase I *Coxiella burnetii* infections. Infect Immun 72:6666–6675

Brooks DL, Ermel RW, Franti CE, Ruppanner R, Behymer DE, Williams JC (1986) Q fever vaccination of sheep: challenge of immunity in ewes. Am J Vet Res 47:1235–1238

Brown GL (1973) Clinical aspects of "Q" fever. Postgrad Med J 49:539–541

Burnet FM, Freeman M (1937) Experimental studies on the virus of "Q" fever. Med J Aust 2:299–305

Carrieri MP, Tissot-Dupont H, Rey D, Brousse P, Renard H, Obadia Y, Raoult D (2002) Investigation of a slaughterhouse-related outbreak of Q fever in the French Alps. Eur J Clin Microbiol Infect Dis 21:17–21

Chen C, Bouman TJ, Beare PA, Mertens K, Zhang GQ, Russell-Lodrique KE, Hogaboam JP, Peters B, Felgner PL, Brown WC, Heinzen RA, Hendrix LR, Samuel JE (2009) A systematic approach to evaluate humoral and cellular immune responses to *Coxiella burnetii* antigens. Clin Microbiol Infect 15:156–157

Chen C, Dow C, Wang P, Sidney J, Read A, Harmsen A, Samuel JE, Peters B (2011) Identification of CD4+ T cell epitopes in *C. burnetii* antigens targeted by antibody responses. PLoS One 6:e17712

Coleman SA, Fischer ER, Cockrell DC, Voth DE, Howe D, Mead DJ, Samuel JE, Heinzen RA (2007) Proteome and antigen profiling of *Coxiella burnetii* developmental forms. Infect Immun 75:290–298

Dellacasagrande J, Ghigo E, Raoult D, Capo C, Mege JL (2002) IFN-gamma-induced apoptosis and microbicidal activity in monocytes harboring the intracellular bacterium *Coxiella burnetii* require membrane TNF and homotypic cell adherence. J Immunol 169:6309–6315

Denison AM, Massung RF, Thompson HA (2007a) Analysis of the O-antigen biosynthesis regions of phase II isolates of *Coxiella burnetii*. FEMS Microbiol Lett 267:102–107

Denison AM, Thompson HA, Massung RF (2007b) IS1111 insertion sequences of *Coxiella burnetii*: characterization and use for repetitive element PCR-based differentiation of *Coxiella burnetii* isolates. BMC Microbiol 7:91–98

Deringer JR, Chen C, Samuel JE, Brown WC (2011) Immunoreactive *Coxiella burnetii* Nine Mile proteins separated by 2D electrophoresis and identified by tandem mass spectrometry. Microbiology 157:526–542

Derrick EH (1937) "Q" fever, a new entity: clinical features, diagnosis and laboratory investigation. Med J Aust 2:281–299

Dilbeck PM, McElwain TF (1994) Immunohistochemical detection of *Coxiella burnetii* in formalin-fixed placentas. J Vet Diagn Invest 6:125–127

Domingues P, Palkovic P, Toman R (2002) Analysis of phospholipids from *Coxiella burnetii* by fast atom bombardment mass spectrometry. A rapid method for differentiation of virulent phase I and low virulent phase II cells. Acta Virol 46:121–124

Dupont HT, Thirion X, Raoult D (1994) Q fever serology: cut off determination for microimmunofluorescence. Clin Diagn Lab Immunol 1:189–196

Engelhard VH (1994) Structure of peptides associated with class I and class II MHC molecules. Annu Rev Immunol 12:181–207

Enserink M (2010) Infectious diseases. Questions abound in Q-fever explosion in the Netherlands. Science 327:266–277

Fenollar F, Fournier PE, Raoult D (2004) Molecular detection of *Coxiella burnetii* in the sera of patients with Q fever endocarditis or vascular infections. J Clin Microbiol 42:4919–4924

Fernandes I, Rousset E, Dufour P, Sidi-Boumedine K, Cupo A, Thiery R, Duquesne V (2009) Evaluation of the recombinant heat shock protein B (HspB) of *Coxiella burnetii* as a potential antigen for immunodiagnostic of Q fever in goats. Vet Microbiol 134:300–304

Field PR, Santiago A, Chan SW, Patel DB, Dickeson D, Mitchell JL, Devine PL, Murphy AM (2002) Evaluation of a novel commercial enzyme-linked immunosorbent assay detecting *Coxiella burnetii*-specific immunoglobulin G for Q fever prevaccination screening and diagnosis. J Clin Mirobiol 40:3526–3529

Fishbein DB, Raoult D (1992) A cluster of *Coxiella burnetii* infections associated with exposure to vaccinated goats and their unpasteurized dairy products. Am J Tropic Med Hyg 47:35–40

Flores-Ramirez G, Toman R, Sekyova Z, Skultety L (2009) *In silico* prediction and identification of outer membrane proteins and lipoproteins from *Coxiella burnetii* by the mass spectrometry techniques. Clin Microbiol Infect 15:196–197

Fournier PE, Casalta JP, Habib G, Messana T, Raoult D (1996) Modification of the diagnostic criteria proposed by the Duke Endocarditis Service to permit improved diagnosis of Q fever endocoarditis. Am J Med 100:629–633

Fournier P-E, Raoult D (2003) Comparison of PCR and serology assays for early diagnosis of acute Q fever. J Clin Microbiol 41:5094–5098

Frangoulidis D, Schropfer E, Al Dahouk S, Tomaso H, Meyer H (2006) Comparison of four commercially available assays for the detection of IgM phase II antibodies to *Coxiella burnetii* in the diagnosis of acute Q fever. Ann N Y Acad Sci 1078:561–562

Frankel D, Richet H, Renvoise A, Raoult D (2011) Q fever in France, 1985–2009. Emerg Infect Dis 17:350–356

Fries LF, Waag DM, Williams JC (1993) Safety and immunogenicity in human volunteers of a choloform-methanol residue vaccine for Q fever. Infect Immun 61:1251–1258

Geddes AM (1983) Q fever. Br Med J 287:927–928

Genig VA (1968) A live vaccine1/M-44 against Q-fever for oral use. J Hyg Epidemiol Microbiol Immunol 12:265–273

Glazunova O, Roux V, Freylikman O, Sekeyova Z, Fournous G, Tyczka J, Tokarevich N, Kovacova E, Marrie TJ, Raoult D (2005) *Coxiella burnetii* genotyping. Emerg Infect Dis 11:1211–1217

Guatteo R, Beaudeau F, Joly A, Seegers H (2007) Performances of an ELISA applied to serum and milk for the detection of antibodies to *Coxiella burnetii* in dairy cattle. Revue Med Vet 158:250–252

Hackstadt T (1986) Antigenic variation in the phase I lipopolysaccharide of *Coxiella burnetii* isolates. Infect Immun 52:337–340

Hellenbrand W, Breuer T, Petersen L (2001) Changing epidemiology of Q fever in Germany, 1947–1999. Emerg Infect Dis 7:789–796

Hendrix LR, Samuel JE, Mallavia LP (1991) Differentiation of *Coxiella burnetii* isolates by analysis of restriction-endonuclease-digested DNA separated by SDS-PAGE. J Gen Microbiol 137:269–276

Hendrix LR, Mallavia LP, Samuel JE (1993) Cloning and sequencing of *Coxiella burnetii* outer membrane protein gene *com1*. Infect Immun 61:470–477

Hernychova L, Toman R, Ciampor F, Hubalek M, Vackova J, Macela A, Skultety L (2008) Detection and identification of *Coxiella burnetii* based on the mass spectrometric analyses of the extracted proteins. Anal Chem 80:7097–7104

Hoover TA, Culp DW, Vodkin MH, Williams JC, Thompson HA (2002) Chromosomal DNA deletions explain phenotypic characteristics of two antigenic variants, phase II and RSA 514 (Crazy), of the *Coxiella burnetii* Nine Mile strain. Infect Immun 70:6726–6733

Hoover TA, Vodkin MH, Williams JC (1992) A *Coxiella burnetii* repeated DNA element resembling a bacterial insertion sequence. J Bacteriol 174:5540–5548

Howe D, Barrows LF, Lindstrom NM, Heinzen RA (2002) Nitric oxide inhibits *Coxiella burnetii* replication and parasitophorous vacuole maturation. Infect Immun 70:5140–5147

Izzo AA, Marmion BP, Hackstadt T (1991) Analysis of the cells involved in the lymphoproliferative response to *Coxiella burnetii* antigens. Clin Exp Immunol 85:98–108

Jager C, Willems H, Thiele D, Baljer G (1998) Molecular characterization of *Coxiella burnetii* isolates. Epidemiol Infect 120:157–164

Jensen TK, Montgomery DL, Jaeger PT, Lindhardt T, Agerholm JS, Bille-Hansen V, Boye M (2007) Application of fluorescent *in situ* hybridization for demonstration of *Coxiella burnetii* in placentas from ruminant abortions. APMIS 115:347–353

Johnson JW, Eddy GA, Pederson CE Jr (1976) Biologic properties of the M-44 strain of *Coxiella burnetii*. J Infect Dis 133:334–338

Johnson JW, McLeod CG, Stookey JL, Higbee GA, Pedersen CE Jr (1977) Lesions in guinea pigs infected with *Coxiella burnetii* strain M-44. J Infect Dis 135:995–998

Kazar J, El-Najdawi E, Brezina R, Schramek S (1977) Search for correlates of resistance to virulent challenge in mice immunized with *Coxiella burnetii*. Acta Virol 21:422–430

Kazar J, Brezina R, Palanova A, Turda B, Schramek S (1982) Immunogenicity and reactogenicity of Q fever chemovaccine in persons professionally exposed to Q fever in Czechoslovakia. Bull World Health Organ 60:389–394

Kazar J, Votruba D, Propper P, Schramek S (1986) Onset and duration of immunity in guinea pigs and mice induced with different Q fever vaccines. Acta Virol 30:499–506

Klaassen CHW, Nabuurs-Franssen MH, Tilburg JJHC, Hamans MAWM, Horrevorts AM (2009) Multigenotype Q fever outbreak, the Netherlands. Emerg Infect Dis 15:613–614

La Scola B (2002) Current laboratory diagnosis of Q fever. Semin Pediatr Infect Dis 13:257–262

Li Q, Niu D, Wen B, Chen M, Qiu L, Zhang J (2005) Protective immunity against Q fever induced with a recombinant PI antigen fused with HspB of *Coxiella burnetii*. Ann N Y Acad Sci 1063:130–142

Lukacova M, Gajdosova E, Skultety L, Kovacova E, Kazar J (1994) Characterization and protective effect of a 29 kDa protein isolated from *Coxiella burnetii* by detergent Empigen BB. Eur J Epidemiol 10:227–230

Lukacova M, Barak I, Kazar J (2008) Role of structural variations of polysaccharide antigens in the pathogenicity of Gram-negative bacteria. Clin Microbiol Infect 14:200–206

Luoto L, Casey ML, Pickens EG (1965) Q fever studies in Montana: detection of asymptomatic infection among residents of infected dairy premises. Am J Epidemiol 81:356–369

Marmion B (2007) Q fever: the long journey to control by vaccination. Med J Aust 186:164–166

Marmion BP, Storm PA, Ayres JG, Semendric L, Mathews L, Winslow W, Turra M, Harris RJ (2005) Long-term persistence of *Coxiella burnetii* after acute primary Q fever. Q J Med 98:7–20

Meiklejohn G, Reimer LG, Graves PS, Helmick C (1981) Cryptic epidemic of Q fever in a medical school. J Infect Dis 144:107–113

Million M, Thuny F, Richet H, Raoult D (2010) Long-term outcome of Q fever endocarditis: a 26-year personal survey. Lancet Infect Dis 10:527–535

Moos A, Hackstadt T (1987) Comparative virulence of intra- and interstrain lipopolysaccharide variants of *Coxiella burnetii* in the guinea pig model. Infect Immun 55:1144–1150

Morisawa Y, Wakiguchi H, Takechi T, Kurashige T, Nagaoka H (2001) Intractable Q fever treated with recombinant gamma interferon. Pediatr Infect Dis J 20:546–547

Muller HP, Schmeer N, Rantamaki L, Semler B, Krauss H (1987) Isolation of a protein antigen from *Coxiella burnetii*. Zentralbl Bakteriol Mikrobiol Hyg A 265:277–289

Narasaki CT, Mertens K, Samuel JE (2011) Characterization of the GDP-D-mannose biosynthesis pathway in *Coxiella burnetii*: the initial steps for GDP-β-D-virenose biosynthesis. PLoS One 6(10):e25514

Neth OW, Falcon D, Peromingo E, Soledad Camacho M, Rodriguez-Gallego C, Obando I (2011) Successful management of chronic multifocal Q fever osteomyelitis with adjuvant interferon-gamma therapy. Pediatr Infect Dis J 30(9):810–812

Omsland A, Cockrell DC, Fischer ER, Heinzen RA (2008) Sustained axenic metabolic activity by the obligate intracellular bacterium *Coxiella burnetii*. J Bacteriol 190:3203–3212

Omsland A, Cockrell DC, Howe D, Fischer ER, Virtaneva K, Sturdevant DE, Porcella SF, Heinzen RA (2009) Host cell-free growth of the Q fever bacterium *Coxiella burnetii*. Proc Natl Acad Sci U S A 106:4430–4434

Ormsbee RA, Marmion BP (1990) The disease. In: Marrie TJ (ed) Q fever, vol 1. CRC Press, Boca Raton, p 226

Ormsbee RA, Bell EJ, Lackman DB, Tallent G (1964) The influence of phase on the protective potency of Q fever vaccine. J Immunol 92:404–412

Palkovicova K, Ihnatko R, Vadovic P, Betinova E, Skultety L, Frangoulidis D, Toman R (2009) A monoclonal antibody specific for a unique biomarker, virenose, in a lipopolysaccharide of *Coxiella burnetii*. Clin Microbiol Infect 15:183–184

Panning M, Kilwinski J, Greiner-Fischer S, Peters M, Kramme S, Frangoulidis D, Meyer H, Henning K, Drosten C (2008) High throughput detection of *Coxiella burnetii* by real-time PCR with internal control system and automated DNA preparation. BMC Microbiol 8:77–84

Peter O, Dupuis G, Peacock MG, Burgdorfer W (1987) Comparison of enzyme-linked immunosorbent assay and complement fixation and indirect fluorescent-antibody tests for detection of *Coxiella burnetii* antibody. J Clin Microbiol 25:1063–1067

Pierce CY, Barr JR, Woolfitt AR, Moura H, Shaw EI, Thompson HA, Massung RF, Fernandez FM (2007) Strain and phase identification of the U.S. category B agent *Coxiella burnetii* by matrix assisted laser desorption/ionization time-of-flight mass spectrometry and multivariate pattern recognition. Anal Chem Acta 583:23–31

Raoult D, Houpikian P, Tissot Dupont H, Riss JM, Arditi-Djiane J, Brouqui P (1999) Treatment of Q fever endocarditis: comparison of 2 regimens containing doxycycline and ofloxacin or hydroxychloroquine. Arch Intern Med 159:167–173

Redd T, Thompson HA (1995) Secretion of proteins by *Coxiella burnetii*. Microbiology 141:363–369

Richardus JH, Donkers A, Dumas AM, Schaap GJP, Akkermans JPWM, Huisman J, Valkenburg HA (1987) Q fever in the Netherlands: a sero-epidemiological survey among human population groups from 1968-1983. Epidemiol Infect 98:211–219

Russel-Lodrigue KE, Andoh M, Poels MW, Shive HR, Weeks BR, Zhang GQ, Tersteeg T, Masegi T, Hotta A, Yamaguchi T, Fukushi H, Hirai K, McMurray DN, Samuel JE (2009) *Coxiella burnetii* isolates cause genogroup specific virulence in mouse and guinea pig models of acute Q fever. Infect Immun 77:5640–5650

Samoilis G, Aivaliotis M, Vranakis I, Papadioti A, Tselentis Y, Tsiotis G, Psaroulaki A (2010) Proteomic screening for possible effector molecules secreted by the obligate intracellular pathogen *Coxiella burnetii*. J Proteome Res 9:1619–1626

Sawyer LA, Fishbein DB, McDade JE (1987) Q fever: current concepts. Rev Infect Dis 9:935–946

Schmeer N, Muller P, Langel J, Krauss H, Frost JW, Wieda J (1987) Q fever vaccines for animals. Zentralbl Bakteriol Microbiol Hyg A 267:79–88

Schmeer N, Muller H-P, Baumgartner W, Wieda J, Krauss H (1988) Enzyme-linked immunosorbent fluorescence assay and high-pressure liquid chromatography for analysis of humoral immune responses to *Coxiella burnetii* proteins. J Clin Microbiol 26:2520–2525

Schneeberger PM, Hermans MHA, van Hannen EJ, Schellekens JJA, Leenders ACAP, Wever PC (2010) Real-time PCR with serum samples is indispensible for early diagnosis of acute Q fever. Clin Vaccine Immunol 17:286–290

Schramek S, Mayer H (1982) Different sugar compositions of lipopolysaccharides isolated from phase I and pure phase II cells of *Coxiella burnetii*. Infect Immun 38:53–57

Sekeyova Z, Kowalczewska M, Decloquement P, Pelletier N, Spitalska E, Raoult D (2009) Identification of protein candidates for the serodiagnosis of Q fever endocarditis by an immunoproteomic approach. Eur J Clin Microbiol Infect Dis 28:287–295

Sekeyova Z, Kowalczewska M, Vincentelli R, Decloquement P, Flores-Ramirez G, Skultety L, Raoult D (2010) Characterization of antigens for Q fever serodiagnostics. Acta Virol 54:173–180

Shaw EI, Moura H, Woolfitt AR, Ospina M, Thompson HA, Barr JR (2004) Identification of biomarkers of whole *Coxiella burnetii* phase I by MALDI-TOF mass spectrometry. Anal Chem 76:4017–4022

Smadel JE, Snyder MJ, Robbins FC (1948) Vaccination against Q fever. Am J Hyg 47:71

Stein A, Kruszewska D, Gouvernet J, Raoult D (1997) Study of the 16S-23S ribosomal DNA internal spacer of *Coxiella burnetii*. Eur J Epidemiol 13:471–475

Stoker MG, Fiset P (1956) Phase variation of the Nine Mile and other strains of *Rickettsia burnetii*. Can J Microbiol 2:310–321

Svraka S, Toman R, Skultety L, Slaba K, Homan WL (2006) Establishment of a genotyping scheme for *Coxiella burnetii*. FEMS Microbiol Lett 254:268–274

Thiele D, Willems H, Krauss H (1994) The 16S/23S ribosomal spacer region of *Coxiella burnetii*. Eur J Epidemiol 10:421–426

Tilburg JJHC, Melchers WJG, Pettersson AM, Rossen JWA, Hermans MHA, van Hannen EJ, Nabuurs-Franssen MH, de Vries MC, Horrevorts AM, Klaassen CHW (2010) Interlaboratory evaluation of different extraction and real-time PCR methods for detection of *Coxiella burnetii* DNA in serum. J Clin Microbiol 48:3923–3927

Tissot-Dupont H, Vaillant V, Rey S, Raoult D (2007) Role of sex, age, previous valve lesion, and pregnancy in the clinical expression and outcome of Q fever after a large outbreak. Clin Infect Dis 44:232–237

To H, Hotta A, Zhang GQ, Nguyen SV, Ogawa M, Yamaguchi T, Fukushi H, Amano K, Hirai K (1998) Antigenic characteristics of polypeptides of *Coxiella burnetii* isolates. Microbiol Immunol 42:81–85

Varghees S, Kiss K, Frans G, Braha O, Samuel J (2002) Cloning and porin activity of the outer membrane protein P1 from *Coxiella burnetii*. Infect Immun 70:6741–6750

Vigil A, Ortega R, Nakajima-Sasaki R, Pablo J, Molina DM, Chao C-C, Chen H-W, Ching W-M, Felgner PL (2010) Genome-wide profiling of humoral immune response to *Coxiella burnetii* infection by protein microarray. Proteomics 10:2259–2269

Vigil A, Chen C, Jain A, Nakajima-Sasaki R, Jasinskas A, Pablo J, Hendrix LR, Samuel JE, Felgner PL (2011) Profiling the humoral immune response of acute and chronic Q fever by protein microarray. Mol Cell Proteomics 10(10):M110.006304

Vodkin MH, Williams JC (1986) Overlapping deletion in two spontaneous phase variants of *Coxiella burnetii*. J Gen Microbiol 132:2587–2594

Vodkin MH, Williams JC, Stephenson EH (1986) Genetic heterogeneity among isolates of *Coxiella burnetii*. J Gen Microbiol 132:455–463

Waag DM, Bolt CR, Marrie TJ, Williams JC (1991) The biology of *Coxiella burnetii*. In: Williams JC, Thompson HA (eds) Q fever. CRC Press, Boca Raton, pp 164–167

Waag DM, England MJ, Pitt ML (1997) Comparative efficacy of a *Coxiella burnetii* chloroform:methanol residue (CMR) vaccine and a licensed cellular vaccine (Q-Vax) in rodents challenged by aerosol. Vaccine 15:1779–1783

Waag DM, England MJ, Tammariello RF, Byrne WR, Gibbs P, Banfield CM (2002) Comparative efficacy and immunogenicity of Q fever chloroform:methanol residue (CMR) and phase I cellular (Q-Vax) vaccine in cynomolgus monkeys challenged by aerosol. Vaccine 20:2623–2634

Waag DM, England MJ, Bolt CR, Williams JC (2008) Low-dose priming before vaccination with the phase I chloroform-methanol residue vaccine against Q fever enhances humoral and cellular immune responses to *Coxiella burnetii*. Clin Vaccine Immunol 15:1505–1512

Williams JC, Cantrell JL (1982) Biological and immunological properties of *Coxiella burnetii* vaccines in C57BL/10ScN endotoxin-nonresponder mice. Infect Immun 35:1091–1102

Williams JC, Damrow TA, Waag DM, Amano K (1986) Characterization of a phase I *Coxiella burnetii* chloroform-methanol residue vaccine that induces active immunity against Q fever in C57BL/10ScN mice. Infect Immun 51:851–858

Williams JC, Peacock MG, Waag DM, Kent G, England MJ, Nelson G, Stephenson EH (1990) Vaccines against coxiellosis and Q fever. Ann N Y Acad Sci 590:88–111

Wisniewski HJ, Kleiman MW, Lackman DB, Krumbiegel ER (1969) Demonstration of inapparent infection with disease agents common to animals and man. Health Lab Sci 6:173–177

Zhang GQ, Samuel JE (2003) Identification and cloning potentially protective antigens of *Coxiella burnetii* using sera from mice experimentally infected with Nine Mile phase I. Ann N Y Acad Sci 990:510–520

Zhang YX, Zhi N, Yu SR, Li QR, Yu GC, Zhang X (1994) Protective immunity induced by 67K outer membrane protein of phase I *Coxiella burnetii* in mice and guinea pigs. Acta Virol 38:327–332

Zhang GQ, Kiss K, Seshadri R, Hendrix LR, Samuel J (2004) Identification and cloning of immunodominant antigens of *Coxiella burnetii*. Infect Immun 72:844–852

Zhang G, To H, Russell KE, Hendrix LR, Yamaguchi T, Fukushi H, Hirai K, Samuel JE (2005) Identification and characterization of an immunodominant 28-kilodalton *Coxiella burnetii* outer membrane protein specific to isolates associated with acute disease. Infect Immun 73:1561–1567

Zhang G, Russell-Lodrigue KE, Andoh M, Zhang Y, Hendrix LR, Samuel JE (2007) Mechanisms of vaccine-induced protective immunity against *Coxiella burnetii* infection in BALB/c mice. J Immunol 179:8372–8380

Chapter 17
Epidemic Q Fever in Humans in the Netherlands

Wim van der Hoek, Gabriëlla Morroy, Nicole H.M. Renders, Peter C. Wever, Mirjam H.A. Hermans, Alexander C.A.P. Leenders, and Peter M. Schneeberger

Contents

W. van der Hoek (✉)
National Institute for Public Health and the Environment (RIVM), Centre for Infectious Disease Control (Clb), P.O. Box 1 Antonie van Leeuwenhoeklaan 9, 3720 BA, Bilthoven, The Netherlands
e-mail: wim.van.der.hoek@rivm.nl

G. Morroy
Department of Infectious Disease Control, Municipal Health Service 'Hart voor Brabant', 's Hertogenbosch, The Netherlands

Academic Collaborative Centre AMPHI, Department of Primary and Community Care, Radboud University Nijmegen Medical Centre, Nijmegen, The Netherlands

N.H.M. Renders • P.C. Wever • M.H.A. Hermans • A.C.A.P. Leenders • P.M. Schneeberger
Department of Medical Microbiology and Infection Control, Jeroen Bosch Hospital, Henri Dunantstraat 1, P.O. Box 90153, 5200 ME, 's Hertogenbosch, The Netherlands

R. Toman et al. (eds.), *Coxiella burnetii: Recent Advances and New Perspectives in Research of the Q Fever Bacterium*, Advances in Experimental Medicine and Biology 984,
DOI 10.1007/978-94-007-4315-1_17,

Abstract In 2005, Q fever was diagnosed on two dairy goat farms and 2 years later it emerged in the human population in the south of the Netherlands. From 2007 to 2010, more than 4,000 human cases were notified with an annual seasonal peak. The outbreaks in humans were mainly restricted to the south of the country in an area with intensive dairy goat farming. In the most affected areas, up to 15% of the population may have been infected. The epidemic resulted in a serious burden of disease, with a hospitalisation rate of 20% of notified cases and is expected to result in more cases of chronic Q fever among risk groups in the coming years. The most important risk factor for human Q fever is living close (<5 km) to an infected dairy goat farm. Occupational exposure plays a much smaller role. In 2009 several veterinary control measures were implemented including mandatory vaccination of dairy goats and dairy sheep, improved hygiene measures, and culling of pregnant animals on infected farms. The introduction of these drastic veterinary measures has probably ended the Q fever outbreak, for which the Netherlands was ill-prepared.

Keywords Q fever • *Coxiella burnetii* • Netherlands • Dairy goats • Epidemic

17.1 Introduction

Since its first description in abattoir workers in Australia in 1935, Q fever has been considered primarily an occupational disease for abattoir workers, sheep shearers, farmers, and veterinarians. Occasional outbreaks among the general population have been described in different countries but these were mostly confined to small areas and were of short duration. The 2007–2010 epidemic of Q fever in the Netherlands with more than 4,000 notified human cases was unique. We describe the different aspects of this exceptionally large epidemic, primarily from the human health perspective, and provide details of ongoing research that will add considerably to the global knowledge base of Q fever. Topics covered are the surveillance of acute Q fever before and during the epidemic; the challenges in laboratory diagnostics; the long-term effects of Q fever; prevention of severe disease by vaccination; risks for pregnant women; and the drastic veterinary measures on dairy goat and dairy sheep farms that were implemented from 2009 and that have probably played a major role in stopping the epidemic.

17.2 Surveillance of Acute Q Fever and Diagnostic Criteria

17.2.1 Q Fever as a Rarity Before 2007

The diagnosis of Q fever was very rare in the Netherlands before 1977, despite increasing numbers of reported cases from other countries. Extensive studies were carried out between 1951 and 1956 on cattle (n = 524) and on patients with atypical pneumonia (n = 6,000). These studies tested serum samples using complement fixation test (CFT) and used animal (guinea pig) cultures. None of them revealed a positive result (Wolff and Kouwenaar 1954). Then in 1956, just as these studies were being phased out, the first three human cases of Q fever were diagnosed in the Netherlands (Westra et al. 1958; Dekking and Zanen 1958). One patient worked at a slaughterhouse, one was thought to have been infected in Switzerland, and a third had spent time living near sheep. When 28 of these sheep were serologically analysed, one tested positive.

17.2.2 An Increase in Reported Cases

In 1976 Q fever was added to the list of notifiable diseases in the Netherlands. This was quickly followed by a rise in the number of reported cases – an average of 2–3 a year between 1977 and 1980, and an average of 20 a year up until 2007. Thirty-three Q fever cases diagnosed between 1979 and 1983 have been described in more detail (Richardus et al. 1984a). An in-house developed immunofluorescence assay (IFA) was used and IgM phase II ≥1:16 was considered reactive. Apart from the usual clinical presentation, epidemiological analysis to identify possible sources showed that 67% of these patients had acquired Q fever in the Netherlands, while the rest were probably infected in a variety of other European countries. However, a reanalysis of the data shows that the serologic profiles described in the patients diagnosed with acute Q fever were quite heterogeneous, with mismatches of CFT and IFA results. Therefore, the group of patients described was a heterogeneous group with acute, past resolved and chronic infections.

17.2.3 Early Seroprevalence Studies

Within the context of increasing number of cases, extensive serologic studies were conducted in the Netherlands in the period between 1982 and 1985 (Richardus et al. 1984b, 1985, 1987), using CFT and IFA with *Coxiella burnetii* antigen phase II from Virion (Virion Ltd., Zurich, Switzerland). For IFA, IgG antibodies were tested to the phase II antigen, with a cut-off of ≥1:16. The study tested a selection of serum samples from groups of people considered to be at high risk of infection. This approach showed very high seropositive rates among veterinarians working with large domesticated animals (84%) and with small domesticated animals (77%), as well as taxidermists (70%) and wool spinners (58%). However, a range of high seropositivity rates (14–73%) was also found in the control groups. The authors suggest that these high rates were achieved because the IFA used in their study was more sensitive than the CFT used in previous studies. No control experiments with CFT were performed in this study, and the IFA results were not confirmed with additional titrations.

Unfortunately, the original data are not available for statistical reanalysis. If transmission rates in the 1970s and 1980s were high, then serologic evidence of this should be evident in older age groups in recent serological surveys. We speculate that the lack of specificity in the in-house IFA, combined with a low cut-off may have influenced the seroprevalence rates of this study.

17.2.4 The Need for Clearer Analysis Before the 2007 Epidemic

When the Dutch Q fever epidemic began in 2007, some speculated that the disease may have been previously overlooked because of under-diagnosis and misclassification, and that increased awareness had created a pseudo-epidemic through the misclassification of acute infections (Van Knapen, personal communication). Indeed, a considerable amount of misclassification is possible if diagnosis relies on detection of IgM phase II, which can persist for months or even years. Clearly, the pre-epidemic situation needed analysis. In addition a change in laws and regulations regarding infectious diseases in the Netherlands also started to have an impact.

17.2.5 Changing Laws and Regulations

The analysis of national data regarding disease incidence relies on notifications. Regulations that control notification are important as regulatory changes can influence epidemiological curves. When Q fever was added to the list of notifiable diseases in 1976, clinicians were legally required to notify public health authorities of Q fever patients, and municipalities (advised by municipal health services) were obliged to enforce legal actions to curb epidemics when necessary. The use of the Q fever notification system was analysed in 2002 (van Gageldonk-Lafeber et al. 2003). Aggregated data from between 1988 and 2002, retrieved from laboratories, hospital admissions and discharge records, were compared to the national data set of Q fever notifications. This analysis showed that only 50% of diagnosed cases were reported by clinicians because laboratories were not required to provide notification at that time. In 2008, a new law was introduced in the Netherlands to comply with international health regulations. According to this law, laboratories were also obliged to provide notification. This approach was expected to improve notification records – as each case would be notified by both the laboratory and the consulting clinician.

17.2.6 Possible Q Fever Clusters Before 2007 Detected Retrospectively

Clinical Q fever in animals was diagnosed in the Netherlands in 2005 in two dairy goat herds with high abortion rates (Wouda and Dercksen 2007). Van den Wijngaard et al. (2011) speculated that unrecognised outbreaks might have preceded the first recognised outbreak in 2007. With this in mind, they used space-time scan statistics and syndromic surveillance to search for hidden Q fever clusters before and during 2007. Hospitalisation data for lower respiratory, hepatitis and endocarditis infections occurring between 2005 and 2008 were aggregated by week, age group and postal codes. Alternative causes of outbreaks were excluded by reviewing all mandatory notified diseases with similar clinical presentation in the same period. Surveillance data on influenza-like illness were also included to assess whether clusters of hospital admissions for lower respiratory tract infections could be due to influenza. From 2005 to 2008, a total of 20 lower respiratory tract infection clusters and two hepatitis clusters were detected. Scan statistics for space-time clusters detected one specific cluster – the Q fever epidemic in 2007. However, ten other clusters were also detected that could be due to other causes, including a major confirmed Legionella outbreak.

Three clusters which occurred earlier than the recorded outbreak – two in 2005 and one 2006 – could be due to Q fever because there was a Q fever-affected farm nearby and there was no alternative explanation for the cluster. In 2007, a number of clusters of lower respiratory tract infection and one hepatitis cluster were also found, and could be attributed to the actual Q fever epidemic. Three clusters in 2007 could not be attributed to Q fever, because they could not be linked to Q fever abortion waves on farms.

In conclusion, Q fever may have spread unnoticed among humans before 2007, and routine cluster scanning may facilitate earlier detection of comparable epidemics in the future. There is indeed circumstantial evidence of limited clustered spreading of *C. burnetii* among humans before 2007, but this analysis also confirms that the major outbreak started in 2007. However, once the Q fever epidemic was established, it may have resulted in an increased number of diagnoses, influencing epidemic curves (van der Hoek et al. 2010a).

17.2.7 Recent Serosurveillance: The PIENTER Study

In 2006, a population-based seroprevalence study was carried out by the National Institute for Public Health and the Environment to evaluate the Dutch National Immunisation Programme. This programme (PIENTER) has been described in detail by van der Klis et al. (2009). It was a national survey in which participants were asked to donate blood and complete a questionnaire on demographics, health perception and activities related to infectious diseases. Data and sample collection was finalised in June 2007, after which the stored serum samples were used to screen for the presence of *C. burnetii* antibodies (Schimmer et al. 2011).

Given the screening considerations described above, this study used a combined test strategy to measure seroprevalence, using an enzyme-linked immunosorbent assay (ELISA) IgG phase II (Serion Immundiagnostica, Würzburg, Germany) on the study group of 5,654 samples, followed by a confirmation of positives by IFA (Focus Diagnostics, Cypress, California, USA). IFA was also used to estimate the ELISA's false negative rates on 504 randomly chosen ELISA negative samples.

Of the 5,654 samples tested, 85 were positive with the ELISA IgG phase II. Of these, 47 had borderline levels and 15 were negative in an IFA IgG phase II screened with 1:32. This resulted in a seroprevalence of 1.5% using ELISA to screen and IFA to confirm. In the 504 ELISA negative samples tested in IFA, six (1.2%) had titres ranging from 1:32 to 1:128. Using IFA as the 'agreed standard', the adjusted seroprevalence estimate was 2.4%. These results underscore the problems encountered while comparing different seroprevalence studies.

The results from this study yielded a low seroprevalence in the Netherlands before 2007, but the low numbers still represent a considerable amount of exposure. Seropositivity in males was higher than in females and increased with age. No regional differences were observed, even when sheep, goat and cattle densities were examined. However, higher seroprevalence was associated with increasing age, being born abroad (specifically in Turkey), keeping ruminants and having occupational contact with animals. In conclusion, this study supports the concept of the massive localised introduction of *C. burnetii* in the human population from 2007 onwards (Schimmer et al. 2011).

Clearly, Q fever has been circulating at a low level in the Netherlands since the 1950s. There may have been a temporary increase in exposure during the 1980s, but data from the older serosurveys and notifications may lack accuracy. Recent studies have confirmed a massive exposure of Q fever in the Netherlands from 2007 onwards.

17.3 The Dutch Epidemic from 2007 to 2010

17.3.1 Concerns Rise in 2007

Between March and June 2007, six cases of acute Q fever were notified by regional microbiology laboratories to public health authorities in the province of North Brabant in the south of the Netherlands. These patients were admitted with atypical pneumonia to a number of hospitals in the province. Concerned, the regional Municipal Health Service (MHS) analysed the cases in detail, but could not link them epidemiologically. Then a general practitioner from a nearby village reported an excess of patients with pneumonia at his practice. Initially, these patients were mistakenly thought to have *Mycoplasma pneumoniae* infection due to serologic cross-reactions, but they were eventually confirmed as having acute Q fever in July 2007 (Roest et al. 2011; van Steenbergen et al. 2007).

Eventually, a total of 168 human cases were notified in North Brabant in 2007. Dairy goats were identified as the source of the human Q fever cases in North Brabant – the Animal Health Service confirmed a considerable number of Q fever-induced abortions at several farms in the region. The unusually hot and dry weather in the spring of 2007 may have caused airborne transmission of contaminated dust particles. The outbreak seemed to have been concentrated around a single village, but a specific point source could not be identified. A case-control study was performed in the village (Karagiannis et al. 2009) and contact with manure, hay and straw were shown to be risk factors. It was shown that people living in the eastern part of the village close to ruminant farms (one of which had a recent history of abortion problems) were at higher risk than people living in other parts of the village. Contact with animals and the consumption of raw milk products were not significant risk factors in the multivariable analysis. In general acute Q fever seemed not to be related to the working environment but there were reports of incidental cases that occurred after visits to dairy goat farms with abortion problems.

17.3.2 Source of Epidemiological Data

The epidemiological data comes from the national registry of notifiable infectious diseases. Attending physicians and heads of microbiology laboratories have a legal obligation to notify the diagnosis of human Q fever to the MHS, which enters the cases into an anonymous national electronic database ('Osiris') monitored by the Centre for Infectious Disease Control. Since the beginning of 2007, notification criteria for acute Q fever in Osiris have been a combination of clinical presentation matching Q fever, with either a four-fold IgG titre rise or a positive IgM phase II antibody test measured by IFA, ELISA, or CFT. During the course of the outbreaks, certain adaptations were made to the notification criteria. In July 2008, a clinical presentation matching Q fever was further defined as fever, or pneumonia, or hepatitis. In February 2010, an additional laboratory criterion was the detection by polymerase chain reaction (PCR) of *C. burnetii* DNA in serum or respiratory material. However, diagnoses based on PCR were already accepted before that time. Given the above criteria, misclassification was possible when isolated IgM was used as a sole measure, since IgM can be a false positive, or persist for months after a past resolved infection. Moreover, clinical symptoms may be aspecific.

Another important source of epidemiological data was a questionnaire routinely dispatched to notified acute Q fever cases by the MHS, which included questions about environmental risk factors and clinical characteristics. This questionnaire was received from 74% of notified cases with onset of illness in 2007 and from 93% of notified cases with onset of illness in 2008.

17.3.3 2007–2010 Overview: Not an Isolated Incident

As the number of notifications increased from May 2008, it became evident that the 2007 outbreak was not an isolated incident. A total of 3,489 Q fever patients who experienced onset of disease between 2007 and 2009 were notified. Of these, 194 cases had a date of onset in 2007, 982 in 2008, and 2,313 in 2009. The epidemic curve (Fig. 17.1) shows a seasonal pattern, with most cases occurring in spring and early summer. The highest incidences were seen in the south of the country, mainly in the province of North Brabant; the affected area expanded to the north and the south during the epidemic (Fig. 17.2). Patient characteristics from 2007 to 2009 were presented by Schneeberger et al. (2010a). The median age of the confirmed notified patients was 50 years, and over 60% of them were male.

The additional MHS questionnaires showed that only a small proportion of patients lived on a farm or worked in the agriculture or meat processing sectors. However, notified patients frequently reported that they had been in contact with a diverse number of animals and animal products.

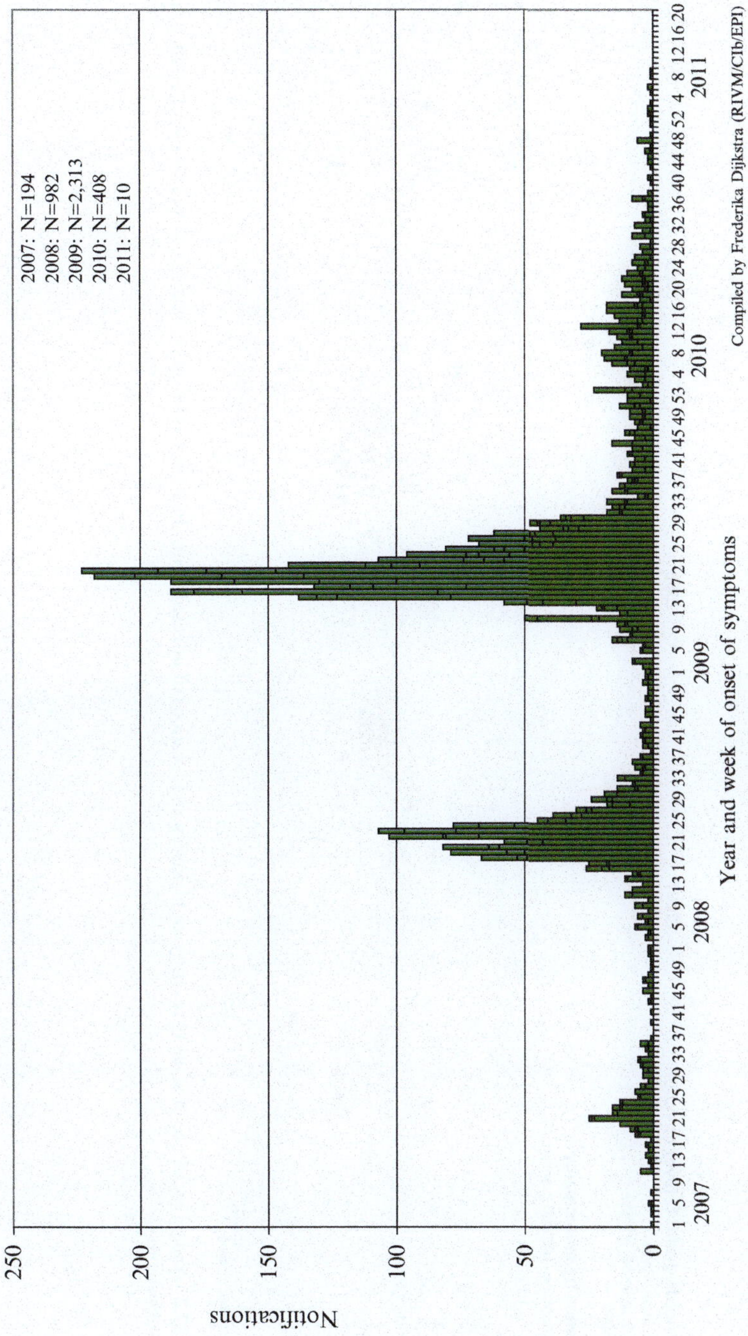

Fig. 17.1 Number of notified acute Q fever patients from 1 January 2007 to 23 March 2011 with known day of onset of illness

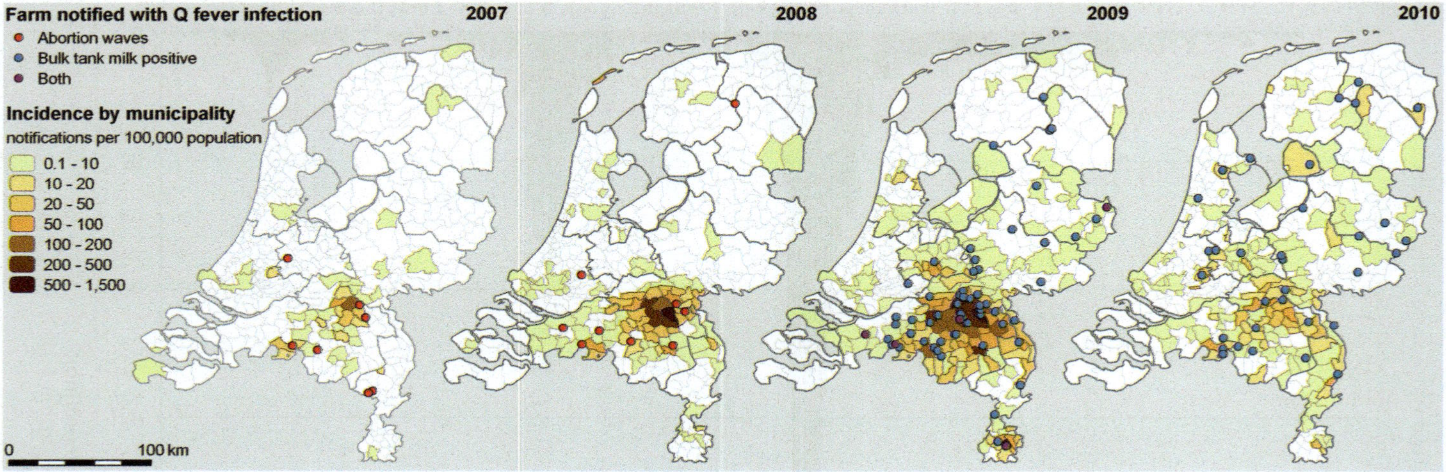

Fig. 17.2 Annual incidence of acute Q fever by municipality and locations of infected farms. Incidence is based on year of onset of illness of notified acute Q fever patients

In 2007, the percentage of hospitalized patients (50%) was largely influenced by active case findings in a retrospective survey among hospitalised cases (van der Hoek et al. 2010b). In 2008 and 2009 it was 20%, still much higher than the 2–5% hospitalisation rate reported in the literature (Raoult et al. 2000). Fever was the most frequently reported symptom (92%), followed by fatigue (78%) and headache (69%). Pneumonia was diagnosed in 62% of patients, while endocarditis (3%) and hepatitis (<1%) were relatively rare. Underlying diseases were frequently reported. Almost 49% of patients smoked, which is relatively high compared to percentages in the general population (30% for males and 24% for females, according to Statistics Netherlands).

17.3.4 Diagnostic Delay and Influence of Influenza A(H1N1)

All notified patients in 2007 for whom additional laboratory data was available were diagnosed either by IFA or CFT. In 2008, 3% of cases were diagnosed by PCR. The most popular method in 2008 was still IFA (75%), although CFT, ELISA, PCR and other methods were also used. In 2009, 79% of notified patients were diagnosed serologically and 20% by PCR. IFA was used in more than half of the cases in 2009, CFT was used in more than a quarter of the cases and ELISA was used in 14% of cases.

The median diagnostic delay (the delay between the date of onset of illness and the date of *C. burnetti* diagnosis) decreased from 82 days in 2007, to 28 days in 2008, to 20 days in 2009 (van der Hoek et al. 2010a). The diagnostic delay was due to lack of awareness by medical staff and the delay in making a definitive diagnosis, as routine diagnostics mainly relied on seroconversion in convalescent serum. Increased awareness and improved routine laboratory services, such as the introduction of IFA, ELISA, and in 2009 PCR, have reduced this diagnostic delay.

Under conditions of high incidence, the positive predictive values of tests are very high. In the autumn of 2009, pandemic influenza A(H1N1) 2009 with more or less similar symptoms interfered with the analysis of the Q fever epidemic. In this third year of the Q fever outbreak, a high background prevalence of antibodies to *C. burnetii*, specifically positive IgM titres, were common in most of the affected areas. This made it much more difficult to determine the exact start of an acute Q fever episode, thus making notifications less reliable. In fact, many patients diagnosed in the laboratory in 2010 had probably experienced clinical signs of acute Q fever much earlier. Under these circumstances, the persistence of IgM makes it difficult to measure the actual decline of the incidence of the disease.

17.3.5 A Link with Goats and Sheep

In May 2008, an outbreak of Q fever occurred in a psychiatric care institution in Nijmegen near the 2007 outbreak area (Koene et al. 2011). At least 28 in-patients, staff, and visitors had laboratory confirmed Q fever illness and several patients in the institution developed atypical pneumonia. It was discovered that these patients had been in close contact with lambs as part of their therapy sessions. Then a large number of goats unexpectedly aborted their offspring on a farm close by, and Q fever was confirmed in the farmer and his wife living there (Roest et al. 2011). An urban cluster identified in 2008 was found to be related to a goat farm with high abortion rates in the area. Patients lived downwind of the goat farm and a house located <2 km from the farm was associated with a higher risk of Q fever infection, compared to a house located at ≥5 km (Schimmer et al. 2010). In 2009, 59% of notified human cases lived within a 5 km zone of a bulk tank milk-positive dairy goat or sheep farm, and 12% (roughly one million people) of the Dutch population lived within such zones (Roest et al. 2011; van der Hoek et al. 2010b). The available evidence in the Netherlands points to dairy goat farms with Q fever-induced abortion problems as the main source of the human outbreaks with a smaller role for dairy sheep and non-dairy sheep.

17.3.6 Transmission from Animals to Humans

Infection of humans is caused by inhalation of contaminated aerosols that can spread over some distance. Especially when infected pregnant small ruminants abort, billions of *C. burnetii* end up in the environment while fewer than 10 organisms are sufficient to seed an infection (Benenson and Tigertt 1956). The organism's ability to persist in the environment may result in a continued risk for infection weeks to months after the birthing event.

The size of the community outbreak in the Netherlands suggests that transmission predominantly takes place through wide-scale environmental contamination or multiple point-source contamination sites. There is strong epidemiological evidence that most human cases are caused by abortion waves on dairy goat farms. People living close to such farms are at risk. Infected farms that have no abortion waves can still be infectious when there is close contact with animals. The transmission route is the same, through inhalation of contaminated aerosols, but the dose is much lower, hence closer contact is required for infection. Based on detailed information from notified patients, occupational exposure can explain only a small proportion of the acute Q fever cases in the Netherlands.

Despite the evidence pointing towards dairy goat farms with Q fever-induced abortion problems, there were a number of such farms without any human cases in the surrounding population. In the 5 km areas around 27 farms with clinical abortion problems, environmental data sets were collected. This showed clear differences between areas with and without transmission to humans in vegetation density and in average groundwater conditions (van der Hoek et al. 2011a). Areas without transmission had higher vegetation densities, based on remotely sensed satellite imagery, and relatively shallow groundwater conditions suggesting that vegetation and soil moisture are relevant factors in the transmission of *C. burnetii* from infected small ruminant farms to humans.

Alternative routes of transmission are unlikely to have played an important role. Q fever is a zoonotic disease with no convincing evidence for human-to-human transmission. Information from notified acute Q fever patients makes it very unlikely that consumption of unpasteurised dairy products has played an important role. In 2008 the manure streams from dairy goat farms were investigated in some detail. Manure was often transported to other parts of the country to be used for example in flower bulb cultivation but in the recipient areas no Q fever cases were reported (unpublished data). It was therefore concluded that manure did not play an important role. More than 2000 ticks have been collected from sheep and the environment but no *C. burnetii* was detected. Dairy goats in the Netherlands are kept indoors in deep litter stables and are not affected by ticks.

17.4 Veterinary Control Measures

17.4.1 The Veterinary Situation at the Beginning of the Outbreak

The world's largest Q fever epidemic recorded to date occurred in an area densely populated with people and domesticated animals, suggesting that animal farming in such areas poses a risk for zoonotic diseases such as Q fever in humans. Initially, the evidence to link the outbreak to goat farming was largely circumstantial, in the absence of DNA fingerprinting techniques for *C. burnetii* that could have matched bacteria from human and animal samples. While there was no sound evidence base for control measures, the subsequent rapid expansion in the scale of epidemic was unforeseen. National and regional public health authorities were largely unprepared for an outbreak of this magnitude, and international literature on smaller outbreaks provided insufficient guidance on several key issues – such as appropriate control measures, the possible effects of the epidemic on pregnant women, the most adequate therapy for acute Q fever, the identification and classification of chronic Q fever, and the use of the Australian human vaccine for Q fever. The most affected province, North Brabant, has a surface area of 5,100 km^2 and currently houses 2.4 million people and 6.4 million livestock (80,000 sheep, 135,000 goats, 660,000 cows and 5.5 million pigs) (Statistics Netherlands 2011), with a goat density that increased five-fold between 1990 and 2007. In retrospect, abortion waves due to *C. burnetii* infection among the goat population were reported from 2005 onwards, although they were not recognized as such at that time.

17.4.2 Veterinary Measures in Response to the 2008 Outbreak

The widespread pattern of the outbreak in 2008 was alarming and pointed to several clusters with multiple sources. In June 2008, the government announced the mandatory notification of Q fever on dairy goat and sheep farms with >5% abortions due to *C. burnetii* infection, and introduced appropriate hygiene measures.[1] During a period of 90 days following the detection of Q fever at a farm, a manure removal ban and visiting restrictions were implemented.[1] However, no restrictions on the transport of animals from infected farms were imposed, and other possible veterinary measures to contain the outbreak, such as a breeding ban, were not included.

Then, in October 2008, the Dutch government authorised the voluntary vaccination of animals on large dairy goat and sheep and recreational farms using the non-registered Coxevac® vaccine (Ceva Santé Animale, France). From November 2008, goat and sheep at smaller farms were also vaccinated. However because of limited vaccine availability – just 80,000 doses – vaccinations could only be provided within a 45 km radius of the outbreak source. In February 2009, a nationwide hygiene protocol became mandatory for all dairy goat and sheep farms, whether infected or not.[1] The eradication of vermin became compulsory, the cleaning of stables during lambing season and for 30 days afterwards was forbidden, and manure had to be stored and covered for at least 90 days before use.

17.4.3 Veterinary Measures in Response to the 2009 Outbreak

Despite these measures, the outbreak was still far from contained. Over 2,000 new human acute Q fever cases were notified from late March 2009 onwards, in a larger area than in 2008. In response, in April 2009, the government extended the vaccination campaign to include a compulsory vaccination programme.[1] Farms with a public function and dairy goat and sheep farms with more than 50 animals in the epidemic centre had to vaccinate their animals before 2010. All Q fever infected farms outside of the area were also obliged to vaccinate their animals. Vaccination of animals on farms in the rest of the Netherlands was still on a voluntary basis. In July 2009, the pasteurization of manure for a minimum of 1 h at 70°C was permitted instead of a decomposition period of 90 days.[1] Restrictions on incoming and outgoing animal transport on *C. burnetii* infected farms were imposed from October 2009 onwards.[1]

17.4.4 Bulk Tank Milk Monitoring

Also from October 2009 onwards, the government set out a new strategy to identify infected farms that did not have an abortion rate above 5%. Farms with more than 50 dairy goats or sheep were obliged to participate in Q fever bulk tank milk

[1] Information from official documents of the Ministry of Agriculture, Nature and Food Quality. Available at http://overheid-op.sdu.nl/cgi/login

monitoring.[1] Bulk milk tanks were sampled once every 2 months (and later on once every 2 weeks) and tested for the presence of *C. burnetii* DNA using a real-time PCR by the Animal Health Service. To separate infected farms from non-infected farms, a cycle threshold (Ct) of 36 as detected by a-real-time PCR test targeting IS1111 was used. This threshold, which was set arbitrarily, is close to the detection limit of real-time PCR tests, implying that the outcome of the test in the lower range (from Ct 34 to 36) is determined stochastically. Positive samples were forwarded for confirmatory testing to the Central Veterinary Institute, and confirmed farms were declared infected. Infected farms were identified on the basis of a positive PCR outcome only, as information on background values of *C. burnetii* DNA load in goat bulk tank milk samples was unavailable. This approach may well have resulted in farms being declared infected when they posed no threat to human health.

17.4.5 Drastic Veterinary Measures

In December 2009, Zembla, a current affairs television programme co-produced by the Dutch Broadcasting Association and a Dutch public newscaster, raised critical concerns about the role of the Dutch government in containing the Q fever outbreak. The Dutch government responded to the increasing concerns by administrators, professionals, and the public by making the location of the 55 *C. burnetii*-infected farms public, announcing a breeding ban on infected farms, and increasing the frequency of tank milk monitoring from bi-monthly to bi-weekly.[1] The decision was taken to implement the most radical measure possible, the pre-emptive culling of all pregnant goats on infected farms.[1] Male goats on infected farms were also culled, as they could supposedly transmit the disease via semen.

A total of 50,355 goats and sheep were culled from 21 December 2009 to June 2010 on 89 bulk tank milk positive farms. Of 517 culling-workers, involved, 17.5% seroconverted for antibodies to *C. burnetii* despite use of personal protective equipment (Whelan et al. 2011). Seroprevalence of *C. burnetii* in workers before the culling activity was 13%, which is similar to findings among blood donors residing in the high-incidence area in the Netherlands in 2009 and in similar high-risk occupational groups internationally (Anderson et al. 2011). Symptomatic infection was recorded in 31% of the seroconverters. A strong dose-response relationship was shown between risk of seroconversion and number of hours worked on the farms and working inside the stable (in close proximity to the animals). In other settings internationally, a risk-gradient has also been shown with close direct and indirect animal contact over time (Porten et al. 2006; Casolin 1999). Given the high risk of infection during culling activities, additional preventive measures should be taken. The Health Council of the Netherlands (2010) has already issued first advice on risk groups suitable for human vaccination against Q fever. However this advice does not extend to culling workers.

17.4.6 Effect of Veterinary Measures in 2010

Exponential spread of Q fever did not occur during the spring of 2010, as feared. Still, almost 400 new Q fever patients were diagnosed during that year. By June 2010, all dairy goats and sheep had been vaccinated twice with Coxevac, and on July 15th the breeding ban for non-infected farms was lifted. The reason for the approximately five-fold decrease in the number of human infections between 2009 and 2010 is mainly attributed to the culling of pregnant goats and sheep, and the vaccination programme. Other factors could have contributed such as hygiene measures, climate, and increasing immunity among the general population.

17.4.7 Identification of the Source of the Outbreak

Recent genotyping studies point to a multi-strain bacterium in both livestock and humans (Huijsmans et al. 2011) as the cause of the Dutch Q fever outbreak and not simply one highly virulent *C. burnetii* strain. Several conditions may have favoured the introduction and rapid spread of *C. burnetii* among livestock in the Netherlands since 2005. In general, goats are kept in large herds in 'deep litter stables' – stables on concrete floors with pits. Straw is regularly added to these deep litter stables, which allows for relatively unhygienic conditions as potentially infected excreta such as urine, faeces and birth products are not regularly removed. Furthermore, the straw is often bought from countries such as France and Germany, which might be a source of *C. burnetii*. Relatively high quantities of *C. burnetii* DNA were measured in samples of stocked straw that had not yet been used in deep litter stables (unpublished observation of M.H.A. Hermans and P.C. Wever). Newborn goats are often fed raw cow colostrum. Colostrum is the highly nutritious milk produced by mammals just before giving birth, and cow colostrum may contain large quantities of *C. burnetii* DNA (unpublished observations of M.H.A. Hermans and P.C. Wever). Therefore, the role of straw and colostrum are interesting for further investigation as potential sources of multi-strain *C. burnetii* infection in goats. When the pits in deep litter stables are full – which happens two or three times a year – everything (including manure and birth products) is removed and spread over the fields or transported elsewhere. Furthermore, the open air stables allow wind to blow through and take *C. burnetii*-infected dust into the environment. April 2007, May and June of 2008, and April of 2009 were unusually dry by Dutch standards – this type of weather condition combined with wind has been documented to play an important role in other Q fever outbreaks (Hawker et al. 1998).

17.5 Laboratory Issues

17.5.1 The Optimization of Acute Q Fever Diagnostics in the Netherlands During an Outbreak

Prior to recognition of the Q fever outbreak in 2007, Q fever diagnostics were performed by a limited number of Dutch microbiology laboratories. The National Institute for Public Health and the Environment functioned as a reference laboratory using the IFA as a reference method, while other regional microbiology laboratories used the CFT. Following recognition of the scale of the outbreak in mid 2007, a number of microbiology laboratories in the epidemic area began to offer Q fever diagnostics. The laboratories chose to use either IFA or CFT, depending in part on the type of serologic assays already in use by them. Seroconversion can be detected earlier by IFA compared to CFT (between 10 and 15 days after infection, versus 2–3 weeks, respectively). In addition, IFA allows the separate analysis of IgM and IgG antibodies against *C. burnetii* phase II and phase I antigens (IgM-II, IgG-II, IgM-I and IgG-I antibodies). Analysis of these four antibodies makes it possible to identify various stages in the acute Q fever infection. The presence of solitary IgM-II antibodies – the first antibody to appear in the serologic response – indicates early acute infection, whereas the presence of IgG-II, IgM-I and IgG-I antibodies reflects a later stage of acute infection. CFT, in contrast, has been reported to be less prone to false positive results than IFA, especially with detection of IgM-II antibodies. Both methods are labour-intensive, non-automated and subject to inter- and intraobserver variation (Maurin and Raoult 1999; Wegdam-Blans et al. 2010).

An important drawback to the serological diagnosis of acute Q fever is the lag phase in antibody response of up to 3 weeks after the onset of clinical symptoms. In 2008, the ongoing outbreak resulted in the development of real-time PCR assays targeting the multicopy IS1111 insertion element by several microbiology laboratories. These assays were used for the detection of *C. burnetii* DNA in serum, respiratory samples, urine specimens, tissues and amniotic fluids. Subsequently, an interlaboratory evaluation of different DNA extraction and real time PCR methods for the detection of *C. burnetii* DNA in serum was conducted. Overall, a low degree of variation was observed in the sensitivity of the evaluated real time PCR assays, although assays amplifying short DNA fragments yielded better results than those producing a large DNA fragment (Tilburg et al. 2010). Next, performance of one of the PCR was evaluated retrospectively on serum samples of acute Q fever patients at various stages of the serological response. *C. burnetii* DNA was detected in serum from 98% of seronegative acute Q fever patients and in 90% of patients with solitary IgM-II antibodies. Ultimately, the PCR became negative as the serological response to *C. burnetii* further developed, with subsequent appearance of IgG-II, IgM-I and IgG-I antibodies (Schneeberger et al. 2010a, b).

17.5.2 Increasing Diagnostic Demands

To cope with the surge in diagnostic demands, which occurred during 2009 (e.g. over 18,000 requests for Q fever diagnostics were received by one single laboratory), several microbiology laboratories in the epidemic area began using algorithms to provide accurate, fast, cost-effective and standardized acute Q fever diagnostics. One such algorithm used an ELISA for IgM-II antibodies, performed on an automated processing system as an initial screening step. In the case of a positive or dubious ELISA result, IFA was performed as a confirmation step. PCR was performed after a negative ELISA result and when the serum sample was either acquired ≤14 days after onset of disease or referred by a hospital physician (noted for their lack in providing a date of onset of disease). When acute Q fever diagnostics on the first serum sample were non-conclusive, a second serum sample was requested after 14 days. Overall, this diagnostic approach led to a significant reduction in the number of labour-intensive, non-automated IFA tests performed, with an increased diagnostic yield on first serum samples due to the introduction of PCR techniques. Likewise, an algorithm was introduced using the ELISA for IgM-II antibodies as a screening step followed by CFT as a confirmation step (Wegdam-Blans et al. 2009).

17.5.3 A Consensus on Q Fever Diagnosis

In 2010, the National Institute for Public Health and the Environment and the Dutch Society for Medical Microbiology formed a working group to develop a consensus on the microbiological diagnosis of acute Q fever. In September 2010, a consensus document was published recommending the use of algorithms in which PCR, ELISA for IgM-II antibodies and either IFA or CFT were incorporated (Wegdam-Blans et al. 2010). The diagnosis 'confirmed acute Q fever' is established by a single positive PCR result with an appropriate clinical presentation or an IgG-II seroconversion or a four-fold or higher increase in IgG-II titre detected by IFA or CFT (requiring multiple serum samples). The diagnosis of 'possible acute Q fever' is made by a positive IgM-II result (in the presence or absence of IgG-II, IgM-I, IgG-I antibodies) in a single serum sample with an appropriate clinical presentation and should be confirmed by either an IgG-II seroconversion or a four-fold or higher increase in IgG-II titre in a follow-up serum sample. The introduction of the term 'possible acute Q fever' was the result of an increasing number of patients with past resolved Q fever and persisting antibody titres against phase II antigens in particular, as well as reported false-positive IgM-II results from IFA and ELISA.

17.5.4 Cross-reactions in Serologic Tests for *Coxiella burnetii*

While screening methods should be very sensitive and can be less specific, several screening methods for infectious diseases have shown cross-reactions with other infections. For Q fever, most cross-reactions described in the literature are those with other agents, which cause pulmonary infections such as *Legionella pneumophila* (Dwyer et al. 1988; Finidori et al. 1992).

In the Netherlands, patients with a pulmonary infection are usually screened for several pathogens. This approach recognises cross-reactions with other agents that cause pulmonary infections. The Q fever epidemic in the Netherlands was first thought to be caused by *Mycoplasma pneumoniae*, because CFT results of several patients showed low titres against this organism. Therefore, we investigated cross-reactions between sera from patients with high Mycoplasma titres and the ELISA screening assay for Q fever. No cross-reactions were found.

Cross-reactions with other pathogens are less likely to be recognised. We investigated cross-reactivity in sera taken from patients with recent Epstein-Barr virus (EBV) and Cytomegalovirus (CMV) infections (IgM positive with low avidity). In a *Coxiella* screening ELISA (IgM phase II), 16/72 EBV IgM positive sera reacted while 7/33 CMV IgM positive sera reacted. In both EBV IgM- and CMV IgM-positive patients, the results in the Q fever ELISA were generally low positive. When performing an IFA phase I and II assay on the ELISA positive samples, all EBV IgM positive patients became negative. However, 40% of CMV IgM positive patients also had positive IFA tests and we concluded that these patients probably had two infections going on at around the same time.

When screening patients with a proven seroconversion for *C. burnetii*, we found very few cross-reactions in tests for EBV IgM (2%) but somewhat more in tests for CMV IgM (8%). Unfortunately, we were unable to follow these patients to see if these were true double infections or not. We concluded that recent EBV and CMV infections can cause cross-reactive antibodies against *C. burnetii* in ELISA tests, but not in IFA tests. It is possible that a recent Q fever infection can also cause cross-reactions in the test for CMV IgM. Therefore, screening with ELISA, followed by confirmation with IFA, is a good way to exclude false positive tests for *C. burnetii* due to recent EBV and CMV infections.

17.5.5 Seroprevalence Surveys

Strategies for diagnosing acute and chronic Q fever in individual patients differ from population-based seroprevalence surveys. Acute Q fever is diagnosed mainly by the detection of antibodies of IgM and IgG subclasses against phase II of *C. burnetii*. The diagnosis of chronic Q fever relies on high titres of phase I IgG antibodies; the sole presence of phase II IgG antibodies against *C. burnetii* indicates a previous infection (Tissot Dupont et al. 1994; Waag et al. 1995). IgG-II antibody levels remain constantly high for almost a year and then slowly decrease, remaining detectable for years after first detection (Dupuis et al. 1985). Therefore, the study of seroprevalence relies on the detection of IgG-antibodies against phase II of *C. burnetii* in serum samples– antibodies can be detected by CFT, IFA or ELISA. What we need to establish is: which test works best?

17.5.6 *The IFA/ELISA Debate: The Need for a Standard*

Studies in the late 1980s and early 1990s reported that the ELISA was a more sensitive and specific method than either the IFA or CFT (Cowley et al. 1992; Peter et al. 1988). Field et al. (2002) evaluated the performance of an ELISA IgG kit (Panbio) against an in-house IFA. The two tests had moderate (53%) agreement; the ELISA had a sensitivity of 71% and a specificity of 96%. A number of different in-house IFAs were evaluated against ELISA, using different methods and cut-offs (D'Harcourt et al. 1996; Setiyono et al. 2005) and the results are illustrative of the difficulties in comparing studies of the serodiagnosis and seroprevalence of Q fever. Nevertheless, IFA has been proclaimed the gold standard reference method in the literature (Fournier et al. 1998), although this method is laborious when compared to the ELISA (which is easier to automate and more suitable for testing large sample numbers). Commercial ELISA and IFA tests are both in current use for the diagnosis of acute and chronic Q fever and in seroprevalence studies globally (Anderson et al. 2009; Gilsdorf et al. 2008; McCaughey et al. 2008). The uncertainty regarding a valid standardised test, coupled with insufficient knowledge about the specific fate of antibodies against *C. burnetii* makes the seroepidemiology of Q fever a difficult undertaking.

Serum samples from a case-control study conducted in 2007 in the Netherlands to investigate the source and routes of transmission in this outbreak were used to evaluate the performance of one commercially available ELISA (Serion Immundiagnostica, Würzburg, Germany) and one IFA (Focus Diagnostics, Cypress, California, USA). Four hundred and eighty-seven human sera were evaluated in terms of sensitivity, specificity and kappa value. The sensitivity and specificity of ELISA for the detection of IgG phase II antibodies were 59% and 97%, respectively (Blaauw et al. 2011). Seroprevalence varied depending on the method used; it was 12.7% when tested with IFA and 6.2% when tested with the ELISA. When measuring IgM antibodies to phase II *Coxiella* antigen, the two tests were comparable, with a kappa value of 0.89. Sensitivity was 82% and specificity 100%. These results support the concept that the ELISA performs reasonably well when diagnosing acute Q fever. However, in past infections, as defined by the sole presence of IgG antibodies, low positive samples have been missed by ELISA. More longitudinal studies using different test systems are needed to measure the levels of antibodies to *C. burnetii* in human serum. A single standard must be agreed on and defined in order to be able to easily compare results from various serosurveys.

17.6 Q Fever and Pregnancy During the 2007–2010 Q Fever Outbreaks in the Netherlands

17.6.1 *The International Literature*

When it became apparent that Q fever had become a major problem in the Netherlands from 2007 onwards, discussion arose about the health threat to pregnant women (Schimmer et al. 2009; van der Hoek et al. 2010b). An estimated 90% of acute Q fever infections in pregnancy present without clinical signs, which is much higher than among non-pregnant persons. International literature suggests that untreated acute Q fever infection during pregnancy may result in adverse pregnancy outcomes in up to 81% of cases (Tissot Dupont et al. 2007; Carcopino et al. 2007; Langley et al. 2003). These outcomes include abortion or intra-uterine foetal death and premature delivery or low birth weight. Furthermore, the risk of developing chronic Q fever infection is reported to be higher in pregnant women (Maurin and Raoult 1999; Carcopino et al. 2009). The only way to detect subclinical Q fever is through screening of serum for antibodies to *C. burnetii*.

17.6.2 The First Year of the Epidemic

In 2007 the Q fever outbreak was confined to a relatively small area. In July 2007, the Outbreak Management Team of the Netherlands decided to offer all pregnant women living in that area a screening test for Q fever. This decision was based on the fact that a policy based on signs and or symptoms was not possible (Parker et al. 2006) as asymptomatic infections could carry the same risk for adverse pregnancy outcome and chronic infection as symptomatic cases. Testing of all pregnant women in outbreak situations was also common policy in other countries. In France, the recommendation was to treat pregnant women testing positive and for those testing negative, to repeat testing on a monthly basis until delivery (Tissot Dupont et al. 2007). Public health practitioners in the south of the Netherlands tried to identify all women who were pregnant or who had recently delivered in the affected area. They were contacted by letter and offered the test. Out of 29 women identified through midwives and obstetricians working in the area, 19 responded, were interviewed and underwent serological testing with IFA (Focus Diagnostics) in which a titre of 1:64 was considered positive. None of these women experienced or had experienced signs or symptoms of Q fever. Two women however had serological evidence of a recent infection and one of an older infection (Meekelenkamp et al. 2009). The two women with serological evidence of recent infection were treated with cotrimoxazole for the duration of the pregnancy, as recommended in the literature (Carcopino et al. 2007). Both delivered under strict hygiene measures and both pregnancies and deliveries were without complications. Birth products tested by PCR all were negative. In none of the neonates there was serological or PCR evidence of vertical transmission of Q fever.

17.6.3 2008–2010

In the following years, the epidemic spread to a much larger geographic area with a total population of almost two million people. This raised the question whether screening of pregnant women for acute Q fever infection was necessary or feasible. The adverse effects from untreated Q fever infections were compared with the possible side effects of long-term antibiotic treatment during pregnancy. In July 2008 an international meeting organized by the Centre for Infectious Disease Control and the Health Council of the Netherlands met to discuss the feasibility of screening of all pregnant women for recent Q fever infection.

In all reported studies, there are a limited number of pregnant women with Q fever for whom pregnancy outcomes (adverse or otherwise) have been reported (<100 women for all studies combined). Most reports concern retrospectively collected data, which don't allow quantification of the risk for an adverse outcome of an infection during pregnancy. Based on this information and the fact that the epidemic had spread to a larger geographical area, the recommendation was that screening of all pregnant women for recent Q fever was not recommended (Health Council of the Netherlands 2008). Several retrospective and prospective studies were set up in response to the urgent need for better quantification of the risk of adverse pregnancy outcome among pregnant women with acute Q fever in early pregnancy.

17.6.4 A Population-Based Retrospective Follow-up Study

In a retrospective study, the presence of antibodies against *C. burnetii* during pregnancy was determined by testing sera that had routinely been collected in the Prenatal Screening for Infectious Diseases and Erythrocyte Immunization (PSIE) programme (http://www.rivm.nl/pns_en/). This screening programme for hepatitis B, syphilis, and HIV is offered to all pregnant women in the Netherlands at around the 12th week of pregnancy. Samples were available in a high percentage of pregnancies as they are often stored for a period of 1 year after initial analysis. Sera were analysed using an IFA for detection of IgG and IgM antibodies. A recent infection was defined as the presence of anti-phase II IgM and anti-phase II IgG antibodies with a titre of ≥1:64. A possible infection was defined as a solitary IgM II ≥1:64 and a past infection as the presence of anti-phase I and II IgG antibodies without IgM being present. Information on pregnancy outcome was obtained from the Netherlands Perinatal Registry (PRN) – a database that represents the joint efforts of the professional organizations of midwives, gynaecologists, obstetrically-trained general practitioners and paediatricians in the Netherlands. The PRN contains perinatal data from 16 weeks of gestation onwards for 96% of all births in the Netherlands. It was estimated that 60% of pregnant women were included in the study during the study period, based on the registered number of births. In this study, almost 4.5% of women had a recent or past infection. The presence of antibodies against *C. burnetii* was not significantly associated with an adverse pregnancy outcome as measured by: preterm delivery (gestational age below 37 weeks), low birth weight <2,500 g, birth weight for gestational age <10th percentile, foetal or neonatal mortality, congenital malformation and 5-min Apgar score <7 (van der Hoek et al. 2011b).

17.6.5 A Prospective Screen and Treat Study

In 2010 a clustered randomized controlled trial among pregnant women within the area of high transmission was started in the Netherlands (Munster et al. 2010a). The study participants were recruited by the midwives in these high risk areas. The midwife centres were randomized to recruit pregnant women from the control group or the intervention group. When taking part in the intervention group, blood samples were taken and tested immediately for Q fever. Patients were referred to a hospital for further pregnancy monitoring and long-term bacteriostatic treatment, if found positive for acute or chronic Q fever. In the control arm, blood samples were stored and analyzed for Q fever only after delivery. If tested positive for Q fever after pregnancy, antibiotics were started if needed as part of regular health care. The objective of the study was to measure differences in obstetric or maternal complications in Q fever positive women between screened and control group. Because the outbreak of Q fever in the Netherlands was successfully managed, relatively few pregnant women included in this study experienced a recent infection with *C. burnetti*. By September 2010, 815 samples had been examined, showing an overall seroprevalence of 15%, but with only 4% having a serologic profile suggesting recent infection (Munster et al. 2010b). The final results of this study are not yet available.

17.6.6 No Evidence of Adverse Effects on Pregnancy Outcome in the Netherlands

Data from the literature on the effects of Q fever infection in pregnant women are limited. Currently the best available evidence with regard to adverse pregnancy outcomes comes from a large case series and from several case reports documenting one to two cases (Carcopino et al. 2007; Denman and Woods 2009; Jover-Diaz et al. 2001; Stein and Raoult 1998; Rey et al. 2000). Case reports and case series have methodological limitations and selective publication of severe outcomes cannot be ruled out. In contrast, in the Dutch outbreak the presence of antibodies against *C. burnetii* in early pregnancy was not associated with adverse pregnancy outcome. This might be explained by a possible difference in pathogenicity of different bacterial strains or because we were not able to include early miscarriages in the study. We have to conclude that in the Dutch outbreak of Q fever, no evidence was found for adverse effects on pregnancy outcome among pregnant women with an asymptomatic Q fever infection in early pregnancy. Based on this, there is insufficient basis for recommending large-scale screening of pregnant women in high incidence areas.

17.7 Long-Term Effects of Acute Q Fever

17.7.1 From Acute to Chronic Illness

According to the literature, 60% of infected Q fever patients are asymptomatic, while 20% of patients develop mild symptoms (CSL 2009). The remaining 20% present with more severe symptoms including high fever, severe headache, night sweating, nausea, diarrhoea, pneumonia, hepatitis, pericarditis, myocarditis, neurological symptoms and weight loss (Mertens et al. 2007). The acute illness spontaneously resolves after 2–6 weeks (Marrie 1990). However, the organism or its partly degraded remains can persist in bone marrow, which can cause future episodes. Chronic illness after acute Q fever can express itself in different forms (Karakousis et al. 2006; Wildman et al. 2002). Classic Q fever endocarditis may take 10–15 years to develop and presents with cardiac vegetations that contain viable *Coxiella* bacteria. Recrudescent granulomatous infections can also occur. Patients with these two forms present with elevated levels of antibodies and persistent presence of viable *C. burnetii*. Another long-term effect of Q fever is QFS (post-Q fever fatigue syndrome). Contrary to the first two forms, QFS may present while there are no viable *Coxiella* and antibody levels are low or negligible. This is confusing for clinicians and patients alike.

17.7.2 Laboratory Diagnosis of Chronic Q Fever

Acute Q fever may develop into chronic Q fever in 2% of patients, a potentially lethal disease with endocarditis as the main presentation (ECDC 2010). Patients with previous cardiac valve pathology, aneurysms or vascular grafts, the immunocompromised and women who are infected during pregnancy are at risk of chronic Q fever (Maurin and Raoult 1999). An IFA IgG phase I antibody titre ≥1:800 is considered highly predictive for chronic Q fever (Tissot Dupont et al. 1994, 2007; Landais et al. 2007). The final diagnosis of chronic Q fever is made when a suspect serologic profile is combined with a positive PCR (Fenollar et al. 2004). However, considerable uncertainties exist about the value of serology to identify chronic cases, and the value of a positive PCR is not completely clear. At the regional laboratory of Jeroen Bosch Hospital ('s-Hertogenbosch, the Netherlands), located at the epicentre of the Dutch outbreak, we evaluated the serologic profiles of 686 patients diagnosed with acute Q fever in 2007 and 2008 at 3, 6 and 12 months after diagnosis (van der Hoek et al. 2011c). Our results differ from data provided by others, as high IgG phase I antibody titres at a 3-month follow-up were not predictive for chronic Q fever and IgG phase I antibody titres greater than IgG phase II antibody titres were rarely seen. An IgG phase I ≥1:1,024 at 6 months seemed to have the highest sensitivity for detecting chronic Q fever, but the probability that cases with this profile actually had chronic Q fever is low. Chronic Q fever cases show a persistently high (≥1:1,024) or increasing IgG phase I antibody titre, combined with a persistently high (≥1:4,096) IgG phase II antibody titre. A serologic cut-off at ≥1:1,024 (or at the previously proposed ≥1:800) provides adequate sensitivity and positive predictive value. The study confirmed that IgG phase I is a good screening test, in our case with a cut-off of ≥1:1,024, at a follow-up of between 6 and 12 months after the acute Q fever episode. A more stringent follow-up scheme is required for patients with clinical risk factors. Based on the experience gained since 2007, the serologic follow-up strategy is now one analysis at 9 months after an episode of acute Q fever. For patients with specific risk factors, the follow-up strategy at 3, 6 and 12 months is maintained, with serology combined with PCR. The diagnosis of chronic Q fever and the decisions about treatment were made by a multidisciplinary team of medical specialists, based on serologic profile, PCR results, the presence of clinical risk factors, clinical presentation, and other patient characteristics. Of the 686 acute Q fever cases that were followed up, 1.6% converted to a classic chronic case with microbiological evidence (van der Hoek et al. 2011c). In the epidemic in the Netherlands, we found that the antibody titre of IgG phase I ≥1:1,024 is not useful for immunocompromised patients and every follow-up serum sample must be tested by PCR independently of the serological profile. In endocarditis patients, we concluded that the PCR in a minority of patients is negative despite having vegetations on echocardiography. Almost every vascular patient has a chronic serological profile and a positive PCR.

A Dutch consensus on chronic Q fever was recently formulated (Wegdam-Blans et al. 2011). A distinction is made between 'proven', 'probable', and 'possible' chronic Q fever. Proven chronic Q fever requires (1) a positive PCR in tissue or blood in the absence of an acute Q fever infection; or (2) an IFA phase I IgG titre ≥1:1,024 and evidence of endocarditis; or (3) an IFA phase I IgG titre ≥1:1,024 and evidence of vascular infection by radiologic imaging.

17.7.3 Fatigue in Q Fever Patients

Following acute Q fever, up to 60% of patients may experience post-infection fatigue symptoms. These symptoms can persist for 6–12 months, after which they spontaneously resolve (Ayres et al. 1998). Post-infection fatigue also occurs after other infectious diseases such as Lyme disease (Marques 2008). In 10–15% of Q fever patients, fatigue can last from 5 to 10 years (Marmion et al. 1996) and is then often referred to as QFS, with a symptom presentation similar to chronic fatigue syndrome (CFS). Some studies state that cytokine deregulation and immunomodulation due to the persistence of *C. burnetii* may be responsible for prolonged fatigue, but others contradict this (Penttila et al. 1998). An impaired or deregulated immune response or the long-term persistence of the bacteria or its antigens and the immune response may also play a role.

17.7.4 A Typical Q Fever Patient

'Jan Verkerk' is a 48-year-old self-employed male. He ran a small, family-owned bicycle shop and did most of the work himself. He was also an active sportsman, running 15 km three times a week and cycling daily. He had no known underlying physical or psychological diseases. In May 2007 he developed Q fever and visited his general practitioner (GP) for the first time in years, presenting with high fever and pneumonia. These acute symptoms disappeared during ensuing weeks, but 1 year later and despite his best efforts, he still hadn't resumed running at his normal level. He was feeling constantly tired and was struggling to manage his business. He did not sleep well due to night sweating, he was unable to concentrate, and he suffered muscle and joint pains. He visited his GP many times, but several blood tests revealed nothing. He felt misunderstood by his GP, who seemed unable to help him, and he worried that if he did not recover he might not be able to manage his shop any longer. After speaking with other patients who told a similar story, he called the Department of Infectious Diseases of the Municipal Health Service to find out if this was a normal experience, whether others had similar problems, what further investigations could be done, and how he could be treated for his persisting symptoms.

17.7.5 The Health Status of Q Fever Patients After Long-Term Follow-up

In response to the many signals and questions about persisting symptoms, particularly fatigue in Q fever patients from the 2007 cohort, the collaborative multidisciplinary study Q-Quest I was started in 2008 (Morroy et al. 2011). A validated questionnaire, the Nijmegen Clinical Screening Instrument (NCSI), was used to obtain a detailed assessment of the health status in Q fever patients 12–26 months after the onset of their illness. This study is the largest and longest follow-up study of Dutch Q fever patients from the 2007 and 2008 outbreaks. In 2009, we asked 870 Q fever patients from the 2007 and 2008 outbreaks to complete the questionnaire based on an empirical definition of health status (Vercoulen et al. 2008), covering physiological functioning, symptoms, functional impairment and quality of life (QoL) as the main domains. These domains were subdivided into eight sub-domains: subjective symptoms, dyspnoea emotions, fatigue, behavioural impairment, subjective impairment, general QoL, health related quality of life, and satisfaction with relations (Peters et al. 2009). We compared the NCSI scores of these Q fever patients with normal data from healthy individuals and patients with severe chronic obstructive pulmonary disease (COPD).

Our findings demonstrate that in comparison to healthy individuals, Q fever patients – especially those that were hospitalized – present 12–26 months after the onset of illness with more severe clinically relevant subjective symptoms, functional impairment and impaired quality of life. The long-term health status of two-thirds of Q fever patients was severely affected for at least one sub-domain. Year of illness onset, level of education and smoking behaviour had no significant influence on sub-domain mean scores. Published data on the health status and its sub-domains of Q fever patients are scarce. Hatchette et al. (2003) reported that 52% of Q fever patients were symptomatic and had an impaired QoL 27 months after infection, using the 36-Item Short Form Health Survey (SF-36), with significantly lower scores, compared to non-infected controls in the domains of physical pain, function and role, emotional role and social function.

In Q-Quest I, the sub-domains 'general QoL' and 'fatigue' were severely and clinically impaired, compared to the reference group. More than half, 59% of patients had abnormal (mild to severe) fatigue, similar to other publications, which indicate that 60% of patients reported protracted fatigue (CSL 2009) and up to 69% fatigue (Ayres et al. 1998) 5 years after infection. A small study on Dutch patients that measured a 1-year follow-up and also used the NCSI reported a higher rate of 53% of patients with severe fatigue (Limonard et al. 2010), whereas the Q-Quest I study reported 44%.

Health status can be impaired after pneumonia regardless of the causative organism. Dutch pneumonia patients had significantly affected SF-36 scores 18 months after pneumonia on the subscales 'physical function' and 'general health status' (El Moussaoui et al. 2006). Survivors of a Legionnaire's Disease outbreak in the Netherlands 17 months after infection reported severely impaired SF-36 domains: 'physical role function', 'general health' and 'vitality' (Lettinga et al. 2002). Up to 75% of patients reported fatigue. In Q-Quest I hospitalization in the acute phase was significantly related to long-term behavioural impairment (OR 2.8, 95% CI 1.5–5.1), poor health-related quality of life (OR 2.3, 95% CI 1.5–4.0), and subjective symptoms (OR 1.9, 95% CI 1.1–3.6) (Morroy et al. 2011). Severity of initial illness generally has a negative influence on long-term QoL (Lowry and Pakenham 2008; Testa and Simonson 1996). Similarly, the severity of the acute Q fever symptoms predicts long-term symptoms (Hickie et al. 2006). Hospitalization can be seen as an indicator of the severity of the initial infection. We conclude that Q fever patients with severe acute illness are more likely to experience long-term impaired QoL. Lung or heart disease, depression and arthritis also significantly affected the long-term health status of Q fever patients. Other authors state that underlying heart (de Leon et al. 2009; Juenger et al. 2002) or lung disease (Daudey et al. 2010), arthritis (Garip et al. 2010), depression (Beard et al. 2010) and diabetes (Glasgow et al. 1997) all have a negative effect on the health status in different sub-domains. In Q-Quest I this effect was also found for all underlying conditions, except for diabetes. It was not possible to compare data with existing studies as most of these studies focus on specific diseases (such as COPD) and grades of severity.

17.7.6 The Q Fever Patient Society

In 2007 and 2008 the Q fever outbreaks in the province of North Brabant did not receive much media attention. In 2009, the outbreaks expanded to a larger area outside Brabant and patient numbers rose to over 2,000. The number of patients presenting with long-term effects grew, the precautionary veterinary measures were stepped up and media attention increased as a result. This fed the public interest. Then, at the height of the Q fever epidemic in 2009, most of the media attention switched to the influenza pandemic. Some general practitioners and other medical doctors and public health officials felt that the concurrent Q fever outbreak received insufficient attention. At the same time, GPs and patients increasingly reported long-term complaints. In November 2009, with help and financial support from the Province of North Brabant, a Q fever Patient's Society was founded. This society offers patients a platform to meet and express concerns and needs such as on treatment options.

Several hospitals now run Q fever out-patient departments for follow-up of Q fever patients but care in these centres is not standardized. In June 2010, the Patient's Society requested the Minister of Health to focus attention on patients with long-term complaints after acute Q fever infection. The National Institute for Public Health and the Environment was asked to draft guidelines on the treatment of long-term complaints after acute Q fever, and the product of a multidisciplinary working group is expected in 2012.

17.7.7 An Opportunity for More Research and Understanding

Many questions on the late effects of Q fever remain unanswered, such as the effectiveness of treatment of QFS (Rimes and Chalder 2005) with cognitive behavioural treatment and graded exercise therapy. The outbreaks in the Netherlands offer a unique opportunity for prospective research (the Q-Quest II study) on the long-term health outcomes in Q fever patients. With more than 4,000 acute Q fever cases reported up to November 2010 and symptoms that can last for 10 years or more, a considerable burden of disease in coming years is expected for patients and the affected communities. GPs and other medical doctors should be aware that Q fever patients may present with long-term symptoms, especially if they have been hospitalized or have co-morbidity (heart or lung disease, or depression). Ongoing research on the treatment and recovery of Q fever patients should offer a better understanding of the delayed and long-term effects of this zoonosis. There is a particular need for randomised clinical trials to test the effectiveness of treatment options.

17.8 Q Fever Vaccination in the Netherlands

17.8.1 Vaccination Decisions During the Q Fever Epidemic

The annual Q fever epidemics that began in 2007 prompted Dutch policy makers to consider introducing a human vaccination programme to protect people at risk for severe outcomes of the disease. However, early live attenuated and sub-unit vaccines were abandoned because of low efficacy and safety concerns, leaving just one human Q fever vaccine. This whole-cell vaccine was developed and registered in Australia and is licensed under the name Q-vax. It is not registered in the Netherlands or in any other European country. There were logistical and legal constraints to introducing a non-registered vaccine that required extensive testing of subjects before vaccination. Human vaccination can play no role in controlling the epidemic but the increasing number of reports of long-term effects in patients with chronic Q fever eventually caused both professionals and decision makers to reconsider the introduction of the vaccine in the Netherlands. At the same time, hospitals and public health services were confronted with an increasing number of worried Q fever patients, both acute and chronic, some of whom travelled to Australia at their own expense to be vaccinated. In 2010, the Government asked the Health Council of the Netherlands to advise on the possible use of the vaccine.

17.8.2 The Q-Vax Vaccine

Q-vax consists of formaline inactivated *C. burnetii* and was developed by CSL limited (CLS Biotherapies). It has been licensed in Australia to protect at-risk slaughterhouse employees and veterinary professionals (Ackland et al. 1994). In this respect, the vaccine was quite successful and is still in use (Gilroy et al. 2001; Marmion et al. 1984).

Analysis of the vaccine's efficacy in selected groups of professionals with a potentially high attack rate shows a protection rate of 97% (Gefenaite et al. 2011). Vaccinating subjects without a measurable immune response to *C. burnetii* is safe, but does commonly result in mild local reactions (33–48%) or mild systemic reactions (9%) such as headache (Marmion et al. 1990). Between 2002 and 2006, a large campaign in Australia saw the vaccination of 50,000 patients, resulting in eight serious adverse events requiring hospital admission and one life-threatening event. No deaths have ever been recorded after vaccination (Gidding et al. 2009). It is noteworthy that this data comes from a specific group of young and predominantly healthy males, the vaccine is only given to subjects over 15 years of age, and it is not administered to pregnant women.

Data is not available on the effectiveness of the vaccine in persons other than healthy workers. Furthermore, the vaccine can only be given to those not previously in contact with *C. burnetii,* as vaccinating subjects that have already mounted an immunological response may lead to serious adverse reactions such as sterile abscesses and systemic symptoms of inflammation. To prevent this, serology and skin testing must be performed to identify those who have previously had contact with *C. burnetii*. Although these tests are not complicated *per se*, they can be difficult to organise and require specific skills such as administering and interpreting of the skin test. To further complicate matters, laboratory tests are not standardised, and different serologic tests systems and cut-off values are used.

17.8.3 Target Groups for Vaccination

Patients affected with Q fever come mainly from specific areas in the south of the Netherlands. However, considerable differences occur within the affected area, and people living near affected farms may be especially affected. Nevertheless, localised mass vaccination has never been considered.

Preliminary data indicate that high numbers of professionals have been infected with *C. burnetii* – studies performed among goat and sheep farmers and veterinarians showed seroprevalence figures of up to 80%. Since the majority of these risk groups had already been exposed with a limited burden of disease, it was decided not to vaccinate them. However, those just starting out in a high-risk career, such as veterinarians, could be considered as candidates for vaccination.

The Q fever vaccine could be of use for population groups with underlying disease that make them at risk for long-term effects. Although these long-term effects are quite rare, they can be very serious and include endocarditis and the infection of large blood vessels (Landais et al. 2007; Botelho-Nevers et al. 2007). The treatment of chronic Q fever requires long-term (>1.5 years) antibiotic treatment and sometimes cardiovascular surgical interventions.

Patients with pathologic heart valves or blood vessels are particularly at risk. However, most of the studies in this area have been performed retrospectively and suffer from considerable selection bias. This means that the true contribution and magnitude of the risk associated with pre-existent factors is not known. It is also unclear whether minor valve or vessel pathology could develop into serious pathology during chronic Q fever. Furthermore, little is known about the incubation period of serious long-term effects of chronic Q fever.

Once the decision has been made to vaccinate patients at risk, these uncertainties matter and must be considered. For example, the screening of all acute Q fever patients for heart defects with echocardiography (as advised in the international literature) was not feasible during the large-scale Dutch outbreak (Limonard et al. 2010; Botelho-Nevers et al. 2007). Similar screening options for aneurysms in a given population may also not be feasible.

For these reasons, defining and selecting patient groups for vaccination is not a simple matter. In 2010, the Health Council of the Netherlands (2010) identified the following groups as eligible for vaccination:

- Patients who have had endocarditis in the past
- Patients with artificial heart valves
- Patients with significant congenital heart anomalies, including those that required repair with grafts
- Patients with structural defects of the aortic or mitral valve
- Patients with known aneurysm of the aorta
- Patients with vascular grafts
- Patients with severe peripheral vascular disease (such as Buerger's disease)

17.8.4 Deciding to Vaccinate

Even though the use of the vaccine in certain groups has been advocated, this vaccine is not licensed in the Netherlands and its administration will not be part of a nationally steered programme. However, it is considered part of health care under the responsibility of the treating physician. Together with the patient, the physician must weigh the potential benefits and disadvantages of vaccine administration. These decisions need to be made with full awareness of the medical and the epidemiological risks involved.

Vaccine administration can only be carried out after a professional skin test reading and serology result analysis. This requires a standardised process with similar cut-off titres and specificity tests, as well as centralised vaccination to realise standardised quality of care. Therefore, even though the vaccine has not been added to the national vaccine programme, its introduction in 2011 was coordinated by the National Institute of Public Health and the Environment, in collaboration with a commercial partner, regional public health departments and local physicians. In the vaccination campaign, early 2011, 1,354 people were vaccinated, all from the defined high risk groups.

17.9 Conclusion

Between 2007 and 2009, the Netherlands experienced an unprecedented series of seasonal outbreaks of Q fever. Dairy goats are clearly implicated in these outbreaks. In 2010 there were a much lower number of notified acute Q fever cases than in 2009, probably due to the drastic veterinary interventions such as culling of pregnant goats on infected farms, vaccination, and hygiene measures. But the risk of Q fever outbreaks and possibly other zoonotic diseases remains high because of the cohabitation of 2.4 million inhabitants with 6.4 million animals in the province of North Brabant. A great deal of knowledge has been generated in the past few years but many questions remain. Ongoing research, including 20 PhD projects, is expected to significantly advance the knowledge base. Attention is now shifting from acute Q fever to the problem of long-term effects of Q fever, the extent of which is not yet known and which poses important challenges for diagnosis and treatment.

References

Ackland JR, Worswick DA, Marmion BP (1994) Vaccine prophylaxis of Q fever. A follow-up study of the efficacy of Q-Vax (CSL) 1985–1990. Med J Aust 160:704–708

Anderson AD, Kruszon-Moran D, Loftis AD et al (2009) Seroprevalence of Q fever in the United States, 2003–2004. Am J Trop Med Hyg 81:691–694

Anderson AD, Baker TR, Littrell AC et al (2011) Seroepidemiologic survey for *Coxiella burnetii* among hospitalized US troops deployed to Iraq. Zoonoses Public Health 58:276–283

Ayres JG, Flint N, Smith EG et al (1998) Post-infection fatigue syndrome following Q fever. QJM 91:105–123

Beard C, Weisberg RB, Keller MB (2010) Health-related quality of life across the anxiety disorders: findings from a sample of primary care patients. J Anxiety Disord 24:559–564

Benenson AS, Tigertt WD (1956) Studies on Q fever in man. Trans Assoc Am Physicians 69:98–104

Blaauw GJ, Notermans DW, Schimmer B et al (2011) The application of an enzyme-linked immunosorbent assay or an immunofluorescent assay test leads to different estimates of seroprevalence of *Coxiella burnetii* in the population. Epidemiol Infect. doi:10.1017/S0950268811000021

Botelho-Nevers E, Fournier PE, Richet H et al (2007) *Coxiella burnetii* infection of aortic aneurysms or vascular grafts: report of 30 new cases and evaluation of outcome. Eur J Clin Microbiol Infect Dis 26:635–640

Carcopino X, Raoult D, Bretelle F et al (2007) Managing Q fever during pregnancy: the benefits of long-term cotrimoxazole therapy. Clin Infect Dis 45:548–555

Carcopino X, Raoult D, Bretelle F et al (2009) Q fever during pregnancy: a cause of poor fetal and maternal outcome. Ann N Y Acad Sci 1166:79–89

Casolin A (1999) Q fever in New South Wales Department of Agriculture workers. J Occup Environ Med 41:273–278

Cowley R, Fernandez F, Freemantle W et al (1992) Enzyme immunoassay for Q fever: comparison with complement fixation and immunofluorescence tests and dot immunoblotting. J Clin Microbiol 30:2451–2455

CSL (2009) A guide to Q-fever and Q-fever vaccination. CSL Biotherapies, Parkville, ABN 66120398067

D'Harcourt SC, Soto AB, Burgos VC et al (1996) Comparison of immunofluorescence with enzyme immunoassay for detection of Q fever. Eur J Clin Microbiol Infect Dis 15:749–752

Daudey L, Peters JB, Molema J et al (2010) Health status in COPD cannot be measured by the St George's Respiratory Questionnaire alone: an evaluation of the underlying concepts of this questionnaire. Respir Res 11:98

de Leon CF, Grady KL, Eaton C et al (2009) Quality of life in a diverse population of patients with heart failure: baseline findings from the heart failure adherence and retention trial (HART). J Cardiopulm Rehabil Prev 29:171–178

Dekking F, Zanen HC (1958) Q-fever in the Netherlands. Ned Tijdschr Geneeskd 102:65–68

Denman J, Woods M (2009) Acute Q fever in pregnancy: report and literature review. Intern Med J 39:479–481

Dupuis G, Peter O, Peacock M et al (1985) Immunoglobulin responses in acute Q fever. J Clin Microbiol 22:484–487

Dwyer DE, Gibbons VL, Brady LM et al (1988) Serological reaction to *Legionella pneumophila* group 4 in a patient with Q fever. J Infect Dis 158:499–500

ECDC (2010) Risk assessment on Q fever. European Centre for Disease Prevention and Control, Stockholm. doi:10.2900/28860

El Moussaoui R, Opmeer BC, de Borgie CA et al (2006) Long-term symptom recovery and health-related quality of life in patients with mild-to-moderate-severe community-acquired pneumonia. Chest 130:1165–1172

Fenollar F, Fournier PE, Raoult D (2004) Molecular detection of *Coxiella burnetii* in the sera of patients with Q fever endocarditis or vascular infection. J Clin Microbiol 42:4919–4924

Field PR, Santiago A, Chan SW et al (2002) Evaluation of a novel commercial enzyme-linked immunosorbent assay detecting *Coxiella burnetii*-specific immunoglobulin G for Q fever prevaccination screening and diagnosis. J Clin Microbiol 40:3526–3529

Finidori JP, Raoult D, Bornstein N et al (1992) Study of cross-reaction between *Coxiella burnetii* and *Legionella pneumophila* using indirect immunofluorescence assay and immunoblotting. Acta Virol 36:459–465

Fournier PE, Marrie TJ, Raoult D (1998) Diagnosis of Q fever. J Clin Microbiol 36:1823–1834

Garip Y, Eser F, Bodur H (2010) Health-related quality of life in rheumatoid arthritis: comparison of RAQoL with other scales in terms of disease activity, severity of pain, and functional status. Rheumatol Int. doi:10.1007/s00296-009-1353-1

Gefenaite G, Munster JM, van Houdt R et al (2011) Effectiveness of the Q fever vaccine: a meta-analysis. Vaccine 29:395–398

Gidding HF, Wallace C, Lawrence GL et al (2009) Australia's national Q fever vaccination program. Vaccine 27:2037–2041

Gilroy N, Formica N, Beers M et al (2001) Abattoir-associated Q fever: a Q fever outbreak during a Q fever vaccination program. Aust N Z J Public Health 25:362–367

Gilsdorf A, Kroh C, Grimm S et al (2008) Large Q fever outbreak due to sheep farming near residential areas, Germany, 2005. Epidemiol Infect 136:1084–1087

Glasgow RE, Ruggiero L, Eakin EG et al (1997) Quality of life and associated characteristics in a large national sample of adults with diabetes. Diabetes Care 20:562–567

Hatchette TF, Hayes M, Merry H et al (2003) The effect of *C. burnetii* infection on the quality of life of patients following an outbreak of Q fever. Epidemiol Infect 130:491–495

Hawker JI, Ayres JG, Blair I et al (1998) A large outbreak of Q fever in the West Midlands: windborne spread into a metropolitan area? Commun Dis Public Health 1:180–187

Health Council of the Netherlands (2008) Advisory letter: meeting on Q fever in the Netherlands. The Hague: Health Council of the Netherlands, publication no. 2008/28. [in Dutch, summary in English] Accessible at: http://www.gezondheidsraad.nl/sites/default/files/201008_r.pdf

Health Council of the Netherlands (2010) Human vaccination against Q fever. The Hague: Health Council of the Netherlands, publication no. 2010/08E. Accessible at: http://www.gezondheidsraad.nl/sites/default/files/201008E.pdf

Hickie I, Davenport T, Wakefield D et al (2006) Post-infective and chronic fatigue syndromes precipitated by viral and non-viral pathogens: prospective cohort study. BMJ 333:575

Huijsmans CJ, Schellekens JJ, Wever PC et al (2011) Single-nucleotide-polymorphism genotyping of *Coxiella burnetii* during a Q fever outbreak in The Netherlands. Appl Environ Microbiol 77:2051–2057

Jover-Diaz F, Robert-Gates J, Andreu-Gimenez L et al (2001) Q fever during pregnancy: an emerging cause of prematurity and abortion. Infect Dis Obstet Gynecol 9:47–49

Juenger J, Schellberg D, Kraemer S et al (2002) Health related quality of life in patients with congestive heart failure: comparison with other chronic diseases and relation to functional variables. Heart 87:235–241

Karagiannis I, Schimmer B, van Lier A et al (2009) Investigation of a Q fever outbreak in a rural area of The Netherlands. Epidemiol Infect 137:1283–1294

Karakousis PC, Trucksis M, Dumler JS (2006) Chronic Q fever in the United States. J Clin Microbiol 44:2283–2287

Koene RP, Schimmer B, Rensen H et al (2011) A Q fever outbreak in a psychiatric care institution in The Netherlands. Epidemiol Infect 139:13–18

Landais C, Fenollar F, Thuny F et al (2007) From acute Q fever to endocarditis: serological follow-up strategy. Clin Infect Dis 44:1337–1340

Langley JM, Marrie TJ, Leblanc JC et al (2003) *Coxiella burnetii* seropositivity in parturient women is associated with adverse pregnancy outcomes. Am J Obstet Gynecol 189:228–232

Lettinga KD, Verbon A, Nieuwkerk PT et al (2002) Health-related quality of life and posttraumatic stress disorder among survivors of an outbreak of Legionnaires disease. Clin Infect Dis 35: 11–17

Limonard GJ, Peters JB, Nabuurs-Franssen MH et al (2010) Detailed analysis of health status of Q fever patients 1 year after the first Dutch outbreak: a case-control study. QJM 103:953–958

Lowry TJ, Pakenham KI (2008) Health-related quality of life in chronic fatigue syndrome: predictors of physical functioning and psychological distress. Psychol Health Med 13:222–238

Marmion BP, Ormsbee RA, Kyrkou M et al (1984) Vaccine prophylaxis of abattoir-associated Q fever. Lancet ii:1411–1414

Marmion BP, Ormsbee RA, Kyrkou M et al (1990) Vaccine prophylaxis of abattoir-associated Q fever: eight years' experience in Australian abattoirs. Epidemiol Infect 104:275–287
Marmion BP, Shannon M, Maddocks I et al (1996) Protracted debility and fatigue after acute Q fever. Lancet 347:977–978
Marques A (2008) Chronic Lyme disease: a review. Infect Dis Clin North Am 22:341–348
Marrie TJ (1990) Acute Q fever. In: Marrie TJ (ed) Q fever. The disease, vol 1. CRC Press, Boca Raton
Maurin M, Raoult D (1999) Q fever. Clin Microbiol Rev 12:518–553
McCaughey C, McKenna J, McKenna C et al (2008) Human seroprevalence to *Coxiella burnetii* (Q fever) in Northern Ireland. Zoonoses Public Health 55:189–194
Meekelenkamp JCE, Notermans DW, Rietveld A et al (2009) Seroprevalence of *Coxiella burnetii* in pregnant women in the province of Noord-Brabant in 2007. Infectieziekten Bull 20:57–61 [in Dutch]. http://www.rivm.nl/cib/publicaties/bulletin/jaargang_20/bull2002/prevalentie-coxiella-burnetii.jsp. Accessed May 2011
Mertens K, Samuel JE, Raoult D et al (2007) Bacteriology of *Coxiella*. In: Raoult D, Parola P (eds) Rickettsial diseases. Informa Healthcare, New York
Morroy G, Peters JB, van Nieuwenhof M et al (2011) The health status of Q-fever patients after long-term follow-up. BMC Infect Dis 11:97. doi:10.1186/1471-2334-11-97
Munster JM, Leenders AC, van der Hoek W et al (2010a) Cost-effectiveness of a screening strategy for Q fever among pregnant women in risk areas: a clustered randomized controlled trial. BMC Womens Health 10:32
Munster JM, Leenders ACAP, Aarnoudse JG et al (2010b) A screening strategy for Q fever during pregnancy. Poster presentation at the 2010 European Scientific Conference on Applied Infectious Disease Epidemiology (ESCAIDE) – 11–13 Nov 2010, Lisbon, Portugal
Parker NR, Barralet JH, Bell AM (2006) Q fever. Lancet 367:679–688
Penttila IA, Harris RJ, Storm P et al (1998) Cytokine dysregulation in the post-Q-fever fatigue syndrome. QJM 91:549–560
Peter O, Dupuis G, Bee D et al (1988) Enzyme-linked immunosorbent assay for diagnosis of chronic Q fever. J Clin Microbiol 26:1978–1982
Peters JB, Daudey L, Heijdra YF et al (2009) Development of a battery of instruments for detailed measurement of health status in patients with COPD in routine care: the Nijmegen Clinical Screening Instrument. Qual Life Res 18:901–912
Porten K, Rissland J, Tigges A et al (2006) A super-spreading ewe infects hundreds with Q fever at a farmers' market in Germany. BMC Infect Dis 6:147
Raoult D, Tissot-Dupont H, Foucault C et al (2000) Q fever 1985–1998. Clinical and epidemiologic features of 1,383 infections. Medicine (Baltimore) 79:109–123
Rey D, Obadia Y, Tissot-Dupont H et al (2000) Seroprevalence of antibodies to *Coxiella burnetti* among pregnant women in South Eastern France. Eur J Obstet Gynecol Reprod Biol 93:151–156
Richardus JH, Schaap GJ, Donkers A et al (1984a) Q fever in the Netherlands; a description of 33 case reports observed between 1979 and 1983. Ned Tijdschr Geneeskd 128:2253–2258
Richardus JH, Donkers A, Schaap GJ et al (1984b) Serological study of the presence of antibodies against *Coxiella burnetti* and *Brucella abortus* in veterinarians in the Netherlands. Tijdschr Diergeneeskd 109:612–615
Richardus JH, Dumas AM, Huisman J et al (1985) Q fever in infancy: a review of 18 cases. Pediatr Infect Dis 4:369–373
Richardus JH, Donkers A, Dumas AM et al (1987) Q fever in the Netherlands: a sero-epidemiological survey among human population groups from 1968 to 1983. Epidemiol Infect 98:211–219
Rimes KA, Chalder T (2005) Treatments for chronic fatigue syndrome. Occup Med (Lond) 55:32–39
Roest HI, Tilburg JJ, van der Hoek W et al (2011) The Q fever epidemic in The Netherlands: history, onset, response and reflection. Epidemiol Infect 139:1–12
Schimmer B, Dijkstra F, Vellema P et al (2009) Sustained intensive transmission of Q fever in the south of the Netherlands, 2009. Euro Surveill 14(19)

Schimmer B, ter Schegget R, Wegdam M et al (2010) The use of a geographic information system to identify a dairy goat farm as the most likely source of an urban Q-fever outbreak. BMC Infect Dis 10:69

Schimmer B, Notermans DW, Harms MG et al (2011) Low seroprevalence of Q fever in The Netherlands prior to a series of large outbreaks. Epidemiol Infect. doi:10.1017/S0950268811000136 1-9

Schneeberger P, Schimmer B, Dijkstra F et al (2010a) Epidemiology and control of Q fever in the Netherlands, 2007–2009 (Abstract 2151) European Congress of Clinical Microbiology and Infectious Diseases, Vienna Austria, 10–13 Apr 2010

Schneeberger PM, Hermans MH, van Hannen EJ et al (2010b) Real-time PCR with serum samples is indispensable for early diagnosis of acute Q fever. Clin Vaccine Immunol 17:286–290

Setiyono A, Ogawa M, Cai Y et al (2005) New criteria for immunofluorescence assay for Q fever diagnosis in Japan. J Clin Microbiol 43:5555–5559

Statistics Netherlands (CBS) STATLINE: landbouw; gewassen, dieren, grondgebruik naar region. http://www.cbs.nl/nl-NL/menu/home/default.htm. Accessed May 2011

Stein A, Raoult D (1998) Q fever during pregnancy: a public health problem in southern France. Clin Infect Dis 27:592–596

Testa MA, Simonson DC (1996) Assesment of quality-of-life outcomes. N Engl J Med 334: 835–840

Tilburg JJ, Melchers WJ, Pettersson AM et al (2010) Interlaboratory evaluation of different extraction and real-time PCR methods for detection of *Coxiella burnetii* DNA in serum. J Clin Microbiol 48:3923–3927

Tissot Dupont H, Thirion X, Raoult D (1994) Q fever serology: cutoff determination for microimmunofluorescence. Clin Diagn Lab Immunol 1:189–196

Tissot Dupont H, Vaillant V, Rey S et al (2007) Role of sex, age, previous valve lesion, and pregnancy in the clinical expression and outcome of Q fever after a large outbreak. Clin Infect Dis 44:232–237

van den Wijngaard CC, Dijkstra F, van Pelt W et al (2011) In search of hidden Q-fever outbreaks: linking syndromic hospital clusters to infected goat farms. Epidemiol Infect 139:19–26

van der Hoek W, Dijkstra F, Wijers N et al (2010a) Three years of Q fever in the Netherlands: faster diagnosis. Ned Tijdschr Geneeskd 154:A1845

van der Hoek W, Dijkstra F, Schimmer B et al (2010b) Q fever in the Netherlands: an update on the epidemiology and control measures. Euro Surveill 15(12):pii–19520

van der Hoek W, Hunink J, Vellema P et al (2011a) Q fever in the Netherlands: the role of local environmental conditions. Int J Environ Health Res. doi:10.1080/09603123.2011.574270

van der Hoek W, Meekelenkamp JCE, Leenders ACAP et al (2011b) Antibodies against *Coxiella burnetii* and pregnancy outcome during the 2007–2008 Q fever outbreaks in the Netherlands. BMC Infect Dis 11:44

van der Hoek W, Versteeg B, Meekelenkamp J et al (2011c) Follow-up of 686 acute Q fever patients and detection of chronic infection. Clin Infect Dis 52(12):1431–1436

van der Klis FR, Mollema L, Berbers GA et al (2009) Second national serum bank for population-based seroprevalence studies in the Netherlands. Neth J Med 67:301–308

Van Gageldonk-Lafeber AB, Koopmans MPG, Bosman A et al (2003) Het vóórkomen van Q-koorts in Nederland. Infectieziekten Bulletin [in Dutch] 14:173–177

van Steenbergen JE, Morroy G, Groot CA et al (2007) An outbreak of Q fever in The Netherlands–possible link to goats. Ned Tijdschr Geneeskd 151:1998–2003

Vercoulen JH, Daudey L, Molema J et al (2008) An Integral assessment framework of health status in chronic obstructive pulmonary disease (COPD). Int J Behav Med 15:263–279

Waag D, Chulay J, Marrie T et al (1995) Validation of an enzyme immunoassay for serodiagnosis of acute Q fever. Eur J Clin Microbiol Infect Dis 14:421–427

Wegdam-Blans M, Verduin C, Wulf M et al (2009) The role of complement fixations tests in the diagnosis of acute Q fever: is PCR always necessary? Ned Tijdschr Med Microbiol 17:28

Wegdam-Blans MC, Nabuurs-Franssen MN, Horrevorts AM et al (2010) Laboratory diagnosis of acute Q fever. Ned Tijdschr Geneeskd 154:A2388

Wegdam-Blans MCA, Kampschreur LM, Nabuurs-Franssen MH et al (2011) Dutch consensus chronic Q fever. Tijdschrift voor Infectieziekten 6:71–73

Westra SA, Lopes CE, ten Berg J (1958) The first cases of Q-fever in the Netherlands. Ned Tijdschr Geneeskd 102:69–72

Whelan J, Schimmer B, Schneeberger P et al (2011) Q fever among culling workers, the Netherlands, 2009–2010. Emerg Infect Dis 17:1719–1723

Wildman MJ, Smith EG, Groves J et al (2002) Chronic fatigue following infection by *Coxiella burnetii* (Q fever): ten-year follow-up of the 1989 UK outbreak cohort. QJM 95:527–538

Wolff JW, Kouwenaar W (1954) Investigation on occurrence of Q fever in Netherlands. Ned Tijdschr Geneeskd 98:2726–2732

Wouda W, Dercksen DP (2007) Abortion and stillbirth among dairy goats as a consequence of *Coxiella burnetii*. Tijdschr Diergeneeskd 132:908–911

Chapter 18
Coxiella-like Endosymbionts

Jianmin Zhong

Contents

Abstract In the past two decades, many *Coxiella*-like bacteria have been found in hard ticks and soft ticks as well as in vertebrate hosts. It is interesting to note that many ticks harbor *Coxiella*-like bacteria with high prevalence. *Coxiella*-like bacteria and virulent *Coxiella burnetii* have high homology to each other; they form a monophyletic clade based on 16S rRNA sequence data and subsequent phylogenetic tree analyses. In this chapter, methods of detection, phylogeny, prevalence and density, distribution in tick organs, transmission routes, bacteria-host interactions, and putative functions of the *Coxiella*-like bacteria are reviewed.

Keywords Bacteria-host interactions • *Coxiella*-like bacteria • Phylogeny • Ticks • Vertebrate host • Endosymbiont • Transovarial transmission • *Rickettsia* • Transstadial transmission • Homologous

J. Zhong (✉)
Department of Biological Sciences, Humboldt State University,
1 Harpst Street, Arcata, CA 95521, USA
e-mail: jz15@humboldt.edu

R. Toman et al. (eds.), *Coxiella burnetii: Recent Advances and New Perspectives in Research of the Q Fever Bacterium*, Advances in Experimental Medicine and Biology 984,
DOI 10.1007/978-94-007-4315-1_18,

18.1 Introduction

The genus *Coxiella* contains only one validly described species, *C. burnetii* (Derrick 1937; Parker et al. 2006). *C. burnetii* is an obligate intracellular bacterium (Maurin and Raoult 1999) and the aetiological agent of human Q fever, a zoonotic disease that is distributed world-wide and causes acute or chronic disease with symptoms including fever, hepatitis, and respiratory complications (Raoult 1993; Waag 2007).

Studies of *Coxiella* have largely focused on pathogenic strains of *C. burnetii*, which have been identified in arthropods, birds, fish, mammals, and reptiles (Waag 2007; Angelakis and Raoult 2010). Recently, many *Coxiella*-like bacteria have been reported in ticks. Ticks infected with *Coxiella*-like bacteria include *Amblyomma americanum* (Jasinskas et al. 2007; Klyachko et al. 2007; Clay et al. 2008), *Carios capensis* (Reeves et al. 2005), *Haemaphysalis longicornis* (Noda et al. 1997; Lee et al. 2004), *Ixodes ricinus* (Schabereiter-Gurtner et al. 2003), *I. woodi* (Kurtti et al. 2002), *Ornithodoros moubata* (Noda et al. 1997), *Rhipicephalus sanguineus* (Noda et al. 1997; Bernasconi et al. 2002), and *R. turanicus* (Bernasconi et al. 2002).

18.2 Phylogenetic Classification of *Coxiella*-like Bacteria

Coxiella-like bacteria are similar but not identical to *C. burnetii*. Based on 16S rRNA nucleotide sequences, all *Coxiella*-like bacteria belong to the genus of *Coxiella* in the γ-subgroup of the *Proteobacteria*. The sequence identity of the 16S rRNA gene among the *Coxiella*-like bacteria varies from species to species. When the 16S rRNA nucleotide sequences are compared with each other using *C. burnetii* as a reference genome, a maximal and a minimal nucleotide identity of 98% and 91%, respectively, is observed (Table 18.1).

The current definition of *Coxiella*-like bacteria is mainly based on phylogenetic analyses of 16S rRNA sequences (Noda et al. 1997; Bernasconi et al. 2002; Kurtti et al. 2002; Jasinskas et al. 2007; Heise et al. 2010). All *Coxiella*-like bacteria, with nucleotide sequence identity of the 16S rRNA gene between 91% and 98%, are subclusters in the cluster of the genus *Coxiella*. Figure 18.1 is a phylogram of the 16S rRNA sequences of all *Coxiella*-like bacteria that are available in GenBank. In the phylogram, *C. burnetii*, *C. cheraxi*, *A. americanum* bacterium, *O. moubata* bacterium, *H. longicornis* bacterium, *R. sanguineus* bacterium, a *Coxiella* species in golden mantle rosella and hawk-headed parrot, and marine dinoflagellate bacterium, are monophyletic. Specifically, phylogenetic analysis suggests that all are more closely related to each other than to other bacteria, such as *Rickettsiella* species, *Aquicella siphonis*, *Legionella* species, *Methylococcus capsulatus*, and *Escherichia coli*. Some portions of the tree are weakly supported by bootstrapping analysis, such as the clade of *C. burnetii*, *H. longicornis* bacterium, and *O. moubata* symbionts. The sequences of the 16S rRNA gene are too conserved to strongly support the lineage of these taxonomic units. More variable regions are needed to precisely define phylogenetic relatedness.

Table 18.1 Pairwise nucleotide sequence identities of the 16S rRNA gene of *Coxiella*-like bacteria

GenBank nucleotide accession numbers[a]															
16S rRNA	EU143670	EU143669	AY939824	NC002971	AB001521	AB001519	D84559	AB058918	NC006369	AY744776	NC002977	NC000913	AF383621	U97547	AY359283
EU143670	100%	98%	92%	96%	95%	95%	96%	93%	86%	87%	89%	84%	86%	86%	85%
EU143669		100%	91%	96%	95%	95%	96%	92%	86%	87%	89%	84%	86%	87%	85%
AY939824			100%	93%	93%	91%	91%	91%	86%	86%	88%	85%	86%	86%	85%
NC002971				100%	98%	95%	95%	93%	87%	87%	89%	85%	86%	87%	84%
AB001521					100%	95%	95%	94%	87%	87%	89%	85%	88%	87%	86%
AB001519						100%	94%	92%	85%	86%	88%	83%	85%	86%	83%
D84559							100%	91%	85%	86%	88%	84%	85%	86%	84%
AB058918								100%	87%	87%	89%	84%	86%	86%	85%
NC006369									100%	98%	87%	84%	86%	86%	84%
AY744776										100%	88%	84%	86%	85%	84%
NC002977											100%	86%	86%	86%	86%
NC000913												100%	83%	83%	82%
AF383621													100%	97%	87%
U97547														100%	86%
AY359283															100%

Homologous sequences were obtained from GenBank and were aligned by ClustalX. A total of 1,218–1,249 base pairs of the 16S rRNA gene were obtained after trimming unaligned nucleotide sequences at 5′- and 3′-ends. The nucleotide sequence identity between two 16S rRNA sequences was compared by BLAST analysis

[a]Sequences included in the pairwise comparisons were hawk-headed parrot bacterium (EU143669), golden mantle rosella bacterium (EU143670), *Amblyomma americanum* bacterium (accession number AY939824), *Coxiella burnetii* (NC_002971), *Ornithodoros moubata* bacterium (AB001521), *Haemaphysalis longicornis* bacterium (AB001519), *Rhipicephalus sanguineus* bacterium (D84559), a marine dinoflagellate bacterium (AB058918), *Legionella pneumophila* (NC_006369), *Legionella anisa* (AY744776), *Methylococcus capsulatus* (NC_002977), *Escherichia coli* K-12 (NC_000913), a *Rickettsiella* sp. of *Ixodes woodi* (AF383621), *Rickettsiella grylli* (U97547), and the protozoan-associated bacterium *Aquicella siphonis* (AY359283)

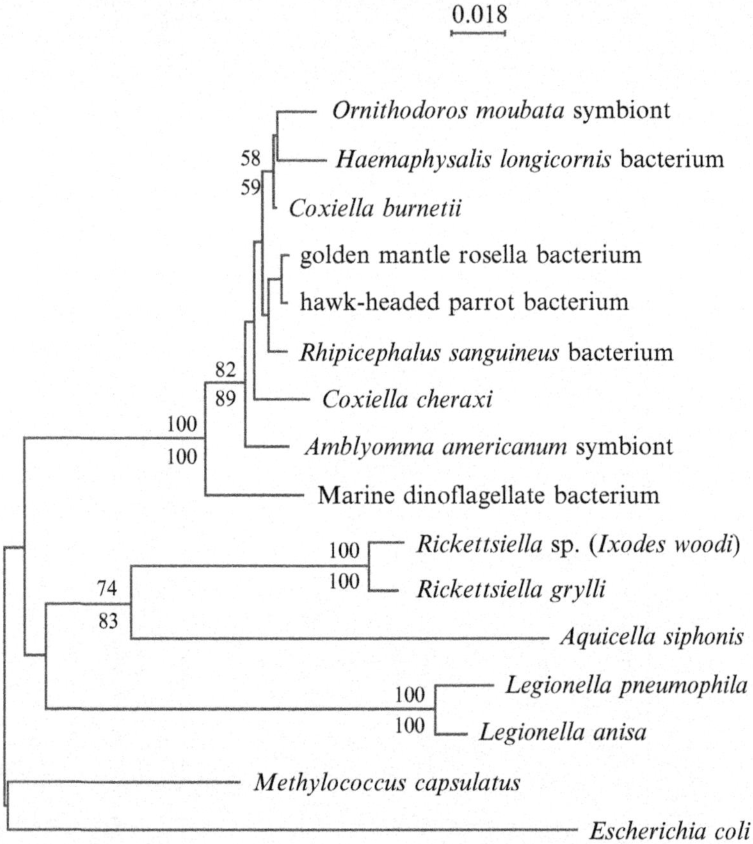

Fig. 18.1 Phylogram of partial 16S rRNA genes of the *Coxiella*-like bacteria and selected γ-proteobacteria. *Ornithodoros moubata* bacterium (AB001521), *Haemaphysalis longicornis* bacterium (AY342035), *Coxiella burnetii* bacterium (NC_002971), golden mantle rosella bacterium (EU143670), hawk-headed parrot bacterium (EU143669), *Rhipicephalus sanguineus* bacterium (D84559), *Coxiella cheraxi* (EF413063), *Amblyomma americanum* bacterium (accession number AY939824), a marine dinoflagellate bacterium (AB058918), a *Rickettsiella* sp. of *Ixodes woodi* (AF383621), *Rickettsiella grylli* (U97547), the protozoan-associated bacterium *Aquicella siphonis* (AY359283), *Legionella pneumophila* (NC_006369), *Legionella anisa* (AY744776), *Methylococcus capsulatus* (NC_002977), and *Escherichia coli* K-12 (NC_000913). Nucleotide positions with gaps in the aligned sequences were excluded. Bootstrap values of nodes with 50% values by maximum likelihood (500 replicates; shown above the line) or neighbor-joining (1,000 replicates; shown below the line) distance criteria are indicated and were estimated using PHYLO_WIN phylogenetic analysis software (http://pbil.univ-lyon1.fr/software/phylowin.html). Bar, nucleotide distance

In addition to the widely used 16S rRNA gene, phylogenetic analysis based on multiple housekeeping genes (*fusA*, *rpsF*, and *rpsG*) confirmed that the *Coxiella*-like bacterium of *A. americanum* belongs to the genus of *Coxiella* (Jasinskas et al. 2007). Among the housekeeping genes, the result of the phylogenetic

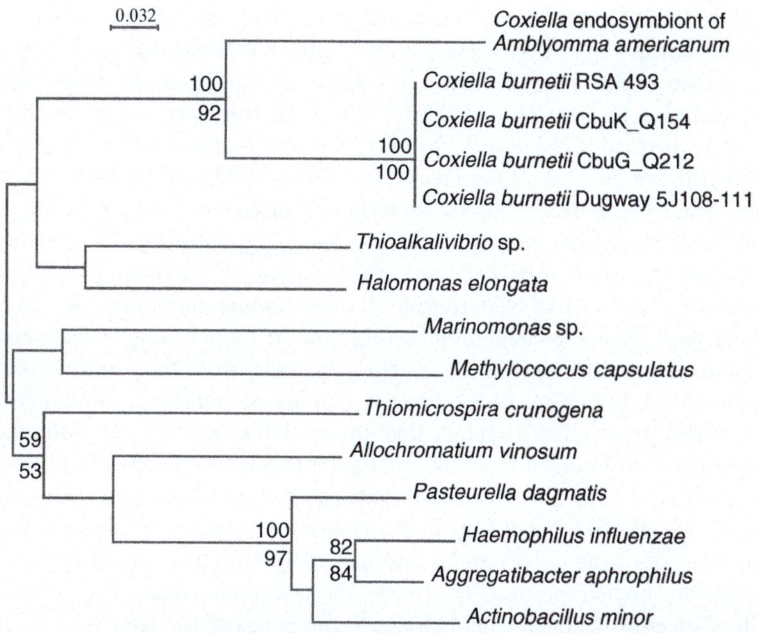

Fig. 18.2 Phylogram of partial *fusA* genes of *Coxiella*-like bacteria and selected other γ-proteobacteria *Coxiella* endosymbiont of *Amblyomma americanum* (DQ908900), *Coxiella burnetii* CbuK_Q154 (NC_011528), *C. burnetii* RSA 493 (NC_002971), *C. burnetii* CbuG_Q212 (NC_011527), *C. burnetii* Dugway 5J108-111 (NC_009727), *Thioalkalivibrio* sp. HL-EbGR7 (NC_011901), *Halomonas elongata* DSM 2581 (NC_014532), *Marinomonas* sp. MWYL1 (NC_009654), *Methylococcus capsulatus* str. Bath (NC_002977), *Thiomicrospira crunogena* XCL-2 (NC_007520), *Allochromatium vinosum* DSM 180 (NC_013851), *Pasteurella dagmatis* ATCC 43325 (NZ_ACZR01000006), *Haemophilus influenzae* PittHH (NZ_AAZH01000030), *Aggregatibacter aphrophilus* NJ8700 (NC_012913), and *Actinobacillus minor* 202 (NZ_ACFT01000107). Nucleotide and protein databases were searched with BLASTN and BLASTX algorithms. Amino acid sequences of FusA were first aligned by ClustalX. Then DNA sequences of *fusA* were aligned on a codon-basis by reference to the corresponding FusA amino acid sequences using CodonAlign, version 2.0 (http://www.sinauer.com/hall/index.php). Aligned *fusA* DNA sequences were used to estimate the phylogeny in PHYLO_WIN phylogenetic analysis software using neighbor-joining (1,000 replicates; shown above the line) and maximum likelihood (500 replicates; shown below the line). Minimal bootstrap values of at least 50% are indicated. Bar, nucleotide distance

analysis based on the 438 base pair coding sequence of *fusA* is shown in Fig. 18.2. Phylogenetic results based on other housekeeping genes are not included in this chapter, since the number of *rpsF* and *rpsG* genes is far less than the number of *fusA* genes from related species in GenBank. The *fusA* gene encodes a translational elongation factor G protein (Sharer et al. 1999). Both *fusA* and 16S rRNA are essential for the function of cells, thus their gene sequences variations were used to study evolutionary relationships between *Coxiella*-like bacteria and *C. burnetii*. Phylogeny of the *fusA* gene shows that *C. burnetii* CbuK_Q154, *C. burnetii* RSA 493,

C. burnetii CbuG, and *C. burnetii* Dugway 5J108-111 are clustered together with strong bootstrap support. However, both neighbor-joining and maximum likelihood analysis of the *fusA* gene failed to support monophyly of the genus *Coxiella*, which includes *C. burnetii* and the *Coxiella*-like bacterium in *A. americanum* (Fig. 18.1). Instead, *Coxiella*-like bacteria form a sister taxon to *C. burnetii*, and together this clade plus *Halomonas elongata* and *Thioalkalivibrio* sp. forms another sister taxon to a group of bacteria that includes members of the genera *Actinobacillus*, *Aggregatibacter*, *Haemophilus*, *Pasteurella*, *Allochromatium*, *Thiomicrospira*, *Methylococcus*, and *Marinomonas*. Consistent with a distinct taxon of the *Coxiella*-like bacterium of *A. americanum*, phylogenetic analysis of the *fusA* gene shows considerable divergence in branch lengths between this bacterium and *C. burnetii* (Fig. 18.2). The conflict between the two trees (*fusA* vs. 16S rRNA) might be due to the limited number of homologous sequences of the *fusA* gene from related species that are available in GenBank. Although the phylogenetic tree inferred from the *fusA* gene is not completely congruent with the tree inferred from the 16S rRNA gene, neighbour joining analysis showed that the *Coxiella*-like bacterium in *A. americanum* is in one clade with other members in the genus of *Coxiella* and that the conflicting clades were weakly supported by bootstrapping (Fig. 18.2). These findings support the conclusion that the *Coxiella*-like bacterium belongs to the genus of *Coxiella*.

18.3 Detection and Prevalence of *Coxiella*-like Bacteria in Ticks

Molecular methods that do not rely on culture isolation, especially 16S rRNA sequencing, play a pivotal role in identification of unculturable bacterial species isolated from environmental microhabitats (Fox et al. 1977; Gupta et al. 1983; Janda and Abbott 2007). The assays allow researchers to determine the spectrum of bacteria in a particular microhabitat without culture media (Schmidt and Relman 1994). Many of these methods, including polymerase chain reaction (PCR), real-time quantitative PCR, 16S rRNA gene cloning and sequencing, direct 16S rRNA sequencing, electron microscopy, and molecular probing, have been used to detect the presence of the *Coxiella*-like bacteria in ticks. Noda et al. identified a *Coxiella*-like bacterium in *R. sanguineus* and *H. longicornis* using PCR and direct sequencing of 16S and 23S rRNA genes (Noda et al. 1997). Bernasconi et al. detected the presence of a *Coxiella* species in *R. sanguineus* ticks by direct sequencing of PCR fragments of 16S rDNA (Bernasconi et al. 2002). Later, Heise et al. reported a *Coxiella*-like bacterium by sequencing of 16S rRNA clones of *A. americanum* (Heise et al. 2010). Last, Lee et al. identified a *Coxiella*-like bacterium in *H. longicornis* by PCR of the 16S rRNA gene (Lee et al. 2004).

In addition to *Rhipicephalus* and *Haemaphysalis* ticks, *Coxiella*-like bacteria are present in other tick species. Kurtti et al. reported a *Coxiella*-like bacterium in *I. woodi* by PCR, cloning of the 16S rRNA gene, and electron microscopic examination (Kurtti et al. 2002). By using a 16S rDNA clone library, PCR, denaturing gradient gel electrophoresis followed by direct sequencing, Schabereiter-Gurtner et al. reported a *Coxiella*-like bacterium in *I. ricinus* (Schabereiter-Gurtner et al. 2003). In 2007, Jasinskas et al. constructed a cDNA library using RNA extracted from the midgut of *A. americanum*. Sequencing analysis of 500 cDNA clones revealed that 20 (4%) clones were homologous to nucleotide sequences of *C. burnetii* (Jasinskas et al. 2007). The same *Coxiella*-like bacterium was confirmed by Klyachko et al. and Clay et al. using PCR, direct sequencing, transmission electron microscopy, and fluorescence *in situ* hybridization of the 16S rRNA gene in laboratory-reared and field-collected *A. americanum* (Klyachko et al. 2007; Clay et al. 2008). In summary, both hard ticks and soft ticks are infected with *Coxiella*-like bacteria.

The prevalence of *Coxiella*-like bacteria varies widely among species of ticks in North America and Europe, ranging from 6.25% in *R. sanguineus* to 100% in *A. americanum* (Bernasconi et al. 2002; Jasinskas et al. 2007). Jasinskas et al. reported that the prevalence of the *Coxiella*-like bacterium in wild unfed *A. americanum* is 100% by real-time quantitative PCR (Jasinskas et al. 2007). The ubiquity and prevalence rate of the *Coxiella*-like bacterium in wild unfed *A. americanum* was confirmed by standard PCR (Klyachko et al. 2007). A similar prevalence rate (96.7%) was reported in wild unfed *A. americanum* by nested PCR (Heise et al. 2010). For colony-reared unfed *A. americanum*, Jasinskas et al. reported that the prevalence rate of the *Coxiella*-like bacterium is 100% by real-time quantitative PCR (Jasinskas et al. 2007). Heise et al. reported a slightly lower prevalence rate (89%) in the colony-reared unfed *A. americanum* by nested PCR (Heise et al. 2010). Surprisingly, the prevalence rate of the *Coxiella*-like bacterium in fed *A. americanum* is only 5.1–11.7% as detected by nested PCR (Heise et al. 2010). In contrast, a significantly higher prevalence rate (100%) was reported in fed *A. americanum* ticks by real-time quantitative PCR (Zhong et al. 2007). The difference in prevalence rates was attributed to the diagnostic method used. Whereas Heise et al. used nested PCR to detect the *Coxiella*-like bacterium, Jasinskas et al. used real-time quantitative PCR, which could be more sensitive than the nested PCR method (Drago et al. 2004; Farcas et al. 2004; Keen et al. 2007; Genc et al. 2010).

In addition to being present in flat and fed *A. americanum*, *Coxiella*-like bacteria were identified in all stages of *A. americanum* including adult, egg, larva, and nymph (Jasinskas et al. 2007; Klyachko et al. 2007). It is interesting to note that the prevalence rate of *Coxiella*-like bacteria is lower in other flat ticks. Bernascon et al. reported that 3 out of 48 (6.25%) *R. sanguineus* and 5 out of 48 *R. turanicus* (10.42%) were infected with *Coxiella*-like bacteria (Bernasconi et al. 2002). Lee et al. reported that the prevalence rate of the *Coxiella*-like bacterium in *H. longicornis* is 2% (Lee et al. 2004). However, it is not known if the lower prevalence rates in *R. sanguineus*, *R. turanicus*, and *H. longicornis* are due to the conventional PCR method used for detection. Future studies using more sensitive real-time quantitative PCR or nested PCR techniques will provide insight into this question.

18.4 Distribution of *Coxiella*-like Bacteria in Tick Organs

Coxiella-like bacterial endosymbionts of ticks do not have a specific tissue location. They are found across diverse tick organs, although commonly infected organs include ovaries and Malpighian tubules. Studies of the localization of *Coxiella* species in certain organs of ticks occurred two decades ago. Weyer reported that infection with *C. burnetii* primarily occurs in gut epithelial cells of *O. moubata* (Weyer 1975). Using PCR, Noda et al. detected a *Coxiella*-like bacterium in the ovaries and Malpighian tubules of *H. longicornis*, *R. sanguineus*, and *O. moubata* (Noda et al. 1997). The presence of a *Coxiella*-like bacterium in the ovarian tissues and Malpighian tubules of *I. woodi* was reported by Kurtti et al. using PCR and direct sequencing (Kurtti et al. 2002). Later in 2007, Jasinskas et al. detected a *Coxiella*-like bacterium in the midgut, ovaries, and salivary glands of *A. americanum* using real-time PCR (Jasinskas et al. 2007). The infection sites of the *Coxiella*-like bacterium in *A. americanum* were also reported by Klyachko et al., who used fluorescence *in situ* hybridization, PCR, and transmission electron microscopy to localize a *Coxiella*-like bacterium to the midgut, Malpighian tubules, ovaries, and salivary glands (Klyachko et al. 2007).

In summary, it appears that *Coxiella*-like bacteria exist in many organs of ticks. However, few studies have investigated the burden of *Coxiella*-like bacteria in each organ. Jasinskas et al. reported mean bacterial/tick ratios of 381 in the midgut, 79 in ovaries, and 12 in salivary glands. The results of Jasinskas et al. confirmed that the majority of the *Coxiella*-like bacterium exists in the midgut and ovaries of *A. americanum* (Jasinskas et al. 2007). The primary distribution of *Coxiella*-like bacteria in the reproductive organs of tick hosts suggest they use transovarial transmission as an important route of transmission route during their life cycle.

18.5 Transmission Routes of *Coxiella*-like Bacteria and *C. burnetii* in Ticks and Vertebrate Hosts

Little is known about the transmission cycle of *Coxiella*-like bacteria in nature. It appears that transmission of the bacteria in nature relies on tick bites, rather than through inhalation of contaminated aerosols and/or contacting contaminated dust particles, as is the case of *C. burnetii* (De et al. 1950). Although it is not known if *Coxiella*-like bacteria are transstadially transmitted by ticks, it is clear they are transovarially transmitted. Using PCR, Reeves et al. demonstrated that a *Coxiella*-like bacterium is transmitted transovarially by *Carios capensis* (Reeves et al. 2005). Later, Clay et al. reported that the frequency of transovarial transmission of a *Coxiella* symbiont by *A. americanum* is 100% by PCR and sequencing analysis (Clay et al. 2008). The high rate of the transovarial transmission of *Coxiella*-like bacteria indicates that ticks act as reservoirs, thus maintaining a population of the bacteria in nature. In contrast, transovarial transmission of virulent *C. burnetii* rarely occurs in nature. Daiter reported that transovarial transmission of *C. burnetii* by *Hyalomma asiaticum* decreased significantly from generation to generation. It was estimated that second- and third-generation females of *H. asiaticum* have a 9- and 20-fold lower frequency of transovarial transmission than first- and the second-generation females, respectively (Daiter 1977).

C. burnetii is also spread and maintained in nature by vertebrate hosts. The Q-fever agent is a zoonotic pathogen, and rodents are major reservoirs in nature, although cattle, goats, and sheep are the main sources of human infection (Aitken 1989; Woldehiwet 2004). Despite numerous reports of infections of *C. burnetii* in vertebrate hosts over the past decades, there are very few reports of infections of vertebrates by *Coxiella*-like bacteria. In the first report of a *Coxiella*-like bacterial infection of vertebrate hosts, Shivaprasad et al. showed that the hawk-headed parrot and golden mantle rosella are infected with a *Coxiella* species by amplifying the organism's 16S rRNA from the livers of these animals (Shivaprasad et al. 2008). In addition, transmission electron microscopy revealed intracellular bacteria with a similar morphology and size as *Coxiella*-like bacteria in ticks. Electron micrographs of liver and lungs of infected birds revealed bacteria that varied in shape (pleomorphic to elongated) and size (0.45 μm wide to more than 1.0 μm long). Despite the identification of the *Coxiella*-like bacteria in birds, it is unclear if birds, such as hawk-headed parrot and golden mantle rosella, serve as reservoirs for *Coxiella*-like bacteria. Future studies are needed to determine if vertebrates are persistently infected with *Coxiella*-like bacteria, and if they truly serve as reservoirs of the organism in nature.

18.6 *Coxiella*-like Bacterium-Host Interactions

All species in the genus *Coxiella* are intracellular bacteria (Raoult 1993; Waag 2007). *Coxiella*-like bacteria share tick host and geographic distribution with *C. burnetii* (Parola and Raoult 2001; Childs and Paddock 2003). However, *C. burnetii* is harbored in 0–37.6% of ticks sampled, a prevalence rate that is significantly lower than that of *Coxiella*-like bacteria (Mediannikov et al. 2010). For example, the prevalence rate of the *Coxiella*-like bacterium in *A. americanum* is 100% (Jasinskas et al. 2007; Klyachko et al. 2007; Heise et al. 2010). However, a high overall variation in prevalence rates of *Coxiella*-like bacteria in ticks suggests different transmission routes.

Sequencing of a 9-kb chromosomal region of the *Coxiella*-like bacterium in *A. americanum* revealed a reduced genome that lacks corresponding genes encoding hypothetical proteins in the genome of *C. burnetii* (Jasinskas et al. 2007). Zhong et al. studied transovarial and transstadial transmissions of this bacterium by detecting its presence in adults, egg masses, larvae, and nymphs using real-time quantitative PCR. Amplification of elongation factor G and macrophage inhibitory factor genes was used to detect copy numbers of the *Coxiella*-like bacterium and *A. americanum* tick cells, respectively. It was found that median values of the *Coxiella*-like bacterium were low in eggs and larvae, but the median values were 10-folder higher in nymphs and adults (Zhong et al. 2007). In 2008, Clay et al. used real-time PCR to confirm that the *Coxiella*-like bacterium was present in engorged *A. americanum*, as well as tick eggs and larval clutches (Clay et al. 2008). Both of the above reports demonstrated that the *Coxiella*-like bacterium is transmitted in *A. americanum* by the transovarial and transstadial transmission.

The *Coxiella*-like bacterium has intimate and persistent relationships with *A. americanum* (Zhong et al. 2007). Zhong et al. studied potential functions of this relationship by comparing engorged females injected with tetracycline or rifampicin, with PBS buffer-injected controls. Antibiotic treatment cured the *Coxiella*-like bacterium of *A. americanum* to result in ticks with longer oviposition, fewer eggs, greater egg weight loss, lower egg hatching rates, and lower numbers of viable larvae per tick. Thus, antibiotic injection results show an association between the *Coxiella*-like bacterium and reproductive fitness of *A. americanum* (Zhong et al. 2007).

Although little is known about functions of symbiotic *Coxiella*-like bacteria in ticks, it has been hypothesized that they synthesize various nutrients that are required by arthropod hosts (Moran et al. 2003; Douglas 2007; Zhong et al. 2007; Baumann et al. 1995; Faria e Silva et al. 1994). Ticks that feed entirely on blood are a virtual breeding ground for symbiotic partnerships because vertebrate blood rarely contains sufficient quantities of amino acids and B-vitamins. Therefore, the diet of blood feeding arthropods is often complemented with bacterially synthesized vitamins. The partnership is thought to contribute to the fitness of the tick host (Beard et al. 2001; Akman et al. 2002; Wu et al. 2006).

18.7 Other Bacterial Endosymbionts of Ticks

Symbiotic association of bacteria with ticks is widespread in nature. Species of notable bacterial symbionts that are vertically transmitted in ticks are *Coxiella*, *Francisella*, and *Rickettsia* (Noda et al. 1997; Scoles 2004). The *Francisella* endosymbionts have been identified from both argasid and ixodid ticks in the past decades. In 1973, Burgdorfer et al. reported that *Dermacentor andersoni* harbors a *D. andersoni* symbiont (Burgdorfer et al. 1973). The genus of the bacterium was later identified as *Francisella* (Niebylski et al. 1997a). Other *Francisella*-like bacteria were described in *A. maculatum* (Scoles 2004), *A. triste* (Venzal et al. 2008), *D. albipictus* (Scoles 2004). *D. andersoni* (Burgdorfer et al. 1973; Niebylski et al. 1997a; Sun et al. 2000; Scoles 2004), *D. hunteri* (Scoles 2004), *D. nitens* (Scoles 2004), *D. occidentalis* (Scoles 2004), *D. variabilis* (Sun et al. 2000; Scoles 2004; Goethert and Telford 2005), and *Ornithodoros moubata* (Noda et al. 1997). The nucleotide sequence identity of 16S rRNA between *Francisella* endosymbionts in *Dermacentor* ticks and *F. tularensis* is within a range of 87.6–98.1% (Scoles 2004). Based on 16S rRNA gene sequences, the *Francisella*-like bacteria were phylogenetically most closely related to virulent *F. tularensis*. All of the *Francisella* endosymbionts, along with virulent *F. tularensis*, clearly belong to a monophyletic group (Niebylski et al. 1997a; Scoles 2004; Goethert and Telford 2005).

PCR analysis indicates high prevalence rates of *Francisella*-like bacteria in ticks. The prevalence rate of the *Francisella* endosymbiont in *D. andersoni* is 94.4% (Niebylski et al. 1997a). Prevalence rates of two *Francisella*-like species of *D. variabilis* are 100% and 55%, respectively (Goethert and Telford 2005). The high prevalence rates demonstrate widespread endemicity of *Francisella* endosymbiont in ticks. *Francisella* endosymbionts are transmitted transovarially, rather than horizontally (Niebylski et al. 1997a; Goethert and Telford 2005). The rate of transovarial transmission was reported as high as 95.6% (Niebylski et al. 1997a). Results of morphological studies confirmed transovarial transmission of *Francisella* endosymbionts. By using electron microscopy, Reinhardt et al. reported *Francisella* endosymbionts in the ovaries and Malpighian tubules of *O. moubata* (Reinhardt et al. 1972). Using 16S rRNA gene, Noda et al. and Niebylski et al. detected the location of *Francisella* species in the ovaries and Malpighian tubules of *O. moubata* (Noda et al. 1997) and *D. andersoni* (Niebylski et al. 1997a).

Most rickettsial endosymbionts have been identified from Ixodid ticks, such as *A. americanum* (Weller et al. 1998), *D. andersoni* (Simser et al. 2001; Baldridge et al. 2004), *I. ricinus* (Simser et al. 2002), and *I. scapularis* (Weller et al. 1998). Other *Rickettsia* species were initially reported as endosymbionts or nonpathogenic; however, after years of research some appear to be pathogenic for lab animals. An example is *R. massiliae*, which was discovered in *Rhipicephalus* spp. ticks in 1993 and considered nonpathogenic. The bacterium is now considered a virulent *Rickettsia* species (Vitale et al. 2006). In contrast to the less than 1% prevalence rate of virulent *R. rickettsii*, nonpathogenic *Rickettsia* species are detected in 10–100% of tick hosts (Stromdahl et al. 2008; Dergousoff et al. 2009; Labruna 2009).

The offspring of ticks acquire spotted fever group rickettsiae (SFG) via transovarial transmission (Burgdorfer and Brinton 1975; Macaluso et al. 2001, 2002; Goddard 2003; Loftis et al. 2005; Matsumoto et al. 2005; Freitas et al. 2009). Another transmission route of SFG rickettsiae in nature is through transstadial transmission (Goddard 2003; Loftis et al. 2005; Matsumoto et al. 2005; Freitas et al. 2009). In addition to the transovarial and transstadial transmission, vertebrate hosts play a role as reservoirs in maintaining virulent spotted fever group in nature (Rovery and Raoult 2008). The transmission of endosymbiotic rickettsiae is poorly defined. However, endosymbiotic *R. peacockii* multiplies in *D. andersoni* and is transovarially transmitted (Niebylski et al. 1997b). Zhong et al. have demonstrated that a rickettsial endosymbiont in *I. pacificus* is transmitted by transovarial and transstadial transmission (unpublished data).

High prevalence rates and transovarial transmission are two trademarks of symbiotic rickettsial species. In addition to the above criteria, Paddock (2009) summarized three additional features of rickettsial endosymbionts in ticks, using *R. peacockii* as an example: (1) endosymbiotic rickettsial species are nonpathogenic or mildly pathogenic for lab rodents, (2) they reside in organs other than salivary glands in tick hosts, and (3) they have a nonfunctional *ompA* gene (encoding an outer membrane protein) because of three premature stop codons in its coding sequence (Baldridge et al. 2004).

Endosymbiotic rickettsial species are believed to play an essential role in host fitness. These intracellular bacteria reside in a variety of arthropod species and are generally associated with reproductive manipulation, such as male-killing (von der Schulenburg et al. 2001; Werren et al. 1994), parthenogenesis (Giorgini et al. 2010), and other reproduction manipulations (Sakurai et al. 2005; Perotti et al. 2006). Recently, it has been reported that endosymbiotic *Rickettsia* species play a role in increasing susceptibility to insecticides and microbe infections (Maudlin and Ellis 1985; Kontsedalov et al. 2008). Other studies have reported that rickettsial species in ticks exert competitive inhibition in the transovarial transmission of highly virulent *R. rickettsii* (Burgdorfer et al. 1981; Price 1953; Macaluso et al. 2002).

18.8 Conclusions

Tick-associated *Coxiella*-like bacteria are closely related to the Q fever pathogen, *C. burnetii*. Though many *Coxiella*-like bacteria have been reported in hard and soft ticks, only the *Coxiella*-like bacterium in *A. americanum* has been demonstrated to be a bacterial symbiont in the tick. The bacterium has a reduced genome, is transmitted by transovarial and transstadial transmission, and contributes to the fitness of the tick host. Future studies investigating the intimate relationship between *Coxiella*-like bacteria and tick hosts will provide needed insight into the nature of the symbiosis. An important research goal in this field is to elucidate the roles bacterial symbionts play in the life strategies of important arthropod pests, with implications for improving pest control.

References

Aitken ID (1989) Clinical aspects and prevention of Q fever in animals. Eur J Epidemiol 5:420–424

Akman L, Yamashita A, Watanabe H et al (2002) Genome sequence of the endocellular obligate symbiont of tsetse flies, *Wigglesworthia glossinidia*. Nat Genet 32:402–407

Angelakis E, Raoult D (2010) Q fever. Vet Microbiol 140:297–309

Baldridge GD, Burkhardt NY, Simser JA et al (2004) Sequence and expression analysis of the *ompA* gene of *Rickettsia peacockii*, an endosymbiont of the Rocky Mountain wood tick, *Dermacentor andersoni*. Appl Environ Microbiol 70:6628–6636

Baumann P, Baumann L, Lai CY et al (1995) Genetics, physiology, and evolutionary relationships of the genus *Buchnera*: intracellular symbionts of aphids. Annu Rev Microbiol 49:55–94

Beard CB, Dotson EM, Pennington PM et al (2001) Bacterial symbiosis and paratransgenic control of vector-borne Chagas disease. Int J Parasitol 31:621–627

Bernasconi MV, Casati S, Peter O et al (2002) *Rhipicephalus* ticks infected with *Rickettsia* and *Coxiella* in Southern Switzerland (Canton Ticino). Infect Genet Evol 2:111–120

Burgdorfer W, Brinton LP (1975) Mechanisms of transovarial infection of spotted fever Rickettsiae in ticks. Ann N Y Acad Sci 266:61–72

Burgdorfer W, Brinton LP, Hughes LE (1973) Isolation and characterization of symbiotes from the Rocky Mountain wood tick, *Dermacentor andersoni*. J Invertebr Pathol 22:424–434

Burgdorfer W, Hayes SF, Mavros AJ (1981) Nonpathogenic rickettsiae in *Dermacentor andersoni*: a limiting factor for the distribution of *Rickettsia rickettsii*. In: Burgdorfer W, Anacker RL (eds) Rickettsiae and rickettsial diseases. Academic, New York

Childs JE, Paddock CD (2003) The ascendancy of *Amblyomma americanum* as a vector of pathogens affecting humans in the United States. Annu Rev Entomol 48:307–337

Clay K, Klyachko O, Grindle N et al (2008) Microbial communities and interactions in the lone star tick, *Amblyomma americanum*. Mol Ecol 17:4371–4381

Daiter AB (1977) Transovarial and transspermal transmission of *Coxiella burneti* by the tick *Hyalomma asiaticum* and its role in the ecology of Q-rickettsiosis. Parazitologiia 11:403–411

De LP, Lennette EH, Deome KB (1950) Q fever in California; recovery of *Coxiella burnetii* from naturally-infected air-borne dust. J Immunol 65:211–220

Dergousoff SJ, Gajadhar AJ, Chilton NB (2009) Prevalence of *Rickettsia* species in Canadian populations of *Dermacentor andersoni* and *D. variabilis*. Appl Environ Microbiol 75:1786–1789

Derrick EH (1937) Q fever, a new fever entity: clinical features, diagnosis and laboratory investigation. Med J Aust 2:281

Douglas AE (2007) Symbiotic microorganisms: untapped resources for insect pest control. Trends Biotechnol 25:338–342

Drago L, Lombardi A, De Vecchi E et al (2004) Comparison of nested PCR and real time PCR of Herpesvirus infections of central nervous system in HIV patients. BMC Infect Dis 4:55

Farcas GA, Zhong KJ, Mazzulli T et al (2004) Evaluation of the RealArt Malaria LC real-time PCR assay for malaria diagnosis. J Clin Microbiol 42:636–638

Faria e Silva PM, Fiorini JE, Soares MJ et al (1994) Membrane-associated polysaccharides composition, nutritional requirements and cell differentiation in *Herpetomonas roitmani*: influence of the endosymbiont. J Eukaryot Microbiol 41:55–59

Fox GE, Magrum LJ, Balch WE et al (1977) Classification of methanogenic bacteria by 16S ribosomal RNA characterization. Proc Natl Acad Sci U S A 74:4537–4541

Freitas LH, Faccini JL, Labruna MB (2009) Experimental infection of the rabbit tick, *Haemaphysalis leporispalustris*, with the bacterium *Rickettsia rickettsii*, and comparative biology of infected and uninfected tick lineages. Exp Appl Acarol 47:321–345

Genc A, Eroglu F, Koltas IS (2010) Detection of *Plasmodium vivax* by nested PCR and real-time PCR. Korean J Parasitol 48:99–103

Giorgini M, Bernardo U, Monti MM et al (2010) *Rickettsia* symbionts cause parthenogenetic reproduction in the parasitoid wasp *Pnigalio soemius* (Hymenoptera: Eulophidae). Appl Environ Microbiol 76:2589–2599

Goddard J (2003) Experimental infection of lone star ticks, *Amblyomma americanum*, with *Rickettsia parkeri* and exposure of guinea pigs to the agent. J Med Entomol 40:686–689

Goethert HK, Telford SR 3rd (2005) A new *Francisella* (Beggiatiales: Francisellaceae) inquiline within *Dermacentor variabilis* say (Acari: Ixodidae). J Med Entomol 42:502–505

Gupta R, Lanter JM, Woese CR (1983) Sequence of the 16S Ribosomal RNA from *Halobacterium volcanii*, an Archaebacterium. Science 221:656–659

Heise SR, Elshahed MS, Little SE (2010) Bacterial diversity in *Amblyomma americanum* (Acari: Ixodidae) with a focus on members of the genus *Rickettsia*. J Med Entomol 47:258–268

Janda JM, Abbott SL (2007) 16S rRNA gene sequencing for bacterial identification in the diagnostic laboratory: pluses, perils, and pitfalls. J Clin Microbiol 45:2761–2764

Jasinskas A, Zhong J, Barbour AG (2007) Highly prevalent *Coxiella* sp. bacterium in the tick vector *Amblyomma americanum*. Appl Environ Microbiol 73:334–336

Keen J, Farcas GA, Zhong K et al (2007) Real-time PCR assay for rapid detection and analysis of PfCRT haplotypes of chloroquine-resistant *Plasmodium falciparum* isolates from India. J Clin Microbiol 45:2889–2893

Klyachko O, Stein BD, Grindle N et al (2007) Localization and visualization of a coxiella-type symbiont within the lone star tick, *Amblyomma americanum*. Appl Environ Microbiol 73:6584–6594

Kontsedalov S, Zchori-Fein E, Chiel E et al (2008) The presence of *Rickettsia* is associated with increased susceptibility of *Bemisia tabaci* (Homoptera: Aleyrodidae) to insecticides. Pest Manag Sci 64:789–792

Kurtti TJ, Palmer AT, Oliver JH Jr (2002) *Rickettsiella*-like bacteria in *Ixodes woodi* (Acari: Ixodidae). J Med Entomol 39:534–540

Labruna MB (2009) Ecology of rickettsia in South America. Ann N Y Acad Sci 1166:156–166

Lee JH, Park HS, Jang WJ et al (2004) Identification of the *Coxiella* sp. detected from *Haemaphysalis longicornis* ticks in Korea. Microbiol Immunol 48:125–130

Loftis AD, Gill JS, Schriefer ME et al (2005) Detection of *Rickettsia*, *Borrelia*, and *Bartonella* in *Carios kelleyi* (Acari: Argasidae). J Med Entomol 42:473–480

Macaluso KR, Sonenshine DE, Ceraul SM et al (2001) Infection and transovarial transmission of rickettsiae in *Dermacentor variabilis* ticks acquired by artificial feeding. Vector Borne Zoonotic Dis 1:45–53

Macaluso KR, Sonenshine DE, Ceraul SM et al (2002) Rickettsial infection in *Dermacentor variabilis* (Acari: Ixodidae) inhibits transovarial transmission of a second *Rickettsia*. J Med Entomol 39:809–813

Matsumoto K, Ogawa M, Brouqui P et al (2005) Transmission of *Rickettsia massiliae* in the tick, *Rhipicephalus turanicus*. Med Vet Entomol 19:263–270

Maudlin I, Ellis DS (1985) Association between intracellular rickettsial-like infections of midgut cells and susceptibility to trypanosome infection in *Glossina* spp. Z Parasitenkd 71:683–687

Maurin M, Raoult D (1999) Q fever. Clin Microbiol Rev 12:518–553

Mediannikov O, Fenollar F, Socolovschi C et al (2010) *Coxiella burnetii* in humans and ticks in rural Senegal. PLoS Negl Trop Dis 4:e654

Moran NA, Plague GR, Sandstrom JP et al (2003) A genomic perspective on nutrient provisioning by bacterial symbionts of insects. Proc Natl Acad Sci U S A 100(Suppl 2):14543–14548

Niebylski ML, Peacock MG, Fischer ER et al (1997a) Characterization of an endosymbiont infecting wood ticks, *Dermacentor andersoni*, as a member of the genus *Francisella*. Appl Environ Microbiol 63:3933–3940

Niebylski ML, Schrumpf ME, Burgdorfer W et al (1997b) *Rickettsia peacockii* sp. nov., a new species infecting wood ticks, *Dermacentor andersoni*, in western Montana. Int J Syst Bacteriol 47:446–452

Noda H, Munderloh UG, Kurtti TJ (1997) Endosymbionts of ticks and their relationship to *Wolbachia* spp. and tick-borne pathogens of humans and animals. Appl Environ Microbiol 63:3926–3932

Paddock CD (2009) The science and fiction of emerging rickettsioses. Ann N Y Acad Sci 1166:133–143

Parker NR, Barralet JH, Bell AM (2006) Q fever. Lancet 367:679–688

Parola P, Raoult D (2001) Ticks and tickborne bacterial diseases in humans: an emerging infectious threat. Clin Infect Dis 32:897–928

Perotti MA, Clarke HK, Turner BD et al (2006) *Rickettsia* as obligate and mycetomic bacteria. FASEB J 20:2372–2374

Price WH (1953) Interference phenomenon in animal infections with rickettsiae of Rocky Mountain spotted fever. Proc Soc Exp Biol Med 82:180–184

Raoult D (1993) Treatment of Q fever. Antimicrob Agents Chemother 37:1733–1736

Reeves WK, Loftis AD, Priestley RA et al (2005) Molecular and biological characterization of a novel *Coxiella*-like agent from *Carios capensis*. Ann N Y Acad Sci 1063:343–345

Reinhardt C, Aeschlimann A, Hecker H (1972) Distribution of *Rickettsia*-like microorganisms in various organs of an *Ornithodorus moubata* laboratory strain (Ixodoidea, Argasidae) as revealed by electron microscopy. Z Parasitenkd 39:201–209

Rovery C, Raoult D (2008) Mediterranean spotted fever. Infect Dis Clin North Am 22:515–530, ix

Sakurai M, Koga R, Tsuchida T et al (2005) *Rickettsia* symbiont in the pea aphid *Acyrthosiphon pisum*: novel cellular tropism, effect on host fitness, and interaction with the essential symbiont *Buchnera*. Appl Environ Microbiol 71:4069–4075

Schabereiter-Gurtner C, Lubitz W, Rolleke S (2003) Application of broad-range 16S rRNA PCR amplification and DGGE fingerprinting for detection of tick-infecting bacteria. J Microbiol Methods 52:251–260

Schmidt TM, Relman DA (1994) Phylogenetic identification of uncultured pathogens using ribosomal RNA sequences. Methods Enzymol 235:205–222

Scoles GA (2004) Phylogenetic analysis of the *Francisella*-like endosymbionts of *Dermacentor* ticks. J Med Entomol 41:277–286

Sharer JD, Koosha H, Church WB et al (1999) The function of conserved amino acid residues adjacent to the effector domain in elongation factor G. Proteins 37:293–302

Shivaprasad HL, Cadenas MB, Diab SS et al (2008) *Coxiella*-like infection in psittacines and a toucan. Avian Dis 52:426–432

Simser JA, Palmer AT, Munderloh UG et al (2001) Isolation of a spotted fever group *Rickettsia*, *Rickettsia peacockii*, in a Rocky Mountain wood tick, *Dermacentor andersoni*, cell line. Appl Environ Microbiol 67:546–552

Simser JA, Palmer AT, Fingerle V et al (2002) *Rickettsia monacensis* sp. nov., a spotted fever group *Rickettsia*, from ticks (*Ixodes ricinus*) collected in a European city park. Appl Environ Microbiol 68:4559–4566

Stromdahl EY, Vince MA, Billingsley PM et al (2008) *Rickettsia amblyommii* infecting *Amblyomma americanum* larvae. Vector Borne Zoonotic Dis 8:15–24

Sun LV, Scoles GA, Fish D et al (2000) *Francisella*-like endosymbionts of ticks. J Invertebr Pathol 76:301–303

Venzal JM, Estrada-Pena A, Portillo A et al (2008) Detection of alpha and gamma-proteobacteria in *Amblyomma triste* (Acari: Ixodidae) from Uruguay. Exp Appl Acarol 44:49–56

Vitale G, Mansuelo S, Rolain JM et al (2006) *Rickettsia massiliae* human isolation. Emerg Infect Dis 12:174–175

von der Schulenburg JH, Habig M, Sloggett JJ et al (2001) Incidence of male-killing *Rickettsia* spp. (alpha-proteobacteria) in the ten-spot ladybird beetle *Adalia decempunctata* L. (Coleoptera: Coccinellidae). Appl Environ Microbiol 67:270–277

Waag DM (2007) *Coxiella burnetii*: host and bacterial responses to infection. Vaccine 25:7288–7295

Weller SJ, Baldridge GD, Munderloh UG et al (1998) Phylogenetic placement of rickettsiae from the ticks *Amblyomma americanum* and *Ixodes scapularis*. J Clin Microbiol 36:1305–1317

Werren JH, Hurst GD, Zhang W et al (1994) Rickettsial relative associated with male killing in the ladybird beetle (*Adalia bipunctata*). J Bacteriol 176:388–394

Weyer F (1975) Observations on the behaviour of *Coxiella burneti* in the argasid tick *Ornithodoros moubata*. Tropenmed Parasitol 26:219–231

Woldehiwet Z (2004) Q fever (coxiellosis): epidemiology and pathogenesis. Res Vet Sci 77:93–100

Wu D, Daugherty SC, Van Aken SE et al (2006) Metabolic complementarity and genomics of the dual bacterial symbiosis of sharpshooters. PLoS Biol 4:e188

Zhong J, Jasinskas A, Barbour AG (2007) Antibiotic treatment of the tick vector *Amblyomma americanum* reduced reproductive fitness. PLoS One 2:e405

Chapter 19
Molecular Typing of *Coxiella burnetii* (Q Fever)

Robert F. Massung, Sally J. Cutler, and Dimitrios Frangoulidis

Contents

Abstract Although we live in the age of genomics and the availability of complete genome sequences of *Coxiella burnetii* has increased our understanding of the genomic diversity of the agent, it is still somewhat a "query" microorganism. The epidemiology of Q fever is complex due to the worldwide distribution, reservoir and vector diversity, and a lack of studies defining the dynamic interaction between these factors. In addition *Coxiella* is an agent that could be used as a bioterror weapon. Therefore, typing methods that can discriminate strains and be used to

R.F. Massung • S.J. Cutler • D. Frangoulidis (✉)
Bundeswehr Institute of Microbiology, Neuherbergstr. 11,
80937 Munich, Germany
e-mail: DimitriosFrangoulidis@Bundeswehr.org

R. Toman et al. (eds.), *Coxiella burnetii: Recent Advances and New Perspectives in Research of the Q Fever Bacterium*, Advances in Experimental Medicine and Biology 984,
DOI 10.1007/978-94-007-4315-1_19, © Springer Science+Business Media Dordrecht 2012

trace back infections to their source are of paramount importance. In this chapter we provide an overview of historical and current typing methods and describe their advantages and limitations. Recently developed techniques such as MLVA and SNP typing have shown promise and improved the discrimination capacity and utility of genotyping methods for molecular epidemiologic studies of this challenging pathogen.

Keywords Q fever • *Coxiella burnetii* • Typing • Molecular epidemiology

19.1 Introduction

Q fever is a zoonosis caused by the Gram negative coccobacillus *Coxiella burnetii* belonging to the class gammaproteobacteria. Despite the worldwide distribution of the agent, Q fever is rarely reported, probably as a result of both clinical and diagnostic challenges presented by *C. burnetii* infection compounded by the lack of requirement for reporting cases in many countries (Fournier et al. 1998). Despite this, it persists with a large reservoir among multiple species, regularly leading to outbreaks.

The organism is highly infectious, with experimental estimates suggesting an infectious dose of less than ten organisms. Furthermore, they are highly resistant against both heat and desiccation, have a ubiquitous distribution, are relatively easy to cultivate and remain infectious over several kilometers when aerosolized. These characteristics resulted in the inclusion of *C. burnetii* among agents tested in the old biological weapon programs of the USA and the former Soviet Union. Similarly, this was also the justification for its inclusion on the CDC-list as a category B potential bioterrorism agent.

An overview and historical reflection, together with actual and future bioweapon potential was provided in 2003 (Madariaga et al. 2003). Although estimates of the impact and implications following an intentional release are somewhat speculative, the threat and consequences can be immense. Besides control and other measurements the bio-forensic aspects have gained importance in the recent years. The anthrax letters in 2001 serve as an important example of how an infectious agent can be misused (Jernigan et al. 2002) and how a limited number of infections from a bioterror event can have immense public, political, and economic impact. Subsequent to an intentional release, molecular detection and diagnostic methods will be important for identifying the agent, and molecular typing methods for forensic analyses to identify the composition and source of the biologic agent.

Against this background, recent attention focused upon *C. burnetii* has highlighted our limited understanding of its epidemiology, population diversity and basic biology. Huge improvements in diagnostic capabilities have been achieved with the introduction of molecular techniques, although serological methods still play a vital role in diagnosing human infections and for population screening. Whole genome sequence data has demonstrated the genetic diversity among strains of

C. burnetii, and this diversity has enabled the design and application of high-resolution molecular typing systems that can be useful for epidemiologic investigations of infectious episodes.

In addition, these methods have the potential to differentiate between natural and deliberate release of the pathogen. Another important aspect in this special field is the so-called background level of this pathogen in the environment. Due to the ubiquitous occurrence of *Coxiella* throughout the world, relevant traces of this pathogen could complicate interpretation of bio-forensic investigations.

Molecular epidemiological investigations can be hindered by diagnostic delays resulting from the very low amounts of DNA often seen in clinical human material, consequently trace back analysis of cases are rarely performed. In recent years some promising publications presented data that could make it possible to do such epidemiological analysis as seen in other diseases (e.g. tuberculosis, salmonellosis) in the future (Driscoll 2009; Malorny et al. 2008).

In this chapter we will give an overview of the different typing methods that have been applied to *C. burnetii* and will discuss their merits and importance. Furthermore, we discuss the discriminatory power, robustness, reproducibility and limitations of these methods (Table 19.1).

19.2 Typing Methods Used Before 2005

Given the highly infectious nature of *C. burnetii* and the variety of potential sources for natural infection from proximity of livestock, consumption of their products, exposure to infected arthropods, or indeed following potential deliberate release, it is of paramount importance to be able to investigate the source of any outbreak. To this end a variety of different genotyping methods have been applied for epidemiological investigation of both naturally occurring outbreaks, or those from a suspected bioterroristic act. Initial typing systems described were based on plasmid types, restriction fragment length polymorphisms (RFLP) analyzed with SDS-PAGE, pulsed-field gel electrophoresis (PFGE), and sequence studies of single genome targets like 16s/23s, *com1*, *mucZ*, and *icd*. They showed different levels of discriminatory power and epidemiological significance, but all suffered with problems of inter- and intra-laboratory reproducibility, hampering their wide use.

19.2.1 Plasmid Types

C. burnetii strains appear with five different plasmid types, independent from the LPS-associated phases: four different plasmids (QpH1, QpRS, QpDV, and QpDG) and one type with a chromosomal plasmid-homologous sequence (Thiele et al. 1994; Valková and Kazár 1995; Willems et al. 1997; Lautenschläger et al. 2000; Jäger et al. 2002). The characterization of these led to a classification into five genomic

Table 19.1 History of *C. burnetii* typing methods

Method/target region	Marker/discriminatory power	Reference (year)	Notes
RFLP	4 pattern, 6 isolates	Vodkin et al. (1986)	
RFLP SDS-PAGE	6 groups, 32 isolates	Hendrix et al. (1991)	
16s, *16s/23s*-ITS	2 "groups", 6 isolates non-groupable, 22 isolates	Stein et al. (1993, 1997)	
Com1 gene	4 groups, 21 isolates	Zhang et al. (1997)	19 nucleotide differences, 10 amino acid changes of 1,060 bp
RFLP-PFGE	20 pattern, 80 isolates	Jäger et al. (1998)	
Com1 and *mucZ* gene	5 groups (com1), 4 groups (mucZ), 27 isolates	Sekeyova et al. (1999)	
icd gene, PCR-RFLP	3 groups (icd), 2 groups (PCR-RFLP), 19 isolates	Nguyen and Hirai (1999)	
MST	10 marker, 30 groups, 173 isolates	Glazunova et al. (2005)	Biggest, worldwide strain collection tested, sequence analysis mandatory
Micro-array	7 groups, 24 isolates	Beare et al. (2006)	2,103 ORFs of Nine Mile strain, large and expensive approach
MLVA/IRS-PCR	7 marker, 9 types in 5 clusters (MLVA), 4 different IRS-PCRs, 6 pattern (IRS-PCR), 16 isolates	Arricau-Bouvery et al. (2006)	
MLVA	17 marker, 36 groups, 42 isolates	Arricau-Bouvery et al. (2006)	Capillary sequencer or gel-based, harmonisation needed
IS1111 PCR	5 groups, 21 isolates	Denison et al. (2007)	Similar results than RFLP SDS-PAGE, simple, PCR based method
MLVA	3 marker, 11 samples, 4 genotypes	Klaassen et al. (2009)	Modified marker from Arricau-Bouvery et al. (2006), "direct" typing of clinical material

groups. Some plasmid types could be associated with various geographic regions. A formerly hypothesised correlation of these genomic groups with the virulence or the clinical appearance could not be confirmed in later studies (Thiele and Willems 1994).

19.2.2 RFLP/PFGE

Restriction fragment length polymorphism analysis using SDS-PAGE following *Eco*RI- and *Bam*HI-digestion of total chromosomal DNA lead to a classification of *C. burnetii* strains in six groups (I-VI) concordant with that of plasmid typing (Hendrix et al. 1991). Analysis of *Not*I and *SFi*I digested DNA restriction fragments by PFGE confirmed four of these six DNA fragment patterns, representing isolates from the genomic groups I, IV, V and VI (Heinzen et al. 1990) and use of further restriction endonucleases were subsequently used to map the *Coxiella* genome (Frazier et al. 1991). Beside these groups, Thiele and co-researchers reported five additional DNA banding patterns after endonuclease restriction with *Not*I (Thiele et al. 1993). By using larger collections of *C. burnetii* isolates derived from animals and humans in Europe, USA, Africa and Asia, 20 different restriction patterns could be distinguished after *Not*I restriction of *C. burnetii* DNA and PFGE (Jäger et al. 1998).

19.2.3 com1, mucZ and icd-Gene

In 1997, the *com*1-gene was first used to analyze 21 isolates from different human and animal sources (Zhang et al. 1997). Although the gene was highly conserved, it showed variation in sequence permitting four groups to be defined. Interestingly a cluster of isolates derived from chronic Q fever cases could be identified. These results were partially confirmed by another study where 37 isolates could be divided into five groups by *com*1-gene sequencing (Sekeyová et al. 1999). Additionally, *muc*Z-gene variation showed four different groups, but without correlation with geographical origin or disease (acute/chronic).

The isocitrate dehydrogenase (*icd*) gene of *C. burnetii* was studied 1999 (Nguyen and Hirai) and differentiated 19 isolates from different backgrounds into three groups. Zhang et al. (1997) used the *com*1-gene and similarly observed that isolates from chronic cases clustered together. Although these studies produced promising data at the time, these methods were not widely used in subsequent years as a result of their relatively poor discriminatory power and unconvincing epidemiological validity.

19.2.4 16/23S-Gene

In common with other microbial typing, 16S and 23S-genes have been assessed for their ability to delineate *Coxiella* from other closely related species. In a small study from 1993 (Stein et al. 1993) six different isolates from either acute or chronic cases and with dispersed origins from all over the world, were compared using a 1,418 bp fraction of the 16S-rRNA, revealing only very slight variation. Three isolates differed from the others showing base substitutions at three positions but failed to show any epidemiological or clinical correlation. Another study using an internal transcribed 16S–23S rDNA spacer (ITS) region to evaluate 22 different *C. burnetii* isolates similarly failed to reveal any correlation with regard to epidemiology, virulence or taxonomy (Stein et al. 1997).

19.3 High Resolution Post-2005 Typing

Availability of complete genomic sequences has enabled application of a range of highly discriminatory typing approaches including multispacer Sequence Typing (MST), positioning of *IS1111*-elements, multiple loci variable number of tandem repeat analysis (VNTR or MLVA), and single nucleotide polymorphism (SNP) typing.

19.3.1 MST

This method is based on the analysis of the intergenic regions of genomes. These non-coding regions (= "spacer") are used for typing because they are not constrained by selective pressure to encode functional proteins, as are coding genes. Variations of the sequenced regions like Single Nucleotide Polymorphism (SNPs), deletions or insertions formed the basis for discrimination of genotypes. This technique was first introduced for *Coxiella* in 2005, and could discriminate 173 isolates into 30 genotypes and furthermore showed concordance with both geographical and plasmid clusters previously described (Glazunova et al. 2005). Use of ten markers had the advantage of high discriminatory power and transferability, as this is a sequence-based method unlike the preceding gel-based methods. This makes the results easier to compare between different laboratories and offers also the ability to identify new alleles and variations inside the studied regions. Analysis of results is facilitated through use of an internet-based data-link that permits comparison with those strain profiles banked (http://ifr48.timone.univ-mrs.fr).

19.3.2 MLVA-Typing

Interrogation of genomic sequence data by software such as tandem repeat finder, enabled identification of small repeat units that often vary between isolates. This has become an established typing technique from other bacteria over recent years. MLVA typing schemes were established for *C. burnetii* independently by two laboratories in 2006. Up to 17 different genomic target-regions could be used for differentiation of *Coxiella* strains. The discriminatory power is higher than that of MST (42 isolates yielded 37 genotypes), but comparison of results from different laboratories remains problematic. Recent comparative studies have highlighted the mandatory requirement for adjustment to standards arising from use of different platforms (gel-based versus capillary-sequencer-based) to accurately determine the length of the PCR products. In addition, exact definitions of consensus sequences of the single targets are still missing making harmonization of the assigned repeat number difficult (Arricau-Bouvery et al. 2006; Svraka et al. 2006).

Often diagnosis of Q fever is delayed through its reliance upon serology. Recently, aided by a particularly large and protracted outbreak in the Netherlands (2007–2010), PCR techniques have been used more frequently as primary diagnostics for suspected acute Q fever cases. This leads to the possibility of directly typing *Coxiella* from clinical material with its obvious advantages of timesaving and rapid instigation of appropriate control measures. The major challenge arises from the very low amounts of *Coxiella*-specific DNA in acute cases that make it very difficult to get reliable results. This can be partially overcome by shortening primer lengths to improve sensitivity. This is not applicable to all known markers, but a reduced panel of three markers has been used successfully in the Dutch outbreak to analyze clinical samples from acute and chronic Q fever cases with various samples (swabs, urine, sputum, plasma) derived from both humans and animals (Klaassen et al. 2009). Four different genotypes could be clearly identified demonstrating the value of direct typing.

Fluctuations in the stability of MLVA/VNTR results have been observed among other bacterial like *Escherichia coli*, *Mycobacterium leprae*, *Salmonella* associated with cultivation, environmental and therapeutic stresses (Cooley et al. 2010; Xing et al. 2009; Malorny et al. 2008). The selected loci are deemed to be relative stable supported by the fact that isolates of the reference strain Nine Mile (RSA493) from different laboratories with widely varied cell culture passages showed no differences (Svraka et al. 2006; Frangoulidis unpublished data).

In addition, like the MST-method, a web based data repository has been established facilitating comparative analysis with known strain repeat patterns (http://minisatellites.u-psud.fr/).

19.3.3 Infrequent Restriction Site-(IRS)-PCR

An alternative approach uses IRS-PCR. This was applied to 14 *Coxiella* isolates using four different IRS PCR's resolving these into six different groups (Arricau-Bouvery et al. 2006). This method was compared with MLVA (see Sect. 19.3.2) showing good agreement. Nevertheless this method is not easy to establish in different laboratories and is prone to inter-laboratory variation. Despite these reservations, IRS-PCR has proved to be valuable for differentiation of two strains that could not be differentiated by MLVA.

19.3.4 Microarray-Based Whole-Genome Comparisons/Typing

The availability of complete genome sequences for *C. burnetii* has allowed whole genome comparisons utilizing microarray technology (Beare et al. 2006, 2009; Seshadri et al 2003). Genomes of a panel of 24 different strains were hybridized to a high-density custom Affymetrix GeneChip containing all open reading frames (ORFs) of the Nine Mile phase I (NMI) reference isolate. Isolates contained up to 20 genomic polymorphisms consisting of 1–18 ORFs each. These were mostly complete ORF deletions, although partial deletions, point mutations and insertions were also identified. A total of 139 chromosomal and plasmid ORFs were polymorphic among all *C. burnetii* isolates, representing circa 7% of the NMI coding capacity (Beare et al. 2006). In summary, these findings confirmed the relatedness of former RFLP-grouped strains and established new distinct groups. Eight different ORF-polymorphism types could be identified among the 24 strains, with two new major groups (VII and VIII) with the plasmid type QpDV.

19.3.5 IS1111-Typing

The *IS1111*-insertion sequence, coding for a transposase, is seen in up to 56 copies in *C. burnetii* genomes. As a consequence, this element is often used as a specific target providing sensitive diagnostic PCRs. In 2007, Denison et al. established a genotyping system based on four of these insertion sequence regions, which were analyzed by PCR, using an antisense primer binding inside of the *IS1111*-elements in combination with an upstream sense primer specific for each element (Denison et al. 2007). The algorithm proposed by Denison et al. (Fig. 19.1) allowed a classification into genomic groups I–V, according to the six clusters derived by different authors described above using plasmid profiles, PCR or restriction-endonuclease digested DNA. One advantage of this method is its ease to perform inside and between laboratories, when comparing up to five different PCR results embedded in the above mentioned decision tree. An improved version of this method using variation of repeat units within the upstream region of *IS1111* elements, thus enhancing discriminatory power was recently published by Hanczaruk et al. (2009).

19.3.6 Single Nucleotide Polymorphism (SNP)-Typing

Sequence mismatches determined by previous MST genotyping (Glazunova et al. 2005) have been used to develop a SNP-based method for genotyping *C. burnetii* using real-time PCR that is rapid and inexpensive compared to most methods (Keim et al. 2004; Papp et al. 2003). Either TaqMan dual probe assays or SYBR Green Mismatch Amplification Mutation Assays (MAMA) are used to determine genotypes

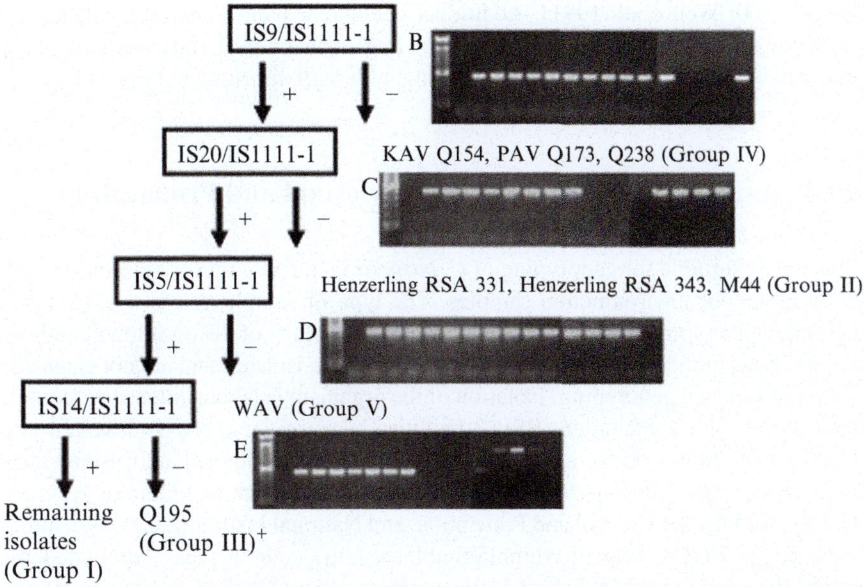

Fig. 19.1 Proposed algorithm flowchart and agarose gel photographs of corresponding PCR products (**a**) To discriminate isolates of *C. burnetii* using the algorithm shown, PCR is performed using the primer pairs depicted in boxes. Positive PCR products are subjected to subsequent PCR reactions as shown in the flowchart, while negative PCR reactions allow for the discrimination of isolates into the groups shown (**b**) PCR products using primer pairs IS 9 and IS1111-1 (**c**) PCR products using primer pairs IS 20 and IS1111-1. (**d**) PCR products using primer pairs IS 5 and IS1111-1 (**e**) PCR products using primer pairs IS 14 and IS1111-1. (**b–e**) Lanes 1, 100 bp ladder; 2, negative control; 3, 9Mi/I; 4, 9Mi/II; 5, RSA 514; 6, Nine Mile Baca; 7, Scottish; 8, Ohio; 9, Australian QD; 10, Q195; 11, Henzerling RSA 343; 12, Henzerling RSA 331; 13, M44; 14, KAV Q154; 15, PAV Q173; 16, Q238; 17, WAV. The star serves to denote that KAV Q154, PAV Q173, Q238, Henzerling RSA 331, Henzerling RSA 343, M44, and WAV also lack PCR products to IS 14, but only Q195 will give positive products in all other PCR reactions

based on SNP's within MST loci (Li et al. 2004). A recent report utilized 14 SNP assays to genotype over 40 isolates of *C. burnetii* and to differentiate these isolates into 8 distinct genogroups (Priestley et al. 2009). This method showed good correlation relating to geographic origin, infection characteristics, and plasmid types of isolates. A recent American goat isolate clustered among genogroup IV human heart valve isolates (K-Q154, P-Q173, F-Q228) and a previous American goat isolate (Priscilla-Q177) (Samuel et al. 1985). Isolates grouping together with Nine Mile phase I have a worldwide distribution, including isolates from North America (AD California, Idaho Q, Dyer, RSA329, Ohio 314), Australia (Australia QD) and Europe (Scottish, Cypriot). Isolates Poker Cat, Arsenaula Dog Uterus, G-Q212 and Ko-Q229 grouped together, all of which were isolated in Nova Scotia. Likewise, four eastern European isolates (Cb 48, Kmen L, Kmen S1, Florian) (Jäger et al. 1998) formed a unique cluster. Interestingly, Chinese isolates (Qi-Yi and Xingiao) clustered with two Italian isolates, Henzerling and M44, from World War II (Brezina 1958;

Genig 1960; Wen et al. 1991). Additional genome sequence analyses will lead to further development of SNP assays as shown by Huijsmans et al. (2011) most recently that will be able to provide rapid genotyping with high discriminatory power.

19.4 Impact of Sample, Sampling Method and Preparation

Specimens suitable for genotyping of *C. burnetii* include human or veterinary clinical samples, or environmental samples. The type of sample that is available will determine the typing method that can be employed. Many of the traditional methods require large numbers of purified bacteria grown from isolates, and are not amenable for rapid forensic genotyping. Isolation of the organism is labour-intensive, requires specialized high-containment BSL-3 facilities because the agent is classified as a Category B bioterrorism agent (Biosafety in Microbiological and Biomedical Laboratories, 5th Edition, 2007, U.S. Department of Health and Human Services, Centers for Disease Control and Prevention, and National Institutes of Health, http://www.cdc.gov/OD/ohs/biosfty/bmbl5/bmbl5toc.htm), and this has stimulated the recent development of PCR-based genotyping methods. PCR-based methods have proven useful for the detection of the agent in clinical samples collected from naturally occurring outbreaks (Amitai et al. 2010; Bamberg et al. 2007; Schneeberger et al. 2009; Tissot-Dupont et al. 2004; Turra et al. 2006) and in environmental samples (Kersh et al. 2010; Schulz et al. 2005). Human whole blood, serum, or the buffy coat fraction collected during the acute phase of infection, prior to antibiotic therapy, can be useful for genotyping using PCR-based methods or for isolation. Heart valves obtained following valve replacement surgery in patients with chronic Q fever are often heavily infected and provide a good substrate for all methods. Clinical samples may be used for genotyping following use of DNA extraction methods optimized for these samples and subsequent molecular diagnostic assays (Panning et al. 2008; Tilburg et al. 2010).

The samples most commonly used for veterinary testing are milk, faeces, urine, and vaginal discharge. Among domesticated animals known as reservoirs of *C. burnetii* (cattle, sheep, goats) the routes by which bacteria are shed varies and will influence optimal sample collection. While cattle and goats shed primarily in milk, sheep shed the bacteria mainly in faeces and vaginal fluids (Rodolakis et al. 2007). Both goats and sheep can shed large numbers of *C. burnetii* in the placenta and other birthing products during parturition. Although many species of wildlife can be infected, the shedding route or tissues that are optimal for bacterial isolation are not well characterized.

Environmental samples are often problematic as they contain large quantities of organic material making DNA extraction difficult and can inhibit PCR amplification. Consequently, these should be tested for inhibition prior to initiating genotype analysis. The ubiquitous nature of *C. burnetii* in the environment could make the forensic evaluation of a suspected intentional release difficult. Environmental samples collected from various regions across the USA identified *C. burnetii* DNA in 23.8% of the more than 1,600 samples tested (Kersh et al. 2010). Any analysis of environmental samples to investigate a suspected introduction of *C. burnetii* by natural (reservoir movement, wind, other methods of spread) or intentional means must include an evaluation of the naturally occurring background level, and genotypic composition and diversity. Ideally, a highly sensitive PCR-based detection assay should be employed in conjunction with genotyping assays, the former to identify the agent and the latter to differentiate introduced *C. burnetii* from pre-existing background bacteria.

19.5 Implications Due to Epidemiology, Forensics and Bioterrorism

Recent studies have confirmed the ubiquitous worldwide distribution of *C. burnetii* among animal species, within the environment, and even in foods. Q fever is considered endemic in every country except New Zealand and Antarctia. Although positive DNA results are frequently seen in milk or in different environmental settings, isolation and/or typing to verify these findings remain elusive. Consequently, molecular typing and epidemiology is largely dependent upon on human or animal derived isolates. The most comprehensive analysis to date was reported by Glazunova et al. in 2005. Using MST data of 173 isolates from various geographical locations and from different host species, they revealed different clusters correlating with plasmid type and geographical regions. Interestingly, no correlation was apparent with either host species or associated clinical presentation. MLVA/VNTR confirmed these findings when analyzing a smaller but similarly heterogenous strain collection (Arricau-Bouvery et al. 2006). Forty-two isolates could be allocated to regions and plasmid types. Regional clusters were more precise than those assessed using MST. Others confirmed these findings with MLVA disclosing clusters from closed locations or countries, although isolate numbers were small (Chmielewski et al. 2009). Data from the Dutch 2007–2009 outbreak confirmed the power of MLVA/VNTR for regional epidemiological analysis (Klaassen et al. 2009; Roest personal communication 2010).

The predominant role of livestock, particularly during parturition, in transmission of human Q fever is irrefutable. Despite this, the ubiquitous nature of *C. burnetii* has raised the intriguing possibility that transmission might also occur through other mechanisms. The possibility of arthropod-borne cases has often been raised. Indeed, support was given through the isolation of the NM strain from a *Dermacentor andersoni* tick, with virulence of this isolate later confirmed through laboratory infection, however, to date, no human cases have been directly attributed to tick bites. Similarly the role of flies has been raised as potentially transmitting *C. burnetii* to livestock rather than directly to humans (Nelder et al. 2008).

Others have questioned whether Q fever could be acquired through a food-borne route? This is particularly a concern for milk as shedding of viable organisms in milk is well established (Berri et al. 2007; Barlow et al. 2008). This is obviously of importance for transmission among livestock, but whether it might pose an infection risk for humans is an area of debate. The majority of milk used in the food chain is pasteurized, a process whose efficacy was assessed using *C. burnetii* as a result of the resilience of the small cell variant (Cerf and Condron 2006). Several studies have reported detection of *C. burnetii* by PCR of bulk milk tank samples with alarming frequency up to 94% from the USA and generally lower incidence in Europe (4.7% Switzerland; 3.5% Turkey). Although consumption of contaminated milk has been associated with seroconversion among recipients, no clinical cases have ensued. Risk assessments are complicated by the fact that up to 60% of those becoming infected with *C. burnetii* will remain asymptomatic.

Unpasteurized milk is available for human consumption and still utilized for a variety of soft cheeses, for example Italian buffalo cheeses (mozzarella di bufala) and goat's cheese in Peru. The likely resistance of the small cell variant to manufacturing processes is a cause for concern. Similarly, shell eggs and even mayonnaise have been incriminated as potential sources for human infection with 4.2% of eggs and 17.6% of mayonnaise from Japan testing positive (Tatsumi et al. 2006). Later Swiss studies were unable to corroborate these findings (Fretz et al. 2007). A recent study demonstrated the presence of viable *C. burnetii* in unpasteurized milk being sold for human consumption in the USA (Loftis et al. 2010).

The resilient nature of *C. burnetii* and consequently its ability to withstand hostile environmental conditions for prolonged periods of time, has led to its incrimination in causing some outbreaks not directly associated with animal contact. One such outbreak was associated with structural renovations in a cardboard box-manufacturing factory in Newport, UK. Compressed straw had been used for wall construction and was hypothesised as the likely source for 95 human cases that occurred when these walls were pulled down after some 50 years (van Woerden et al. 2004).

Although a plethora of typing methods have been applied to a representative number and distribution of *Coxiella* isolates, it is still a challenge to draw relevant conclusions about the molecular epidemiology of *C. burnetii*. Some of the challenges mentioned above, including delayed clinical recognition, limited period of bacteremia and typically low organism burden, hinder epidemiological investigations. Those studies giving epidemiological information identified a wide range of diversity. Failure to find correlation of geno- or phenotype patterns with clinical appearance might arise from use of typing methods based upon mobile chromosome regions that might show a quite random genesis/evolution independent from regional or host factors. For this reason, the old plasmid-based nomenclature still has relevance and is often used as a comparative standard for newer methods. Deployment of SNP–based typing systems might overcome some of the above-mentioned limitations resulting from the stability of these targets due to the very low mutation rates seen among intracellular bacteria (10^{-9} to 10^{-11} per base pair per generation) (Drake et al. 1998; Jordan et al. 2002). To date, no "classical" virulence factors like toxins or pore-forming proteins, have been identified among *Coxiella*. Our lack of understanding the virulence mechanisms of this organism makes it difficult to develop bioforensic tools capable of recognizing modified/manipulated organisms. Application of discriminating methods like MLVA or *IS1111*-typing that produce an isolate "signature", hold promise for partially overcoming our knowledge gaps and providing a means for differentiation of isolates.

The idealistic way to maximize data of potential bioterroristic release is whole genome sequencing, as seen following the anthrax release in 2001 (Chen et al. 2010). This in itself can be problematic for this intracellular organism. Consequently, we are reliant upon robust epidemiological information gained during an outbreak situation, as seen in the Netherlands since 2007 (see above), where molecular typing methods were used to identify the probable infection source.

19.6 Conclusions

The last 5 years have seen a rapid expansion of molecular typing techniques for *C. burnetii* and their relevance assessed during outbreak situations like in the Netherlands (see above and Chap. 17). Given that scientific work on *C. burnetii* is reduced to some small groups in the world, the development and deployment of such methods is not as fast as in other bacteria like *E. coli*, *S. aureus* or *Mycobacterium tuberculosis*. Therefore a true typing gold standard is still missing and harmonization of promising techniques like MLVA or MST is a challenge. Although significant efforts in whole genome analysis have been undertaken, there are more queries rather than established facts. Several methods now show promise for delivery of reproducible, portable, highly discriminatory typing that could fulfill the requirements for an accepted standard. All of them work well with isolates, but face the same challenge of low bacterial numbers in human clinical samples. Thus for reliable typing, we are still dependent upon obtaining isolates with the associated difficulties of handling this containment level 3 pathogen and reliant upon any selective pressures that cultivation methods might exert. Another important challenge for these methods is harmonization and comparison of results between different laboratories. The growing technical power of whole genome sequencing has shown that it as the potential to replace all other loci-/target based methods/techniques (Chen et al. 2010). As with many other features of *C. burnetii*, the intricacies of its genomic information remain a "query" although next-generation sequencing technology has the potential to rapidly expand the *Coxiella* genome information database. Likewise, developments in SNP and array-based technologies hold promise for the rapid and high-resolution genotyping of *C. burnetii* strains involved in natural outbreaks or a suspected intentional release.

References

Amitai Z, Bromberg M, Bernstein M, Raveh D, Keysary A, David D, Pitlik S, Swerdlow D, Massung R, Rzotkiewicz S, Halutz O (2010) A large Q fever outbreak in an urban school in Central Israel. Clin Infect Dis 50:1433–1438

Arricau-Bouvery N, Hauck Y, Bejaoui A, Frangoulidis D, Bodier CC, Souriau A, Meyer H, Neubauer H, Rodolakis A, Vergnaud G (2006) Molecular characterization of *Coxiella burnetii* isolates by infrequent restriction site-PCR and MLVA typing. BMC Microbiol. doi:10.1186/1471-2180-6-38

Bamberg WM, Pape WJ, Beebe JL, Nevin-Woods C, Ray W, Maguire H, Nucci J, Massung RF, Gershman K (2007) Outbreak of Q fever associated with a horse-boarding ranch, Colorado, 2005. Vector Borne Zoonotic Dis 7:394–402

Barlow J, Rauch B et al (2008) Association between *Coxiella burnetii* shedding in milk and subclinical mastitis in dairy cattle. Vet Res 39:23

Beare PA, Samuel JE, Howe D, Virtaneva K, Porcella SF, Heinzen RA (2006) Genetic diversity of the Q fever agent, *Coxiella burnetii*, assessed by microarray-based whole-genome comparisons. J Bacteriol 188:2309–2324

Beare PA, Unsworth N, Andoh M, Voth DE, Omsland A, Gilk SD, Williams KP, Sobral BW, Kupko JJ 3rd, Porcella SF, Samuel JE, Heinzen RA (2009) Comparative genomics reveal extensive transposon-mediated genomic plasticity and diversity among potential effector proteins within the genus *Coxiella*. Infect Immun 77:642–656

Berri M, Rousset E et al (2007) Goats may experience reproductive failures and shed *Coxiella burnetii* at two successive parturitions after a Q fever infection. Res Vet Sci 83:47–52

Brezina R (1958) A change in the virulence of the Henzerling strain of *Coxiella burneti* during passage in white mice. Acta Virol 2:220–227

Cerf O, Condron R (2006) *Coxiella burnetii* and milk pasteurization: an early application of the precautionary principle? Epidemiol Infect 134:946–951

Chen PE, Willner KM, Butani A, Dorsey S, George M, Stewart A, Lentz SM, Cook CE, Akmal A, Price LB, Keim PS, Mateczun A, Brahmbhatt TN, Bishop-Lilly KA, Zwick ME, Read TD, Sozhamannan S (2010) Rapid identification of genetic modifications in *Bacillus anthracis* using whole genome draft sequences generated by 454 pyrosequencing. PLoS One 5(8): pii: e12397

Chmielewski T, Sidi-Boumedine K, Duquesne V, Podsiadly E, Thiéry R, Tylewska-Wierzbanowska S (2009) Molecular epidemiology of Q fever in Poland. Pol J Microbiol 58:9–13

Cooley MB, Carychao D, Nguyen K, Whitehand L, Mandrell R (2010) Effects of environmental stress on stability of tandem repeats in *Escherichia coli* O157:H7. Appl Environ Microbiol 76:3398–3400

Denison A, Thompson H, Massung RF (2007) IS1111 insertion sequences of *Coxiella burnetii*: characterization and use for repetitive element PCR-based differentiation of *Coxiella burnetii* isolates. BMC Microbiol. doi:10.1186/1471-2180-7-91

Drake JW, Charlesworth B, Charlesworth D, Crow JF (1998) Rates of spontaneous mutation. Genetics 148:1667–1686, Review

Driscoll JR (2009) Spoligotyping for molecular epidemiology of the *Mycobacterium tuberculosis* complex. Methods Mol Biol 551:117–128

Fournier P, Marrie T, Raoult D (1998) Diagnosis of Q fever. J Clin Microbiol 36:1823–1834

Frazier ME, Heinzen RA, Stiegler GL, Mallavia LP (1991) Physical mapping of the *Coxiella burnetii* genome. Acta Virol 35(6):511–518

Fretz R, Schaeren W et al (2007) Screening of various foodstuffs for occurrence of *Coxiella burnetii* in Switzerland. Int J Food Microbiol 116:414–418

Genig VA (1960) Attenuated variant "M" of Rickettsia burneti as a possible live vaccine against Q-fever. Vestn Akad Med Nauk SSSR 15(2):46–57 [In Russian]

Glazunova O, Roux V, Freylikman O, Sekeyova Z, Fournous G, Tyczka J, Tokarevich N, Kovacava E, Marrie T, Raoult D (2005) *Coxiella burnetii* genotyping. Emerg Infect Dis 11:1211–1217

Hanczaruk M, Meyer H, Frangoulidis D (2009) A genotyping system for *Coxiella burnetii* based on IS1111-elements. Paper presented at the 61rd annual meeting of the German Society of Hygiene and Microbiology (DGHM), University of Göttingen, Germany, 20–23 Sept 2009

Heinzen R, Stiegler GL, Whiting LL, Schmitt SA, Mallavia LP, Frazier ME (1990) Use of pulsed field gel electrophoresis to differentiate *Coxiella burnetii* strains. Ann N Y Acad Sci 590:504–513

Hendrix L, Samuel J, Mallavia L (1991) Differentiation of *Coxiella burnetii* isolates by analysis of restriction-endonuclease-digested DNA separated by SDS-PAGE. J Gen Microbiol 137:269–276

Huijsmans CJJ, Schellekens JJA, Wever PC et al (2011) Single-nucleotide-polymorphisms genotyping of *Coxiella burnetii* during a Q fever outbreak in the Netherlands. Appl Environ Microbiol 77:2051–2057

Jäger C, Willems H, Thiele D, Baljer G (1998) Molecular characterization of *Coxiella burnetii* isolates. Epidemiol Infect 120:157–164

Jäger C, Lautenschläger S, Willems H, Baljer G (2002) *Coxiella burnetii* plasmid types QpDG and QpH1 are closely related and likely identical. Vet Microbiol 89:161–166

Jernigan DB, Raghunathan PL, Bell BP, Brechner R, Bresnitz EA, Butler JC, Cetron M, Cohen M, Doyle T, Fischer M, Greene C, Griffith KS, Guarner J, Hadler JL, Hayslett JA, Meyer R, Petersen LR, Phillips M, Pinner R, Popovic T, Quinn CP, Reefhuis J, Reissman D, Rosenstein N, Schuchat A, Shieh WJ, Siegal L, Swerdlow DL, Tenover FC, Traeger M, Ward JW, Weisfuse I, Wiersma S, Yeskey K, Zaki S, Ashford DA, Perkins BA, Ostroff S, Hughes J, Fleming D, Koplan JP, Gerberding JL (2002) Investigation of bioterrorism-related anthrax, United States, 2001: epidemiologic findings. Emerg Infect Dis 8:1019–1028

Jordan IK, Rogozin IB, Wolf YI, Koonin EV (2002) Microevolutionary genomics of bacteria. Theor Popul Biol 61:435–447

Keim P, Van Ert MN, Pearson T, Vogler AJ, Huynh LY, Wagner DW (2004) Anthrax molecular epidemiology and forensics: using the appropriate marker for different evolutionary scales. Infect Genet Evol 4:205–213

Kersh GJ, Wolfe TM, Fitzpatrick KA, Candee AJ, Oliver LD, Patterson NE, Self JS, Priestley RA, Loftis AD, Massung RF (2010) Presence of *Coxiella burnetii* DNA in the environment of the United States, 2006 to 2008. Appl Environ Microbiol 76:4469–4475

Klaassen CHW, Nabuurs-Franssen MH, Tilburg JJHC, Hamans MAWM, Horrevorts AM (2009) Multigenotype Q fever outbreak, the Netherlands. Emerg Infect Dis 15:613–614

Lautenschläger S, Willems H, Jäger C, Baljer G (2000) Sequencing and characterization of the cryptic plasmid QpRS from *Coxiella burnetii*. Plasmid 44:85–88

Li B, Kadura I, Fu D, Watson DE (2004) Genotyping with TaqMAMA. Genomics 83:311–320

Loftis AD, Priestley RA, Massung RF (2010) Detection of *Coxiella burnetii* in commercially available raw milk from the United States. Foodborne Pathog Dis 7(12):1453–1456

Madariaga M, Rezai K, Trenholme GM, Weinstein RA (2003) Q fever: a biological weapon in your backyard. Lancet Infect Dis 3:709–721

Malorny B, Junker E, Helmuth R (2008) Multi-locus variable-number tandem repeat analysis for outbreak studies of *Salmonella enterica* serotype Enteritidis. BMC Microbiol. doi:10.1186/1471-2180-8-84

Nelder MP, Lloyd JE et al (2008) *Coxiella burnetii* in wild-caught filth flies. Emerg Infect Dis 14:1002–1004

Nguyen SV, Hirai K (1999) Differentiation of *Coxiella burnetii* isolates by sequence determination and PCR-restriction fragment length polymorphism analysis of isocitrate dehydrogenase gene. FEMS Microbiol Lett 180:249–254

Panning M, Kilwinski J, Greiner-Fischer S, Peters M, Kramme S, Frangoulidis D, Meyer H, Henning K, Drosten C (2008) High throughput detection of *Coxiella burnetii* by real-time PCR with internal control system and automated DNA preparation. BMC Microbiol 19:77

Papp AC, Pinsonneault JK, Cooke G, Sadee W (2003) Single nucleotide polymorphism genotyping using allele-specific PCR and fluorescence melting curves. Biotechniques 34:1068–1072

Priestley RA, Hornstra HM, Pearson T, Keim P, Massung RF (2009) The state of the SNP: using real-time PCR to genotype *Coxiella burnetii*. Abstract #34, 23rd meeting of the American Society for Rickettsiology, Hilton Head Island, SC

Rodolakis A, Berri M, Héchard C, Caudron C, Souriau A, Bodier CC, Blanchard B, Camuset P, Devillechaise P, Natorp JC, Vadet JP, Arricau-Bouvery N (2007) Comparison of *Coxiella burnetii* shedding in milk of dairy bovine, caprine, and ovine herds. J Dairy Sci 90:5352–5360

Samuel JE, Frazier ME, Mallavia LP (1985) Correlation of plasmid type and disease caused by *Coxiella burnetii*. Infect Immun 49:775–779

Schneeberger PM, Hermans MH, van Hannen EJ, Schellekens JJ, Leenders AC, Wever PC (2009) Real-time PCR on serum samples is indispensable for early diagnosis of acute Q-fever. Clin Vaccine Immunol 17:286–290

Schulz J, Runge M, Schröder C, Ganter M, Hartung J (2005) Detection of *Coxiella burnetii* in the air of a sheep barn during shearing. Dtsch Tierarztl Wochenschr 112:470

Sekeyová Z, Roux V, Raoult D (1999) Intraspecies diversity of *Coxiella burnetii* as revealed by com1 and mucZ sequence comparison. FEMS Microbiol Lett 180:61–67

Seshadri R, Paulsen IT, Eisen JA, Read TD, Nelson KE, Nelson WC, Ward NL, Tettelin H, Davidsen TM, Beanan MJ, Deboy RT, Daugherty SC, Brinkac LM, Madupu R, Dodson RJ, Khouri HM, Lee KH, Carty HA, Scanlan D, Heinzen RA, Thompson HA, Samuel JE, Fraser CM, Heidelberg JF (2003) Complete genome sequence of the Q-fever pathogen *Coxiella burnetii*. Proc Natl Acad Sci U S A 100:5455–5460

Stein A, Saunders NA, Taylor AG, Raoult D (1993) Phylogenic homogeneity of *Coxiella burnetii* strains as determinated by 16S ribosomal RNA sequencing. FEMS Microbiol Lett 113:339–344

Stein A, Kruszewska D, Gouvernet J, Raoult D (1997) Study of the 16S–23S ribosomal DNA internal spacer of *Coxiella burnetii*. Eur J Epidemiol 13:471–475

Svraka S, Toman R, Skultety L, Slaba K, Homan WL (2006) Establishment of a genotyping scheme for *Coxiella burnetii*. FEMS Microbiol Lett 254:268–274

Tatsumi N, Baumgartner A et al (2006) Detection of *Coxiella burnetii* in market chicken eggs and mayonnaise. Ann N Y Acad Sci 1078:502–505

Thiele D, Willems H (1994) Is plasmid based differentiation of *Coxiella burnetii* in 'acute' and 'chronic' isolates still valid? Eur J Epidemiol 10:427–434

Thiele D, Willems H, Köpf G, Krauss H (1993) Polymorphism in DNA restriction patterns of *Coxiella burnetii* isolates investigated by pulsed field gel electrophoresis and image analysis. Eur J Epidemiol 9:419–425

Thiele D, Willems H, Haas M, Krauss H (1994) Analysis of the entire nucleotide sequence of the cryptic plasmid QpH1 from *Coxiella burnetii*. Eur J Epidemiol 10:413–420

Tilburg JJ, Melchers WJ, Pettersson AM, Rossen JW, Hermans MH, van Hannen EJ, Nabuurs-Franssen MH, de Vries MC, Horrevorts AM, Klaassen CH (2010) Interlaboratory evaluation of different extraction and real-time PCR methods for the detection of *Coxiella burnetii* DNA in serum. J Clin Microbiol 48(11):3923–3927

Tissot-Dupont H, Amadei MA, Nezri M, Raoult D (2004) Wind in November, Q fever in December. Emerg Infect Dis 10:1264–1269

Turra M, Chang G, Whybrow D, Higgins G, Qiao M (2006) Diagnosis of acute Q fever by PCR on sera during a recent outbreak in rural south Australia. Ann N Y Acad Sci 1078:566–569

Valková D, Kazár J (1995) A new plasmid (QpDV) common to *Coxiella burnetii* isolates associated with acute and chronic Q fever. FEMS Microbiol Lett 125:275–280

van Woerden HC, Mason BW et al (2004) Q fever outbreak in industrial setting. Emerg Infect Dis 10:1282–1289

Vodkin MH, Williams JC, Stephenson EH (1986) Genetic heterogeneity among isolates of *Coxiella burnetii*. J Gen Microbiol 132:455–463

Wen BH, Yu SR, Yu GQ, Li QJ, Zhang X (1991) Analysis of proteins and lipopolysaccharides from Chinese isolates of *Coxiella burnetii* with monoclonal antibodies. Acta Virol 35:538

Willems H, Ritter M, Jäger C, Thiele D (1997) Plasmid-homologous sequences in the chromosome of plasmidless *Coxiella burnetii* Scurry Q217. J Bacteriol 179:3293–3297

Xing Y, Liu J, Sakamuri RM, Wang Z, Wen Y, Vissa V, Weng X (2009) VNTR typing studies of *Mycobacterium leprae* in China: assessment of methods and stability of markers during treatment. Lepr Rev 80:261–271

Zhang GQ, To H, Yamaguchi T, Fukushi H, Hirai K (1997) Differentiation of *Coxiella burnetii* by sequence analysis of the gene (com1) encoding a 27-kDa outer membrane protein. Microbiol Immunol 41:871–877

Index

R. Toman et al. (eds.), *Coxiella burnetii: Recent Advances and New Perspectives in Research of the Q Fever Bacterium*, Advances in Experimental Medicine and Biology 984, DOI 10.1007/978-94-007-4315-1, © Springer Science+Business Media Dordrecht 2012

The manufacturer's authorised representative in the EU is Springer Nature Customer Service Centre GmbH, Europaplatz 3, 69115 Heidelberg, Germany. If you have any concerns regarding our products, please contact ProductSafety@springernature.com

Printed and bound by CPI Group (UK) Ltd, Croydon, CR0 4YY
22/09/2026
02220946-0015